DIAGNOSIS AND IDENTIFICATION OF PLANT PATHOGENS

Developments in Plant Pathology

VOLUME 11

Diagnosis and Identification of Plant Pathogens

Proceedings of the
4th International Symposium of the European Foundation for Plant Pathology,
September 9–12, 1996, Bonn, Germany

Edited by

H.-W. DEHNE,
G. ADAM,
M. DIEKMANN,
J. FRAHM,
A. MAULER-MACHNIK
AND
P. VAN HALTEREN

KLUWER ACADEMIC PUBLISHERS
DORDRECHT / BOSTON / LONDON

Library of Congress Cataloging-in-Publication Data

A C.I.P. Catalogue record for this book is available from the Library of Congress.

ISBN-13:978-94-010-6508-5 e-ISBN- 13:978-94-009-0043-1
DOI: 10.1007/978-94-009-0043-1

Published by Kluwer Academic Publishers,
P.O. Box 17, 3300 AA Dordrecht, The Netherlands.

Sold and distributed in the U.S.A. and Canada
by Kluwer Academic Publishers,
101 Philip Drive, Norwell, MA 02061, U.S.A.

In all other countries, sold and distributed
by Kluwer Academic Publishers,
P.O. Box 322, 3300 AH Dordrecht, The Netherlands.

Softcover reprint of the hardcover 1st edition 1997

Editorial board

Contents

ORAL PRESENTATIONS

Session 1: General Aspects

Session 2: Taxonomy and Differentiation

Session 4: Nucleic Acid Based Hybridisation Techniques

Session 5: Disease Assessment and Quantification

Session 6: Disease Prognosis and Expert Systems

POSTER PRESENTATIONS

Taxonomy and Differentiation

Serological Assays

Nucleic Acid Based Hybridisation Techniques

Disease Assessment and Quantification

Preface

The diagnosis and identification of plant pathogens provides the basis of plant pathology and phytomedicine. The Executive Committee of the EUROPEAN FOUNDATION FOR PLANT PATHOLOGY (EFPP) had no problem to identify this actual topic as topic for the 4th Symposium, which was held from September 8th to the 12th at the University of Bonn. It was suggested to have introductory papers and papers on actual research on recently identified topics.

The development of diagnosis and pathogen identification is very important to keep plants healthy and to provide a successful and efficient disease control. On the other hand the most important task of the EUROPEAN FOUNDATION FOR PLANT PATHOLOGY is to improve the international communication, especially in the European hemisphere. Another important duty is to provide the contact between all associated societies – of specific importance seems to be the contact to societies and colleagues from eastern European countries. Times have changed and gratefully we are obliged to hold the contact to our colleagues from the east. During the last meeting we could hold this contact to a certain extent and this should be a premise for the future.

During 1998 the EUROPEAN FOUNDATION FOR PLANT PATHOLOGY will join the 7th International Congress of Plant Pathology held at Edinburgh from August 9-14, 1998. The 5th Symposium of the EUROPEAN FOUNDATION FOR PLANT PATHOLOGY will be arranged by our Italian colleagues.

Many thanks to those colleagues, who joined the symposium at Bonn, we hope to meet you at Edinburgh in 1998 and in Taormina in the year 2000 -

Heinz-W. Dehne

Acknowledgments

A meeting like the 4th Symposium of the European Foundation for Plant Pathology on the "Diagnosis and Identification of Plant Pathogens", which has been held form September 9th to the 12th at the University of Bonn, can only be successful if certain, necessary requirements are fulfilled: Most important is the participation of scientists, willing to share their results and to discuss the actual status in this particular field - many thanks to those that joined the Symposium. Additionally it needs an active Organizing Committee to establish and arrange an interesting program. The members of the Organizing Committee also reviewed the papers for the Proceedings of this Symposium and served as editors. Therefore my thanks go to Günther Adam, Hamburg, Marlene Diekmann, Rome, Johann Frahm, Münster, and Paul van Halteren, Wageningen. The symposium was also assisted by the joint meeting of the COST-actions 821 ("Arbuscular mycorrhizas") and 823 ("Nucleic acid based differentiation of plant feeding insects and their vector status") that was held right before and in connection with the EFPP-Symposium at the University of Bonn.

The success of a meeting is furthermore depending on financial assistance – I want to express my thanks to the National Science Foundation (Deutsche Forschungsgemeinschaft) and the Ministry of Foreign Affairs (Auswärtiges Amt) for providing substantial support to the symposium. Additionally the following companies attributed to the symposium financially – therefore many thanks go to AgrEvo, BASF, Bayer, Bioreba, Ciba-Geigy, Cyanamid, Leica, Sanofi Pasteur, Zeiss and Zeneca.

I want to thank also the City of Bonn, especially our mayor, Mrs. Paß-Weingarts, who presented an interesting welcome address to all participants during a reception at the Old Cityhall (Altes Rathaus). My deep thanks go to all colleagues and members of our Institute for Plant Diseases, the Faculty of Agronomy and the University of Bonn that made the meeting possible by personal engagement and their individual support.

Heinz-W. Dehne

Session 1:

General Aspects

THE IMPORTANCE OF DIAGNOSIS AND PROGNOSIS IN INTEGRATED PLANT PROTECTION

Head of Division Dr. Ralf Petzold,
Federal Ministry of Food, Agriculture
and Forestry, Bonn

Mr. President, Ladies and Gentlemen,

First of all, I would like to convey to you the best wishes of the Federal Ministry of Food, Agriculture and Forestry for this 4th International Symposium.

As you know for sure, in 1986, thus ten years ago already, the German Parliament legally prescribed integrated plant protection. as the guideline for plant protection. Since 27 July 1993, this concept which is no longer questioned by experts has applied to all Member States of the European Union. The relevant passage in the Council Directive 91/414/EEC of 15 July 1991 concerning the placing of plant protection products on the market (Article 3, para. 3) reads as follows:

"Member States shall prescribe that plant protection products must be used properly. Proper use shall include compliance with the conditions stipulated in the authorisation and specified on the labelling, and the application of the principles of good plant protection practice, as well as, whenever possible, the principles of integrated control."

Incidentally, at a global level, too, integrated plant protection represents a norm for sustainable agriculture (UN Climate Conference in Rio de Janeiro 1992, Agenda 21, Chapters 19 and 14).

Even if the contents of integrated plant protection vary to some degree due to the lack of an internationally accepted definition, a pragmatic distinction can still be made in that integrated plant protection compared with good plant protection practice comprises a plant protection concept of a higher quality. Non-chemical measures, also including the clean start in the enterprise, and chemical measures are conceptionally linked in integrated plant protection. To do what is economically justifiable only or, to be more precise, to target one's actions better also means to refrain from any unnecessary impact on the agricultural ecosystem.

What is the significance of diagnosis and prognosis in this higher-quality plant protection concept?

H.-W. Dehne et al. (eds.), Diagnosis and Identification of Plant Pathogens, 3–8.

The answer is actually very simple. Without a clean diagnosis and a highly accurate prognosis, the concept cannot be put into practice. The early and reliable detection of harmful organisms and the forecast of occurrence, infestation curve and level of damage with the utmost probability constitute the basis of all decisions which have to be made. Mistakes and underestimations owing to a lack of certainty in the diagnosis or prognosis can lead to substantial yield and quality losses or to wrong decisions imposing unnecessary strains on the agricultural ecosystem.

In practice, however, the actual difficulty lies in the fact that a different degree of accuracy is required for different purposes and the possibilities have to be geared to the availabilities.

Let us turn to the diagnosis. Depending on the objective, a distinction should be made between

1. Diagnosis for practical operating purposes, that is the knowledge of the harmful organism; for example, are we dealing with a harmful organism already known in the holding or with an introduced harmful organism appearing for the first time?

2. Diagnosis for extension purposes. Validation of the on-farm diagnosis, development of new control concepts which are even better adapted.

3. Diagnosis in the scientific field; exact diagnostics, information on diagnostic differences, further development of diagnosis methods, etc.

Apart from this, further aims of diagnostic efforts arise with an indirect reference to integrated plant protection derived from legal provisions as, for example

4. Diagnosis with respect to notification procedures regarding the occurrence of new or more virulent harmful organisms

5. Diagnosis within the framework of supervisory measures

6. Diagnosis in case of dispute settlements.

<u>As to 1</u>

The diagnosis at plant level requires a sound training and further education on the part of the farm manager. Combined with extension work, almost all problems are solvable today. With regard to a stepped-up introduction of integrated plant protection, however, it should be noted that for a certain transitional period the improved diagnosis will entail an increased use of plant protection agents until the actual relevance of the respective infestation can be more effectively restricted. Afterwards, the benefits of a better diagnosis and prognosis will have a positive effect. It is important that the diagnosis methods are methods which are as simple as possible and do not take up much time. The current farm sizes, in general, do not allow time-consuming diagnostics on the farm. The ratio between the costs for the use of plant protection agents and the yield and quality risk clearly shows that uncertainties tend to be compensated by the use of plant protection agents today. Frequently, the results of counting, measuring and assessing are scientifically corroborated and available for application. Yet, it is only after a considerable delay that they are put into practice. Furthermore, to what extent harmful organisms have adapted or shifts in the population or species inventory have taken place is not systematically monitored at the farm. As a rule, we only gain insights when resistances against specific plant protection agents occur. The currently prevailing stability rests essentially on the good co-operation between practice and extension work. This brings us round to the second field.

As to 2

Even though in the field of extension work, the diagnosis forms an integral part of the concept of integrated plant protection, this field is coming under pressure due to the increasing cost-cutting constraints in the public agencies of many countries. The practical implementation of the broad introduction of integrated plant protection which is, for example, stipulated as a political aim in the USA, will not least be s question of financially securing in the long term the necessary support of farmers and gardeners provided by the extension agencies. Moreover, with the increasing privatisation of extension work, the danger increases that the benefits of targeted control measures as part of integrated plant protection are superseded by more prophylaxis to prevent crop damage at all events. Wrong assessments which have led to severe yield losses are leaving clear marks. The willingness to take risks is frequently impaired for years. All the same, extension work constitutes a special transmission belt for diagnostics. Scientific innovations have to prove their worth and stand the test in practice. Optimisations are possible through the co-operation of extension services with scientific institutes.

As to 3

The diagnostics in the scientific field represents, in principle, the preliminary stage of an accurate diagnosis in all subsequent fields of application. Innovations are in demand here. This concerns the entire spectrum of diagnostic questions with the modern instruments of molecular-genetic, biochemical, serological, electron- microscopical technologies and molecular-biological processes. The whole range of today's scientific approaches manifests itself in the contributions to this international symposium of the European Foundation for Plant Pathology. It is quite amazing what great advances have been made in the past few years. A very large share of which is directly or. indirectly related to integrated plant protection, whether via the improved, more precise and reliable diagnosis of harmful organisms or whether with a view to the diagnosis of antagonists or to the species inventory of agricultural ecosystems or in the context of genetic engineering with respect to resistance breeding.

As to 4

The International Plant Protection Convention and regional plant protection organisations such as, for example, the EPPO or NAPPO as well as supranational organisations like the European Union demand that the individual states monitor their national territory as to the occurrence of new harmful organisms. If this is the case, there is a duty to report the occurrence. This is designed to protect the domestic production and, since trade operates on a worldwide basis today, other countries, too. The early detection of harmful organisms which can trigger off great damage as well as targeted actions mean a real chance for prevention; they can at least considerably delay the control which would otherwise be necessary. Let me illustrate this on the basis of two examples.

The first example refers to *Lymantria dispar* which had been a serious problem in the south-west of Germany in 1993 and 1994. The situation became particularly explosive when a threat through introduction from Germany was established by the USA and Canada on the grounds that the females have a better ability to fly which otherwise only Asian strains have. That the American and Canadian colleagues are afraid of a renewed introduction is understandable since they run large-scale and costly programmes to protect their forests and speak from bitter experience. Thus, the point in question was whether the hypothesis is correct that Asian strains have successively crossed westwards. The diagnosis results to date do not present a uniform picture. Due to the low population density last year and this year, the problems have eased a little. Yet, the demand is still made that each consignment from Germany should be free from *Lymantria dispar*. Incidentally, the US-Army has separately inspected and declared each part, also in the private sphere, during the return of its units, e.g. from Frankfurt.

The second example is *Diabrotica virgifera*, a maize pest. It has been introduced via the war to Yugoslavia and steadily spreads to the north. It remained at first undiscovered owing to the war turmoil. The diagnosis is simple, the problem is to stop this harmful organism which plays a big role in the US from spreading any further. This should serve as an example that it is also vital to carry out the diagnosis at the right time.

<u>As to 5</u>

In contrast to the diagnosis with respect to the occurrence of new harmful organisms with their special requirements, diagnosis in the context of control tasks consists of routine tasks to be carried out at minimum cost and which should not be time-consuming. The testing for freedom from viruses falls under this category, but also diagnostic work with respect to the control of protected areas or for the confirmation that no infestation with *Clavibacter* is at hand in case of. potatoes. Here, further efforts are required to introduce and to optimise modem methods.

<u>As to 6</u>

Finally, the diagnostics in the context of settlement of disputes. The next speaker, Mr. Paul v. Halteren, will supply convincing arguments on this issue. Contrary to the diagnosis in case of the control tasks, in this case diagnosis methods do not have to be equally optimised in economic terms. What is expected of them is freedom of doubt in the results. Such methods of diagnosis should be developed and generally accepted at least for the harmful organisms subject to a passport which are regulated in the Community Directive. From today's perspective, there is also a backlog in this case.

Let us turn to the prognosis now.

It is no doubt true that without a sound prognosis a re-orientation of plant protection to targeted measures will not be possible. If it is a question of integrated plant protection, the methods of prognosis are in different stages of development. Typical examples are the integrated plant protection in fruit growing, where many years of experience are at hand regarding the prognosis, but also the economic damage thresholds in weed control. The prognosis becomes more complicated with respect to one-year crops, when the weather plays a big role or the development of harmful organisms depends on several parameters which are interactive. In these cases, too, the development of individual prognosis methods has been successful in the past such as, for example, the well-known scab prognosis or in case of the hop mildew. The rapid development in data processing has given a new impetus to the development of prognosis methods. The possibility of processing large amounts of data also rendered possible the modelling of biological processes and to relate them to the meteorological data. Within the past ten years, it has been possible in the Federal Republic of Germany, even though at a cost of about DM 15 million to develop nine further methods of prognosis in total and to press ahead with them until they are ready for practical application. In particular, we are dealing with

- the calculation of the intensity of attack in case of *Cercosporella*,
- the negative prognosis with respect to cereal mildew,
- the calculation of the degree of danger in case of *Rhynchosporium*,
- the population dynamics of the grain aphid,
- the prognosis on the beginning of an epidemic in case of potato blight,
- observation of the blight epidemic,
- prognosis of the potato beetle,
- the population dynamics of the cabbage fly,
- the population dynamics of the carrot fly.

The institute in Bonn hosting this event today has played a prominent part in these efforts. Hopefully, these methods will soon become part of the standard repertory and also encourage other working groups to develop prognosis methods for other harmful organisms, too. Let me just go into one thing, in particular. The actual difficulty did not lie in the modelling of the biological processes, even though new ground had to be broken here, too, to some extent or in the continuous provision of meteorological data related to the necessity in question, but in the question how the methods which have been developed can be secured for the future and adapted if need be. This has now been safeguarded after a grim struggle through the co-operation of the *Laender* which are competent for the implementation of plant protection in the Federal Republic of Germany.

Ladies and Gentlemen,
I intended to present to you with my observations the various direct and indirect references of diagnosis and prognosis to integrated plant protection. From the variety of tasks on the one hand and the scarcity of resources on the other hand inevitably result the approach to the setting of priorities, co-operation and thus division of labour. How such a "network" can be organized and financed, however, is currently completely open. However, supraregional bodies have started considering the first approaches. It turned out very soon in the process that many states are facing serious difficulties in incurring new financial obligations binding them in the medium term or even long term. Yet, this should not keep us from seeking and achieving co-operation wherever and in whatever way possible. This symposium illustrates this very well.

Thank you for your attention.

PHYTOSANITARY IMPLICATIONS OF DIAGNOSTIC ACTIVITIES AND VICE VERSA

DR. P. VAN HALTEREN
Plant Protection Service
P.O. Box 9060, 6700 GW Wageningen
The Netherlands

ABSTRACT

The fourth International EFPP-Symposium focuses on status, recent developments and future trends in the field of diagnostics and identification of plant pathogens and parasites. It attains even more weight if it is placed in a broader spectrum of interests, more in particular phytosanitary subjects, quarantine problems and political objectives. In this paper seven subjects are being dealt with, ranging from innocent to major economic events:

1. new findings, 2. *Phoma* spp, 3. *Meloidogyne chitwoodi* and *M. fallax*, 4. *Thrips palmi*, 5. Mediterranean fruitfly, 6. Brown rot of potatoes, 7. Mad cow's disease(BSE).

Diagnostic activities and research bear phytosanitary implications, and in return phytosanitary and political feelings and facts affect diagnostic activities, including research. It is a two-way lane.

In Europe, trade barriers imposed by protectionism are exceptional and in any case either hidden or denied. Pest Risk Analysis, Diagnostic Protocols and good terminology are in an advanced stage of development. This is, however, not the case with jurisprudence and evidential value of modern biochemical techniques.

The official title of this EFPP Symposium is Diagnosis and Identification of Plant Pathogens, at least according to the first and second circular. In the accompanying text is said "pathogens and parasites" which widens the scope and objectives of this conference already. Normally, diagnostics is considered as detection and identification. With this the whole field is opened up sufficiently to elaborate on the interaction between diagnostics and phytosanitary actions.

Detection is almost always done with a particular objective in mind; this is less the case with identification. Identification and studying identification methods is often done more or less as a hobby or by scientists to further science. This does not mean

H.-W. Dehne et al. (eds.), Diagnosis and Identification of Plant Pathogens, 9–13.

that this is a waste. Certainly not! Quite often after some time, for purposes not thought of at the time of identification, the information becomes very useful.

Plant Protection Services all over the world must take care that their phytosanitary actions are scientifically sound, economically and politically justified, and legally correct. Indeed no small matter!

Scientifically sound means amongst others that sampling and detection procedures have to follow statistical requirements and that for identification generally accepted diagnostic techniques are applied.

The interaction between phytosanitary requirements and diagnostic techniques is very important. No proper phytosanitary and quarantaine measures are possible without a solid diagnostic basis. But also a considerable part of diagnostic research should be done with the purpose of pursuing and backing phytosanitary and quarantaine measures.

The Organising Committee of this Conference thought it of value if by means of a number of practical examples this relation between diagnostics and phytosanitary activities would be illustrated.

The **first example** is the finding of insects, nematodes, fungi, bacteria and viruses new in an area, on a new host plant or new to science. These are often not or not yet pests or pathogens of a plant, crop or habitat. Such discoveries or descriptions are the order of the day and somewhere reported in scientific or general literature. They most often do not give rise to phytosanitary activities.

The Diagnostic Centre of the Dutch Plant Protection Service mentions anywhere between 50 and 100 of those organisms every year in its Annual Report.

The **second example** is one that stems from the difficulties of keeping the different *Phoma* species apart. Many *Phoma* species are plant pathogens and it is important to know the host plant range of each species and pathotypes. This information may have a bearing on possible quarantine measures possibly imposed by countries. Keeping the species apart may be done by normal morphological or morphometrical characteristics, or as will be presented later this week by Man in 't Veld by isozyme analyses.

The interaction is best described by a need to discriminate between several pathogenic forms of *Phoma* spp. and the newest scientific techniques provide the opportunity. Or vice versa, the techniques are becoming available and these are applied in a field were there was a need for them.

The **third example** are *Meloidogyne chitwoodi* and *Meloidogyne fallax*, two plant parasitic root-knot nematodes. The first originates from the United States and was apparently distributed the world over around 1930, without having been described as a species at that time. After gentle but persistent prodding by APHIS, the U.S. Plant Protection Service, eventually several crops were found to be attacked by this nematode in the Netherlands. Although the nematode was found a few years ago in Dutch material collected in 1933, it has never caused a problem and lived unnoticed in

Western Europe for over at least half a century. Admittedly, *M. chitwoodi* is now spreading in the Netherlands and is becoming a problem in certain areas.

Because the Americans created as it were a problem that was not there, the Dutch Plant Protection Service, started research and discovered that *M. chitwoodi* was associated with the extraordinary intensive farming system, and on the way discovered an entirely new root-knot nematode, recently described as *Meloidogyne fallax*.

The theory is that *M. chitwoodi* has been distributed from the Americas all over the world but is a problem only in light sandy soils of temperate climates with an intensive cropping system and that *M. fallax* originates from Western Europa and lives there virtually unnoticed.

This is an example were an unimportant and imaginary phytosanitary problem has lead to the discovery of an entirely new root-knot nematode and a better understanding of the biology and characteristics of a thus far poorly known species.

The **fourth example** is the insect *Thrips palmi*. This thrips was found at import inspections in the Netherlands and elsewhere and at the suggestion of the Netherlands the insect was added to the EU and EPPO quarantine lists about 5 years ago. Unfortunately, in 1991 it was found on *Ficus* plants in the Netherlands and a phytosanitary action resulted in the destruction of almost 3 million ECU worth of plants. Much to the dismay of the importers which in turn resulted in a legal battle that is still going on.

A discussion started over the presence of this thrips in the U.S. and Central America and whether *Ficus* spp. are host plants. In fact the Americans claimed that this thrips has never been found on *Ficus benjamini*, whereas in the Netherlands the insect was only found on imported *Ficus* from the Americas and on imported orchids from South East Asia.

The conflict went so far that in 1995 American scientists from a university in Florida made a survey, paid by the *Ficus* grower with which the authorities fought that legal battle, to investigate the presence of *Thrips palmi* in the Netherlands. To cut the story short, the Dutch government will probably win this court case and earlier this year the Service finally intercepted *Thrips palmi* on *Ficus benjamini* imported from Florida.

In this case there was a phytosanitary problem without a taxonomic problem. *Thrips palmi* is a well-described species and there are no differences of opinion concerning the taxonomic status. There were, however, differences of opinion whether some plant species, *Ficus* spp., were indeed host plants of this quarantine organism and about the geographic distribution.

Example number 5 is the Mediterranean fruitfly, *Ceratitis capitata*. The presence of the fly was recorded in the Netherlands around 1960 and which was published at that time. This information has been enough for the Japanese Plant Protection Service to prohibit the import of sweet pepper, tomatoes and other fruit vegetables grown in greenhouses in the glasshouse district of Western Holland. It took the Dutch Plant

Protection Service over 10 years to convince Japan that it is unrealistic to hamper trade based on such evidence.

The result as it stands is that the Service monitors the presence of the medfly in hundreds of pheromone traps all over the area. This, including the harbour area of Rotterdam where indeed an occasional medfly is caught. This of course because of the enormous citrus imports through this harbour from Mediterranean countries. It is certainly walking on a phytosanitary tightrope!

Had the Service not published the catching of the medfly in the thirties, than it would have been unlikely that the Japanese had prohibited the import of fruit vegetables from greenhouses of Holland some 15 years ago. This indicates the importance of what was for a long time considered to be an innocent publication.

The economic importance is very clear in the **sixth example**, brown rot of potatoes caused by *Pseudomonas solanacearum*. Worldwide some 5 pathotypes are known, of which one, race three, is common on potato in the Mediterranean area. The diagnosis of this bacterium followed the method developed by the European and Mediterranean Plant Protection Organisation (EPPO) for ringrot in potato. There was no diagnostic problem with the detection and identification of *Pseudomonas solanacearum* in potatoes until it was found in seed potatoes in the Netherlands in 1995 and almost overnight thousands of samples, especially for the export of seed potatoes, had to be processed. How to carry out this efficiently? And all of a sudden scientists were discussing the merits of shaking versus macerating the tissues, the quality of a selective medium, the reliability of immunofluorescence and Elisa, the value of testing 200 tubers out of 25 tons, the survival in process water and the importance of *Solanum dulcamara*.

The discussion was not only created by a phytosanitary problem, but by the sheer size of the more than 50,000 samples that had to be processed within a few months.

The **last example** is that of the mad cow disease: the less than effective handling of an existing problem, an unlikely but not impossible transmission to man and a discussion about the possible agent of the disease, an infectious protein without a nucleic acid core. A type of agent that probably none of us had ever heard of before.

Nevertheless with devastating consequences, economically and politically, for the British community and Europe as a whole.

In Europe ordinary protective trade barriers have become rare. Every now and than an outright, but unjustified, denial that a plant pathogenic agent is present in a given country occurs and of course the tightening of phytosanitary measures a bit more than called for are still frequent. On the whole, Europe is on the right track, however.

To keep on track it is of paramount importance that internationally agreed diagnostic protocols are developed for all EU and EPPO quarantine organisms. It is equally important that an internationally accepted Pest Risk Analysis Scheme is finalised within short.

International cooperation is the order of the day and will probably continue to grow.

There is, however, one aspect that continues to scare. That is jurisprudence and evidential value of modern techniques or the lack there-off. When applying the techniques that are being discussed during this conference on populations of pathogens, differences will be found and recorded.

I am convinced that within the near future several of our scientists will stand in court to explain that a particular electrophoretic band or fatty acid profile does or does not mean a sufficiently distinctive difference with an official quarantine organism and thus that the agent under discussion indeed does or does not pose a considerable phytosanitary risk and justifies a phytosanitary action.

It was the intention to enlighten the interactions between phytosanitary and diagnostic activities. Hopefully, this has been successful.

GENERAL ASPECTS ON DIAGNOSTIC METHODS: VIRUSES, MICRO-ORGANISMS

H. HUTTINGA
DLO Research Institute for Plant Protection (IPO-DLO)
P.O. Box 9060, 6700 GW Wageningen
The Netherlands

'Preventing a disease is better than curing it' is a good leading principle in plant protection. And in order to prevent the outbreak of a disease in a crop it makes sense to start with healthy material: healthy plant material and non-contaminated substrate in which the plants will be grown.

Detection methods, and not necessary diagnostic methods, are needed to select healthy starting material. Application of detection methods in plant protection in fact is checking whether or not a specified pathogen is present. It fundamentally differs from diagnosis, the act or process of deciding the nature of a diseased condition by examination of the symptoms.

Before detection methods can be designed for a pathogen, this pathogen has to be identified and characterized. After that one can look for a specific feature which can be used to detect the pathogen amidst the surrounding pool of non- pathogenic organisms and maybe even other pathogens. There should be a strict correlation between the characteristic used for detection of a pathogen and its pathogenicity.

Characteristics used for detection can be divided in two groups: i) biological properties related to the interaction of the pathogen with its host and/or vector, and ii) intrinsic properties of the pathogen itself. Methods based on biological properties, like symptomatology, can be very sensitive and decisive: one particle of potato virus Y may cause a completely diseased potato plant, so in this way one can detect 80×10^{-18} g of virus. However, the use of symptomatology certainly also has many disadvantages, e.g. adequate plants sometimes are not available throughout the year. Detection methods based on intrinsic properties of pathogens are well known, e.g. ELISA is based on the intrinsic antigenic properties of for instance the outer surface of virus particles. For viroids one can use the unique feature of the internal base-pairing of these single-stranded circular RNAs to detect them by bi-directional electrophoresis. Many modern molecular biological methods, like molecular hybridization, PCR, NASBA and AFLP, are based on the properties of the genetic material of viruses and microorganisms.

The effectiveness of a detection method is highly influenced by the way (how, where, when) samples for testing are collected. Results can often be improved by increasing the physiological activity of the material to be tested. Modern and more

H.-W. Dehne et al. (eds.), Diagnosis and Identification of Plant Pathogens, 15–16.

sensitive methods require other ways of handling and preparation than older less sensitive tests do. Especially for methods with a very high sensitivity there is a high risk of carry-over from one sample to another. Furthermore the spread of material to be detected by aerosols should carefully be prevented.

Good detection methods should have a high signal/noise ratio. The signal in a test can be increased by e.g. stimulating the multiplication of a specific bacterium on a (s)elective medium or by e.g. improving the conditions in a plant like breaking dormancy of potato tubers, resulting in an increase in virus titers. In fact PCR and NASBA are also ways of specifically increasing the amount of (part of the) genetic material of the pathogen one is looking for. Note, however, that PCR nor NASBA is a diagnostic or a detection method. It has to be followed by a detection method for the expected amplicons, e.g. by a TaqmanR system or by using molecular beacons. Noise in a test can be reduced e.g. by preparing highly specific antibodies, primers and probes. The use of AFLP can facilitate the finding of nucleic acid regions specific for a pathogen, and therefore in designing specific sequences for probes and primers.

Each detection test is a compromise between sensitivity and specificity. It is very difficult to devise tests in which both sensitivity and specificity are high. Depending on the material, the nature of the pathogen (whether quarantine organism or not) and the interests of the consumer and the producer one has to make a choice on which kind of test to use. It is wise to combine detection methods in those cases that really matter.
Diagnosis and detection as such can not make a healthy product. This should be achieved by a mix of many efforts during the whole process of producing a crop. Detection methods can only be used to check whether or not all those efforts led to the desired result: a healthy and safe agricultural product.

Reference

Huttinga, H. (1996) Sensitivity of indexing procedures for viruses and viroids. Advances in Botanical Research (including Advances in Plant Pathology) 23, 59-71.

FUNGAL TAXONOMY IN CRISIS

W. GAMS
Centraalbureau voor Schimmelcultures
3840 AG Baarn, The Netherlands

The negative-sounding word "crisis" should not be used light-heartedly for a highly valued science. But the two meanings of the word crisis, derived from Greek *krinein* = to discern, both apply all too precisely to the components of fungal taxonomy:

- a turning point, especially of a disease, *i.e.* a drastical change of functioning
- a moment of danger or suspense, *e.g.* in politics, commerce; in our case a disastrous decline in interest for a very important discipline.

Reliable fungal identifications and thus taxonomy are the indispensible basis for all other mycological work. Insufficient numbers of experts are available to do this kind of work and this has major repercussions on the quality of mycological (and phytopathological) research, especially in verification, communication and utilization of the results. Further, consider people talking about organisms of which they do not even know the structure and the closest relatives.

Fungal taxonomy has acquired a bad reputation in recent years, for which several factors are responsible.

- Continuous changes of classifications and names. People are frightened in any case by the enormous number of fungal taxa. (There are estimated to be six times more fungi than green plants.)
- A vicious circle then arises: fewer and fewer institutions and universities teach fungal taxonomy, and few students allow themselves to become fascinated by this exciting discipline. As a consequence, very few positions are open to fungal taxonomists in higher education and research.

Nevertheless, progress in fungal taxonomy has never been as fast as it is now.

Fungal taxonomy as a scientific discipline is in a critical situation that requires a redefinition and restatement of its position in a period when molecular biology is sweeping over all previous knowledge.

So far fungal taxonomy has generally been morphology-based. In contrast with bacteria, filamentous fungi have morphological features that very often allow a correct identification from a glance (or careful study) through the microscope or even macroscopically. There are, however, many less clearly differentiated fungi where

H.-W. Dehne et al. (eds.), Diagnosis and Identification of Plant Pathogens, 17–21.

morphological classification is difficult if not impossible. Moreover, a robust phylogenetic backbone of fungal taxonomy has so far been completely lacking.

Chemotaxonomic approaches (secondary metabolites) have contributed very much to the taxonomy of *Penicillium*, *Aspergillus* and related fungi, and lichen-forming fungi. This method is not successful in groups that are poor in secondary compounds. Electrophoretic isoenzyme patterns are also a rapid and efficient tool to characterize and differentiate critical fungi.

In the last ten years molecular biology has gained the field, sometimes, regrettably, at the expense of morphological expertise. It is particularly the analysis of selected genome sequences that allows a reconstruction of phylogeny at various taxonomic levels.

The phylogenetic analysis of the DNA sequences of several genome fractions combined with a careful morphological analysis provides the best basis for a **stable classification.** From such a system further properties of an organism can be deduced. Molecular data have resulted in less drastic changes in taxonomy than feared by many. They contribute a dimension of stability at higher ranks which has not been attained with other methods. If morphological features are then evaluated with this information at hand, it often becomes possible to reach accurate identifications when relying on informative features.

From this kind of work we have learned that the oomycetes (kingdom *Chromista*) are quite unrelated with other fungi (kingdom *Fungi*). We also have learned that the ascomycetes have no affinity whatsoever with red algae as was once suspected. The basidiomycetes and ascomycetes must have moved apart early in the Devonian age. Among the most primitive ascomycetes are organisms like *Taphrina* (causal agent of peach leaf curl, witches' brooms etc.), and the fission yeast *Schizosaccharomyces*; other yeast-like fungi follow somewhat higher up the tree. Next come discomycetous, plectomycetous and many lichenized groups, and later pyrenomycete and loculoascomycete orders. The powdery mildews with globose ascomata are not related to the pyrenomycete group and therefore their unique ascomata are not homologous to either cleistothecia or perithecia. At a finer level, the formerly united ascomycetous groups featuring perithecia with a very long beak are now differentiated into the unrelated genera *Ophiostoma* and *Ceratocystis*, which are also easily distinguished in their asexual, anamorphic forms of sporulation. At the finest level, the recently appearing `aggressive strain' of *Ophiostoma ulmi* has been found to differ sufficiently from this species, particularly at DNA level, to warrant its specific distinction as *O. novo-ulmi*. In *Trichoderma*, molecular taxonomy is now beginning to play a vital role. *T. harzianum* is mainly known as a valuable biocontrol agent of plant-pathogenic fungi. A closely related, morphologically almost indistinguishable, but genetically distinct, highly aggressive fungus now threatens mushroom cultivation in Britain, Canada and the USA. To diagnose this, molecular or chemotaxonomic tools are almost indispensable.

One of the most staggering discoveries of molecular mycology this year is the biogeographic analysis of the *Gibberella fujikuroi* (*Fusarium moniliforme*) complex by O'Donnell *et al.* (1996). Besides the seven species known so far, 21 further species are now recognized (and also morphologically differentiated); these are subdivided into a number of related groups that were originally associated with certain plants in certain continents. The overall connection of these groups suggests a phylogenetic development parallel with that of other biota in the southern hemisphere continents radiating out from Gondwanaland. It seems that we are at last arriving at a stabilized system of real biological species in these problematic groups.

Molecular studies also contribute towards an integration of asexual or conidial fungi into the system of ascomycetes and basidiomycetes. Nevertheless, the recognition of anamorphic (mitosporic) genera will remain indispensable for several decades, at least as long as microscopic examination of a fungus has any meaning and efficacy (Gams, 1995). The cases where known teleomorph connections allow the application of a teleomorph name are still in the minority.

To produce a robust taxonomy, a voluminous material must be carefully studied using different methods. To keep up with developments in the field, a taxonomist must have enough knowledge (or at least understanding) of molecular evolution to be able to collaborate with molecular biologists and experts of other disciplines. Creating a sound fungal taxonomy is becoming more difficult, and problems can often only be solved by the collaboration between morphologists and molecular experts, often also ecologists or pathologists.

A remarkable additional advance due to molecular techniques is the possibility of identification of certain species with rapidity and reliability far superior to techniques used heretofore that was not available so far (but this is beyond the scope of my topic).

Stabilization of nomenclature: Changing names have made many enemies for taxonomy. There is currently a very strong trend to stabilize names whenever possible. Name changes have a threefold origin.

- A reclassification of genera becomes necessary to reflect new discoveries on relationships, and the ensuing name changes should be welcomed as increasing the predictive value of the name.
- Many name changes are introduced by strict application of the rules of nomenclature (particularly the rule of priority). Although historical truth is of high value (Principle III of the International Code of Botanical Nomenclature), a strict adherence is undesirable, particularly in those cases where very old and poorly described names are unearthed and reactivated, particularly by neotypification. There is no criterion by which to judge what is to be regarded as very old. And no rule has been formulated that strictly governs how to solve the problems.
- Misapplied names, where authors have applied a name in a sense differing from the type; this can be due to several reasons, but is particularly confusing.

Two approaches have been envisaged to solve the problem: (1) Lists of "Names in current use" receive a protected status. (2) Individual names of important genera or species that are threatened by older names can be conserved or, vice versa, the disruptive reintroduction of old names can be prevented by their formal rejection. The second possibility has been widely opened since the last International Botanical Congress in Tokyo (1993), while the proposed protection of listed "Names in current use" did not receive sufficient support. Moreover, except in the *Trichocomaceae* (*Penicillium, Aspergillus* and associated teleomorphs), taxonomic mycologists are just not ready to agree on such lists, and waiting for them could delay stabilization by decades. Therefore the conservation/rejection possibility is the immediately available method of choice.

Let me illustrate this: *Fusarium sambucinum* is threatened by *F. roseum*. You know what a mess Snyder & Hansen made of this "species" and its reintroduction would be very confusing. Thus the name *F. sambucinum* will have to be conserved. The name *Rhizoctonia solani* is threatened by another genus, *Moniliopsis*, and another species name, *napaeae*. Conservation of *Rhizoctonia* with the type species *R. solani* and rejection of the name *napaeae* will solve these problems. Important genera conserved include now *Phoma, Phyllosticta, Pleospora, Venturia, Alternaria, Drechslera, Monilia, Oidium*, and species *Penicillium chrysogenum, Aspergillus niger*, etc. Taxonomists should show responsability and avoid changing names without immediate urgency.

Other problems cannot yet be solved by any means. The causal agent of lodging or eye-spot on wheat caused phytopathologists much pain in converting from *Cercosporella* to *Pseudocercosporella herpotrichoides*. Then von Arx (1983) transferred the fungus to *Ramulispora*. The discovery of the teleomorph, *Tapesia yallundae*, confirms its generic segregation from *Pseudocercosporella*, a genus that comprises anamorphs of the loculo-ascomycete genus *Mycosphaerella*. In any case, the teleomorph name should be used preferentially in such a case. Unfortunately, *Tapesia* has now become a nomen rejiciendum since *Mollisia* has been conserved over this genus. The new combination *Mollisia yallundae* has not yet been proposed, but this may eventually be the final resolution of this problem. Whether *Ramulispora* is really the final name for the anamorph, will have to be decided by a molecular comparison between this species and the type species of the genus. Even such complex problems can be solved.

Fungal taxonomy's real crisis, in the second sense of the word, is a reality in the deficient mycological education and shortage of trained experts. Mycological chairs that become vacant are not filled by taxonomists, and no new such positions are being created. This trend prevails in all taxonomic disciplines. As an exemple, in the Netherlands positions for lecturers in taxonomy are decreasing from 1984 to 2000 by 65%, while the overall decrease in such positions for biologists is 20% (van den Broek, 1996). What regards mycology, taxonomic positions have been suspended in the universities of Zürich, Greifswald, Innsbruck, and the Rijksherbarium in Leiden, and probably others. Even worse, it is hardly possible to find qualified candidates to fill mycological taxonomic chairs. In many other universities fungal taxonomy suffers a second- or third-rank existance. In the few remaining institutes where fungal taxonomy is practised and taught, the staff are completely overloaded with work which is often unrelated to the original task of the institute. Such work most frequently involves the obtaining of additional financial sources and the proposal of applied science projects which then detract from the time available for taxonomic work. This is a real crisis that understandably deters students from taking up the challenge of a taxonomic mycological career. Some well-trained experts have been forced to quit universities and go into private business. In view of the scarcity of academic positions and the strong need for their expertise, this kind of activity begins to be quite lucrative. But these experts are lost for taxonomic teaching unless they can be integrated in University courses, at least as external teachers.

I wish to end in a more optimistic tone. Although fungal taxonomy is becoming difficult, even difficult enough to be recognized as "a real science", and often has to rely on the collaboration of various other types of experts, it still is the basic discipline that is indispensable for other aspects of mycology. The morphological examination of fungi will not become superfluous as long as the human intellect is not completely replaced by computers. Voluminous fungal material must be seen and examined, particularly for biodiversity and ecological studies. This can never be handled and sorted effectively by means other than the well-trained eye.

Numerous students love biology because of their affection for morphological diversity. Their enthusiasm should not be deterred by the problems outlined above. On the contrary, time will show the indispensability of well-trained experts in a discipline that is of vital importance to many issues of human concern. Therefore I foresee good professional opportunities for those students who dare to move in a temporarily unpopular direction. This need must be emphasized again and again to decision makers, so that they do not lose sight for the necessary balance of expertise when planning and creating positions.

With thanks to Prof. D.L. Hawksworth and Dr R.K. Summerbell for improvements of a previous draft.

References

Arx, J.A. von (1983). *Mycosphaerella* and its anamorphs. *Proc. K. Ned. Akad. Wet., Ser. C,* **86**, 15–54.

Broek, J. van den (1996). Leerstoelenoverzicht Biologie toont aan: Klassieke biologie onder zware druk. *BioNiews* **1996** (13), 1–3.

Gams, W. (1995). How natural should anamorph genera be? *Can. J. Bot.* **73 (Suppl. 1)**, S747–S753.

O'Donnell, K., Cigelnik, E. and Nirenberg, H. I. (1996). Molecular systematics and phylogeography of the *Gibberella fujikuroi* species complex of *Fusarium. Mycologia* **88** (in press).

PCR-BASED DIAGNOSTICS OF AGRICULTURAL PESTS AND DISEASES

An Introduction to the Basics with Examples

JÜRG E. FREY and BEATRICE FREY
Swiss Federal Research Station for Fruit-Growing, Viticulture and Horticulture, CH-8820 Wädenswil

Since its introduction about a decade ago, the polymerase chain reaction (PCR) has become a widely accepted technique in many agricultural research institutions. By now, a staggering number of variations to the basic PCR amplification process has been described, and even an approximate overview over these would be beyond the scope of this contribution (see Newton and Graham [1994] for a comprehensive review). Rather, we give a short introduction to the principles of PCR and exemplify the benefit of basic methods for agricultural diagnostics.

1. The Polymerase Chain Reaction

PCR is an in vitro amplification of DNA. It is based on the same processes that assure DNA duplication during (or before) cell duplication. The double-stranded DNA molecules in nuclei of eukaryotic cells are separated (denatured) by an enzyme called helicase. The two separated strands with their unpaired nucleotides are then polymerized by strand-specific polymerases. In the polymerase chain reaction, an exponential amplification occurs in a tube in a three-step process including i) denaturation of double stranded DNA, ii) annealing of primers, and iii) polymerisation, i.e., filling up the single stranded DNA with the nucleotides that complement the template DNA (Figure 1). Primers are short sequences of DNA designed to match known DNA regions upstream and downstream of the DNA fragment of interest. In principle, they represent anchors from which the polymerase starts its work.

The aim of the PCR is to produce DNA in amounts that can be analyzed by a variety of diagnostic procedures. Basically, the technique is simple, time-efficient, and robust. Consequently, PCR, in all its diverse applications, has reached wide acceptance in medical and biological research laboratories in a very short time. In our laboratory we apply PCR mainly for molecular diagnostics. The technique allows to identify groups of individuals at any taxonomic level, e.g., assignments to species after establishment of a molecular species identification key. It also enables the identification of characteristics of single individuals or groups, such as carriers of mutations, e.g., resistance of microorganisms to antibiotics, or resistance of pests to pesticides.

H.-W. Dehne et al. (eds.), Diagnosis and Identification of Plant Pathogens, 23–28.

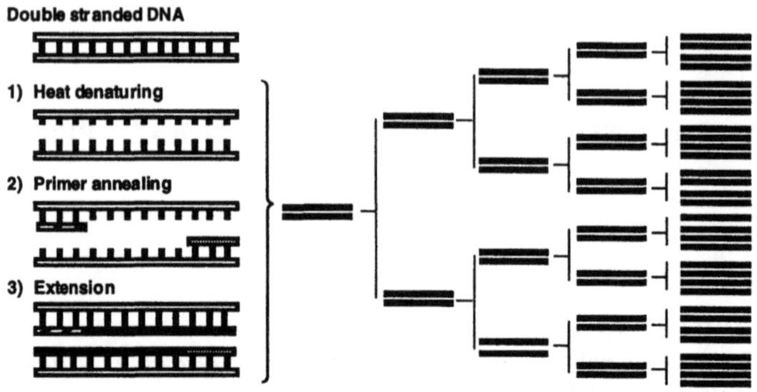

Figure 1: Principle of the polymerase chain reaction: Three-step cyclic amplification.

2. PCR in Molecular Diagnostics

The PCR technique has some very important advantages over other methods for assessing taxon relationships: i) DNA is very stable, thus simplifying handling of samples. Specimen identification based on morphological characters requires fully intact samples. This requirement is often difficult to match if samples have to be taken in the field or are shipped under suboptimal conditions. Biochemical identification keys, i.e., those based on enzyme staining procedures, require that the specimen are alive upon arrival in the laboratory or that the tissue is kept frozen from the time the sample was taken until it is processed in the laboratory. Using DNA as a basis for taxonomic identification of specimen is compatible with a variety of storage methods, e.g., specimen may be frozen, dried, stored in alcohol, or trapped on sticky glue traps. ii) PCR works on minute amounts of DNA. Some assays work with less than 1 nanogram of DNA, making it possible to use very small amounts of tissue, enabling non-invasive sampling, or to use very small animals, such as some hymenopteran egg-parasitoids, or males of some scale insects. iii) PCR is technically robust and rapid, allowing to generate reproducible data with comparatively little effort.

On the other hand, PCR has also some disadvantages. Because very low amounts of DNA are sufficient for successful amplification by PCR (which is one of its advantages), the technique is highly sensitive to contamination and therefore requires sterile conditions for sample preparation. Furthermore, PCR generally requires some sequence information to allow designing of primers. This may be a problem especially in agricultural diagnostics where often identification of organisms is needed for which no molecular information is yet available.

3. RAPD-PCR

The latter problem is avoided by a modification of PCR, the RAPD technique (random amplification of polymorphic DNA). RAPD works with a single short primer of arbitrary sequence (hence "random") of usually 10 base pairs (Figure 2). The primers will attach to inverted repeat sequences on the opposite DNA strands (actually reverse complements) and the polymerase will produce a product if the distance between the primers does not exceed c. 2500 base pairs. Because such inverted repeats may occur more than once within a genome, more than one product can be expected from one reaction with one random primer. These amplification products are generally dominant genetic markers, i.e., the heterozygotic state cannot be assessed which may be a problem for projects with a focus on population genetics. The advantages of RAPD-PCR are that there is no a priori need for DNA sequence information, and that only few steps are involved. Also, it works with very low amounts of template DNA (usually less than 5 nanograms), and it is possible to amplify reproducible fragment patterns from crude DNA extracts without the need of DNA purification.

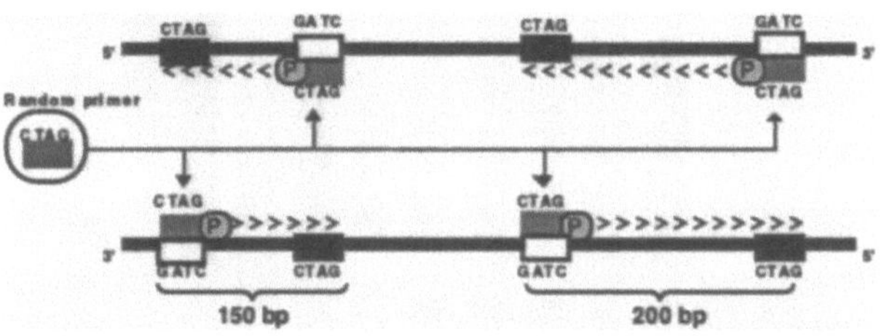

Figure 2: Principle of RAPD-PCR: A single primer of arbitrary nucleotide sequence anneals to reverse complement sequences of the template DNA. Amplification occurs if the distance between inverse annealing sites is less than c. 2500 bp.

We used the RAPD-PCR technique to establish a molecular identification key for six species of scale insects (*Quadraspidiotus* sp.) (Frey and Frey, 1995). The aim of this project was to assess the field-specificity of pheromone traps that are used to monitor the presence of *Q. perniciosus,* a quarantine pest in Switzerland. Because the pheromone analog for the traps was synthesised with compounds identified from the sex pheromone of *Q. perniciosus,* it was thought to be specific for this species. However, mass-catches of *Quadraspidiotus* males occurred on these pheromone traps in several locations in Eastern Switzerland where *Q. perniciosus* was not expected and in fact unlikely to occur. Because no morphological species identification key exists for males of this genus, it was not possible to verify if the specimen attracted by the pheromone in the field were indeed *Q. perniciosus* or if they belonged to another species, indicating that the pheromone was not species specific. We tested 44 random primers (Operon Technologies, California USA) and found four that produced species specific fragment patterns upon electrophoresis (Figure 3). This molecular species identification key enabled us to show that the pheromone used on these traps was also attractive to a second species, *Q. zonatus*, and that the mass-catches in Eastern Switzerland were due to males of this species.

The RAPD-PCR technique has, however, some disadvantages. It is higly sensitive to contamination from diverse sources of DNA, such as that from gut contents, parasitisation, or infection, e.g., by fungi. These foreign DNA sources may produce amplification fragments and may thus obstruct the taxon-specific fragment patterns.

Furthermore, the amplified fragments are completely anonymous with respect to their origin, i.e., there is no information on what the amplified sequences are. Finally, while reproducibility of results within a laboratory is certainly feasible, i.e., using the same chemicals, equipment and operators, it seems to be a challenge to transport RAPD protocols from one laboratory to another. One strategy to overcome this problem is to sequence taxon-specific marker fragments and to design sequence-specific primers that then allow for "regular" PCR.

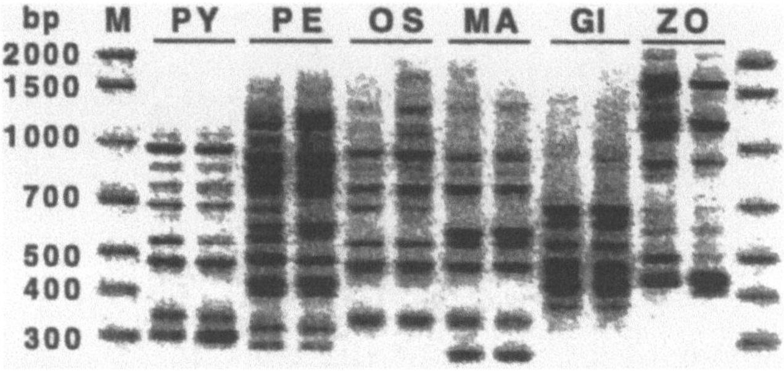

Figure 3: Species diagnostic amplification patterns of two individuals of each of six species of *Quadraspidiotus* scale insects. PY=*Q.pyri*; PE=*Q.perniciosus*; OS=*Q.ostreaeformis*; MA=*Q.marani*; GI=*Q.gigas*; ZO=*Q.zonatus*; bp=base pairs; M=DNA size marker.

4. PCR-RFLP

Some of the problems associated with RAPD-PCR can be avoided using the PCR-RFLP technique (restriction fragment length polymorphism). The amplification step in PCR-RFLP is a "regular" PCR reaction using two different sequence-specific primers (Figure 4). The primers attach to matching nucleotide sequences on both DNA strands bordering the sequence to be amplified. In general, only one amplification product is produced with this approach. The amplification fragment is then analyzed by means of a restriction reaction. The amplification product is digested with a restriction enzyme that recognizes a specific (normally between four to eight bases long) nucleotide sequence. If the amplification product contains the recognition sequence, the restriction enzyme will cut the product into a number of fragments. Provided that the restriction reaction is conducted to completion, i.e., that all existing recognition sequences are cut, the number of restriction fragments always equals that of restriction sites contained in the amplification product, plus one. Incomplete digestion results in additional products, i.e., those that contain a recognition sequence that was not cut. The advantages of this method are primarily its high level of reproducibility that allows for easy adoption in any molecular laboratory. Furthermore, the nature and origin of the amplified fragment is usually precisely known, and the markers obtained after restriction analysis segregate in a codominant fashion. Among the disadvantages are that more experimental steps are involved in PCR-RFLP than in RAPD-PCR, and that the primer sequences have to be known. However, with the increasing number of published primer sequences matching highly conserved genome sites, the latter disadvantage is becoming irrelevant for many studies.

In an ongoing project, we are using PCR-RFLP to establish a molecular species identification key for some economically important thrips species (Thysanoptera: Thripidae). We have been amplifying a piece of mitochondrial DNA and so far, we have digested it with 12 different restriction enzymes, two of which have produced species-specific fragment patterns for four different thrips species (Figure 5). We are now in the process of increasing sample sizes for all species to assess the reliability of the patterns, and we include more species.

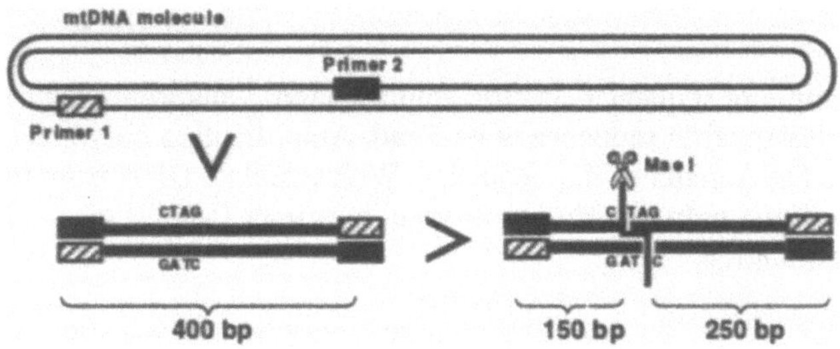

Figure 4: Principle of PCR-RFLP: In the first step, a specific fragment of template DNA is amplified with selective primers. In the second step, the amplified fragment is analyzed by a restriction reaction.

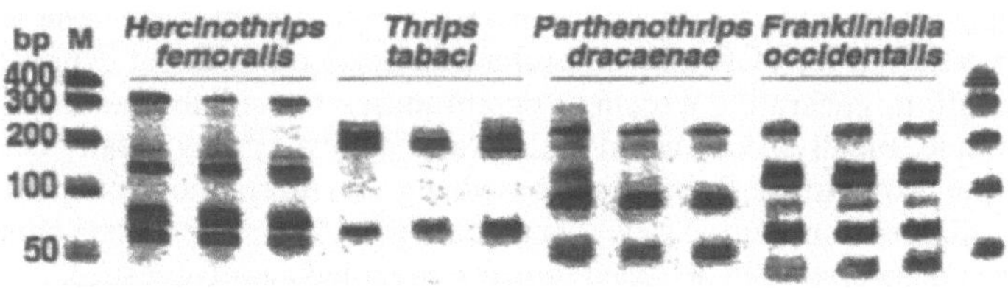

Figure 5: Species diagnostic restriction patterns of three individuals of each of four species of thrips (Thysanoptera: Thripidae) (bp=base pairs; M=DNA size marker).

5. Sequence-specific Amplification

The PCR-RFLP approach did not solve our problems in another project aimed at diagnosing five different species of storage rot fungi (Hilber et al., this issue). We found no restriction enzyme that could differentiate between the amplified ITS-fragments (internal transcribed spacer) of *Botryotinia fuckeliana* and *Monilinia* sp., respectively. Sequencing of the amplification products revealed only very little differentiation between these two species. However, we found a sequence with a three base pair mismatch and designed species-specific primers matching this sequence (Figure 6). With these primers it is now possible to differentiate between *B. fuckeliana* and *Monilinia* sp. (see Figure 3 in Hilber et al., this issue).

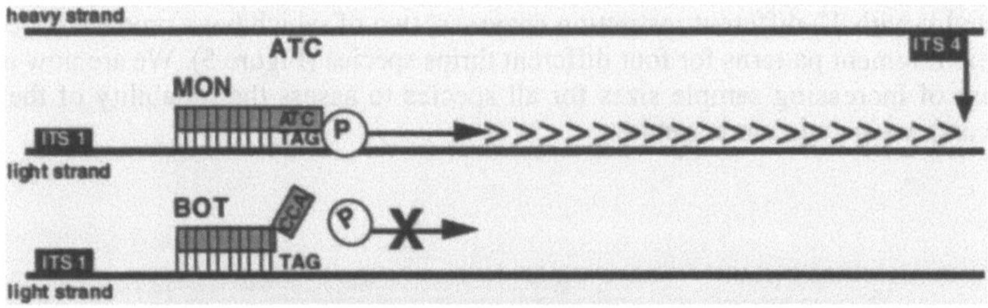

Figure 6: Principle of sequence-specific amplification: One primer is provided with a species-specific sequence at its 3'end. Amplification only occurs if the template DNA matches this sequence. With template DNA from the storage rot *Monilinia*, only the *Monilinia*-specific primer (MON) results in successful amplification.

6. Conclusions

These examples show that comparatively simple PCR techniques can be highly beneficial to the field of agricultural diagnostics. Being based on DNA, PCR can be used on any form of life. We have used PCR to address questions in various species and developmental stages of insects, in nematodes, plants, fungi, and bacteria. The techniques and expertise required for PCR may be exchanged between different departments of a research institution, generating excellent opportunities for collaborative projects that use this technique. Finally, PCR-based molecular diagnostics may greatly benefit from well-established taxonomical frameworks based, e.g., on morphological characters, and vice-versa. Collaboration with classical taxonomists and/or systematists should therefore always be considered in PCR-based projects on molecular diagnostics.

Acknowledgements

We thank Urs Hilber, Benno Graf, and Robert Baur for their helpful comments.

References

Frey, J.E. and Frey, B. (1995) Molecular identification of six species of scale insects (*Quadraspidiotus* sp.) by RAPD-PCR: assessing the field-specificity of pheromone traps. *Molecular Ecology* **4**, 777-780.

Hilber, U.W., Hilber-Bodmer, M., Schüepp, H., and Frey, J.E. (1997) Diagnosis and identification of pathogens causing apple storage rots, by using PCR-RFLP and allele-specific amplification. (this issue).

Newton, C.R. and Graham, A. (1994) *PCR*, BIOS Scientific Publishers, Oxford. (German translation: *PCR*, Spektrum Akademischer Verlag, Heidelberg)

Session 2:

Taxonomy and Differentiation

Electron microscopic methods currently used for diagnosis and characterization of plant viruses

Lesemann, Dietrich-E. and Riedel, Dietmar
Institut für Biochemie und Pflanzenvirologie, Biologische Bundesanstalt für Land- und Forstwirtschaft, Messeweg 11/12, D-38104 Braunschweig, Germany

Currently used electron microscopy (EM) methods which provide effective and versatile applications for several aspects of plant virus research are briefly reviewed based mainly on experimental data from the authors' lab:

1. Routine detection and identification of plant viruses
Unknown virus infections of plants are in many cases most effectively detected and identified by EM. Characteristic particle populations can be recognized in crude tissue extracts whose constituents are nonspecifically adsorbed to support films. The morphology of detected viral particles can be defined in respect of size and shape characteristics and special structural details like flexibility of elongated particles, internal canals, surface fine structures or specific preparation artifacts. The particle size can be measured. Mixed infections by viruses differing in particle morphology can most easily be recognized by EM (Milne 1984).

The typical morphology dictates the choice of antisera for further specific identification by immunoelectron microscopical (IEM) decoration and/or immunosorbent electron microscopy (ISEM) methods. The specific and very conspicuous coating of virion surfaces or parts of these with homologous and to some extent with heterologous antibodies is designated decoration. Mixtures of morphologically identical viruses can be most easily and effectively revealed by decoration. For several genera of elongated viruses a partial 'heterologous' encapsidation has been observed when taxonomically related viruses occur in mixed infections. On particles which appear morphologically intact only parts can be decorated with the antiserum homologous to one of the viruses. Other parts of such particles are not decorated by the same antiserum because they have been built up from coat protein of the serologically distinct second virus of the mixture (Milne 1984).

ISEM, i.e. the specific trapping of virions on antiserum coated grids gives on the one hand clues on the identity of trapped virions and on the other hand dramatically increases sensitivity of detection of known viruses. This is very important if these viruses reach low concentrations only. The increase in particle numbers trapped on antiserum coated grids can be more than 1,000-fold compared to uncoated grids. At appropriate virus concentrations, the number of virions trapped on the antiserum-coated grids is proportional to the virus concentration in the sample. Thus, the determination of relative virus concentrations is possible by ISEM. Several viruses (e.g. luteoviruses, furoviruses and closteroviruses) could not be detected by EM until ISEM became available. With ISEM sensitivity levels similar to those of highly sensitive ELISA formats can be obtained. From mixtures of viruses occurring at very divergent concentrations, the lower concentrated component may only be detected with certainty using ISEM (Milne and Lesemann 1984).

H.-W. Dehne et al. (eds.), Diagnosis and Identification of Plant Pathogens, 31–34.

In conclusion IEM methods offer superior possibilities for a quick detection and identification of plant viruses. Several advantages of IEM, compared to other serological routine methods should be realized: a) direct results, b)no problem with host reactions of antisera, c) high sensitivity of detection, d) analysis of crude plant extracts is possible almost irrespective of their source or composition, e) minimal plant sample sizes, f) minimal antiserum consumption, g) crude and low-titred antisera as well as antisera with host specificity can be effectively used, h) results are rapidly obtained, i) relative broad specificity of the tests allows work with serological variants and even heterologous viruses and j) unexpected or contaminant viruses can be detected.

2. Characterization of fine structure of capsids

EM of negatively stained virions from suspensions permits high resolution analyses of virion fine structures. Examples are the elucidation of the complicated internal structure of rhabdovirus particles (Lesemann and Doraiswamy 1975), the determination of helix parameters of the helically constructed protein capsids of elongated virus particles and the computerized analysis, starting from micrographs, of the clustering of protein subunits on the surface of isometric virions (Milne 1994).

3. Morphology of nucleic acid molecules

Morphological analysis of viral nucleic acid molecules can be performed by using spreading techniques in order to supplement the biochemical characterization of viral nucleic acids. For example the size and linear structure of dsRNA fragments associated with groundnut rosette virus and the size and circular nature of ssDNA components of faba bean necrotic yellows virus could be elucidated by EM (Katul *et al.*, 1993).

4. Determination of serological relationships

Decoration and ISEM techniques can not only be used for the rapid identification of viruses but also permit a simple and direct (semi)quantification of serological relationships between viruses in terms of decoration titers (Vetten *et al.*, 1992 a and b) or relative trapping efficiencies, respectively . Also with low concentrated viruses in crude extracts decoration results can be obtained, even when only a single virion can be visualized. Since the reactions of virus-specific antibodies are directly assessed on the virion surface, clearcut results can be obtained also when antisera exhibit host specificity. Therefore, problems of non-specific reactions often associated with other serological methods using indirect vizualization of reactivity are not encountered. Extremely high titered antisera or especially cross-reactive antisera may cause uncertainty in the distinction by decoration of relatively closely related viruses. In such cases the decoration titers may exceed dilutions of more than 1:10,000, i.e. antiserum dilutions at which the few still reacting antibodies cannot be visualized anymore. The ISEM technique allows for a very sensitive, though laborious, assessment of serological distinctness or serological reactivities (Leseman *et al.*, 1980, 1983; Lesemann 1982; Porth *et al.*, 1987) on the basis of counting particle numbers trapped per standard area.

5. Localization of specific antigens with antisera or monoclonal antibodies (Mab) and epitope analyses

The use of specific antisera or especially Mabs in decoration experiments allows the detection of particular antigens within the protein capsid of virions or of all those epitopes which are exposed at the outer surface of particles, respectively. Reaction sites which exist on small portions of the virion only or which are exposed at very small surface areas only will react with very few antibody molecules. In such cases the antibodies bound to these spots might not be readily recognized by direct visual examination. Immunogold labelling (Louro and Lesemann, 1984) allows the direct visualization of such reaction sites. Examples for such localizations by immunogold labelling are the detection of a second coat protein of beet yellows closterovirus in a short segment of one particle end ('rattlesnake' structure) (Agranosvsky *et al.*, 1995) or the demonstration with specific Mab of glycoproteins of tomato spotted wilt tospovirus on the outer surface of the particles and of nucleoprotein in internal structures of the particles (Adam *et al.*, 1991). Contributions of the technique of immunogold-labelling of Mab reaction sites for epitope analyses have been possible for example with tobacco mosaic virus (Dore *et al.*1988), beet necrotic yellow vein furovirus (Lesemann *et al.*, 1990), beet soil-borne furovirus (Barbarossa *et al.*,1992), bean common mosaic potyvirus (Hasselmann *et al.*, 1993).

6. Cytological alterations of infected cells

Virus-induced alterations of the cellular fine structure can readily be demonstrated in ultrathin sections of fixed and resin-embedded tissues. Virus-specific alterations are represented mainly by the following phenomena: 1) the localization and arrangement of the virions which sometimes can be found in large crystalline arrays., 2) the structure of other cellular inclusions consisting of nonstrucural proteins, 3) the alterations of the cellular membrane system comprizing hypertrophied, loosely or tightly convoluted or regularly arrayed cytoplasmic membranes and 4) the formation of virus-specific membrane-associated or free vesicles containing dsRNA (Francki *et al.* 1985; Lesemann 1988; Lesemann 1995).

7. Immunocytological localization of antigens on ultrathin sections

For analysing virus-host cell interactions immunogold-labelling techniques provide a highly sensitive and specific tool for locating and serologically identifying virions and nonstructural viral proteins (NSP) *in situ* on ultrathin sections of embedded tissues. Conventional antisera to viral coat proteins (CP) as well as antisera to bacterially expressed CP and NSP can be used. For example the detection of CP in virion aggregates formed by Argentine plantago potexvirus was possible on Epon- and on LR-Gold-resin-embedded tissues. Antisera to four bacterially expressed NSP of plum pox potyvirus (PPV) and to its CP have been used in an extensive study to analyse the cellular inclusions induced by PPV and by several other potyviruses (Riedel, D. and Lesemann, D.-E., manuscript in preparation). Results of this study were: 1) All antisera to expressed proteins exhibited relatively strong reactions with host cell components, necessitating absorption of the antisera with healthy plant extracts. 2) Antisera to PPV CP and NSP revealed a broad heterologous reactivity with virions and inclusions, respectively, of some but not all potyviruses tested. 3) All virus-specific inclusions tested were shown to consist of one or several NSP. 4) The location of NSP in cellular inclusions varied from virus to virus. 5) Apart from pinwheel inclusions which consisted only of cylindrical inclusion protein, the other inclusions often contained mixtures of up to four NSP.

8. Localization of nucleic acids on ultrathin sections

Most recently *in situ* hybridization techniques were adapted for the detection of viral (+) and (-) sense RNA by EM. Digoxigenin-labelled specific probes consisting of c. 400 bases of (-) or (+) sense RNA of PPV were successfully used for the localization of plus- and minus-sense viral RNA, respectively, in PPV-infected cells by immunogold labelling (Riedel, D, unpublished results).

9. Perspectives

The examples of applications in the authors' lab have demonstrated the wide spectrum of EM methods in use for the identification and characterization of plant viruses. The inherent limitations of EM will restrict its use for routine diagnosis also in future. For example the techniques are labor-intensive and depend on skill and experience for successful application. Thus they are not likely to be automatized. For future studies of virus-host interactions current limitations are the relatively low sensitivities of detection of antigens and nucleic acids on ultrathin sections. New technical developments should be persued in respect of improved techniques of tissue preservation, embedding and labelling. Sensitivity and topological specificity of immuno-reagents and nucleic acid probes have to be improved for EM applications.

10. References

Adam, G., Lesemann, D.-E. and Vetten, H.J. (1991) Monoclonal antibodies to tomato spotted wilt virus: characterization and application, Ann. appl. Biol. **118**, 87-104.

Agranovsky, A.A., Lesemann, D.-E., Maiss, E., Hull, R. and Atabekov, J.G. (1995) "Rattlesnake" structure of a filamentous plant RNA virus built of two capsid proteins, Proc. Natl. Acad. Sci. USA **92**, 2470-2473.

Barbarossa, L., Vetten, H.J., Kaufmann, A., Lesemann, D.-E. and Koenig, R. (1992) Monoclonal antibodies to beet soil-borne virus, Ann. appl. Biol.**121**, 143-150.

Dore, I., Weiss, E., Altschuh, D. and Van Regenmortel, M.H.V. (1988) Visualization by electron microscopy of the location of tobacco mosaic virus epitopes reacting with monoclonal antibodies in enzyme immunoassay. Virology **162**, 279-289.

Francki, R.I.B., Milne, R.G. and Hatta, T. (1985) Atlas of plant viruses, Vol. 1 and 2, CRC Press Inc., Boca Raton, Florida, 222 and 282 pp.

Hasselmann, D., Lesemann, D.-E. and Vetten, H.J. (1993) Epitope analysis of the capsid proteins of three related potyviruses using monoclonal antibodies, Sixth Internat. Congress of Plant Pathology, Montreal, Abstract No. 17.1.1.

Katul, L., Vetten, H.J., Maiss,E., Makkouk, K.M., Lesemann, D.-E. and Casper, R. (1993) Characterization and serology of virus-like particles associated with faba bean necrotic yellows, Ann. appl. Biol. **123**, 629-647.

Leseman, D.-E. (1982) Advances in virus identification using immunosorbent electron microscopy, Acta Horticulturae **127**,159-173.

Lesemann, D.-E., Makkouk, K.M., Koenig, R. and Samman, E.N. (1983) Natural infection of cucumbers by zucchini yellow mosaic virus in Lebanon, Phytopath. Z. **108**, 304-313.

Lesemann, D.-E. (1988) Recent aspects of the use of cytopathology in virus identification, Acta Horticulturae **234**, 289-298.

Lesemann, D.-E. (1995) Intracellular symptoms, in: Loebenstein, G., Lawson, R.H. and Brunt, A.A. (edts.) Virus and virus-like diseases of bulb and flower crops, John Wiley & Sons, Chichester, pp. 99-116.

Lesemann, D. E., Bozarth, R.F. and Koenig, R. (1980) The trapping of tymovirus particles on electron microscope grids by adsorbtion and serological binding, J. Gen. Virol. **48**, 257-264.

Lesemann, D. and Doraiswamy, S. (1975) Nachweis und elektronenmikroskopische Charakterisierung von "langen", bullet-förmigen Viruspartikeln in kultivierten Orchideen, Phytopath. Z. **84**, 201-214.

Lesemann, D.-E., Koenig, R., Torrance, L., Buxton, G., Boonekamp, P.M., Peters, D. and Schots, A. (1990) Electron microscopical demonstration of different binding sites for monoclonal antibodies on particles of beet necrotic yellow vein virus. J. Gen. Virol. **71**, 731-733.

Louro, D. and Lesemann, D.-E.(1984) Use of protein A-gold complex for specific labelling of antibodies bound to plant viruses I. Viral antigens in suspensions. J. virological Methods **9**, 107-122.

Milne, R.G. (1984) Electron microscopy for the identification of plant viruses in *in vitro* preparations, Methods in Virology, Vol.**VII**,87-120.

Milne, R.G. and Lesemann, D.-E. (1984) Immunosorbent electron microscopy in plant virus studies, Methods in Virology Vol. **VIII**, 85-101.

Porth, A., Lesemann, D.-E. and Vetten, H.J. (1987) Characterization of potyvirus isolates from West African yams (*Dioscorea spp*), J. Phytopathology **120**, 166-183.

Vetten, H.J., Green, S.K. and Lesemann, D.-E. (1992a) Characterization of peanut stripe virus isolates from soybean in Taiwan, J. Phytopathology **135**, 107-124.

Vetten, H.J., Lesemann, D.-E. and Maiss, E. (1992b) Serotype A and B strains of bean common mosaic virus are two distinct potyviruses, Arch. Virol. (**Suppl. 5**), 415-431.

EXPERIENCES WITH THE ISOLATION OF PLANT PATHOGENIC FUNGI

D. AMELUNG
Department of Phytomedicine, Faculty of Agriculture
University of Rostock, Satower Str. 48
D 18051 Rostock (Germany)

The daily work of mycologists is the isolation of the fungi, but sometimes also difficult. Therefore some proven method for the isolation of plant pathogenic fungi are presented.

Two things are essentially for the success: the method of sterilization and the medium. During the sterilisation it is important, that the entire surface is covered with the solution and no air is included. These places are not sterilised otherwise.

Sterilisation solution on sublimate $HgCl_2$ is based. But sublimate is not soluble easily in water. First of all a storage solution must be made: 20 g $HgCl_2$ become to 100 ml conc. HCl given. Thereby the solubility is given and there develop chlorocomplexes. By these storage solution 5 ml to 995 ml alcohol (70%) are given. That is the solution of sterilisation. It works synergistic through the complete solution of sublimat, the development of chlorocomplexes, alcohol and the pH of 1. Also go under the plant segments in this solution immediately.

As first medium for the isolation water agar with 300 ppm chloramphenicole serves to prevent the development of bacteria. Chloramphenicole has be heat stable and can therefore with the agar boiled. Furthermore is the water agar transparent. Small pieces can be observed under the microscope.

For the sterilisation the sample is cut with the scissors in small pieces (approximately 2 x 2 mm). They are done in a nylon screen. The screen is put now in a petri dish with the sterilisation solution. The pieces go under immediately. The time of sterilisation is dependent on the thickness of the pieces. 30 - 60 seconds are sufficient. The screen is taken after this time from the solution. The chosen time of sterilisation can be observed thereby exactly. The screen is given follow in sterile water for the washing. The pieces swim now at the surface. They move quickly, since the physical conditions change suddenly. Thereby the washing is put in addition supports. The pieces become now individually with a fine tweezers on water agar with chloramphenicole. To the sterilisation of the tweezers their tip is dived in alcohol (70%) and flame up. To five pieces can be put on a dish. The cover of the dish is marked with the sample number. On the floor the number of the pieces (1, 2, 3, 4, 5 clockwise) is written.

H.-W. Dehne et al. (eds.), Diagnosis and Identification of Plant Pathogens, 35–36.

After an incubation period of 3 days first hyphae have developed. One can see them with a magnifying glass or microscope. One can recognise the development also at the changed structure of the agar surface by tilting the petri dish. So early as possible the mycel will transfer to another dish. To it is a fine needle, which is selfmade forged and sharpened in form of a scalpel, helpful. The disks are daily to check.

It is to be isolated also possibly fungi with spores. To it must be the spores large or occur in masses, how for example in cirri. That can make one under the microscope with a needle. The best goes that with a thread of glass, which one pull in a desired intensity. Caution, the thread may not be hot. It is better, to cool the thread through dipping in the agar. Thereby the thread is moistened. The spores can stick so better. Through cross crannies with the thread the surface of agar at the places were the spores put down, are to be marked these for the observation. They are otherwise heavy to find.

Propylenoxide can be taken also to the sterilisation. In a 500 ml glass are poured 2,5 ml propylenoxide on cotton and locked the glass. After few minutes the propylenoxide has evaporated. Now can become a glass with the material inside. For the sterilisation are 5 to 15 minutes sufficient. The pieces can be explained subsequently without further treatment on agar.

For the isolation of *Fusarium* from the soil is a bait method successful. Barley or wheat grains are cooked in a water suspension with 50 - 100 ppm Rovral (50% iprodione). The grains are used then as bait. After the incubation in the soil the bait are taken and washed with tap water. Following are put them on filter paper, which is moistened with a water suspension by 50 ppm Rovral,. After the incubation under near UV - light of 4 - 5 days have developed mycelium and spores. Of it can be isolated then.

Isolation of roots are especially difficult, since they are damaged often greatly. For example at *Gaeumannomyces graminis* must be concentrated first of all the fungus. To it a small plastic tube (approximately 20 x 6 mm) with root parts filled. Above a germinated wheat is put. The small tube is stuck in damp sand. The roots increase through and are infected. By these freshly attacked root the fungus can be isolated well.

DETECTION OF DIFFERENT BACTERIAL AND FUNGAL PATHOGENS IN APPARENTLY HEALTHY STRAWBERRY PLANTS

ELISABETH BOSSHARD AND MAYA SCHWINDT
Swiss Federal Research Station, CH-8820 Wädenswil

1. Abstract

The pathogens *Phytophthora fragariae*, *Colletotrichum spp.* and *Xanthomonas fragariae* may be present in the roots or in the system of strawberry plants showing none or atypical symptoms. All three pathogens are known to cause serious losses in some varieties under certain climatic conditions in Switzerland.
For the sanitary control of imported young strawberry plants as green plants, cold-store plants or unrooted stolon-tip plantlets, random samples are drawn at the border and are sent to our lab.
In April and March strawberry fields used for the propagation of young plants are monitored for red stele root rot (*Phytophthora fragariae*) and other root diseases, and fields planted in August for fruit production in the following year are monitored for latent infections with *X. fragariae*, *Colletotrichum* or *Gnomonia* in late fall and spring.
It is required that pathogens are detected and identified in the shortest time possible, specially in sanitary control samples. Methods had also to be adapted to the fact that, depending on the kind of plants to be tested, different plant parts are available for the diagnosis.
In our diagnostic laboratory serological methods are used for the detection and identification of *X. fragariae* and *P. fragariae.* Serological methods may be not very sensitive and not very specific, but they are fast and great numbers of samples can be easily handled. It is assumed that by including control samples and by practicing a carefull interpretation of the results, ELISA-tests can be used to qualify plant material if more accurate methods are not suited for routine control of large numbers of samples, or if they are too expensive.
The Paraquat test or isolation on PDA-dishes are used to detect *Colletotrichum spp* and other pathogens, like *Gnomonia comari* or *Hainesia lythri,* present in the system of strawberry plants. However, it was found that latent infections by these pathogens are difficult to detect under certain conditions.

2. Introduction

Some fungal and bacterial pathogens of strawberries, colonizing the system of the plants, are difficult to control. Growers are therefore adviced to start with healthy young plants. Every so often this simple advice is very difficult to follow. Young strawberry plants are shipped all over Europe, and their origin and the way they are produced are mostly unknown. For this reason the state of health of imported plant material has to be determined.
Different kinds of strawberry plants are imported by nurseries and growers into Switzerland: cold-store plants, unrooted stolon-tip plantlets, and green plants. The plants are tested for freedom of quarantine organisms which, according to EC guidelines, may not be imported into the EC member countries. On the list are the fungi

H.-W. Dehne et al. (eds.), Diagnosis and Identification of Plant Pathogens, 37–41.

Phytophthora fragariae (red stele root rot) and *Colletotrichum acutatum* (antracnose), and the bacterium *Xanthomonas fragariae* (angular leaf spot).
A set of tests, adapted to the kind and origin of samples taken by plant protection inspectors at the border, has been developed.

3. Material and methods

The following diagnosis methods were used to detect quarantine organisms or pathogens reducing plant quality:

3.1. DAS-ELISA FOR X. FRAGARIAE

Antiserum production: Antibodies were produced in collaboration with BIOREBA AG by injecting *X. fragariae* 131 into rabbits.
Sample preparation:
Antigens: cells of *X. fragariae* were suspended in extraction buffer and adjusted to a density of 70% transmittance at 590 nm (approximately $2x10^8$ cfu/ml). Dilutions of 10^6 to 10^4 cfu/ml were used as standard.
Tissue samples were prepared from fresh or frozen plant parts by homogenizing 0.4g of thin slices of rhizomes or pieces of calyxes in 5ml extraction buffer.
ELISA-test: In all tests the DAS-ELISA described by Clark and Adams (1977) was used.

3.2. MODIFIED ELISA FOR *P. FRAGARIAE* AND *P. CACTORUM*

Conjugated antibodies produced by Sabine Werres were used (Werres, 1988).
Sample preparation:
Antigens: lyophilized mycelium of *Phytophthora* isolates was mixed with a spatula tip of Celite (Serva 535, 10-40µm) and homogenized with coating buffer in a mortar (10 mg mycelium/2ml buffer).Dilutions of 10^6 to 10^4 cfu/ml were used as standard.
Tissue samples were prepared as described above (0.1g roots or thin slices of rhizomes /5 ml coating buffer).
Antigen preparations and tissue samples were applied to microtiter plates. After incubation for 4 h at 30°C the plates were washed and the conjugated antibodies were added. Plates were incubated over night at 4°C. After washing the substrate was added. The plates were read after 5 h at room temperature in a Dynatech microplate reader MR5000 at 405 nm.

3.3. ISOLATION OF *COLLETOTRICHUM SPP.* FROM RHIZOMES

Rhizomes of cold-store plants were thoroughly washed in running tap water. After surface-sterilization in 1% NaOCl/70% aethanol during 4 min rhizomes were rinsed in sterile distilled water and cut longitudinally on a sterile glass plate. Small tissue pieces were taken from different areas of the rhizome and were plated out on PDA + antibiotics (100 mg tetracycline, 100 mg streptomycine/ liter). Plates were checked for growth of *Colletotrichum, Gnomonia or Hainesia* after 7 d at 24°C (Bosshard, 1996).

3.4. PARAQUAT-TEST

Preparation of samples for the paraquat-test (Cook 1993): Petioles of green plants were thoroughly washed in running tap water. After surface-sterilization in 70% aethanol (1 min) and 0.5% NaOCl (4 min) petioles were dipped into paraquat (60 sec), rinsed in sterile distilled water and laid out in a moist chamber. After incubation for 7-10 d at room temperature, the samples were checked for acervuli of Colletotrichum.

4. Results

4.1 ELISA DAS FOR *X. FRAGARIAE*

Serum specificity: Repeated tests showed, that the serum cross-reacted with *X.. campestris pv. begoniae* and *X.c.pv. pelargonii*, and to a lesser extent with *X.c. pv. pruni* and *X. c.pv. campestris*; there was a neglectable reaction with a saprophyte isolated from strawberry leaves. Lopez *et al.* (1987), which made the same observations, mention, that the cross-reaction is not problematic as the pathovars are host-specific and don't grow on strawberries to detectable concentrations.
Tests were performed with thin sections of rhizomes of unrooted stolon-tip plantlets, cold-store plants and green plants imported into Switzerland. Most samples were free of X. fragariae; however, the bacterium was detected in cold-store plants and unrooted stolon-tip plantlets of one nursery. Plants produced from this material developed symptoms of angular leaf spot under favourable conditions.
ELISA DAS for *X. fragariae* was also used to identify the cause of reduced growth and the browning of calyxes in field plants.

4.2 MODIFIED ELISA FOR *P. FRAGARIAE* AND *P. CACTORUM*

The antibodies reacted with all Phytophthora species tested, but not with Pythium species. As the sensitivity of the preparation was not very high, the roots of all samples of cold-store and green plants were investigated for red steles macroscopically, and only suspicious cases were tested with ELISA. It was found that in most cases the results were negative and the plants produced from this material were free of the disease.
The modified ELISA for Phytophthora was also used to distinguish cold injuries from symptoms of crown rot (*P. cactorum*).

4.3 DETECTION OF *COLLETOTRICHUM* AND *GNOMONIA*

After *Colletotrichum spp.* had caused serious problems in some strawberry fields in 1995, dying plants as well as symptomfree plant material had to be tested for this pathogen. It was found that *Colletotrichum spp.* could be isolated from rhizoms of both kind of plants. This method was also used to monitor imported cold-store plants. However, the most suitable test to evaluate the state of health of a big number of plants was the paraquat-test developed by Cook (1993). It was used in our lab for monitoring strawberry fields in fall and spring, and for routine control of imported green plants.
With the methods described a distinction between the different species of *Colletotrichum* is not possible. It seems, that *C. acutatum* is present in Switzerland, but that the main damage is caused by *C. gloeosporioides*, which is widespread on different hosts (Bosshard, 1996).
The isolation method described allows to detect other strawberry pathogens like *Gnomonia coumari, Hainesia lythri* or *Verticillium sp..*

5. Discussion

The European and Mediterranean Plant Protection Organization recommends as good plant protection practice in strawberry production the growing of healthy plant material, if possible deriving from a certification scheme (EPPO 1993). This is also the goal of integrated plant and fruit production in Switzerland, where the use of fungicides in the production of young strawberry plants is not allowed as it might hide an infection by *P. fragariae* or *P. cactorum.*

Strawberry diseases which caused problems in recent years have been brought into and spread in Switzerland with imported, latently infected plant material.

X.fragariae is known in Switzerland since 1992 (Grimm *et al.*, 1993). The pathogen may be present latently in the system of young plants without developing symptoms. Infected young plants might show reduced growth. Under favorable weather conditions symptoms like leaf-spots and blackening of the calyxes suddenly appear and reduce the yield of marketable fruit. It is obvious that the distribution of latently infected young plants increases the danger of the epidemic spread of this disease, which cannot be controlled chemically. With the help of DAS-ELISA for *X. fragariae* we try to improve the quality of young plant material.

Phytophthora fragariae has been detected in Switzerland for the first time in the mid-sixties (Lutz *et al.*, 1980). Each spring strawberry fields with motherplants used for the production of unrooted stolon-tip plantlets are controlled for plants affected by the red stele disease. Stunted plants are tested for infection by *P. fragariae.* Recently the reliable but very time-consuming Duncan test (Duncan 1980) has been replaced by a modified ELISA-test for *Phytophthora.* As motherplants as well as cold-store and green plants should be free of all species of *Phytophthora* the disadvantage of the unspecific antibodies detecting also *P. cactorum* might be an advantage. We found that roots of cold-store-plants might be discoloured due to unfavorable storage conditions. With the modefied ELISA for *Phytophthora spp.* a wrong diagnosis can be prevented.

Colletotrichum gloeosporioides is wide spread in Switzerland on hosts like gooseberries, elderberries and cherries, but the outbreak of strawberry fruit rot or the reduced growth and death of *Colletotrichum*-infected young plants has been observed only since 1994 (Bosshard, 1996). It is assumed that changed weather conditions enable so-called warm-country pathogens like *Colletotrichum spp.* to survive in a originally cold country like Switzerland. Investigations on the presence of *C. acutatum* in Switzerland are under way.

The isolation method used to detect *Colletotrichum spp.* in cold-store-plants is quite time-consuming, it is not specific, and its sensitivity is not known. However, it allows to detect also other strawberry pathogens like *Gnomonia coumari, Hainesia lythri* or *Verticillium sp.*.which are not quarantine organisms, but might influence the quality of the plant material and increase the costs for plant protection. The paraquat test is usefull but rather slow to monitor petioles of green or field plants, or calyxes.

Polyclonal antibodies against *Colletotrichum sp.* have been produced, but they proved to be not specific enough.

In recent years detection methods have been improved considerably, but as they are still not fast enough to be used for testing great numbers of plants presently only plants with macroscopically visible symptoms are rejected at the border.

At the time the use of PCR methods are widely discussed (Stammler et al., 1993; Sreenivasaprasad et al., 1996). It has to be determined if this methods are appropriate to work on large numbers of plants, and if their efficiency and costs meet our demands.

Acknowledgment

We thank Dr. S. Werres for the gift of conjugated antibodies against Phytophthora.

References

Bosshard, E. (1996) Why is *Colletotrichum acutatum* a quarantine organism, and *C. gloeosporioides* and *C. fragariae* are not? 3rd International Strawberry Symposium, Veldhoven NL April 28 - May 3, 1996 (in press)

Clark, M.F. and Adams, A.N. (1977) Characteristics of the microplate method of enzyme-linked immunosorbent assay for the detection of plant viruses. J. gen. Virol. **34**, 475-483.

Cook, R.T.A. (1993) Strawberry black spot caused by Colletotrichum acutatum. Ebbels, D. (ed.), British Crop Protection council monograph No ?. 301-304.

Duncan, J.M (1980) A technique for detecting red stele (Phytophthora fragariae) infection of strawberry stocks before planting. Plant Disease 64, 1023-1025.

European and Mediterranean Plant Protection Organization (1993) Good Plant Protection Practice in strawberry. 93/3841

European and Mediterranean Plant Protection Organization (1993). Scheme for the production of certified pathogen-tested strawberry plants. 93/4060; PCF Point 6.2.

Grimm, R., Lips, T. and Vogelsanger, J. (1993) Eckige Blattfleckenkrankheit der Erdbeere. Schweiz. Z. für Obst- und Weinbau **129**, 130.131.

Lutz, M., Lauber H.P. and Stocker W. (1980) Das Wurzelsterben der Himbeeren. Schweizerische Zeitschrift für Obst- und Weinbau 116, 239-245.

Lopez, M.M., Cambra, M., Aramburu, J.M.and Bolinches, J. (1987) Problems of detecting phyopathogenic bacteria by ELISA. Bulletin OEPP/EPPO Bulletin **17**, 113-117.

Richtlinie 92/103/EWG (1.12.1992), Anhang I, Teil A, Abschnitt II.

Sreenivasaprasad, S., Sharada, K., Brown, A.E. and Mills, P.R. (1996) PCR-based detection of *Colletotrichum acutatum* on strawberry. Plant Pathology 45, 650-655.

Stammler, G. and Seemüller, E. (1993) Specific and sensitive detection of Phytophthora fragariae var. rubi in raspberry roots by PCR amplification. Zetischrift für Pflanzenkrankheiten und Pflanzenschutz 100, 394-400.

Werres, S. (1988) Enzyme-linked immunosorbent assay (ELISA) as a method for detection of Phytophthora fragariae Hickman in strawberry roots. Nachrichtenbl. Deut. Pflanzenschutzdienst **40** (10), 146-150.

THE USE OF NESTED PCR FOR THE DETECTION OF *PHYTOPHTHORA FRAGARIAE* IN STRAWBERRY.

P.J.M. BONANTS[1*], M. HAGENAAR-DE WEERDT[1], M.P.E. VAN GENT-PELZER[1], I. LACOURT[2], D.E.L. COOKE[2] AND J.M. DUNCAN[2]

[1]*Institute for Plant Protection (IPO-DLO), PO BOX 9060, 6700 GW Wageningen, The Netherlands*
[2]*Scottish Crop Research Institute (SCRI), Invergowrie, Dundee DD2 5DA, United Kingdom*
[*] *to whom correspondence should be addressed*

1. Introduction

The fungus *Phytophthora fragariae* Hickman is a quarantine organism and is as well pathogenic on strawberry as on raspberry [1]. It means that all plantmaterial for export has to be tested on the presence of the fungus. A sensitive and specific detection method is therefore necessary. At IPO-DLO in the Department of Detection we work on the development of such a detection method. In the past work was done to solve this problem by serological methods [2,3,4,5]. Unfortunately with limited succes. In this study a detection method is developed using nested PCR.
A severely infected strawberry root looks like a rat tail and can easily be recognized as being infected. But for a quarantine organism, like *P. fragariae*, it is also necessary to detect the fungus in symptomless root material.

2. Results

At the start of this project in september 1993 a probe was obtained from dr. Stammler and prof. Seemuller (BBA, Dossenheim, Germany), who worked on *P. fragariae* in raspberry. From the end of this probe primers were designed and they detected the fungus using PCR in infected raspberry roots [6,7]. The probe was located in the rDNA part of the genome and the primers, P1 and P2, were used in our research and checked for specificity and sensitivity.
DNA was amplified from several *Phytophthora* spp. with primers P1 and P2. DNA from all *P. fragariae* isolates reacted with these primers plus DNA from some other *Phytophthora* spp. , e.g. *P. citrophthora*, *P. nicotianae* and *P. capsici*, which were not tested by Stammler. When DNA, isolated from lightly, moderately, severely and uninfected strawberry roots, was tested with these primers it was only possible to detect the fungus in severely infected root material. For a detection method of a quarantine organism criteria such as specificity and sensitivity should be fulfilled. This was not the case with the PCR primers P1 and P2.

H.-W. Dehne et al. (eds.), Diagnosis and Identification of Plant Pathogens, 43–45.

ITS (internal transcribed spacer) regions of the ribosomal DNA (rDNA) are usefull because of their great variability in sequence to develop species specific PCR primers [8,9]. The use of these specific primers for detection purposes has been proven [10,11,12]. After sequencing of the ITS regions of a lot of *Phytophthora spp.* [13,14] and comparing these sequences two primers (DC1 and B5) have been synthesized, which were specific for *P. fragariae*.
The specificity of primers DC1 and B5 was checked by testing it on DNA from a lot of *P. fragariae* isolates and on DNA from some 20 other *Phytophthora* spp. Only *P. fragariae* gave the expected 750 bp PCR product. *P. porri* gave a greater PCR product (2300 bp). All other *Phytophthora* spp. didn't react with the specific primers. After a nested PCR procedure with P1,P2 as outer primers and DC1,B5 as inner primers, the fungus could be detected in lightly infected strawberry root material, something which was not possible with one round of PCR. Also a positive PCR product was seen with DNA from moderately and severely infected roots as template. DNA from uninfected roots was negative.

Because of the difficulty to isolate DNA from a lot of root samples, we decided to focuss on the detection of zoospores in water samples. DNA was isolated from zoospores and nested PCR experiments were performed. Very few (20) zoospores of the fungus could be detected in this way. Water samples of the NAKB (Dutch Inspection Services) could be screened for the presence of *P. fragariae*. Four water samples were tested with the developed nested PCR method. Two samples appeared to be positive and two negative.

Other non electrophoretic methods to detect PCR products are being investigated at IPO-DLO. Beside the gelelectrophoresis new techniques as PCR-ELISA, DIAPOPS and Taq-man are being studied. Also an internal control has been developed to follow the PCR reaction qualitatively. A part of the rDNA subunit of *P. fragariae*, containing the *P. fragariae* specific primersites, was digested with restriction enzyme Hha1 into three fragments. Fragment 1 and 3, which contained those sites, were ligated and cloned into a plasmid. So the plasmid can be added to the PCR reaction to act as internal control giving a PCR product of 650 bp, about 100 bp smaller as the *P. fragariae* PCR product.

3. Conclusion

As a conclusion we can say that it was possible to detect the fungus *P. fragariae* in a specific and sensitive way with the developed nested PCR method. More research has to be done on the induction of zoospores from infected root material to improve to test.

4. References

1. Hickman, C.J. (1940) The Red Core root disease of the strawberry caused by *Phytophthora fragariae*. *J. Pomol. Hortic. Sci.* **18**: 89-119
2. Werres, S. & Casper, R. (1987) Nachweiss von *Phytophthora fragariae* Hickman in Wurzeln der Erdbeerkultursorte "Tenira " mit Hilfe des enzyme linked immunosorbent assay (ELISA). *J. Phytopathol.* **118**: 367-369

3. Werres, S. (1988) Enzyme-linked immunosorbent assay (ELISA) as a method for detection of *Phytophthora fragariae* Hickman in strawberry roots. *Nachrichtenbl. Deut. Planzenschutzd.* **40**: 146-150
4. Amouzou-Alladaye, E., Dunez, J. & Clerjeau, M. (1988) Immunoenzymatic detection of *Phytophthora fragariae* in infected strawberry plants. *Phytopathology* **78**: 1022-1026
5. Olsson, C.H.B. (1995) Diagnosis of *Phytophthora* infections in raspberry and strawberry plants by ELISA tests. *J. Phytopathology* **143**: 307-310
6. Stammler, G., Seemüller, E. & Duncan, J.M. (1993) Analysis of RFLPs in nuclear and mitochondrial DNA and the taxonomy of *Phytophthora fragariae*. *Mycol. Res.* **97**: 150-156
7. Stammler, G. & Seemüller, E. (1993) Specific and sensitive detection of *Phytophthora fragariae* var. *rubi* in raspberry roots by PCR amplification. *J. of Plant Disease and Protection* **100**: 394-400
8. Lee, S.B. & Taylor, J.W. (1992) Phylogeny of five fungus-like protoctistan *Phytophthora* species, inferred from the internal transcribed spacers of ribosomal DNA. *Mol. Biol. Evol.* **9**: 636-653
9. Lee, S.B., White, T.J. & Taylor, J.W. (1993) Detection of *Phytophthora* species by oligonucleotide hybridization to amplified ribosomal DNA spacers. *Phytopathology* **83**: 177-181
10. Nazar, R.N., Hu, X., Schmidt, J., Culham, D. & Robb, J. (1991) Potential use of PCR-amplified ribosomal intergenic sequences in the detection and differentiation of *Verticillium* wilt pathogens. *Physiol. and Molec. Plant Pathol.* **39**: 1-11
11. Henson, J.M., Goins, T., Grey, W., Mathre, D.E. & Elliott, M.L. (1993) Use of polymerase chain reaction to detect *Gaeumannomyces graminis* DNA in plants grown in artificially and naturally infested soil. *Phytopathology* **83**: 283-287
12. Rollo, F., Salvi, R. & Torchia, P. (1990) Highly sensitive and fast detection of *Phoma tracheiphila* by polymerase chain reaction. *Appl. Microbiol. Biotechnol.* **32**: 572-576
13. Cooke, D.E.L., Kennedy, D.M., Guy, D.C., Russell, J., Unkles, S.E. & Duncan, J.M. (1996) Relatedness of Group I species of *Phytophthora* as assessed by randomly amplified polymorphic DNA (RAPDs) and sequences of ribosomal DNA. *Mycol. Res.* **100**: 297-303
14. Cooke, D.E.L. & Duncan, J.M. (1996) Phylogenetic analysis of *Phytophthora* spp. based on ITS1 and ITS2 sequences of rRNA. *Mycol. Res.* (in press)

rDNA SEQUENCING: AN ESSENTIAL GUIDE TO THE TAXONOMY AND DIAGNOSIS OF COLLETOTRICHUM

Nam et ipsa scientia potesta est

JOHN A. BAILEY
IACR-Long Ashton Research Station
Long Ashton,
Bristol, BS18 9AF,
U.K.

1. Introduction

In the genus *Colletotrichum*, reliance on morphological criteria has led to much taxonomic confusion. In the last few years, analysis of rDNA sequences has resolved many questions and it is now possible to propose a new taxonomy for the genus. Examples of incorrect identification are discussed. Accurate and predictive taxonomy, coupled with a full understanding of the extent of phenotypic variability, is essential for meaningful diagnosis. If diagnosis and taxonomy are correct, then pathogen behaviour can be predicted and its control, avoidance, elimination, etc become much easier.

2. A New Taxonomy for Colletotrichum based on rDNA Sequences.

The taxonomy of *Colletotrichum* has been based on highly variable criteria, such as the shape and size of conidia, setae and appressoria, coupled with knowledge of their host plant. These criteria are not satisfactory. Many isolates with indistinguishable morphologies were given different species names which were based solely on the identity of their host. For *Colletotrichum*, several alternative criteria, including biochemical tests, protein profiles and cytochemical procedures based on lectin and monoclonal specificity were proposed (see Bailey and Jeger, 1992). Molecular methods, including RFLPs and RAPDs, are also possible, but like the other methods these are not easily standardised.

Significant advances in taxonomy, and also in understanding the relationships between pathogens are, however, coming from analyses of DNA sequences. DNA sequences provide definitive data that can be readily exchanged and compared. Most of the data obtained to date are based on different regions of the large ribosomal sub-unit, i.e. domains 1 and 2 of the 28S region and the internally transcribed spacers (ITS-1 and -2). Several regions of the large ribosomal sub-unit are especially useful for distinguishing species. In the genus *Colletotrichum*, Sherriff *et al.* (1994) were the first to show the value

H.-W. Dehne et al. (eds.), Diagnosis and Identification of Plant Pathogens, 47–51.

of the different regions of rDNA for taxonomic purposes. Subsequent studies have concentrated on different regions of rDNA, and through cross-referencing to standard isolates significant advances have been made (Bailey *et al.*, 1996; Sherriff *et al.*, 1995; Sreenivasaprasad *et al.*, 1996). On the basis of analysis of rDNA sequences, the following species are proposed.

Species with ovoid conidia:

C. acutatum: morphologically difficult to distinguish from *C. gloeosporioides*, a species with which it is often confused (Sreenivasaprasad *et al.*, 1994).

C. crassipes: similar to *C. acutatum* and *C. gloeosporioides*, but with large lobed appressoria.

C. gloeosporioides: similar to and often confused with *C. acutatum* and *C. crassipes*.

C. orbiculare: contains several pathogens that were previously considered distinct host specific species, i.e. *C. lindemuthianum*, *C. orbiculare*, *C. malvarum* and *C. trifolii*. Distinguished from other species because conidia stay aseptate when appressoria form.

Glomerella cingulata: some forms producing perithecia are the teleomorph of *C. gloeosporioides*. Other forms have DNA sequences that are different from that of *C. gloeosporioides* (C. Nash, LARS, personal communication).

Species with ovoid/weakly falcate conidia, <20um long

C. destructivum: a species with conidia varying from ovoid to slightly falcate. Contains several forms that were previously considered distinct species, e.g. *C. destructivum*, *C. linicola* (=*C. lini*) and *C. truncatum*.

Species with strongly falcate conidia, >15um long

C. capsici: a species on many different dicotyledonous hosts, which is often wrongly designated as a synonym of *C. truncatum*. Distinguished from *C. truncatum* on the basis of its large fully falcate conidia.

C. graminicola: a species associated with maize (*Zea mays)*: difficult to distinguish morphologically from *C. sublineolum*.

C. sublineolum: a species associated with sorghum (*Sorghum bicolor*): difficult to distinguish morphologically from *C. graminicola*

Further studies on their morphology, cytochemistry and host range have allowed many phenotypic characteristics to be defined. An important conclusion from this work is that once a species is established through sequence analysis, several diagnostic characters can be often be discovered. These can now be used to identify some of the proposed species without the need to resort to DNA sequencing. At present, however, some species, e.g. *C. acutatum* / *C.gloeosporioides* and *C. graminicola* / *C. sublineolum* can only be identified unequivocally after DNA analysis. The status of *C. caudatum*, C. *dematium*, *C. coccodes* and *C. falcatum* have yet to be resolved.

3. Examples of Taxonomic Errors and Corrections

3.1. COLLETOTRICHUM ORBICULARE.

Several pathogens, including *C. lindemuthianum* on bean, *C. orbiculare* on cucumber, *C. malvarum* on *Sida* and *C. trifolii* on alfalfa, that have previously been regarded as distinct species (Sutton, in Bailey and Jeger, 1992) are, in fact, very closely related. All these isolates are biological variants of a single species which show distinct host specificity. Because *C. orbiculare* (Berk. and Mont) von Arx was described first (see Sutton, in Bailey and Jeger, 1992), *C. orbiculare* was adopted as the name for this aggregate species. Host origin/specificity can be indicated by use of *forma specialis* (Sherriff *et al.*, 1994): *C. orbiculare* f. sp. from *Phaseolus* spp (previously *C. lindemuthianum*); *C. orbiculare* f. sp. from *Cucumis sativus* (previously *C. orbiculare, syn C. lagenarium*); *C. orbiculare* f. sp. from *Sida spinosa* (previously *C. malvarum*); *C. orbiculare* f. sp. from *Malva pusilla* (previously *C. malvarum*); *C. orbiculare* f. sp. from *Medicago sativum* (previously *C. trifolii*) (Bailey, *et al.*, 1996). Several phenotypic characteristics are consistent with these conclusions and these now provide a simple and rapid means of species identification without the need to resort to DNA sequencing: conidia remain aseptate during appressoria formation (other *Colletotrichum* species produce at least one septum during germination); spores have affinity for a specific lectin and several monoclonal antibodies (Pain et al., 1992); and spherical infection vesicles are formed in epidermal cells (Bailey *et al.*, 1996).

3.2. THE MYCOHERBICIDE BIOMAL

The BioMal pathogen was isolated from *Malva pusilla* and registered as a mycoherbicide for the control of *M. pusilla* in Canada (see Templeton, in Bailey and Jeger, 1992). It was identified and registered as *C. gloeosporioides* f. sp. *malvae*. However, sequence data revealed that the original identification was incorrect. The rDNA sequences of the BioMal pathogen were almost identical to other forms of *C. orbiculare* and hence must be considered as *C. orbiculare* f. sp. from *Malva pusilla* (Bailey *et al.*, 1996). This mycoherbicide produced characteristic conidia, which stained specifically with lectins/antibodies and produced characteristic spherical intracellular infection structures. Surveys of Biomal's pathogenicity failed to reveal any crops to be susceptible. However, since the pathogen is now known to be a member of the *C. orbiculare* aggregate, which contains pathogens of valuable crops, *e.g.*, beans, cucurbits and lucerne, the specificity and stability of BioMal requires verification.

3.3. COLLETOTRICHUM DESTRUCTIVUM

This species aggregate also consists of several pathogens, previously considered as distinct species, i.e *C. destructivum, C. truncatum and C. linicola* (= *C. lini*). The conidial morphology of this aggregate varies greatly: some forms produce conidia that are ovoid (the cowpea anthracnose pathogen), some are elongated (typical *C. destructivum*), whilst others show varying degrees of being falcate (*C. truncatum* and *C. linicola*). However, their initial infection structures are extremely unusual and totally characteristic. On their respective hosts, they all produce large multilobed intracellular infection vesicles in epidermal cells of their hosts (Bailey *et al.*, in Bailey and Jeger, 1992, O'Connell *et al.*, 1993; Latunde-Dada *et al.* 1996a, Latunde-Dada *et al.*, 1996b) again confirming the conservation of phenotypic characters within species defined by molecular analyses. The cowpea anthracnose pathogen, which has long been regarded as a form of *C. lindemuthianum*, has also been shown to be a form of *C. destructivum* (Latunde Dada *et al.*, 1996a and b).

3.4. COLLETOTRICHUM CAPSICI / C. TRUNCATUM

Colletotrichum capsici and *C. truncatum* are names applied inconsistently to pathogens with falcate conidia that do not attack the Gramineae. Sometimes *C. truncatum* is used to describe species with either small or large conidia and *C. capsici* is reserved for a disease of cotton. On other occasions, the use of *C. capsici* is restricted to pathogens with large conidia, and *C. truncatum* (= *C. destructivum*) is reserved for species with smaller conidia. Sequence data (Sherriff *et al.*, 1994) makes a clear distinction between *C. capsici* and *C. truncatum*, which is fully consistent with the morphology of these pathogens: *C. capsici* has characteristic large falcate conidia (20-35µm), whilst *C. truncatum* has smaller less falcate conidia (10-20µm). Their infection processes are also totally different. *C. capsici* produces subcuticular intramural hyphae, whilst *C. truncatum* produces intracellular multilobed biotrophic hyphae. These species have totally different host ranges: *C. capsici* affects a wide range of different plant species, whilst *C. truncatum* is specific to only a few hosts (Bailey *et al.*, in Bailey and Jeger, 1992).

3.5. THE ANTHRACNOSE PATHOGENS OF MAIZE AND SORGHUM

Colletotrichum graminicola is used universally to describe a species causing anthracnose of cereals, notably maize and sorghum. The pathogen from these crops is considered to be the same species. Again DNA sequence analysis has shown that this is incorrect. Pathogens obtained from maize have different sequences than those of isolates obtained from sorghum (Sherriff *et al.*, 1995). The pathogens from maize should thus be considered as *C. graminicola*, whilst those from sorghum should be called *C. sublineolum*. This conclusion is consistent with previous, though generally neglected, studies (Sutton, in Bailey and Jeger, 1992). This distinction is very important because the two pathogens appear to have clearly different phenotypes. Only *C. graminicola* reproduces sexually, the two fungi exhibit different infection processes and *C. graminicola* has a wide host range, whilst *C. sublineolum* has a restricted host range.

3.6. COLLETOTRICHUM ACUTATUM / C. GLOEOSPORIOIDES

Sreenivasaprasad *et al.* (1994) examined several isolates of *Colletotrichum* that caused anthracnose on lupins. All these isolates had been identified previously as *C. gloeosporioides* on the basis of morphological criteria. When rDNA sequences were obtained these identifications were clearly incorrect. All the isolates were forms of *C. acutatum*.

4. References

Bailey, J. A. and Jeger, M. J. 1992. "Colletotrichum: Biology, Pathology and Control", CAB International, Wallingford, 388 pp.

Bailey, J.A., Nash, C., Morgan, L.M. and O'Connell R.J. 1996. Molecular taxonomy of *Colletotrichum* species causing anthracnose on the Malvaceae. Phytopathology: in press.

Latunde Dada, A. O., O'Connell, R. J,. Nash, C., Pring, R. J., Mendgen, K., Lucas, J. A. and Bailey, J. A. 1996a. The infection process and identity of the hemibiotrophic anthracnose fungus (*Colletotrichum destructivum* O'Gara) from cowpea (*Vigna inguiculata* (L.) Walp). Mycol. Res. 100: 1133-1141.

Latunde Dada, A. O., Lucas, J. A. and Bailey, J.A. 1996b. Infection process of *C. destructivum* (O'Gara) from lucerne (*Medicago sativa* L.). Eur. J. Plant Pathol. in press.

O'Connell, R. J., Uronu, A. B., Waksman, G., Nash, C., Keon, J. P. R. and Bailey, J. A. 1993. Hemibiotrophic infection of *Pisum sativum* by *Colletotrichum truncatum*. Plant Pathol. 42: 774-783.

Pain, N. A., O'Connell, R. J., Bailey J. A. and Green, J. R. (1992). Monoclonal antibodies which show restricted binding to four *Colletotrichum* species: *C. lindemuthianum, C. malvarum, C. orbiculare and C. trifolii.* Physiol. Mol. Plant Pathol., 41, 111-126.

Sherriff, C., Whelan, M. J., Arnold, G. M., Lafay, J-F., Brygoo, I. and Bailey, J. A. 1994. Ribosomal DNA sequence analysis reveals new species groupings in the genus Colletotrichum. Exper. Mycol. 18: 121-138.

Sherriff, C., Whelan, M. J., Arnold G. M., and Bailey, J. A. 1995. rDNA sequence analysis confirms the distinction between *Colletotrichum graminicola* and *C. sublineolum*. Mycol. Res. 99: 475-478.

Sreenivasaprasad, S., Mills, P.R. and Brown, A.E. 1994. Nucleotide sequence of the rDNA spacer 1 enables identification of isolates of *Colletotrichum* as *C. acutatum*. Mycol. Res. 186-188.

Sreenivasaprasad, S., Mills, P.R., Meehan, B.M. and Brown, A.E. 1996. Phylogeny and systematics of *Colletotrichum* species based on ribosomal DNA spacer analysis. Genome 39: 499-512.

ANASTOMOSIS GROUPING AND MOLECULAR GENETIC TYPING OF *RHIZOCTONIA SOLANI* CAUSING CRATER DISEASE OF WHEAT IN SOUTH AFRICA

L. MEYER[1], L.H. NEL[1] & D.E. CARLING[2]

[1]*Department of Microbiology and Plant Pathology, University of Pretoria, Pretoria, 0002 South Africa.*

[2] *Agricultural and Forestry Experimental Station, University of Alaska Fairbank, 533 East Fireweed, Palmer, Alaska 99645.*

Crater disease (CD) of wheat (*Triticum aestivum* L.) on the Springbok Flats, South Africa, and bare patch disease of cereals occurring elsewhere in the world, are both caused by *Rhizoctonia solani* Kühn (Samuel & Garrett, 1932; Deacon & Scott, 1985), albeit by different anastomosis groups of the pathogen. Bare patch is ascribed to root infection by anastomosis group (AG) 8 (Neate & Warcup, 1985), whereas recent evidence (Carling *et al.*, 1996) assigned the CD *R. solani* to AG–6. However, unlike other isolates of AG–6 (and all other AGs of *R. solani* for that matter), the CD AG–6 also produces bridging reactions with *R. solani* AG–1 IA (Carling *et al.*, 1996). Furthermore, AG–6 has previously been found only in non–pathogenic association with orchid roots (Uetake *et al.*, 1988), turfgrass (Zhang & Dernoeden, 1995) and as inhabitant of uncultivated soil (Kuninaga *et al.*, 1979), whereas the CD *R. solani* causes a destructive disease. Considering the polymorphic and phylogenetically–diverse nature of AG–6 (Kuninaga & Yokosawa, 1984a) it seems reasonable to assume that the CD *R. solani* could constitute a separate subgroup within this AG.

Advances in molecular techniques, particularly the utilisation of genetic

H.-W. Dehne et al. (eds.), Diagnosis and Identification of Plant Pathogens, 53–57.

markers, have provided answers to basic questions regarding genetic variation and systematic relationships in various groups of phytopathogenic fungi (Sreenivasaprasad *et al.*, 1992; Duncan *et al.*, 1993; Ward & Akrofi, 1994; Liu *et al.*, 1995). Vilgalys & Gonzalez (1990) accordingly indicated a clear differentiation between the various *R. solani* AGs by using restriction fragment length polymorphisms (RFLPs) of rDNA, whereas Vilgalys (1988) showed isolates within particular ISGs to share a high degree of DNA sequence complementarity among their genomes. Considering the applicability of RFLPs in the systematics of *Rhizoctonia* and many other fungal taxa (Vilgalys & Gonzalez, 1990; Kim, *et al.*, 1991), the present investigation resorted to anastomosis and genetic typing of the CD pathogen by RFLP for elucidating its relationship with the various AGs and ISGs of *R. solani*.

CD *R. solani* were acquired from soil and wheat roots on the Springbok Flats in South Africa. Anastomosis group affinities with the various *R. solani* AG tester strains were established for six CD isolates according to a modification (Carling *et al.*, 1987) of the method described by Parmeter *et al.* (1969). Genomic DNA was extracted from the various isolates and ribosomal DNA amplified. For restriction analysis each PCR product was digested with *Cfo*I, *Hpa*II, *Sau*3AI and *Taq*1. Similarity matrices were constructed from the RFLP patterns to provide a quantitative measure of relatedness between isolates.

Inter-isolate anastomosis showed the various CD *R. solani* isolates to be similar regardless of geographic origin. Isolates varied somewhat in respect of anastomosis reaction with the different AG tester strains, but the only pairing of CD *R. solani* resulting in a C2 anastomosis response nevertheless was with the two ISGs of AG–6. In accordance with Carling *et al.* (1996), bridging (C1 reaction) occurred with AG–1 IA. The present investigation also showed bridging between CD isolates and AG–4/AG–4 HGI, a phenomenon that has not been observed before. Although bridging indicates a real but distinct relationship between *R. solani* isolates (Vilgalys & Cubeta, 1994), it is subservient to C2 anastomosis reactions for demonstrating authentic associations.

Placement of CD *R. solani* within AG–6 (Carling *et al.*, 1996) is therefore endorsed. The fact that AG–6 bridges with AG–8 and AG–BI (Sneh *et al.*, 1991), whereas CD isolates did so with AG–1 IA and AG–4 instead, nevertheless point at differences between the two groups. Since rDNA genes contain numerous regions of high sequence conservation and variability over their entire length, it is possible to amplify and study similarity and variation occurring between PCR primer sites at almost any desired taxonomic level (Hillis & Davis, 1986; Vilgalys & Gonzales, 1990; Vilgalys & Cubeta, 1994). In *Rhizoctonia*, RFLP fingerprints are generally consistent with the anastomosis status of a particular isolate (Vilgalys & Gonzales, 1990; Cubeta *et al.*, 1991; Duncan *et al.*, 1993), and the LR7 - LR7–R section of the 25S region in rDNA studied here has previously been found sufficiently variable to separate binucleate *Rhizoctonia* anastomosis groups (Cubeta *et al.*, 1991). Application thereof to determine the phylogenetic relatedness of multinucleate *R. solani* isolates therefore seemed justified.

In this investigation, the major groups that could be recognised according to size of amplified 25S rDNA products (1.4 and 1.8 kb), corresponded with the different size classes of rDNA PCR amplicons reported by Cubeta *et al.* (1991) for binucleate *Rhizoctonia*, and are thought to represent different length–variants of the rRNA gene. Although not accordant with the anastomosis affinities of *R. solani* proposed by Sneh *et al.* (1991), the differentiation was nevertheless considered sufficiently significant for separating isolates yielding a 1.4 kb amplicon from those producing a 1.8 kb amplicon. This apparent unrelatedness of the two groups is substantiated by differences in restriction fragment sizes observed between them after enzymatic digestion with particularly *Hpa*II and *Sau*3AI, as well as separate clustering of the different size groups by Cubeta *et al.* (1991).

The CD isolates proved to be homologous, with all of them yielding a single size PCR amplicon exhibiting similar patterns with each of the four enzymes used. They shared among them a unique and exclusive 610 bp fragment after digestion with *Hpa*II, a characteristic which could in future studies facilitate accurate identification of isolates as CD *R. solani*. Clustering of AG–1 IA and AG–1 IB with AG–6 is in accordance with the report by Liu *et al.* (1995), who analysed the 18S rDNA region of *R. solani* by means of a PCR–based restriction mapping procedure. However, the closer affinity of CD *R. solani* to AG–1 IA and AG–1 IB than to AG–6 does not correspond with the anastomosis reactions showing CD isolates to anastomose with AG–6 and to bridge with AG–1 IA (Carling *et al.*, 1996). Vilgalys & Gonzales (1990), showed AG–1 IA and IB to be highly variable regarding their rDNA restriction patterns, whereas Sneh *et al.* (1991) indicated a low base sequence homology between them. In view of this, greater variation should have been expected in the present investigation between the AG–1 ISGs. To resolve the matter more testers and enzymes obviously need to be screened.

Anastomosis grouping remains the accepted criterion for classification of *R. solani* (Ogoshi, 1987; Vilgalys & Cubeta, 1994) and Carling *et al.* (1996) accordingly placed the CD pathogen in AG–6. Considering its anastomosis and bridging reactions, morphological characteristics, and unique mode of pathogenicity, it is proposed that the pathogen actually represents a distinct ISG within AG–6. *R. solani* AG–6 is a genetically polymorphic group composed of strains with partial phylogenetic relatedness (Kuninaga & Yokosawa, 1984b). Genomic DNA hybridisation divided AG–6 into two genetically different ISGs, with AG–6 HG being a homogeneous group and AG–6 GV consisting of genotypic variable isolates (Kuninaga & Yokosawa, 1984b). Molecular genetic typing done here supports this separation and also links the CD *R. solani* closer to AG–6 GV than to AG–6 HG. It furthermore showed the CD pathogen to differ molecularly from all other *R. solani* AGs associated with bare patch disease of cereals, viz. AG–8, AG–3, AG–4, AG–5, AG–9 and AG–10 (Ogoshi *et al.*, 1990).

5. References

Carling, D.E., Leiner, R.H. & Kebler, K.M., 1987. Characterization of a new anastomosis group (AG–9) of *Rhizoctonia solani*. *Phytopathology* **77**, 1609–1612.

Carling, D.E., Meyer, L., & Brainard, K., 1996. Crater disease of wheat caused by *Rhizoctonia solani* AG–6. *Plant Disease* (In press).

Cubeta, M.A., Echandi, E., Abernethy, T. and Vilgalys, R., 1991. Characterisation of anastomosis groups of binucleate *Rhizoctonia* species using restriction analysis of an amplified ribosomal RNA gene. *Phytopathology* **81**, 1395–400.

Deacon, J.W. and Scott, D.B., 1985. *Rhizoctonia solani* associated with crater disease (stunting) of wheat in

South Africa. *Transactions of the British Mycological Society* **85**, 319–327.

Duncan, S., Barton, J.E. and O'Brien, P.A., 1993. Analysis of variation in isolates of *Rhizoctonia solani* by random amplified polymorphic DNA assay. *Mycological Research* **97**, 1075–1082.

Hillis, D.M. and Davis, S.K., 1986. Evolution of ribosomal DNA: Fifty million years of recorded history in the frog genus *Rana*. *Evolution* **40**, 1275–1288.

Kim, D.H., Martyn, R.D. and Magill, C.W., 1991. Mitochondrial DNA restriction length fragment length polymorphisms in *Fusarium oxysporum* f.sp. *niveum*. *Phytoparasictica* **19**, 211–223.

Kuninaga, S., Yokosawa, R. & Ogoshi, A., 1979. Some properties of anastomosis group 6 and BI in *Rhizoctonia solani* Kühn. *Annals of the Phytopathological Society of Japan* **45**, 207–214.

Kuninaga, S. and Yokosawa, A., 1984a. DNA base sequence homology in *Rhizoctonia solani* Kühn. IV. Genetic relatedness within AG–4. *Annals of the Phytopathological Society of Japan* **50**, 322–330.

Kuninaga, S. and Yokosawa, A., 1984b. DNA base sequence homology in *Rhizoctonia solani* Kuhn. IV. Genetic relatedness within AG–6. *Annals of the Phytopathological Society of Japan* **50**, 346–352.

Liu, Z.L., Domier, L.L. amd Sinclair, J.B., 1995. Polymorphism of genes coding for nuclear 18S rRNA indicates genetic distinctiveness of anastomosis group 10 from other groups in the *Rhizoctonia solani* species complex. *Applied and Environmental Microbiology* **61**, 2659–2664.

Neate, S.M. & Warcup, J.H., 1985. Anastomosis grouping of some isolates of *Thanatephorus cucumeris* from agricultural soils in South Australia. *Transactions of the British Mycological Society* **85**, 615–620.

Ogoshi, A., 1987. Ecology and pathogenicity of anastomosis and intraspecific groups of *Rhizoctonia solani* Kühn. *Annual Review of Phytopathology* **25**, 125–143.

Ogoshi, A., Cook, R.J., and Bassett, E.N., 1990. *Rhizoctonia* species and anastomosis groups causing root rot of wheat and barley in the Pacific Northwest. *Phytopathology* **80**, 784–788.

Parmeter, J.R., Sherwood, R.T. & Platt, W.D., 1969. Anastomosis grouping among isolates of *Thanatephorus cucumeris*. *Phytopathology* **59**, 1270–1278.

Samuel, G. and Garrett, S.D., 1932. *Rhizoctonia solani* on cereals in South Australia. *Phytopathology* **22**, 827–836.

Sneh, B., Burpee, L. and Ogoshi, A., 1991. Identification of *Rhizoctonia* species, APS Press, St. Paul.

Sreenivasaprasad, S., Brown, A.E. and Mills, P.R., 1992. DNA sequence variation and interrelationships among *Colletotrichum* species causing strawberry antrachnose. *Physiological and Molecular Plant Pathology* **41**, 265–281.

Uetake, Y., Obayashi, K., Ogoshi, A. & Tsutsui, K., 1988. Identification of *Rhizoctonia* species isolated from orchids. *Annals of the Phytopathological Society of Japan* **54**, 114.

Vilgalys, R. and Gonzalez, D., 1990. Ribosomal DNA restriction fragment length polymorphisms in *Rhizoctonia solani*. *Phytopathology* **80**, 151–158.

Vilgays, R. and Cubeta, M.A., 1994. Molecular systematics and population biology of *Rhizoctonia*. *Annual Review of Phytopathology* **32**, 135–155.

Zhang, M. & Dernoeden, P.H., 1995. Facilitating anastomosis grouping of *Rhizoctonia solani* isolates from coolseason turfgrasses. *HortScience* **30**, 1260–1262.

IDENTIFICATION OF PHYSIOLOGIC RACES AND FORMAE SPECIALES OF *FUSARIUM OXYSPORUM*

R.P. BAAYEN and C. WAALWIJK
DLO Research Institute for Plant Protection, P.O. Box 9060, 6700 GW Wageningen, The Netherlands

1. Introduction

The anamorphic genus *Fusarium* harbours many plant pathogens. An economically very important species is *F. oxysporum*, which belongs to section *Elegans* and lacks a known teleomorph. Sequence analysis of the 28S rRNA gene indicates that it is closely related to the section *Liseola* [10], with teleomorphs in *Gibberella*. A second species in section *Elegans* is *F. redolens*. Its distinction from *F. oxysporum* has been questioned [3, 8, 15, 18, 19]. No teleomorph has been described for *F. redolens* either: both species propagate in a clonal manner. Genetic diversity is nevertheless considerable, particularly for *F. oxysporum* which comprises more than one hundred formae speciales subdivided in races, and a plethora of saprophytic strains. Such diversity can arise by mutations, transpositions, and the parasexual cycle [13].

Genetic diversity in *F. redolens* is more restricted, with only two formae speciales described thus far: *F. redolens* f.sp. *spinaciae* and *F. redolens* f.sp. *dianthi* [8]. Morphological distinction of the latter from *F. oxysporum* f.sp. *dianthi* is very difficult, and the same disease symptoms are provoked by both taxa [3]. Recent studies show that isolates of *F. oxysporum* f.sp. *dianthi* and *F. redolens* f.sp. *dianthi* can be distinguished unambiguously from one another, however, on basis of the internal transcribed spacer (ITS) region of the ribosomal DNA (rDNA) [2, 17, 18].

Genetic diversity in *F. oxysporum* is traditionally described in terms of host specificity (formae speciales) and cultivar specificity (physiologic races). During the past decade a wealth of data has accumulated on the genetic structure of populations of *F. oxysporum* associated with specific host species, examples of which are presented below. The impact of the novel information on population structure in *F. oxysporum* for diagnosis of races and formae speciales is discussed.

2. Population structure in *F. oxysporum* f.sp. *dianthi*

Pathogenic isolates from a world-wide collection of *Fusarium oxysporum* sensu lato from *Dianthus* spp. were classified into ten vegetative compatibility groups (VCGs).

H.-W. Dehne et al. (eds.), Diagnosis and Identification of Plant Pathogens, 59–62.

Table 1. VCGs coincide with races in *F. oxysporum* f.sp. *dianthi*, but not in *F. oxysporum* f.sp. *lycopersici*. VCGs also coincide with races in *F. redolens* f.sp. *dianthi*

Taxon	Vegetative compatibility group	Race
F. oxysporum f.sp. *dianthi*	0020	4
	0021	2
	0022	1, 8
	0025	11
	0027	10
	0028	9
F. oxysporum f.sp. *lycopersici*	0030	1, 2, 3
	0031	1, 2
	0032	1, 2
F. redolens f.sp. *dianthi*	0001	*1*
	0002	*2*
	0003	*3*
	0004	*4*

On basis of differences in the ITS region of the rDNA between *F. oxysporum* and *F. redolens*, six VCGs were classified as *F. oxysporum* f.sp. *dianthi* and four as *F. redolens* f.sp. *dianthi* [2]. The ability to induce wilt disease on carnation has thus developed convergently in two distinct *Fusarium* species.

VCGs belonging to *F. oxysporum* f.sp. *dianthi* have unique restriction fragment length polymorphisms (RFLPs), esterase profiles, and overall virulence patterns for differential carnation cultivars [2], indicating that VCGs represent the progeny of single individuals that once have independently acquired pathogenicity to carnation. One of the VCGs of *F. oxysporum* f.sp. *dianthi* nevertheless comprised more than one race (Table 1). Races within the same VCG have probably arisen by mutations and transpositions in avirulence genes, whereas races in different VCGs seem not to be related at all. The clonal origin of VCGs is a general concept in *F. oxysporum* [9]. Four distinct races, different from those in *F. oxysporum* f.sp. *dianthi*, also exist in *F. redolens* f.sp. *dianthi* [2]. Both in *F. oxysporum* f.sp. *dianthi* and in *F. redolens* f.sp. *dianthi*, diagnostics for physiologic races can be derived from VCG-based assays.

3. Population structure in *F. oxysporum* f.sp. *lycopersici*

The two major VCGs (0030 and 0031) present among isolates of *F. oxysporum* f.sp.

lycopersici can be differentiated from one another by means of RFLP analyses [7] and isozyme polymorphisms [6]. Although in this forma specialis VCGs again seem to represent clonal lineages, correspondence between VCGs and races is not observed (Table 1). The presence of the same race in more than one VCG can be attributed to the use of monogenic resistance in tomato breeding. Introduction of novel resistance genes has provoked adaptation in both VCGs. Similar observations have been made for *F. oxysporum* f.sp. *cubense* [16] and *F. oxysporum* f.sp. *melonis* [11, 12]. The use of polygenic resistance in carnation breeding for resistance to the dominant race 2 of *F. oxysporum* f.sp. *dianthi* [4] has likely prevented such fungal adaptation.

4. Population structure in *F. oxysporum* f.sp. *lilii*

Among isolates of *F. oxysporum* f.sp. *lilii*, causal agent of basal rot of lily, one major VCG was encountered [1]. Isolates belonging to this VCG share a unique RFLP pattern, again supporting the notion that VCGs form clonal lineages. Pathogenic specialization at cultivar level is of little or no importance in *F. oxysporum* f.sp. *lilii* [14]. Diagnostics for *F. oxysporum* f.sp. *lilii* may be based on the single major VCG.

5. Perspectives for the development of diagnostics

Diagnostics for *Fusarium* at species level may be derived from differences in the ITS region of the rDNA [18, 19]. Specific ITS-based primers may be developed for this purpose. This topic is discussed more extensively elsewhere in this volume [5].

For the development of diagnostics at infraspecific level (formae speciales, races) one should consider the mode of evolution within species. While mating populations are the natural genetic entities for species in section *Liseola* with an active sexual cycle, in the asexual fungus *F. oxysporum* VCGs rather than formae speciales or races should be the target of diagnostics development. Sequence characterized amplified regions (SCARs) are likely to be the most convenient approach to differentiate among VCGs, rather than utilizing RFLPs, RAPDs or isozyme profiling.

Diagnostics at VCG level are promising for those formae speciales where VCGs correspond to (groups of) races. The use of monogenic resistance in plant breeding, however, may provoke fungal adaptation in distinct VCGs to the same resistance gene. Races belonging to different VCGs are likely polyphyletic in origin [11].

Molecular genetic studies show that distinct VCGs within the same forma specialis often differ widely from one another [2]. Such VCGs may well have arisen by convergent adaptation to the same host species. If this hypothesis is correct, and VCGs specialized on the same host do not necessarily have a common ancestor, diagnostics at forma specialis level cannot be developed unless a forma specialis coincides with a single VCG.

6. References

1. Baayen, R.P. (1996) Afweermechanismen bij lelie tegen aantasting door *Fusarium oxysporum*, IPO-DLO report 96-03.
2. Baayen, R.P., van Dreven, F., Krijger, M.C., and Waalwijk, C. (1997) Revision of the classification of races and vegetative compatibility groups in *Fusarium oxysporum* f.sp. *dianthi* and *F. redolens* f.sp. *dianthi*. *Eur. J. Plant Pathol.* **103**, submitted.
3. Baayen, R.P. and Gams, W. (1988) The *Elegans* fusaria causing wilt disease of carnation. I. Taxonomy. *Neth. J. Plant Pathol.* **94**, 273-288.
4. Baayen, R.P., Sparnaaij, L.D., Jansen, J., and Niemann, G.J. (1991) Inheritance of resistance in carnation against *Fusarium oxysporum* f.sp. *dianthi* races 1 and 2, in relation to resistance components. *Neth. J. Plant Pathol.* **97**, 73-86.
5. Baayen, R.P., Waalwijk, C., and Gams, W. (1996) *Fusarium* sections *Elegans* and *Liseola*: taxonomy, rDNA internal transcribed spacer sequences and diagnostics. Elsewhere in this volume.
6. Elias, K.S., and Schneider, R.W. (1992) Genetic diversity within and among races and vegetative compatibility groups of *Fusarium oxysporum* f.sp. *lycopersici* as determined by isozyme analysis. *Phytopathology* **82**, 1421-1427.
7. Elias, K.S., Zamir, D., Lichtmann-Pleban, T., and Katan, T. (1993) Population structure of *Fusarium oxysporum* f.sp. *lycopersici*: restriction fragment length polymorphisms provide genetic evidence that vegetative compatibility group is an indicator of evolutionary origin. *Molec. Plant-Micr. Interact.* **6**, 565-572.
8. Gerlach, W. and Nirenberg, H.I. (1982) The genus *Fusarium* - a pictorial atlas. *Mitt. Biol. Bundesanst. Land.- forstw. Berlin-Dahlem* **209**, 1-406.
9. Gordon, T.R. (1993) Genetic variation and adaptive potential in an asex7ual soilborne fungus, in D.R. Reynolds and J.W. Taylor (eds.), *The fungal holomorph: mitotic, meiotic and pleomorphic speciation in fungal systematics*, CAB International, Wallingford, U.K., pp. 217-224.
10. Guadet, J., Julien, J., Lafay, J.F, and Brygoo, Y. (1989) Phylogeny of some *Fusarium* species, as determined by large-subunit rRNA sequence comparison. *Mol. Biol. Evol.* **6**, 227-242.
11. Jacobson, D.J. and Gordon, T.R. (1991) *Fusarium oxysporum* f.sp. *melonis*: a case study of diversity within a forma specialis. *Phytopathology* **81**, 1064-1067.
12. Katan, T., Katan, J., Gordon, T.R., and Pozniak, D. (1994) Physiologic races and vegetative compatibilitygroups of *Fusarium oxysporum* f.sp. *melonis* in Israel. *Phytopathology* **84**, 153-157.
13 Leslie, J.F. (1990) Genetic exchange within sexual and asexual populations of the genus *Fusarium*, in R.C. Ploetz (ed.), *Fusarium wilt of banana*, APS Press, St. Paul, MN, U.S.A., pp. 37-48.
14. Löffler, H.J.M., Straathof, Th.P., Mouris, J.M., and Baayen, R.P. (1995) Durability of resistance in lily to basal rot: evaluation of virulence and aggressiveness among isolates of *Fusarium oxysporum* f.sp. *lilii*. *Eur. J. Plant Pathol.* **101**, 261-271.
15. Nelson, P.E., Toussoun, T.A., and Marasas, W.F.O. (1983) *Fusarium species. An illustrated manual for identification*, Pennsylvania State University Press, University Park, PA, USA.
16. Ploetz, R.C. (1990) Population biology of *Fusarium oxysporum* f.sp. *cubense*, in R.C. Ploetz (ed.), *Fusarium wilt of banana*, APS Press, St. Paul, MN, U.S.A., pp. 63-76.
17. Waalwijk, C. and Baayen, R.P. (1995) Identification of races of *Fusarium oxysporum* f.sp. *dianthi* byDNA fingerprinting and vegetative compatibility. *Bull. O.E.P.P.* **25**, 117-124.
18. Waalwijk, C., Baayen, R.P., De Koning, J.R.A., and Gams, W. (1996) Ribosomal DNA analyses challenge the status of *Fusarium* sections *Elegans* and *Liseola*. *Sydowia* **48**, 90-104.
19. Waalwijk, C., De Koning, J.R.A., Baayen, R.P., and Gams, W. (1996) Discordant groupings of *Fusarium* spop. from sections *Elegans*, *Liseola* and *Dlaminia* based on ribosomal ITS1 and ITS2 sequences. *Mycologia* **88**, 361-368.

FATTY ACID ANALYSIS IN THE IDENTIFICATION, TAXONOMY AND ECOLOGY OF (PLANT PATHOGENIC) BACTERIA

J.D. JANSE
Department of Bacteriology
Plant Protection Service, P.O. Box 9102, 6700HC, Wageningen, the Netherlands.

1. Introduction

Morphological patterns, that are very important in the classification of most organisms, have only limited value for bacteria. Moreover bacterial anatomy can only be reveiled by laborious electron microscopic methods. Therefore physiological, biochemical, serological and pathological reactions, determined using pure cultures, have been and are used to create recognition patterns for bacteria. These patterns were and are used for classification and identification, and a condensation of them in the form of tables and keys can be found in Bergey's Manuals of Determinative and Systematic Bacteriology. A definition of such man-made patterns is: 'the design made by the relations among a set of objects' (Romesburg, 1984). The contrast between the relations determines the information content of the pattern. The above mentioned patterns are subject to several disadvantages: their generation is usually laborious, expensive and time consuming, their interpretation is liable to errors due to mutations, loss of pathogenicity, choice and quality of test media, etc.

The developments in molecular biology of the past 30 years have generated a lot of knowledge about structure and function of microorganisms, especially of bacteria. It was also discovered that structural elements can be arranged in patterns to be used for discrimination between organisms. Subsequently several chemotaxonomic 'finger-print' methods have been developed, using proteins, nucleic acids, sugars, isoprenoid quinones, mureins, lipids and fatty acids (Goodfellow and Minnekin, 1985). For the 'finger-print' patterns the definition given before also holds true. Interpretation of these patterns is by direct (visual) comparison involving intuition and an expert, by computerized numerical analysis or a combination of both. These 'finger-print' methods have the advantages of being rapid, usually cost effective, applicable in a number of cases to non-culturable organisms or museum specimen, less sensitive to mutation or variation and discrimination at low taxonomic level. However, because these new methods are liable to experimental and/or sample error (also true for so-called genetic methods using nucleic acids) they are to be used with caution. When such a 'finger-print' method is automated, using standard libraries, answers from such an identification system are as good as these libraries and present-day taxonomy are (Janse, 1995;

H.-W. Dehne et al. (eds.), Diagnosis and Identification of Plant Pathogens, 63–70.

Sneath, 1989). One of the 'finger-print' methods used and well established for identification and classification of (plant pathogenic) microbes, namely whole cell fatty acid analysis (FAA) will be treated in more detail in this article.

2. Principle of fatty acid analysis

Fatty acid composition, both qualitative and quantitative, has been known to be a useful characteristic for discrimination between micro-organisms (Abel *et al.*, 1963) for more than 30 years. It also appeared to be a stable character as information for fatty acid formation does not occur on plasmids (Van den Boom and Cronan, 1989). More than 300 fatty acids and related compounds such as aldehydes and dimethyl-aldehydes, have been characterized, most of them also by mass-spectrometry, giving an enormous discriminative potential (Sasser, 1990). An average of cellular fatty acids can be most easily and reproducible obtained by analysis in gas-liquid chromatography, GLC. This average is called here 'fatty acid profile' and should be distinguished from 'fatty acid pattern', the latter being only part of the profile. Especially the development of fused silica capillary colums enabled an accurate analysis (Moss *et al.*, 1980; Eder, 1995). Providing that growth conditions, physiological age of the cells and analysis are well standardized, fatty acid profiles are very reproducible. A high degree of standardization has been realized with an automated system, viz. the Microbial Identification System (MIS) from Microbial ID, Newark, Delaware, USA (Sasser, 1990).

3. Fatty acids of bacteria

Bacteria contain lipids in concentrations of 0.2-50%, usually, 5-10% of dry weight. For FAA the lipids containing esterified fatty acids are important. These are mainly 1) phospholipids, structural elements in the cell membrane; 2) glycolipids, structural elements in the cell membrane, less abundant than phospholipids, common in Actinomycetes; 3) lipid A, the lipid part of lipopolysaccharide (LPS) in the outer membrane of Gram-negative bacteria; 4) lipotheichoic acids, cell wall components of Gram-positive bacteria.

Bacterial fatty acids contain usually 9-20 carbons in a chain. They may be linear, branched, cyclic, saturated, unsaturated or containing 2-or 3-hydroxy groups. Unique bacterial fatty acids are the 3-hydroxy, cyclopropane and branched chain fatty acids. In Gram-negative bacteria the main saturated acids are 16:0, 18:0 and often 14:0, most important unsaturated fatty acids are 16:1 cis or trans 9 and 18:1 cis or trans 11 and hydroxy acids (as part of the LPS) are common. Gram positive bacteria usually contain important quantities of branched fatty acids. Coryneform bacteria and Actinomycetes may contain tuberculostearic acids and mycolic acids (Ratledge and Wilkinson, 1989a,b).

4. The Microbial Identification System (MIS)

The MIS consists of the following elements: a) a protocol for standardized growing and harvesting cells and sample treatment (different protocols for aerobes, microaerophils, anaerobes, yeasts, actinomycetes, fungi, etc.); b) Hewlett Packard (HP)5180 (or update) chromatograph with flame-ionisation detector, integrator (or HP chemstation software) computer with printer, hard disk and disk drive; c) calibration mixture of known fatty acids; d) sofware for identification of individual fatty acids, generation of fatty acid patterns, data storage, identification of micro-organisms, generation of 'mean' fatty acid patterns (library entries) to build the database, generation of database and statistical programs (cluster analysis and principal component analysis) for taxonomic and epidemiological/ecological studies.

5. Pattern generation, using the MIS

To obtain a representative average, qualitative and quantitative, of fatty acids present in bacteria belonging to a wide range of genera, the MIS uses the following procedure (Sasser, 1990):

Bacteria are grown in pure culture for 24 or 48 h on Trypticase Soy Broth Agar (TSBA) at 28°C (slow growers, like *Clavibacter* species, sometimes have to be grown for 72 h) and subsequently c. 40 mg wet weight of cells from a quadrant on the TSBA plate showing still confluent growth is harvested. Cells are saponified for 30 min in a NaOH/MeOH/water mixture at 100°C, fatty acids are methylated for 10 min in a MeOH/6N HCl mixture at 80°C, extraction of methyl esters for 10 min in a hexane/ether mixtrue at room temperature, followed by a base wash for 5 min in a dilute NaOH solution. The extract is placed in a chromatography vial in the autosampler of the chromatograph and the rest of the process is automatic. Separation of acids takes place on a 25m x 0.2 mm phenyl methyl silicone fused silica capillary column, first using a calibration mixture and then samples. The calibration is used in order to update the so called 'peak naming table'. This table contains information for the interpretation of raw chromatographic data and identification of (fatty acid) peaks. Than the fatty acid profile is generated, using the updated peak naming table and an algorithm. Peak naming is achieved by calculating the equivalent chain length (ECL). The ECL can be derived as a function of the retention time of a fatty acid in relation to the retention times of a known series of straight chain fatty acids. The fatty acid pattern is generated by cleaning and adapting the fatty acid profile. Disturbing peaks (e.g. caused by electronic disturbances), that cannot be named, are not used. Subsequently the fatty acid pattern is compared with the libraries present and an identification (including a similarity index) is performed. The patterns can also be used for library generation, taxonomic, epidemiological and ecological studies.

6. The use of fatty acid analysis in the microbial world.

Fatty acid patterns have been shown to be especially valuable for the classification and identification of bacteria, but they can be applied succesfully to a large number of micro-organisms (Lechevalier and Lechevalier, 1989). A compilation of fatty acid data of micro-organisms is given by Ratledge and Wilkinson (1989 a and b). Some examples of applications of FAA are given below:

- Determination of endomycorrhizal fungi, using spores, in the presence of contaminating micro-organisms. FAA analysis of roots provided a measure of colonization development (Bentivenga & Morton, 1994; Graham *et al.*, 1995). Characterization of other fungi was reported by Blomquist *et al.* (1992).
- Differentiation of fermenting and spoilage yeasts in wine (Rozes *et al.*, 1992).
- Determination of composition of biofilms (biological slimes) on paper and board machines. Signature fatty acids were described and amount of cells present could be determined (Vaisanen *et al.*, 1994)
- Check on commercially produced insect pathogens (Siegel *et al.*, 1995).
- Community characterization and identification of bacteria present in decaying sandstone of a cathedral (Descheemaker and Swings, 1995).
- Characterization of Mycoplasms (El-Shabiny and McDonough, 1993).
- Predicting of mosquitocidal activity of *Bacillus* strains, where bioassays on mosquito larvae could be reduced (Frachon *et al.*, 1991).
- Classification of non-nodulating strains of *Frankia* (Mirza *et al.*, 1991).
- Characterization of non-culturable forms of *Campylobacter jejuni* (Hazeleger *et al.*, 1995).

7. Fatty acid analysis in the taxonomy and diagnosis of plant pathogenic bacteria

Fatty acid analysis has proven to be a valuable tool in the taxonomy, diagnosis and ecology of plant pathogenic bacteria. It is being used in quite some diagnostic laboratories, including ours since 1988.

Early studies concerning *Pseudomonas* species were from Ikemoto *et al.* (1978). An extensive study on intra- and inter species discrimination by FAA was performed by Stead (1992). In our laboratory, often in collaboration with other labs, studies have been made for example on a) the discrimination of pathovars within the gall forming *P. syringae* subsp. *savastanoi* (Janse, 1991a), b) *P. (Burkholderia/Ralstonia) solanacearum*, the cause of bacterial wilt in many hosts (where we were able to discriminate between pathovars = races of this pathogen and a study on related bacteria, belonging to the same rRNA group showed a close correlation with DNA-DNA hybridization studies (Janse, 1991b), c) the *P. fluorescens/P. marginalis* soft rot group and *P. aeruginosa* (Janse *et al.,* 1992), d) the recognition of the bacterium causing canker of hazelnut as a separate species, *P. avellanae* (Janse *et al.* , 1996). FAA was useful in the identification of *P. andropogonis* from *Vaccinium corymbosum* (Kobayashi *et al.*, 1995). FAA also played a role in recent reclassifications of *Pseudomonas* species belonging to rRNA group 2 as *Burkholderia* or *Ralstonia* (Yabuuchi *et al.* 1995).

FAA has been very valuable in species delineation and in many cases also pathovar discrimination in the genus *Xanthomonas* (Péros *et al.,* 1994; Vauterin *et al.*, 1991, 1992, 1996b; Yang *et al.* 1993; It has also been used in ecological studies of *Xanthomonas* (Jones *et al.*, 1993; Vauterin *et al.,* 1996 a).

The succesful discrimination of *Erwinia* species and subspecies has been reported several times (De Boer and Sasser, 1986; Kori *et al.*, 1992; Persson and Sletten, 1994). Biotypes of *E. chrysanthemi* from river water were determined by Cother *et al.* 1992. *Erwinia amylovora* was studied by Van der Zwet (1985).

A taxonomic study on *Clavibacter/Curtobacterium* species and subspecies is from Henningson and Gudmestad (1991). The use of FAA to directly detect *C. michiganensis* subsp. *sepedonicus*, in diseased potato tissue was described by Jäger *et al.* (1990).

Biovar discrimination within the genus *Agrobacterium* appeared to be possible (Sawada *et al.,* 1992).

FAA was used in a taxonomic study on *Streptomyces scabies* and related Streptomycetes causing deep scab of potato (Bowers *et al.,* 1995; Faucher *et al.,* 1995).

Phyllosphere studies of bacteria e.g. on wheat (Legard *et al.,* 1994) and sugar beet (Rainey *et al.,* 1994) and characterization of bacterial strains involved in oak decline (Scortichini and Stead, 1993) using FAA have been described.

8. Conclusions: possibilities and limitations of FAA

In an evaluation study on plant pathogenic bacteria, using the MIS system for FAA Stead *et al.* (1992) concluded that FAA (as long as it is well standardized) is a rapid and reliable way of identifying and accuracy of identification (at species and intraspecies level) is generally very good, especially when using an own generated library. This finding is fully confirmed in our lab, as well as another conclusion that infraspecific identification for many pathovars of *P. syringae* is difficult. But this is true for any present laboratory method and it reflects the complexity and narrow relations in this group of bacteria. Of course problems may occur when present day classification or nomenclature (reflected in the library) are not sufficient. Standardization and reproducibility are at a high level and interlaboratory studies confirm this (Persson and Sletten, 1994).

When unknown organisms from complex habitats (like soil) are studied, caution is necessary. When for example an organism is not represented in the library it may show similarity with some organism in the library to a fairly high level, but still being different (Haack *et al.,* 1994; McInroy and Kloepper, 1995; Janse, unpublished; Tonso *et al.,* 1995). This means that in diagnosis FAA is a very valuable aid, but interpretation of data by the diagnostician remains necessary and in some cases additional tests have to be performed. A new development is the on-line generation of fatty acid profiles by gas-chromatography-mass spectrometry, where sample preparation is no longer necessary and only microgram amounts of bacteria are used for analysis (Dworzanski *et al.*, 1990).

In many polyphasic studies FAA has been shown to be an important component and a very good congruence between rRNA-DNA, DNA-DNA-hybridization, protein analysis and sequence data has been found (e.g. Janse, 1992; Rainey *et al.*, 1994; Vauterin *et al.*, 1992) It is a cheap and easy method to trace new species and determine relations between taxa, mainly due to the possibility of statistical comparison of profiles and the possibility to generate libraries of a great many strains studied. (Janse *et al.*, 1996; Kämpfer, 1994; Rozes *et al.*, 1992; Romalde *et al.*, 1995)

9. References

Abel,K., de Schmertzing, H., Peterson, J.I. (1963) Classification of micro-organisms by analysis of chemical composition. I. Feasibility of gas-chromatography. *J. Bacteriol.* **85**, 1039-1044.

Bentivenga, S.P. and Morton, J.B. (1994). Stability and heritability of fatty acid methyl ester profiles of glomelan endomycorrhizal fungi. *Mycol. Res.* **98**, 1419-1426.

Blomquist, G., Andersson, B., Andersson, K. and Brondz, I. (1992) Analysis of fatty acids. A new method for characterization of moulds. *J. Microbiol. Methods* **16**, 59-68.

Bowers, J.H., Kinkel, L.L., Jones, R.K. (1995) Classification of pathogenic strains of *Streptomyces* spp. using fatty acid analysis. *Phytopathology* **85**, 1188 (Abstract).

Cother, E.J., Bradley, J.K., Gillings, M.R. and Fahy, P.C. (1992) Characterization of *Erwinia chrysanthemi* biovars in alpine water sources by biochemical properties, GLC fatty acid analysis and genomic DNA. *J. Appl. Bacteriol.* **73**, 99-107.

De Boer, S.H. and Sasser, M. (1986) Differentiation of *Erwinia carotovora* ssp. *carotovora* and *E. carotovora* ssp. *atroseptica* on the basis of cellular fatty acid composition. *Can. J. Microbiol.* **32**, 796-800.

Descheemaker, P. and Swings, J. (1995) The application of fatty acid methyl ester analysis (FAME) for the identification of heterotrophic bacteria in decaying Lede-stone of the St. Bavo Cathedral in Ghent. *Sci. Total Environ.* **167**, 241-247.

Dworzanski, J.P., Berwald, L. and Meuzelaar, H.L.C. (1990) Pyrolytic methylation-gas chromatography of whole bacterial cells for rapid profiling of cellular fatty acids. *Appl. Environ. Microbiol.* **56**, 1717-1724.

Eder, K. (1995) Gas chromatographic analysis of fatty acid methyl esters. *J. Chromatography. B. Biomedical Applications* **671**, 113-131.

El-Shabiny, L.M. and McDonough, P. (1993) Analysis of fatty acids of Mycoplasma cell. *Veterinary Med. J. Giza* **41**, 5-7.

Faucher, E., Paradis, E., Goyer, C., Hodge, N.C., Hogue, R., Stall, R.E. and Beaulieu, C. (1995) Characterization of streptomycetes causing deep-pitted scab of potato in Quebec, Canada. *Int. Syst. Bacteriol.* **45**, 222-225.

Frachon, E., Hamon, S., Nicolas, L. and De Barjac, H. (1991) Cellular fatty acid analysis as a potential tool for predicting mosquitocidal activity of *Bacillus sphaericus* strains. *Appl. Environ. Microbiol.* **57**, 3394-3398.

Graham, J.H., Hodge, N.C. and Morton, J.B. (1995) Fatty acid methyl ester profiles for characterization of glomalean fungi and their endomycorrhizae. *Appl. Environ. Microbiol.* **61**, 58-64.

Goodfellow, M. and Minnekin, D.E. (1985) *Chemical methods in bacterial systematics*. Academic Press, London, pp. 410.

Haack, S.K., Garchow, H., Odelson, D.A., Forney, L.J. and Klug, M.J. (1994). Accuracy, reproducibility and interpretation of fatty methyl ester profiles of model bacterial communities. *Appl. Environ. Microbiol.* **60**, 2483-2493.

Hazeleger, W.C., Janse, J.D., Koenraad, F.J., Beumer, R.R., Rombouts, F.M. and Abee, T. (1995) Temperature-dependent membrane fatty acid and cell physiology changes in coccoid forms of *Campylobacter jejuni*. *Appl. Environ. Microbiol.* **61**, 2713-2719.

Henningson, P.J. and Gudmestad, N.C. (1991) Fatty acid analysis of phytopathogenic coryneform bacteria. *J. Gen. Microbiol.* **137**, 427-440.

Ikemoto, S., Kuraishi, H., Komagata, K., Azuma, R., Suto, T. and Murooka, H. (1978) Cellular fatty acid composition in *Pseudomonas* species. *J.Gen. Appl. Microbiol.* **24**, 199-213.

Jäger, J., Lorenz, D.H. and Eichhorn, K.W. (1990) Ein schneller Nachweis von *Clavibacter michiganensis* ssp. *sepedonicus* syn. *Corynebacterium michiganense* ssp. *sepedonicum* (Cms) im Gewebe von Kartoffelknollen und Auberginenpflanzen mit der Gas-Chromatographie der Ganzzell-Fettsäuren. *Nachrichtenbl. Deut. Pflanzenschutzd.* **42**, 87-90.

Janse, J.D. (1991a) Pathovar discrimination within *Pseudomonas syringae* subsp. *savastanoi* using whole cell fatty acids and pathogenicity as criteria. *System. Appl. Microbiol.* **13**, 79-84.

Janse, J.D. (1991b) Infra- and intraspecific classification of *Pseudomonas solanacearum* strains, using whole cell fatty acid analysis. *System. Appl. Microbiol.* **14**, 335-345.

Janse, J.D., Derks, J.H.J., Spit, B.E. and van der Tuin, W.R. (1992) Classification of fluorescent soft rot *Pseudomonas* bacteria, including *P. marginalis* strains, using whole cell fatty acid analysis. *System. Appl. Microbiol.* **15**, 538-553.

Janse, J.D. (1995) New methods of diagnosis in plant pathology - perspectives and pitfalls. *Bulletin OEPP/EPPO Bulletin* **25**, 5-17.

Janse, J.D., Rossi, M.P., Angelucci,, L., Scortichini, M., Derks, J.H.J. and Akkermans, A.D.L. (1996) Reclassification of *Pseudomonas syringae* pv. *avellanae* as *Pseudomonas avellanae* (spec. nov.), the bacterium causing canker of hazelnut (*Corylus avellana* L.) *System. Appl. Microbiol.* (in press).

Jones, J.B., Chase, A.R., Hodge, N.C. and Stall, R.E. (1992) Xanthomonad diversity on *Geranium* leaves as determined by monoclonal antobodies MABS, carbon source utilization patterns, fatty acid analysis FAME and pathogenicity. *Phytopathology* **82**, 1066 (Abstract).

Kämpfer, P. (1994) Limits and possibilities of total fatty acid analysis for classification and identification of *Bacillus* species. *System. Appl. Microbiol.* **17**, 86-98.

Kobayashi, D.Y., Stretch, A.W. and Oudemans, P.V. (1995). A bacterial leaf spot of highbush blueberry hardwood cuttings caused by *Pseudomonas andropogonis*. *Plànt Disease* **79**, 839-842.

Kori, Y., Furuya, N., Tsuno, K., Matsuyama, N. (1992) Differentiation of *Erwinia chrysanthemi* and *E. carotovora* subsp. *carotovora* by the cellular fatty acid analysis. *J. Facult. Agr., Kyushu Univ.* **37**, 173-178.

Lechevalier, H. and Lechevalier, M.P. (1989) Chemotaxonomic use of lipids - an overview, in C. Ratledge and S.G. Wilkinson (eds.), *Microbial Lipids* Vol. 1, Academic Press, London, pp. 869-902.

Legard, D.E., McQuilken, M.P., Whipps, J.M., Fenlon, J.S., Fermor, T.R., Thompson, I.P., Bailey, M.J. and Lynch, J.M. (1994) Studies of seasonal changes in the microbial populations on the phyllosphere of spring wheat as a prelude to the release of a genetically modified microorganism. *Agric. Ecosystems & Environ.* **50**, 87-101.

McInroy, J.A. and Kloepper, J.W. (1995) Survey of indigenous bacterial endophytes from cotton and sweet corn. *Plant and Soil* **173**, 337-342.

Mirza, M.S., Janse, J.D., Hahn, D. and Akkermans, A.D.L. (1991) Identification of atypical *Frankia* strains by fatty acid analysis. *FEMS Microbiol. Letters* **83**, 91-98.

Moss, C.W., Dees, S.B. and Guerrant, G.O. (1980) Gas-liquid chromatography of bacterial fatty acids with a fused-silica capillary column. *J. Clin. Microbiol.* **12**, 127-130.

Péros, J.P., Girard, J.C., Lombard, H., Janse, J.D. and Berthier, Y. (1994) Variability of *Xanthomonas campestris* pv. *vasculorum* from sugarcane and other Graminae in Reunion Island. Characterization of a different Xanthomonad. *J.Phytopathol.* **142**, 177-188.

Persson, P. and Sletten, A. (1994) Fatty acid analysis for the identification of *Erwinia carotovora* subsp. *carotovora* and *E. carotovora* subsp. *carotovora*. *Bulletin OEPP/EPPO Bulletin* **25**, 151-156.

Rainey, P.B., Bailey, M.J. and Thompson, I.P. (1994) Phenotypic and genotypic diversity of fluorescent pseudomonads isolated from field-grown sugar beet. *Microbiology (Reading)* **140**, 2315-2331.

Ratledge, C. and Wikinson, S.G. (1989a) *Microbial lipids*, vol. 1. Academic Press, London, pp. 963.

Ratledge, C. and Wikinson, S.G. (1989b) *Microbial lipids*, vol. 2. Academic Press, London, pp. 726.

Romalde, J.L., Magarinos, B., Turnbull, K.D., Baya, A.M., Barja, J.M. and Toranzo, A.E. (1995) Fatty acid profiles of *'Pasteurella' piscicida*: comparison with other fish pathogenic gram-negative bacteria. *Arch. Microbiol.* **163**, 211-216.

Romesburg, H.C. (1984) *Cluster analysis for researchers*. Lifetime Learning Publ., Belmont, USA, pp. 333.

Rozes, N., Garcia-Jares, C., Larue, F. and Lonvaud-Funel, A. (1992) Differentiation between fermenting and spoilage yeasts in wine by total free fatty acid analysis. *J. Sci. Food & Agric.* **59**, 351-357.

Sasser, M. (1990) Identification of bacteria by gas chromatography of fatty acids. *MIDI Techn. Note* **101**, pp. 7.

Sawada, H., Takikawa, Y. and Ieke, H. (1992) Fatty acid methyl ester profiles of the genus *Agrobacterium. Ann. Phytopath. Soc. Japan* **58**, 46-51.

Scortichini, M., Stead, D.E. and Rossi, M.P. (1993) Oak decline: aerobic bacteria associated with declining *Quercus cerris* in Central Italy. *Eur. J. Plant Pathol.* **23**, 120-127.

Siegel, J.P., Smith, A.R. and Novak, R.J. (1995) Cellular fatty acid analysis of *Bacillus thuringiensis* ssp. *israelensis. J. Amer. Mosquito Control Assoc.* **11**, 176-185.

Sneath, P.H.A. (1989) Analysis and interpretation of sequence data for bacterial systematics: the view of a numerical taxonomist. *System. Appl. Microbiol.* **12**, 15-31.

Stead, D.E. (1992) Grouping of plant-pathogenic and some other *Pseudomonas* spp. by using cellular fatty acid profiles. *Int. J. Syst. Bacteriol.* **42**, 281-295.

Stead, D.E., Sellwood, J.E., Wilson, J. and Viney, I. (1992) Evaluation of a commercial microbial identification system based on fatty acid profiles for rapid, accurate identification of plant pathogenic bacteria. *J. appl. Bacteriol.* **72**, 315-321.

Tonso, N.L., Matheson, V.G. and Holben, W.E. (1995) Polyphasic characterization of a suite of bacterial isolates capable of degrading 2,4-D. *Microbial Ecology*, **30**, 3-24.

Vaisanen, O.M., Nurmiaho-Lassila, E.L., Marmo, S.A., Salkinoja-Salonen, M.S. (1994) Structure and composition of biological slimes on paper and board machines. *Appl. Environ. Microbiol.* **60**, 641-653.

Van den Boom, T. and Cronan, J.E. (1989) Genetics and regulation of bacterial lipid metabolism. *Ann. Rev. Microbiol.* **43**, 317-343.

Van der Zwet, T. (1985) Characterization of *Erwinia amylovora* through fatty acid profiling. *Phytopathology* **75**, 1281 (Abstract).

Vauterin, L., Yang, P., Alvarez, A., Takikawa, Y., Roth, D.A., Vidaver, A.K., Stall, R.E., Kersters, K. and Swings, J. (1996a) Identification of non-pathogenic *Xanthomonas* strains associated with plants. *System. Appl. Microbiol.* **19**, 96-105.

Vauterin, L., Yang, P., Hoste, B., Pot, B., Swings, J. and Kersters, K. (1992) Taxonomy of xanthomonads from cereals and grasses based on SDS-PAGE of proteins, fatty acid analysis and DNA hybridization. *J. Gen. Microbiol.* **138**, 1467-1477.

Vauterin, L., Yang, P., Hoste, B., Vancanneyt, M., Civerolo, E.L., Swings, J., Kersters, K. (1991) Differentiation of *Xanthomonas campestris* pv. *citri* strains by sodium dodecyl sulfate-polyacrylamide gel electrophoresis of proteins, fatty acid analysis, and DNA-DNA hybridization. *Int. J. Syst. Bacteriol.* **41**, 535-542.

Vauterin, L., Yang, P., Swings, J.M., Nakase, T. and Komagata, K. (1996b) Utilization of fatty acid methyl esters for the differentiation of new *Xanthomonas* species. *Int. J. Syst. Bacteriol.* **46**, 298-304.

Yabuuchi, E., Kosako, Y., Yano, I., Hotta, H. and Nishiuchi, Y. (1995) Transfer of two *Burkholderia* and an *Alcaligenes* species to *Ralstonia* gen. nov.: Proposal of *Ralstonia pickettii* (Ralston, Palleroni & Douderoff 1973) comb.nov., *Ralstonia solanacearum* (Smith 1896) comb. nov. and *Ralstonia eutropha* (Davis, 1969) comb. nov. *Microbiol. Immun.* **39**, 879-904.

Yang, P., Vauterin, L., Vancanneyt, M., Swings, J. and Kersters, K. (1993) Application of fatty acid methyl esters for the taxonomic analysis of the genus *Xanthomonas. System. Appl. Microbiol.* **16**, 41-71.

ASSESSMENT OF MICROBIAL COMMUNITIES BY FATTY ACID ANALYSIS

V. Vilich
Rheinische Friedrich-Wilhelms-Universität Bonn
Institut für Pflanzenkrankheiten, Phytomedizin in Bodenökosystemen
Nussallee 9, D-53115 Bonn

1. Abstract

Because attractance, chemotropism and inhibition of pests and pathogens are influenced by the community structure of rhizosphere bacteria a measurement technique for the assessment of microbial populations in this ecological niche was developed. Samples of subcrown internodes (SCI) of barley were collected in cereal monocultures and rotations in Syria (ICARDA). Measurements of plated SCI-communities on TSA and GSP according to the Microbial Identification System (MIDI, Inc.) identified treatment based differences in fatty acid profiles and characterized microbial sub-populations. The statistical factor analysis of the Cellophane-Colony-Harvest method gave evidence that incubation on different nutrient or selective media resulted in higher factor loadings for so-called 'marker fatty acids'.

2. Introduction

The role that rhizosphere microbial communities play in the improvement of root health is poorly understood. New concepts aimed at management of overall suppressiveness in a soil-ecosystem through management of specific segments of the microbial community are the exception (Sikora, 1992; Vilich and Sikora, in press). This undoubtedly is related to the complex nature of the microbial community in the rhizosphere. Recently developed methodologies for detection of complex community structures however, should improve our knowledge of rhizosphere microbial communities and our ability to use these communities as buffer systems for disease and pest management. Rotation is the most effective short and long term plant protection measure for control of many soil-borne pests and diseases (Rovira et al. 1990; Vilich, 1993). Advances in methodology needed to study microorganisms in the soil has helped demonstrate the large degree of the spacial and functional interrelationships between pathogenic and non-pathogenic microorganisms associated with the root system and the ultimate effects of these interactions on root health.

H.-W. Dehne et al. (eds.), Diagnosis and Identification of Plant Pathogens, 71–74.

3. Materials and Methods

Field experiments were conducted at ICARDA (International Center for Agricultural Research in the Dry Areas, Aleppo, Syria). Samples of barley subcrown-internodes (SCI) were taken at random at growth stage EC 69 in 1996 in barley rotations and monocultures. Treatments consist of barley/lentil rotation (cv. Arabi aswad), barley/vetch rotation (cv. Rihane) and barley monoculture (cv. Rihane). Subsamples of SCI material (1-2 mg) were suspended in Ringer-solution and incubated on Tryptic Soy Agar (TSA) and GSP (Merck) media. Before plating of the suspensions sterile cellophane segments were placed on media. Bacterial colonies were counted after incubation at 25°C.

Using the Cellophane-Colony-Harvest method (CCH) cellophane segments covered with bacterial colonies were transferred to glas vials and fatty acids extracted according to the MIS whole cell analysis by MIDI, Inc. Fatty acids were analysed in a HP5890 ChemStation using the 'Fungi' method. Controls consisted of sterile cellophane segments treated in the same manner. Statistical analysis were done with STATGRAPHICS PLUS by multivariate methods (Factor/Cluster Analysis) and standardized data.

4. Results and Discussion

The cellophane colony harvest method (CCH) exhibited differences in fatty acid distribution (Tab.1). Major fatty acids (FA) extracted from communities were C16:0, C17:0 cyc and C16:1 cis9. GSP-Agar grown populations yielded more ß-hydroxy FA indicating the presence of gram-negative bacteria like pseudomonads.

TABLE 1. Relative fatty acid content (%) of SCI microbial communities incubated on TSA and GSP-Agar (mean of all treatments)

Fatty acid	TSA	GSP
10:0 3OH	1,28	3,63
12:0	2,44	3,88
12:0 2OH	2,52	4,10
12:0 3OH	2,42	5,41
15:0 iso	1,01	
15:0 aniso	1,03	
16:0	22,99	30,47
16:1 cis9	17,33	28,56
17:0	0,83	
17:0 cycl	10,63	6,23
19:0 cycl	1,38	

Factor analysis exhibited different factor loadings for fatty acids (Fig. 1). On TSA factor loadings were highest (I.quadrant) for C15:0, C17:0 and 15:0iso; ß-hydroxy FA showed highest factor loadings on GSP media.

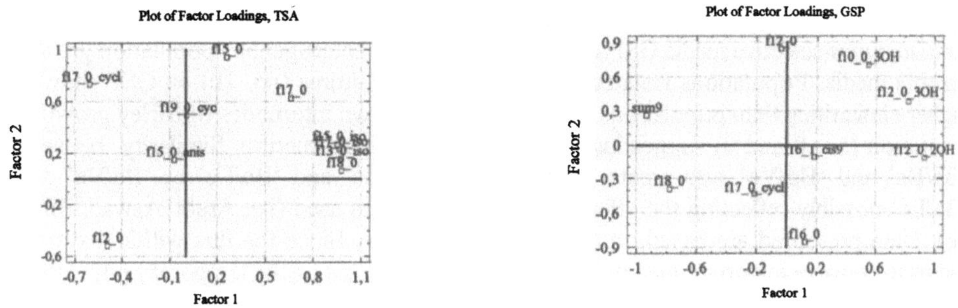

FIGURE 1. Factor loadings of fatty acids of subcrown internode bacterial populations of barley. Incubation on TSA and GSP media

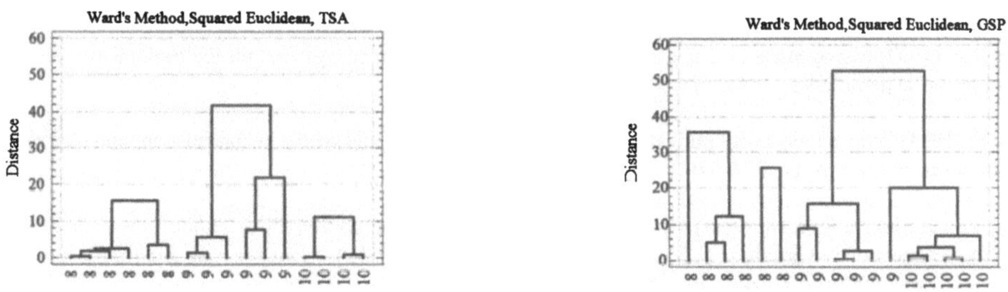

FIGURE 2. Cluster analysis for fatty acids of subcrown internode bacterial populations of barley grown in legume rotation (8: Barley/Lentil; 9: Barley/Vetch) and monoculture (10). Incubation on TSA and GSP media

The result indicated that incubation on nutrient or selective media pointed out marker fatty acids since β-hydroxy FA loaded highstest in GSP-populations and C15:0, C17:0 and - C15:0*iso* loaded highest on TSA media. The β-hydroxy FA have been considered to be biomarkers for gram-negative bacteria (e.g. pseudomonads), C15:0 and C17:0 for bacteria in general and C15:0*iso* for gram-positive bacteria (Zelles et al., 1995).
The Cluster analysis was applied to specific fatty acids. Clustering data showed clear dissimilarities of SCI samples from different barley rotations if the microbial suspensions were incubated on TSA (Fig.2). No differences between treatments occurred using the *Pseudomonas*-selective media GSP which indicated more heterogenous population growth on this media. Populations isolated from barley monocultures (no. 10) on GSP showed higher similarities than populations isolated from subcrown internodes of barley grown in a rotation (no. 8 and 9) suggesting a more homogeneous structure. Similarity between 'Ba/Ba' and 'Ba/Ve' was greater than between 'Ba/Ba' and 'Ba/Le' or 'Ba/Ve' and 'Ba/Le' possibly reflecting the effect of different cultivars used (i.e. Arabi Aswad, Rihane). Data presented are based on a single sampling date. Since the flux within a system seems to be more important than the actual number of species or the temporary population composition (Kennedy and Smith, 1995), additional experiments on simultaneous development of pathogens and microbial communities with time over seasons are under work.

References

Kennedy, A.C. and Smith, K.L. (1995) Soil microbial diversity and the sustainability of agricultural soils, in H.P.Collins, G.P. Robertson and M. J. Klug (eds.), *The significance and regulation of soil biodiversity*, Kluwer Acad. Publ., Netherlands, pp. 75-86

Rovira, A.D., Elliot, L.F. and Cook, R.J. (1990) The impact of cropping systems on rhizosphere organisms affecting plant health, in J.M. Lynch (ed.), *The Rhizosphere*, John Wiley and Sons, Chichester, pp. 389-436

Sikora, R.A. (1992) Management of antagonistic potential in agricultural ecosystems for the biological control of plant parasitic nematodes, *Annu. Rev. Phytopathology* **30**, 245-270

Vilich, V. (1993) Crop rotation with pure stands and mixtures of barley and wheat to control stem and root rot diseases, *Crop Protection* **12**, 373-379

Vilich, V. and Sikora, R.A. (in press) Diversity in soil-borne microbial communities - a tool for biological system management of root health, in G.J. Boland and L.D. Kuykendall (eds.), *Plant Microbe Interactions and Biological Control*, Marcel Dekker, Inc., New York

Zelles, L., Rackwitz, R., Bai, Q.Y., Beck, T. and Beese, F. (1995) Discrimination of microbial diversity by fatty acid profiles of phospholipids and lipopolysaccharides in differently cultivated soils, *Plant and Soil* **170**, 115-122

TESTING VIABILITY AND GERMINATION OF *PLASMODIOPHORA BRASSICAE (WOR.)* RESTING SPORES USING THE FLUORESCENCE MICROSCOPE TECHNIQUE

KATRAIN VENT-SCHMIDT, and H. BOCHOW
Humboldt-University Berlin, Agric.-Hortic. Fac.,
Div. Phytomedicine,
Dorfstr. 9, 13051 Berlin, Germany

Abstract

The usefulness of the fluorescence microscope method for the assessment of resting spore activity of *Plasmodiophora brassicae* (Wor.) was assessed. The influence of root exudates of cruciferous and non-cruciferous plants on the germination of *P. brassicae* resting spores was investigated with a view to the future application for the biocontrol of clubroot.

The stained spores were classed in correspondence with the categories 1 - 7 of DOYLE (1990, unpublished): Non-viable spores when classed as germinated or dead ones, showed a blue sporewall and red fluorescence on the inside, while viable spores showed the blue sporewall alone.

The influence of root exudates from cruciferous and non-cruciferous plants on the germination of *P. brassicae* resting spores is described by the percentage of viable, non-germinated spores.

1. Introduction

The clubroot fungus *Plasmodiophora brassicae (Wor.)* is one of the most dangerous soil-borne root pathogens of crucifers because of the difficulty to control it, especially in soils strongly infested by the long surviving and resistant resting spores. The most common control is by soil liming, increasing the soil pH level up to about 7.0 or 7.2 to inhibit resting spore germination and improve plant resistance with high calcium levels (WEBSTER; DIXON 1991). Useful fungicides are also searched for.

Resting spores of *P. brassicae* (Wor.) will also be influenced by the rhizosphere of growing plants, apparently by root exudates. Root exudates of cruciferous host plants showing clubbing symptoms, were shown to stimulate the germination of resting spores (BOCHOW, 1965; MACFARLANE,1970). Root hair infections were observed in a number of non-host and non-cruciferous plants, indicating that spore germination

H.-W. Dehne et al. (eds.), Diagnosis and Identification of Plant Pathogens, 75–78.

is also influenced by plants in which the biotrophic fungus cannot reproduce (WALTER 1985). SUZUKI et al. (1992) have found the existence of a germination stimulating factor for *P. brassicae* resting spores from host as well as non host plants. These facts and a number of observations in Japan, Scotland, Poland and own experiments suggest the possibility of developing a new biocontrol method of cultivating catch or baiting crops which stimulate resting spore germination but in which the fungus cannot reproduce.

We investigated the influences of root exudates from cruciferous and non-cruciferous plants on the germination of *P. brassicae* resting spores to ascertain if plant specific or non-specific stimulating factors exist to be applied for biocontrol purposes.

Direct observation of spore germination is very difficult and labour intensive and the biotrophic fungus cannot be grown on axenic media. The fluorescence microscopic technique after fluorochrome staining of the spores described by TAKAHASHI and YAMAGUCHI (1988) was found to be useful for the assessment of the activity of resting spores.

2. Materials and methods

Water suspensions of resting spores of *P. brassicae* were prepared from washed, ripe clubroot galls frozen at -20 °C. The galls were grated, centrifuged and the suspension adjusted to 10^7 per ml and kept at 5 °C.

Root exudates were collected from three host plants: Chinese cabbage (*Brassica pekinensis*), Rape seed (*Brassica napus*), Brown mustard (*Brassica juncea*) and three non-host plants: Wheat *(Triticum aestivum)*, Tomato (*Lycopersicon esculentum*), Cucumber (*Cucumis sativus*). Plants were cultivated from surface sterilized seeds in sterile glowed quartz-sand, in 2 l pots with lateral holes for the introduction of ceramic filter candles in the pot center, where roots grow intensively. Root exudates were extracted from the filter candles applying a negative pressure of -0.7 bar, without any disturbance of the plants. Plant free quartz-sand was treated in the same way as control. The collected root exudates were stored at 3 °C for no longer than 10 days.

The effect of the root exudates on *P. brassicae* resting spores were tested by mixing 0.1 ml spore suspension with 0.1 ml root exudates. The mixture was placed on a microscope glass slide in moist chambers (Petri - dishes) to prevent their drying. The Petri-dishes were incubated at 18 °C or 25 °C for three days. The fluorochrome staining solution was prepared by mixing aqueous solutions of calcofluorowhite (100 µg/ml) and ethiumdibromide (30 µg/ml) in a 1:1 ratio (TAKAHASHI and YAMAGUCHI, 1988). A few droplets of staining solution were added to the incubated test slides and staining reactions were examined under the fluorescence microscope with the filter set 05, blue-violet stimulation, 395-440 nm, filter FT 460; LP 470. TAKAHASHI (1991) using this staining method showed that spores with the blue sporewall alone are viable spores and spores with a blue sporewall and red fluorescence on the inside are non-viable spores.

3. Results and discussion

We classed the stained spores in categories 1-7 in correspondence with DOYLE 1990 (unpublished). The described categories are demonstrated in colour pictures, here not illustrateable.

The staining reaction of spores depended on the activity of *P. brassicae* resting spores as described by TAKAHASHI (1991). Spore germination was not reflected by absolute clear staining reactions. Blue fluorescence was detected on the wall of all spores. The germination process is reflected by red fluorescence inside the spores. Similar reactions were seen in empty, damaged or dead spores.

We classed spores with the blue sporewall alone - "blue spores" as viable spores (DOYLEs Category 1); and spores with a blue sporewall and red fluorescence on the inside - "red spores" as non-viable spores. Non-viable spores are germinated or dead ones (DOYLEs categories 3, 4 and 6). Damaged resting spore cases (DOYLEs category 5) and bisquit shaped spores (DOYLEs category 7) were visible but were not counted. DOYLEs category 2, that ist germinating spores, is included in "red spores", because this process is not detectable.

All evaluations were based on the quantitative reduction in percentages of „blue spores“ compared to the control.

The results of infection experiments by WALTER (1985) showed that root exudates cannot, in the context of our experiment damage, destroy or kill resting spores. A decrease in percentages of "blue spores" compared to the control is therefore seen as consequence of spore germination.

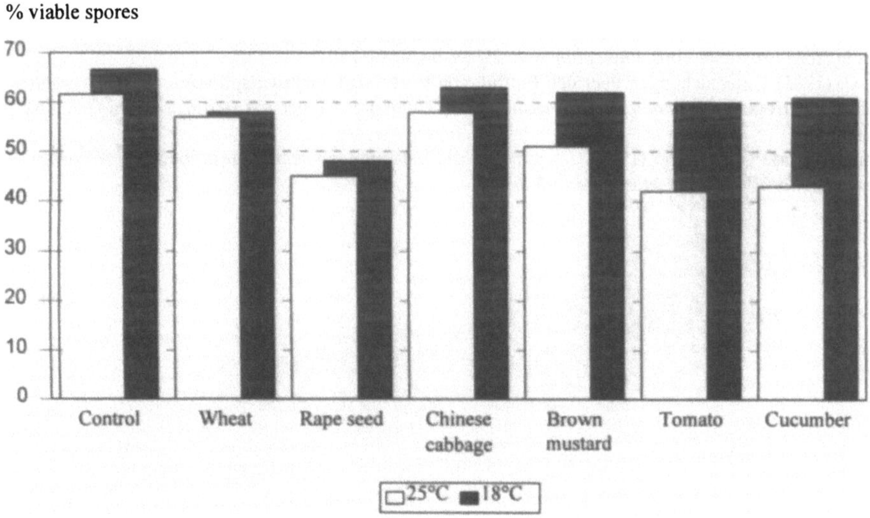

Figure 1: Effect of root exudates of 13 days old plants on resting spores of *P. brassiace (Wor.)*

Root exudates of both host and non-host plants reduced the percentage of viable "blue" *P. brassicae* resting spores (Fig 1). This indicates the existence of a non plant specific root-borne spore germination factor, coinciding with the findings of SUZUKI et al. (1992). We also found indications that temperature, pH and age of the plant influence spore germination.
Presentation and discussion of all these relevant observations will not be given in this methodical paper, but shall be published in the Ph.D. thesis of the junior, lady author.

Aknowledgements
We thank Dr. Doyle from the National University of Ireland, Dublin, for his advice and the permission to use the classification of resting spores elaborated in his hitherto unpublished Ph.D. thesis.

4. References

Bochow, H. (1965) The effect of root diffusates of host and non-host plants on the resting spore germination of *Plasmodiophora brassicae* Wor. In: *Plant Microbes Relationships,* Proceedings of a Symposium. Publishing House of the Czechoslovak Academy of Sciences, Prague 1965, 296-299.
Doyle, O.P. E (1990) Studies on clubroot of Brassicas - disease assessment, control and resting spore viability in *Plasmodiophora brassicae.* Ph.D. thesis, National University of Ireland, 328 pp.
Macfarlane, I. (1970) Germination of resting spores of *Plasmodiophora brassicae. Trans. Br. Myc. Soc.* **55,** 97-112.
Suzuki, K., Matsumiya, E., Ueno, Y. and J. Mizutani (1992) Some properties of germination-stimulating factor from plants for resting spores of *Plasmodiophora brassicae. Ann. Phytopath. Soc. Japan* **58**, 699-705.
Takahashi, K. (1991) Correlation between fluorescent staining reaction and germination in resting spores of *Plasmodiophora brassicae. Ann. Phytopath. Soc. Japan* **57**, 160-164.
Takahashi, K. and T. Yamaguchi (1988) A method for assessing the pathogenic activity of resting spores of *Plasmodiophora brassicae. Ann. Phytopath. Soc. Japan* **54,** 466-475.
Walter, C. (1985) Untersuchungen über den Einfluß von Wirts- und Nichtwirtspflanzen auf das Verhalten des Kohlhernieerregers, *Plasmodiophora brassicae* Wor. Humboldt-Universität Berlin, Sektion Gartenbau, Dissertation A.
Webster, M.A. and G.R. Dixon (1991) Calcium, pH and inoculum concentration influencing colonization by *Plasmodiophora brassicae. Mycol. Res.* **95,** 64-73.

HISTOLOGICAL STUDY OF A SUSCEPTIBLE POTATO CULTIVAR (*Solanum tuberosum* L. cv. Igor) INFECTED WITH POTATO VIRUS Y^{NTN}

JASNA DOLENC[1], BARBARA VILHAR[2], MAJA RAVNIKAR[2] and MARINA DERMASTIA[1,2]
[1]*Department of Biology, Biotechnical faculty, University of Ljubljana and* [2]*The National Institute of Biology, Večna pot 111, 1000 Ljubljana, SLOVENIA*

1. Introduction

Potato cultivar Igor is very susceptible to the potato virus Y^{NTN}, when plants are grown in soil. The symptoms of the potato tuber ring necrotic disease (PTRND) caused by PVY^{NTN} are severely expressed (clear mosaic and brightening on the leaves, severe necrosis on the stems and veins, accelerated leaf senescence and leaf drop; Le Romancer *et al.* 1994). However, systemically PVY^{NTN}-infected plants transferred to stem-node culture exhibit an enhanced tolerance to PVY^{NTN} (Dermastia and Ravnikar, 1996). Subsequent studies on different potato cultivars revealed that susceptibility to PVY^{NTN} was cultivar dependent (Jerman *et al.*, 1996). Moreover, the susceptibility in soil correlated with tolerance in tissue culture and *vice versa*. Thus, the symptoms of PVY^{NTN}-infection were the most prominent in *in vitro* grown plants of cv. Pentland Squire (Jerman *et al.*, 1996), which is known as tolerant cultivar to PVY^{NTN} when grown in soil.

The increased tolerance to PVY^{NTN} in cultivar Igor correlated with increased endogenous concentration of plant hormones - cytokinins, especially biologically active zeatin riboside (Dermastia and Ravnikar, 1996). Historical definition of cytokinins was based on their stimulation of cell division in plant tissue grown *in vitro*. Cytokinins are also involved in control of the cell cycle in apical meristems of roots and shoots, and in secondary meristems of the cambia and phellogen (Jacqmard *et al.*, 1994). The meristems are special groups of cells from which all tissues in a mature plant are ultimately derived. Shoot apical meristems produce the above ground parts of the plant. The shoot apical meristem consists of a small dome of relatively undifferentiated dividing cells. They proliferate and finally differentiate into various cell types, and can regenerate a new meristem if damaged.

In this paper we discuss some histological growth parameters related to the cell cycle of shoot meristem cells, which might explain to some extent the involvement of cytokinins in growth of PVY^{NTN}-infected potato plants of cv. Igor cultivated in stem node culture.

2. Materials and Methods

2.1. PLANT MATERIAL

Single stem node cuttings were excised from 8-week old healthy or systemically infected (PVY^{NTN}) plants (*Solanum tuberosum* L. cv. Igor), which were propagated *in vitro* on the

H.-W. Dehne et al. (eds.), Diagnosis and Identification of Plant Pathogens, 79–84.

Murashige and Skoog (1962) medium adjusted to pH 5.7-5.8 before autoclaving. The cultures were kept at 20±2°C, with a photoperiod of 16 h light / 8 h dark at 50 μmol m^{-2} s^{-1} (Osram LI8W 20 lamps). The stem height and the number of leaves were recorded on day 4, 7, 11, 14, 18, 21, 28 and 35 of culture. Uniform distribution of potato virus Y^{NTN} within all plant organs was confirmed by ELISA (Petrovič *et al.*, 1995).

2.2. PREPARATION OF TISSUE SECTIONS FOR LIGHT MICROSCOPY

The histological studies were conducted on 18- and 35-day-old plants. Shoot tips were fixed in 3:1(v/v) absolute ethanol: glacial acetic acid for at least 24 hours. The tissue was then dehydrated in an alcohol series prior to wax-embedding using standard procedures (Taylor and Francis, 1989). Longitudinal sections (12 μm thicknesses) were cut through the shoot meristem. The slides were rehydrated in an alcohol series, hydrolyzed in 5M HCl (optimum hydrolysis time 30 min, 25°C) and stained with the Feulgen reagent (pararosaniline chloride, BDH, UK) for 2.5 h at 25°C. The sections were then dehydrated, counter-stained by Fast Green (0.1% Fast Green in ethanol) and mounted in DPX.

Dimensions of meristem zones and the number of cells were measured from images portrayed on a TV monitor connected to a light microscope, which was equipped with a video camera. Tracings of zone perimeters were computed into area measurements using a PC with a graphic tablet.

2.3. GROWTH PARAMETERS

2.3.1. *Relative growth rate of the shoot (RGR)*

RGR was calculated as follows:

$$RGR = \frac{ln\ (H2/H1)}{t2 - t1} \tag{1}$$

where *H1* is the stem height at time *t1* and *H2* is the stem height at time *t2*.

2.3.2. *Mean Tissue Doubling Time (MTDT)*

The shoot tip sections of 18- and 35-day-old plants were used to estimate MTDT with the phytomere method. First, the relative growth rate for the shoot apex (*r*) was calculated, from the areas of the apex and (apex + the youngest leaf primordium) on the median longitudinal sections as:

$$r = \frac{ln\ (apex + ylp) - ln(apex)}{t_p} \tag{2}$$

where t_p is the duration of the plastochron (the period between initiation of successive leaf primordia on the shoot apical meristem).

The estimate of the mean tissue doubling time (MTDT) in the shoot apex was calculated as:

$$MTDT = \frac{ln2}{r} \tag{3}$$

3. Results and Discussion

The symptoms of PVY^{NTN}- infection, which are strongly expressed in secondarily infected soil-grown plants were milder after these plants were transferred into the *in vitro* conditions (Dermastia and Ravnikar, 1996). In these plants, PVY^{NTN}-viral particles remained evenly distributed within all plant organs and their concentration in plant tissues was high (Petrovič *et al.*, 1995). Nevertheless, after five weeks of cultivation *in vitro* the morphological characteristics of infected plants did not differ significantly from the healthy controls (Dermastia and Ravnikar, 1996).

While the virus affected the growth rate of systemically infected plants at the beginning of the cultivation (in the first 18 days), the growth was buffered against the infection in the following days (Figure 1). However, after 35 days the stem height of the infected plants was lower than that of the controls due to the slower growth rate in the first stage of culture (data not shown).

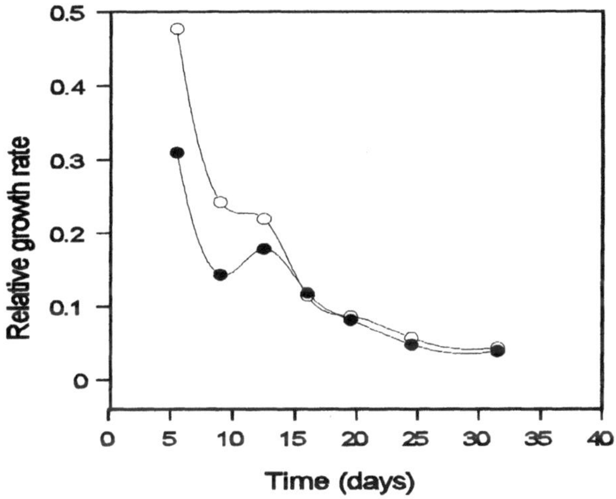

FIGURE 1. Relative growth rate of healthy (○) and PVY^{NTN}-infected (●) plants

After 18 days of culture the total meristem area (apical + subapical region) in infected plants was smaller than in healthy controls, mainly due to the significantly smaller apical region (Table 1). In addition, the total cell number in control meristems was higher than in the infected meristems (Table 1). After 35 days, the meristem size and the number of cells were smaller than on day 18 in both treatments. Nevertheless, no significant differences between the two treatments were detected in areas of meristem zones (Table 1).

Day 18 of the experiment roughly corresponds to the end of the initial period of rapid growth. Before day 18, the control plants grew much faster than the infected plants (Figure 1), therefore on day 18 larger meristems were measured in healthy plants. The general reduction in meristem size correlates with the slower growth rate in both treatments in later stages of culture. When the growth rate slowed down and was similar in both treatments (Figure 1), no significant differences in meristem size were detected (day 35; Table 1).

TABLE 1. Histological characteristics of shoot meristems in healthy and PVY^{NTN}-infected plants

	18 days		35 days	
	Healthy control (n=5)	PVY^{NTN}-infected plants (n=4)	Healthy control (n=5)	PVY^{NTN}-infected plants (n=5)
Height, μm	63.0 ± 2.0	59.0 ± 4.0	58.0 ± 2.0	57.0 ± 3.0
Diameter, μm	129.0 ± 8.0	112.0 ±5.0 *	115.0 ± 10.0	101.0 ± 5.0
Apical region, mm²	3.9 ±2.3	3.0 ±0.2*	3.2 ±0.4	2.8 ±0.2
Subapical region, mm²	2.5 ±0.3	2.2 ±0.3	2.3 ±0.5	1.6 ±0.2
Total no. of cells	99.2 ±9.3	80.5 ±5.2	87.8 ±13.4	75.0 ±4.9

Values are the means of replicates ±SEM. The Mann-Whitney U-test was used for evaluating the levels of significant differences between the healthy control and infected plants.Simbols used in table mean: (*) $P < 0.05$.

The measurements of the mitotic index revealed that at day 18 the mitotic cells were evenly distributed between apical and subapical zones in the control meristems (Table 2). On the other hand, in the infected plants the mitotic index was higher in the subapical region compared to the apical dome (Table 2). In some plant tissues cytokinins stimulate growth and cell division. They increase the growth fraction by releasing cells arrested at a particular stage of the cell cycle (Jacqmard *et al.*, 1994). Therefore, the higher frequency of cells in division in subapical region of infected plants in the early stage of culture might be related to the detected higher cytokinin concentration in infected plants (Dermastia and Ravnikar, 1996).

TABLE 2. Mitotic index (%) in different meristematic zones.

	18 days		35 days	
	Healthy control (n=5)	PVY^{NTN}-infected plants (n=4)	Healthy control (n=5)	PVY^{NTN}-infected plants (n=5)
Total	3.4 ± 0.9	4.3 ± 0.3	1.4 ± 0.4	0.5 ± 0.5
Apical region	3.3 ± 0.7	2.1 ± 0.8	0.8 ± 0.8	0.4 ± 0.4
Subapical region	3.3 ± 1.6	7.4 ± 1.6	1.8 ± 0.8	0.6 ± 0.6

Mitotic index is expressed as a percentage of cells in mitosis in the total population of cells. Values are the means of replicates ±SEM.

Another measure of growth activity is the mean tissue doubling time which can be regarded as an estimate of the mean cell doubling time (Lyndon, 1994). After 18 days of culture the mean tissue doubling time was similar in the shoot apical meristems of the control and of infected plants (Table 3). After five weeks it remained almost the same in plants with virus, but was about 30% longer in the healthy controls (Table 3). High cytokinin concentration in infected plants (Dermastia and Ravnikar, 1996) might at least partly explain this observation. Cytokinins are not only required to initiate cell proliferation in nondividing tissues, but their continued presence is needed to sustain mitotic activity (Jacqmard *et al.*, 1994). In mature plant tissues where the cell cycle is arrested, the level of cytokinins is very low. It might be possible therefore, that in older healthy plants the measured low cytokinin concentration (Dermastia and Ravnikar, 1996) is related to the reduced size of the shoot meristem (Table 1). On the other hand, in PVY^{NTN}-infected plants the concentration of these plant hormones was higher compared to the control plants (Dermastia and Ravnikar, 1996). Thus, the high cytokinin concentration may maintain the steady mean cell doubling time in the shoot meristem and support the growth in the upper vegetative plant parts.

TABLE 3. Mean tissue doubling time

	18 days			35 days		
	Healthy control (n=5)	PVY^{NTN}infected (n=4)	P	Healthy control (n=5)	PVY^{NTN}infected (n=5)	P
MTDT	1.1 ± 0.1	0.9 ± 0.1	NS	1.4 ± 0.1	1.1 ± 0.0	0.01

Values are the means of replicates ±SEM. The Mann-Whitney U-test was used for evaluating the levels of significant differences between the healthy control and infected plants.

The genetics of tolerance to plant virus diseases are not understood, neither is its relationship, if any, to resistance. Tolerant plants exist in most host-virus combinations, probably because of their specific physiological traits (Agrios, 1988). The potato plants of the cultivar Igor grown *in vitro* are obviously susceptible to PVY^{NTN}, but they tolerate the virus and show little damage resulting from the infection. Therefore in this cultivar, transfer of plants from soil to *in vitro* conditions seems to cause physiological changes that lead to tolerance. Alteration of the cytokinin level might be one of the factors involved in this process.

3. References

Agrios, G. N. (1988) *Plant Pathology*, Academic Press, Inc., San Diego, New York, London, Sydney, Tokyo, Toronto.

Dermastia, M. and Ravnikar, M. (1996) Altered cytokinin pattern and enhanced tolerance to potato virus Y^{NTN} in the susceptible potato cultivar (*Solanum tuberosum* cv. Igor) grown *in vitro*, *Physiol. Mol. Plant Pathol.* **48**, 65-71.

Petrovič, N., Gruden, K. and Ravnikar, M. (1995) Purification of potato virus Y^{NTN} from different plant material, *Acta Chim. Slov.* **42** (4), 425-430.

Le Romancer, M., Kerlan, C. and Nedellec, M. (1994) Biological characterization of various geographical isolates of potato virus Y inducing superficial necrosis on potato tubers, *Plant Pathol.* **43**, 138-144.

Lyndon, R. F. (1994) Control of organogenesis at the shoot apex, *New Phytol.* **128**, 1-18.

Jerman, S., Ravnikar, M. and Kovač, M. (1996) Photosynthetic pigments in healthy and PVY^{NTN} -infected potato (*S. tuberosum*) grown in vitro, *Phyton*, in press.

Jacqmard, A., Houssa, C. and Bernier, G. (1994) Regulation of the cell cycle by cytokinins, in D. W. S. Mok and M. C. Mok (eds), *Cytokinins*, CRC Press, Boca Raton, Ann Arbor, London, Tokyo, pp. 197-215.

Murashige, T. and Skoog, F. (1962) A revised medium for rapid growth and bio assays with tobacco cultures. *Physiol. Plantarum* **15**, 473-497.

Taylor, M. and Francis, D. (1989) Cell cycle changes in the shoot apex of *Silene coeli-rosa* during the second and third days of floral induction, *Ann. Bot.* **64**, 625-633.

ASSESSMENT OF CHARACTERISTICS FOR DETERMINING PATHOGEN POPULATIONS OF *RHIZOCTONIA SOLANI* IN RICE-BASED CROPPING SYSTEMS

S. BANNIZA, P.D. BRIDGE, M. HOLDERNESS,
M.A. RUTHERFORD AND S.A. SIMONS
International Mycological Institute, Egham, Surrey, United Kingdom

1. Introduction

Rhizoctonia solani is the causal agent of rice sheath blight, a major constraint to rice production in many SE Asian countries. *R. solani* is considered to be a 'collective species' and genetic diversity within the species is represented by anastomosis groups (AGs) defined from hyphal anastomosis behaviour. Within the AGs other intraspecific groups (ISGs) have been defined on the basis of morphology, pathology, pectinase isoenzymes and DNA sequence homology (Vilgalys & Cubeta, 1994). Isolates of *R. solani* in tropical rice-based cropping systems are generally found to belong to AG-1 1A, but there is little information about diversity among these isolates. Such information is of vital importance for resistance breeding and the development of strategies for disease management.

Research at the International Mycological Institute has concentrated on integrating morphological, biochemical and molecular techniques for the characterisation of the diversity of the pathogen in the rice agro-ecosystem. Within this work, two molecular techniques, the analysis of restriction fragment length polymorphisms of mitochondrial

H.-W. Dehne et al. (eds.), Diagnosis and Identification of Plant Pathogens, 85–88.

DNA (mtDNA-RFLPs) and PCR fingerprinting with primers targeted to variable number tandem repeats (VNTR-PCR) have been used to study the genetic diversity of the isolates.

2. Materials and Methods

Twenty-six isolates of *R. solani* were selected from an extensive collection of isolates from tropical rice-based cropping systems (Table 1), and grown in shaken culture in GYM (Mugnai et al., 1989) for 4 days. The mycelium was freeze dried and DNA was extracted according to Raeder and Broda (1985) and Vilgalys and Hester (1990).

For the analysis of mtDNA RFLP patterns, DNA samples were digested with *Hae*III (Gibco BRL) following the supplier's instruction, and fragments were separated by electrophoresis on a 1% (w/v) agarose gel at a constant voltage of 5v cm^{-1} in TAE buffer.

For PCR, three VNTR-derived oligonucleotides, MR, RY and GF were used to obtain fingerprinting patterns of a subset of 18 isolates (Bridge *et al.*, 1996). Amplification products were separated by electrophoresis in 1.5 % (w/v) agarose as above.

Gels were stained in ethidium bromide (0.5μg ml^{-1}) and visualised under UV light. Presence or absence of characteristic fragments were recorded for each isolate. The binary data matrices were analysed with Jaccard's Coefficient and clustered by UPGMA (MVSP, Plus-Version 2.0, Kovach, 1980-1990).

3. Results and discussion

Fifteen out of 26 isolates of *R. solani* produced mtDNA RFLPs which varied around one pattern of 5 fragments, suggesting this was characteristic for isolates from rice-based cropping systems (mtDNA RFLP group 1 in Table 1). A second group (group 2) of three isolates (359620, 360021, 360023) were clustered close to group 1 with a very similar pattern, whilst the remaining eight isolates had distinct RFLPs. Among those, four isolates (360016, 360336, 360298, 369367) did not anastomose with tester strains of AG-1, AG-2 1, AG-2 2 or AG-4, whereas all other isolates proved to belong to AG-1 (Table 1). Analysis by VNTR-PCR gave very similar results; 13 isolates showed similar but distinct patterns while 5 isolates, including the three non-AG-1 strains, had very different fragments. The major VNTR-PCR group included isolates with RFLP patterns of type 1, 2 and 3 (Figure 1) .

Mitochondrial DNA is thought to be inherited uniparentally and appears to be relatively conserved in fungi, therefore allowing the study of variation within relatively closely related species or sub-species groups. VNTR sequences, in contrast, are chromosomally inherited and show variation within populations (Kendrew, 1994). The mtDNA RFLP analysis in this study revealed that the majority of isolates had similar banding patterns, suggesting that they are relatively closely related and may represent typical populations of the sheath blight pathogen. The isolates with more distinct mtDNA RFLPs, most of which came from non-rice host plants in rice-based cropping systems, appear to represent background populations. This overall structure was also obtained by VNTR PCR analysis where individual isolates were distinguishable within these groups.

4. References

Bridge, P.D., Pearce, D.A., Rivera, A., and Rutherford, M.A. (1996) VNTR derived oligonucleotides as PCR primers for population studies in filamentous fungi. *Letters in Applied Microbiology* (in press).

Kendrew, J. (1994) *The Encyclopedia of Molecular Biology*, Blackwell Science, Oxford.

Mugnai, L., Bridge, P.D., and Evans, H.C. (1989) A chemotaxonomic evaluation of the genus *Beauveria*. *Mycological Research*. **92**, 199-209.

Raeder, U., and Broda, P. (1985) Rapid preparation of DNA from filamentous fungi. *Letters of Applied Microbiology*. **1**, 17-20.

Vilgalys, R., and Hester, M. (1990) Rapid genetic identification and mapping of enzymatically amplified ribosomal DNA from several *Cryptococcus* species. *Journal of Bacteriology*. **172**, 4238-4246.

Vilgalys, R., and Cubeta, M.A. (1994) Molecular systematics and population biology of *Rhizoctonia*. *Annual Review of Phytopathology*. **32**, 135-155.

TABLE 1: Host plant origin, country origin, anstomosis groups and mt DNA RFLP groups of isolates of *Rhizoctonia solani* from rice-based cropping systems (NI=Anastomosis group not identified)

IMI number	Host	Country	AG group	mt DNA RFLP group	mt DNA RFLP subgroup
360337	Grass	Vietnam	1	1	1
360302	Grass	Vietnam	1	1	1
358755	*Oryza sativa*	Philippines	1	1	1
360045	*Oryza sativa*	Vietnam	1	1	1
303152	*Oryza sativa*	Japan	1	1	1
360046	Unknown	Japan	1	1	1
361174	*Oryza sativa*	Nigeria	1	1	1
360305	*Oryza sativa*	Vietnam	1	1	1
361179	*Oryza sativa*	Nigeria	1	1	2
360366	*Oryza sativa*	Vietnam	1	1	3
360378	*Oryza sativa*	Nigeria	1	1	3
360321	*Zea mays*	Vietnam	1	1	3
369671	*Oryza sativa*	Nigeria	1	1	4
361189	*Oryza sativa*	Malaysia	1	1	5
361190	*Oryza sativa*	Malaysia	1	1	5
359620	*Oryza sativa*	Vietnam	1	2	-
360023	*Oryza sativa*	Côte d'Ivoire	1	2	-
360021	*Commelina*	Côte d'Ivoire	1	2	-
360037	*Oryza sativa*	Côte d'Ivoire	1	3	-
360038	*Oryza sativa*	Côte d'Ivoire	1	3	-
369673	Soil	Benin	NI	4	-
360314	*Oryza sativa*	Ghana	1	5	-
360016	*Vigna ung.*	Benin	NI	6	-
360377	*Lycopersicon*	Vietnam	1	7	-
360336	*Brassica oler.*	Vietnam	NI	8	-
360298	*Brassica chin.*	Vietnam	NI	8	-

FIGURE 1: Dendrogram obtained from Jaccard's coefficient after UPGMA clustering of data generated by VNTR-PCR of *Rhizoctonia solani* (Primers: MR, RY, GF)

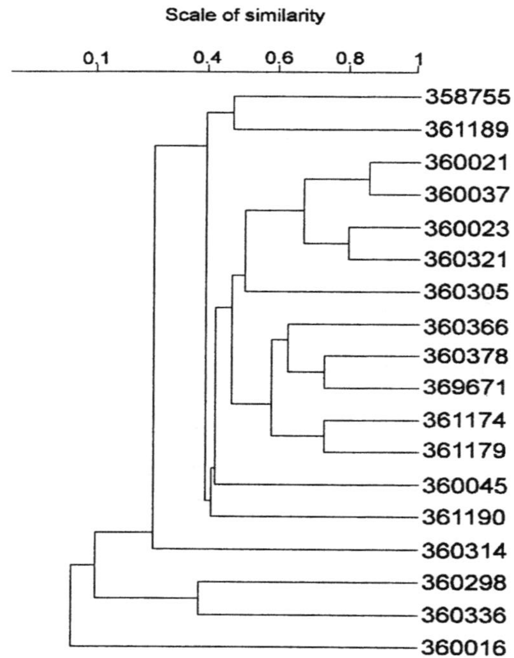

Session 3:

Serological Assays

MONOCLONAL ANTIBODY-BASED ASSAYS FOR DETECTION AND QUANTIFICATION OF *BOTRYTIS CINEREA*

FRANCES M. DEWEY AND LOUISE COLE*
Department of Plant Sciences, University of Oxford, South Parks Rd., Oxford, OX1 3RB
**School af Biological Sciences, Oxford Brookes University, Oxford, OX3 OBP*

1. Introduction

Botrytis cinerea Pers is responsible for the widespread grey mould disease of plants throughout temperate climates. It can infect an amazing diverse variety of plants but it is most noted for the serious losses that it causes in the fresh fruit, vegetable and cut flower industries The air-borne conidia germinating on plant surfaces either cause an immediate damaging, necrotrophic rot in growing plants or, they remain localized in the outer tissues until harvest and shipment where upon the organism then becomes invasive causing considerable post harvest losses. Although losses can be minimised by repeated applications of fungicides throughout the growing season there are increasing environmental and economic concerns about the prophylactic use of fungicides particularly with respect to fresh fruits and vegetables where fungicide residues have been detected. In order to target fungicide applications to the periods of highest risk and thereby reduce the number of spray applications during the growing season, we need to be able to measure not only the environmental factors that have the greatest effect on the spread of the conidia and development of the disease but also quantify the numbers of air-borne spores present at high risk times so that disease forecasting models can be made more accurate. Information on levels of air-borne conidia would also be invaluable in determining the growth stage and route by which quiescent infections are initiated. Furthermore, in order to minimize post harvest losses we also need to be able to detect symptomless infections before shipment of produce. We have addressed these various needs by developing immunological assays using the *Botrytis* - monoclonal antibody BC-KH4 raised by Bossi and Dewey (1992).

The hybridoma cell line, BC-KH4, was raised from splenocytes of a mouse immunized with surface washings from a mycelial culture of the fungus grown on solid media. BC-KH4 is an IgM antibody that recognizes, by plate-trapped antigen enzyme ELISA and immunofluorescence, all species of *Botrytis,* and some species of of unrelated fungi such as *Aspergillus, Verticillium dahliae* but not other species commonly found on the surfaces of fruits and vegetables. Studies with periodate, heat

H.-W. Dehne et al. (eds.), Diagnosis and Identification of Plant Pathogens, 91–94.

and proteinase treated antigen have shown that BC- KH4 recognizes a remarkably heat stable carbohydrate epitope on a glycoprotein molecule. Competitive ELISA tests indicate that the epitope recognized by the antibody is a galactomannan. Immunolabelling of conidia and hyphae both at the light and at the EM level have shown that the BC-KH4 antigen is present in both the hyphal walls, the cross-septa and in the extracellular matrix and that the antigen is produced both *in vitro* and *in vivo* (Cole *et aL,* 1996).

2. Immunolabelling and enumeration of air-borne spores

The capture of *Botrytis* spores on tape in a Burkhard spore trap and their differentiation from other air-borne spores by immunolabelling has now been made possible by using a modified adhesive on the tape (Dewey, 1996). Furthermore, it is possible to count the immunolabelled spores using a digital image analysis system. We have also collected spores in an eppendorf tube, using a Cylcone surface sampler (Burkhard), and quantified them by ELISA thus eliminating the need for a UV microscope and digital image analysis system (Dewey and Karpovich- Tate, unpublished).

3. Detection of spores germinated on surfaces of fruits and flowers

It is generally believed that some, but not all, of the *Botrytis* spores settling on the surfaces of growing fruits such as plums, nectarines and grapes germinate and penetrate the cuticle. They either remain quiescent at this stage or grow subcuticularly before becoming arrested until the fruit is ripe and harvested. In a collaborative study with UNIFRUCO, (S. Africa), we have been using the BC-KH4 monoclonal antibody as a probe to immunolabel spores on the skins of grapes by immunofluorescence microscopy and to determine the number of viable spores on the surface that are likely to grow through the epidermis and produce necrotrophic infections when the fruits are ripe. To do the latter we have developed an immunoblotting system in which small areas of the skin are carefully removed and sandwiched between 2 layers of polysulphone membrane which, in turn, is overlaid on a plate of malt extract agar and left to incubate overnight. Growth of germ tubes that have penetrated through the epidermis can be seen indirectly on the immunolabelled membranes using the BC-KH4 antibody followed by an anti-mouse immunogold conjugate and silver enhancement (Dewey, Seyess & Evans, unpublished).

4. Detection of *Botrytis cinerea* in Must from grapes and plant tissues

Early studies, recently confirmed, have shown that *B. cinerea* antigens present in must from infected grapes can be detected and quantified by plate trapped antigen ELISA (PTA-ELISA, Bossi *et al.,* 1994; Dewey and Evans, unpublished). However, use of the antibody to detect the antigen by PTA-ELISA in extracts from grape skins or plant tissues, particularly green tissues, has proved more difficult. As yet, unidentified substances present in green plant extracts appear to block the binding of the BC-KH4 antigen to the surfaces of the micro-titre wells (Dewey and Broadbent, unpublished). These problems can be overcome by using competitive ELISAs (c-ELISA) in which wells are coated with a dilution of surface washing of the fungus grown *in vitro*. (Fig. 1).

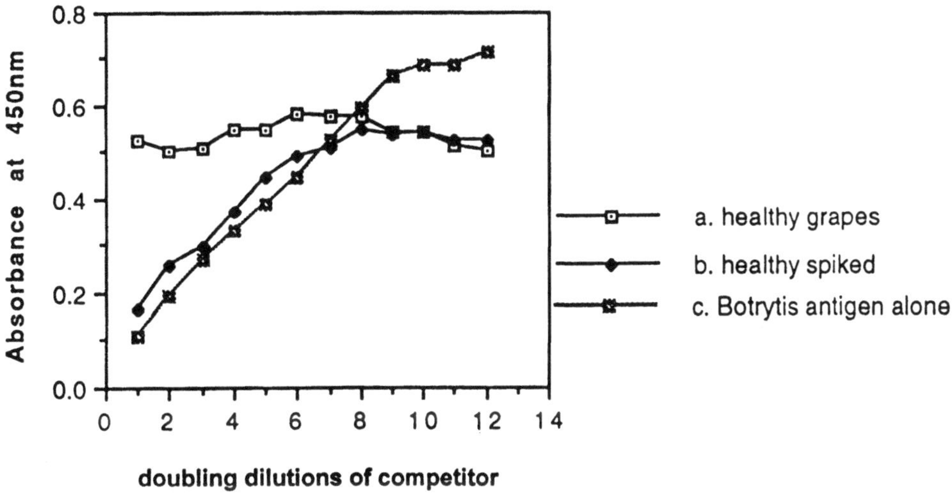

FIGURE 1. Absorbance values from a competition ELISA test of extracts from a) healthy grape skins b) grape skins spiked with *B. cinerea* antigens derived from pure cultures of the fungus c) *Botrytis* antigens only (Dewey and Evans).

Extracts from infected plants are separately incubated with a dilution of the BC-KH4 supernatant for 30 min, centrifuged to remove any precipitate and then incubated with the antigen-coated, pre-blocked plates.

5. Concluding remarks

The immunoassays that we have developed todate for *Botrytis* are proving most useful for research purposes and form the basis from which relatively rapid, user friendly, screening assays can be developed. More than one final assay format will almost certainly have to be developed in order to suit the very different end-user requirements. For assays where specificity is important it will be necessary to raise a species-specific monoclonal antibody and this is in progress.

Acknowledgements

F.M. Dewey wishes to thank the Leverhulme Trust for ongoing support and L. Cole for support from a PCFC grant awarded to C. Hawes, Oxford Brookes University.

6. References

Bossi, R. and Dewey, F.M. (1992). Development of a monoclonal antibody-based immunodetection assay for *Botrytis cinerea. Plant Pathology* **41,** 472-482.

Bossi, R., Cole, L., Spier, A.D. and Dewey, F.M. (1994). Monoclonal antibody-based ELISA for detection of mycelial antigens of *Botrytis cinerea* in fruits and vegetables. In: *Modern assays for Detection of Plant Pathogenic Fungi: Identification, Detection and Quantification* (ed. A. Schots, F.M.Dewey *k* R. Oliver), pp.165-172. CAB International: Wallingford, UK.

Cole, L. Immunocytochemical studies of fungal cell surface molecules and the *Vicia faba. Botrytis* interaction. Ph.D. Thesis, Oxford Brookes University, 1996, pp. 200.

Cole, L., Dewey. F.M. and Hawes, C.R. (1996). Infection mechanisms of *Botrytis* species: prepenetration and preinfection processes of dry and wet conidia. *Mycological Research,* 100. 277-282.

Dewey, F.M. (1996) Production and use of monoclonal antibodies for the detection of fungi. In: *Diagnostics in Crop Production,* pp 85-93, *BCPC Symposium* ***65,*** ed G. Marshall.

SENSITIVE DETECTION OF PLANT PATHOGENS BY USING IMMOBILIZED TARGETS IN TISSUE IMPRINTED MEMBRANES.

M. CAMBRA, A. OLMOS, M.T. GORRIS, N. DURAN, M.P. ROMAN, E. CAMARASA, M.A. DASI.
Instituto Valenciano de Investigaciones Agrarias (IVIA).
Apartado Oficial, 46113, Moncada, Valencia, Spain.

1. Introduction

Extracts preparation limits the number of samples that can be processed. Grinding is time consuming, requires special buffers and equipment, enhances risks of contamination and entails the release of inhibitors of plant origin. In addition, some epitopes and targets are modified when extracts are prepared.

To overcome these disadvantages we investigate the possibilities of simpler and faster preparative methods than conventional extractions or squashes.

The use of membranes to immobilize plant materials by direct tissue imprint, is an alternative to extracts used for detection of plant pathogens. Imprints can be easily prepared in field conditions and is an inexpensive method which do not require buffers nor equipment and small amounts of plant material are needed. In addition, tissue printing allows the recovery of "native" antigens and targets in membranes reducing contamination risks.

The imprinted membranes can be stored and mailed and can be safely used with quarantine of plant pathogens as controls.

Several application of immunoprinting-ELISA, imprint hybridization and the recently described and patented print-capture PCR (PC-PCR), are reviewed.

2. Material and methods

2.1. IMMUNOPRINTING-ELISA

The direct tissue blot of sections of plant material was performed according Lin *et al.* (1990). The printed nitrocellulose (0,45 μ) membranes (Millipore) were developed after a blocking step (1% BSA) of 1 h at room temperature. Direct immunoprinting-ELISA was done by using monoclonal antibodies (MCA) alkaline phosphatase linked for the detection of Citrus tristeza virus (CTV) coat protein (3DF1 and 3CA5, MCAs; Vela *et al.*, 1986) and plum pox virus (PPV) coat protein

H.-W. Dehne et al. (eds.), Diagnosis and Identification of Plant Pathogens, 95–98.

(5B and 4DG5, MCAs; Cambra *et al.*, 1994). The conditions of the tests were previously established (Garnsey *et al.*, 1993; Cambra *et al.*, 1994a). A commercial kit is available for routinary detection of CTV (Plant Print Diagnostics, Valencia, Spain) in which the sampling procedure is indicated. Indirect immunoprinting-ELISA was also assayed for detection of non-structural proteins of PPV (cylindrical inclusions) by using the MCA 11E5H (Cambra *et al.*, 1994b) and for detection of ds-RNA of PPV and CTV in infected woody plants by using an antiserum against the synthetic polynucleotide poly I-poly C (Aramburu *et al.*, 1991). The test was finished by adding goat anti-mouse or anti-rabbit immunoglobulins, alkaline phosphatase labelled.

2.2. IMPRINT HYBRIDIZATION

The method described by Romero-Durban *et al.*, (1995) has been used for detection of viroids in 16 viroid-host combinations including 7 viroids (CEVd, CSVd, HSVd, CVd-IV, CBLVd, CVd-III and CVd-IV) and 7 hosts (chrysanthemum, citron, cucumber, *Gynura*, tomato, peach and avocado). PVDF and Nylon membranes were used. RNA probes were synthesized by a transcription reaction using cloned plasmid DNA, and digoxigenin-labelled UTP. In general, no denaturation of the imprinted membranes was required but enhanced sensitivity was obtained by pretreating the membranes with hydroxy methyl mercury (unpublished results).

2.3. PRINT-CAPTURE PCR AND SPOT-CAPTURE PCR

Print-capture PCR (PC-PCR) or spot-capture PCR (SC-PCR) was done in Whatman 3MM paper. The imprinted paper with sections of plant material or blotted with plant extracts, was analyzed according Olmos *et al.* (1996). The method briefly consisted in three steps: 1. Immobilization of plant material and targets in paper. The membranes were stored or processed. 2. Extraction of targets from squares of paper harboring either the tissue prints or the spotted extracts, by 0.5% Triton X-100, vortex and incubation for 2 min at room temperature, and 3. Capture phase: Triton extracts were transferred to tubes previously coated with one of the following proteins: specific MCA (anti CTV or PPV), normal immunoglobulins (2µg/ml) from non immunized rabbits or skimmed milk (Sveltesse, Nestlé 5%). The extracts were incubated in the coated tubes for 2h at 37°C and washed as in a conventional "immunocapture" phase. The RT-PCR one step protocol was used for PPV detection (Wetzel *et al.*, 1992) or for CTV detection (Cambra *et al.*, in preparation).

2.4. HEMINESTED-PCR AND COLOURIMETRIC PCR-ELISA

Heminested-PCR for PPV detection was performed in two steps, using the PC-PCR (Olmos *et al.*, 1996) followed by a second amplification using an internal primer specific for D or M types of PPV (Olmos *et al.*, unpublished). CTV detection was done by using the same protocol with specific primers (P_A, P_B) and an internal primer (P_A, P_{OD}). The final product was analyzed by electrophoresis in 3% agarose gels and stained by ethidium bromide. The method was assayed for the detection of PPV and CTV in individual aphids (Aphis gossypii) fed in infected plants, as well as with serial dilutions of plant extracts.

A colourimetric PC-PCR ELISA was performed for PPV and CTV following the Boehringer Mannheim method, using streptavidin precoated ELISA plates in which a biotinilated probe (Olmos *et al.*, unpublished) hybridize with amplicons-dUTP-digoxigenin linked. The reaction was visualized by using anti-digoxigenin immunoglobulins peroxidase linked.

3. Results

The use of immunoprinting-ELISA procedure for CTV and PPV allowed the detection of coat protein in woody plants. The method was more sensitive than conventional ELISA. Five stems (two imprints for stem) or ten leaves (one imprint for petiole leave), were necessary for reliable detection. Detection of non-structural proteins of PPV and ds-RNA from PPV or CTV infected plants was more easily obtained by immunoprinting-ELISA than ELISA.

Imprint hybridization was successfully used to detect the viroids in all the pathogen-host combinations assayed. The method is fast and sensitive, and provides additional information on the sites of viroid accumulation. This technique was successfully assayed to follow the movement and accumulation of viroids in infected plants (Romero, 1996).

The PC-PCR method allowed the detection of CTV and PPV targets by using an unespecific capture phase instead the specific capture required in immunocapture-PCR (IC-PCR). The PC-PCR method performed with skimmed milk allowed the detection of PPV and CTV in single aphids as well as in plant materials. The sensitivity of PC-Heminested-PCR allowed the differentiation of isolates of PPV. The sensitivity of colourimetric PCR-ELISA was similar than PC-Heminested-PCR but hundred times more sensitive than PC-PCR, performed with or without a specific capture phase.

The use of direct tissue printing in membranes followed by serological or molecular probes, seems to be an excellent alternative to detect and characterize plant viruses and viroids.

4. References

Aramburu, J., Navas-Castillo, J., Moreno, P. and Cambra, M. (1991) Detection of double stranded RNA by ELISA and immunobinding assay using an antiserum to synthetic polynucleotides, *J. Virol Methods* **33**, 1-11.

Cambra, M., Asensio, M., Gorris, M.T., Pérez, E., García, J.A., Moya, J.J., López-Abella, D., Vela, C., Sanz, A. (1994) Detection of plum pox potyvirus using monoclonal antibodies to structural and non-structural proteins, *Bulletin OEPP/EPPO Bulletin* **24**, 569-577.

Cambra, M., Camarasa, E., Gorris, M.T., Román, M.P., Asensio, M., Pérez, E., Serra, J. and Cambra, M.A. (1994) Detección de proteínas estructurales de virus mediante inmunoimpresión-ELISA y su uso en diagnóstico, *Invest. Agrar.: Fuera de serie* **2**, 221-230.

Garnsey, S.M., Permar, T.A., Cambra, M. and Henderson, C.T. (1993) Direct Tissue Blot Immunoassay (DTBIA) for detection of Citrus Tristeza Virus (CTV), in P. Moreno, J.V. Da Grasa and L.W. Timmer (Eds.), *Proc.12th Conf. Inter. Organ. Citrus Virologitsts IOCV*, Riverside, pp. 39-50.

Lin, N.S., Hsu, Y.H. and Hsu, H.T. (1990) Immunological detection of plant viruses and a mycoplosma like organism by directed tissue blotting on nitrocellulose membranes, *Phytopathology* **80**, 824-828.

Olmos, A., Dasí, M.A., Candresse, T. and Cambra, M. (1996) Print-capture PCR: a simple and highly sensitive method for the detection of Plum pox virus (PPV) in plant tissues, *Nucleic Acid Research* **11**, 2192-2193.

Romero-Durbán, J., Cambra, M. and Durán-Vila, N. (1995) A simple imprint-hybridization method for detection of viroids, *J. Virol. Methods* **55**, 37-47.

Vela, C., Cambra, M., Cortés, E., Moreno, P., Miguet, J.G., Pérez de San Román, C. and Sanz, A. (1986) Production and characterization of monoclonal antibodies specific for citrus tristeza virus and their use for diagnosis. J.Gen. Virol. **67**, 91-96.

Wetzel, T., Candresse, T., Macquaire, G., Ravelonandro, M. and Dunez, J. (1992) A highly sensitive immunocapture polymerase chain reaction method for plum pox potyvirus detection, *J. Virol. Methods* **39**, 27-37.

DEVELOPMENT OF A DOT-IMMUNOBINDING ASSAY FOR THE DETECTION OF *PHYTOPHTHORA* SPP. IN ROOTS

RENATE HAHN[1] and SABINE WERRES[2]

1) *Institute for Plant Diseases and Plant Protection, University of Hannover, Herrenhäuserstr. 2, D - 30419 Hannover*

2) *Federal Biological Research Centre for Agriculture and Forestry, Institute for Plant Protection in Horticultural Crops, Messeweg 11/12, D - 38104 Braunschweig* (corresponding address)

The objective of these studies was to develop a serological method for routine detection of *Phytophthora* spp. in root samples from woody plants.

In all experiments the indirect dot-immunobinding assay (DIBA) and a polyclonal antiserum against an isolate of *Phytophthora cinnamomi* were used. The material to be tested were freeze-dried from liquid culture of the fungus, frozen roots from healthy plants of *Chamaecyparis lawsoniana* 'Ellwooddii' as well as from inoculated and diseased plants of this variety. All plants had been grown in the greenhouse in 5-liter containers. Naphthol-phospat/Fast Blue, -/Fast Red and NBT/BCIP were used as substrates.

One of the biggest problems assiciated with root extracts especially from several-year-old trees and shrubs and the use of polyclonal antibodies is the prevention or reduction of unspecific background reactions caused by root constituents (e.g. phenolic compounds) and unspecific binding of antibodies. To prevent these unspecific reactions and to increase the sensitivity of DIBA, different membranes, blocking solutions, various incubation periods, different washing solutions, buffer additives and conjugate buffers, etc. were tested.

A DIBA method permitting reliable detection of *Phytophthora* spp. in roots of *Chamaecyparis* plants is presented. In this procedure roots from infected plants can be readily distinguished from non-infected ones by a sharply coloured halo around the

H.-W. Dehne et al. (eds.), Diagnosis and Identification of Plant Pathogens, 99–100.

coloured dot on the membrane. Furthermore, the use of Fast Red or Fast Blue as substrates resulted in colour change when samples from infected roots were used.

DETECTION OF RESTING SPORES OF *SPONGOSPORA SUBTERRANEA* F.SP. *SUBTERRANEA* USING SEROLOGICAL TECHNIQUES

J.A. WALSH[1] and U. MERZ[2]
[1]Horticulture Research International, Wellesbourne, Warwick, CV35 9EF, UK, [2]Institute of Plant Sciences, Pathology Group, Federal Institute of Technology, CH-8092 Zurich, Switzerland

A polyclonal antiserum has been produced against spore balls of *Spongospora subterranea* f.sp. *subterranea* prepared from potato tubers. It was able to detect as little as 0.02 spore balls in enzyme-linked immunosorbent assay (ELISA). It detected spore balls of different cultivar origin equally well. It also detected spores from different geographical origins. Western blot analysis revealed that the antiserum detected a number of different proteins, the most distinct of which had a molecular weight of slightly less than 6.5 kDa. In indirect FITC immunofluorescence, discrete bright fluorescent spots were visualised using the specific serum. With the non-specific serum only a very dull background fluorescence was evident.

1. Introduction

Powdery scab of potatoes is caused by infection of tubers by the obligate zoosporic parasite *Spongospora subterranea* f.sp. *subterranea*. The scab-like lesions that are produced on tubers are particularly damaging as they reduce the proportion of marketable crop, carry infection to subsequent crops when the potatoes are used as `seed' and release spore balls that remain viable in field soils for many years. *S. subterranea* is also important as the vector of potato mop-top furovirus which can cause severe damage in some potato cultivars. These diseases have become more important in parts of Europe over the last decade including the U.K., Switzerland, Sweden, Finland, Holland and also other parts of the world including Colombia, USA, Australia, India, Pakistan, New Zealand and Turkey. This increase in importance is thought to be due to the intensification of potato production, shorter rotations, increased irrigation of crops and spread of the disease on seed stocks.

Control of powdery scab is difficult because of the persistence of the resting spores in soil and the difficulty in killing them. The only reliable form of control would be to plant clean seed into uncontaminated land. Sensitive and rapid serological detection of resting spores on tubers would provide a reliable means of identifying contaminated tubers and could allow a certification scheme to be implemented. Harrison *et al.* (1993) described the development of an enzyme-linked immunosorbent assay (ELISA) for the detection and quantification of *S. subterranea* contaminating potato tubers. This paper describes the further development of serological techniques for the detection of *S. subterranea*.

H.-W. Dehne et al. (eds.), Diagnosis and Identification of Plant Pathogens, 101–104.

2. Materials and Methods

2.1 PREPARATION OF SPORE BALL MATERIAL

Spore balls were used as immunogens and in most experiments were obtained from potato tubers cultivar Bintje grown in Switzerland. For additional tests, spore balls were collected from cvs. Indira and Ukama grown in Switzerland, cv. Estima grown in Scotland and an experimental line (NDO 1496-1) grown in the USA.

2.2 PRODUCTION OF ANTISERUM AND ELISA

A polyclonal antiserum to the spore ball stage of *S. subterranea* was produced in a New Zealand White rabbit. Plate trapped antigen (PTA) ELISA was used in all experiments and prepared spore balls were ground with pestles and mortars in 0.05M sodium carbonate buffer (pH 9.6) prior to ELISA.

2.3 SENSITIVITY AND SPECIFICITY OF DETECTION OF RESTING SPORE BALLS

Dilutions containing 4000, 1000, 50, 12.5, 2.5, 0.5 and 0.1 spore balls/ml from potato cv. Bintje were prepared. ELISA was carried out as outlined previously on these samples and scrapings from healthy Bintje tubers. Equal amounts (30 mg) of scrapings of uncontaminated tubers (cv. Bintje) and infected tubers (cv. Agria) taken as before, ground with a pestle and mortar or unground in 2 ml of 0.05M sodium carbonate buffer (pH 9.6) were tested by ELISA to determine the optimum method for preparing samples. The ability of the antiserum to detect resting spores from different potato varieties was also tested. To determine whether the antiserum recognised resting spores of different geographical origin, spore ball preparations from Switzerland, Scotland and USA were quantified, ground with pestles and mortars in carbonate buffer, further diluted in carbonate buffer and then tested by ELISA as described earlier, except that purified γ-globulin was used instead of raw serum.

2.4 STUDIES ON THE ANTIGENS DETECTED BY THE ANTISERUM

To characterise protein antigens, Western blot analysis of total spore ball proteins was carried out (Walsh *et al.*, in press). In order to visualise the position and distribution of antigens detected by the antiserum on and in spore balls, immunofluorescent studies were carried out.

3. Results

3.1 SENSITIVITY AND SPECIFICITY OF DETECTION OF RESTING SPORE BALLS

The prepared and quantified spore balls reacted strongly in PTA ELISA with the raw serum from the final bleed and A_{405} values increased consistently with increased spore ball concentration (Fig. 1). The sample containing 0.02 spore balls (200µl of 0.1 spore balls/ml) gave an A_{405} of 0.04, whereas blank wells gave a value of 0.02. In contrast equivalent weights of scraping from uncontaminated potatoes gave low absorbence values (Fig. 1).

Figure 1. The relationship of the concentration of spore balls of *Spongospora subterranea* prepared from infected potatoes and equivalent concentrations of scrapings from a healthy potato with A_{405} in a PTA ELISA test. (•) resting spores; (o) equivalent weights of scrapings from a healthy potato.

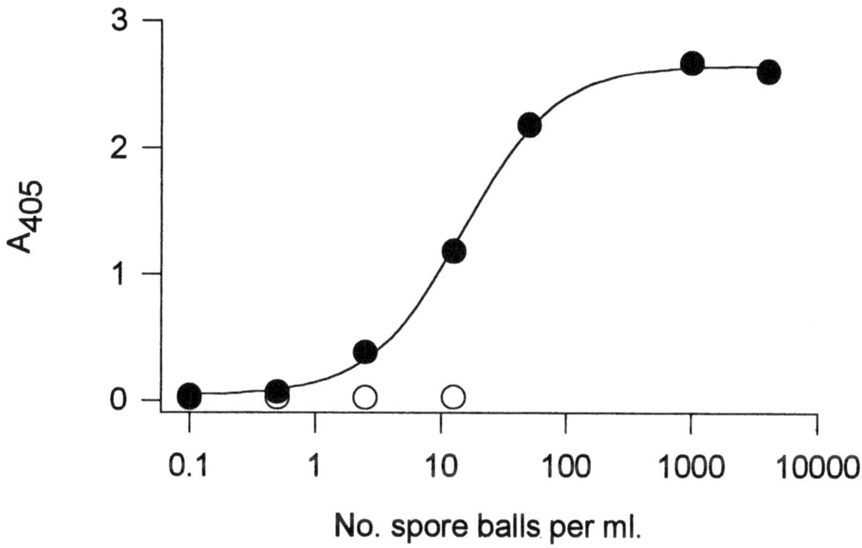

A comparison of the A_{405} obtained in PTA ELISA for equal quantities of ground and unground scrapings showed that although grinding did not affect values for scrapings from uncontaminated potatoes (A_{405} = 0.13 with and without grinding) those of ground scrapings from infected potatoes were almost double those of unground. In similar tests on equal weights of resting spore ball preparations from different potato cultivars there were remarkably no differences in A_{405} values for spores; grinding increased A_{405} values, but to a lesser extent than with scrapings (Table 1).

Table 1. Detection of spore balls of *Spongospora subterranea* obtained from different potato cultivars and the effect of grinding.

Cultivar origin	A_{405}*		
	Bintje	Indira	Ukama
Unground samples	0.48	0.48	0.48
Ground samples	0.55	0.56	0.56

Samples containing scrapings from healthy potatoes gave A_{405} of 0.13 and blank wells gave A_{405} of 0.00.
* Mean value from 2 ELISA plate wells.

Tests on equal quantities of spore balls of different geographical origin demonstrated the ability of raw serum to detect these in PTA ELISA.

3.2 STUDIES ON THE ANTIGENS DETECTED BY THE ANTISERUM

Western blotting revealed a number of bands, the most prominent of which had a molecular weight of slightly less than 6.5 kDa . In the immunofluorescent studies using specific serum, many bright fluorescent spots were seen within spore balls during microscopic examination. In contrast spore balls treated with non-specific serum showed no such fluorescent spots, just a very dim general background fluorescence.

4. Discussion

Further to the earlier description of serological detection of *S. subterranea* (Harrison *et al.*, 1993) this paper describes improvements in the sensitivity of serological detection of spore balls from tubers. The assay detected 0.02 spore balls and was slightly more sensitive than the test described by Harrison *et al.* (1993), perhaps due to differences in antisera titres or assay methods.

As grinding spore ball preparations and scrapings increased A_{405} values in comparison with unground samples, it is recommended that all samples should be ground prior to testing.

The ability to detect spore balls of different potato cultivar or geographical origin suggests that the antiserum has wide application. The antiserum has also been shown to detect spore balls from Pakistan (Merz, pers. comm.) and Australia (Wilson, pers. comm.).

5. Acknowledgements

This work was funded by the Ministry of Agriculture, Fisheries & Food, UK and ETH, Zurich. Collaboration between JAW and UM was made possible through joint British Council and Swiss National Science Foundation funding.

6. References

Harrison, J.G., Rees E.A., Barker, H. and Lowe R. (1993) Detection of spore balls of *Spongospora subterranea* on potato tubers by enzyme-linked immunosorbent assay. *Plant Pathology* **42**, 181-186.

Walsh, J.A., Merz U. and Harrison J.G. (1996) Serological detection of spore balls of *Spongospora subterranea* and quantification in soil. *Plant Pathology* **45**, (in press).

DEVELOPMENT OF AN ENZYME-LINKED IMMUNOSORBENT ASSAY (ELISA) FOR *VERTICILLIUM DAHLIAE* DETECTION IN SOIL

CLAUDIA HEPPNER, RUDOLF HEITEFUSS
Institute for Plant Pathology and Plant Protection,
Georg-August-University Goettingen
Grisebachstr. 6, D-37077 Goettingen, Germany

1. Introduction

In Germany, one of the most important crop rotation diseases of oilseed rape is caused by the fungus *Verticillium dahliae* (Krüger, 1989; Amelung et al., 1996). The pathogen persists in soil for many years with microsclerotia (MS). The inoculum density of *V. dahliae* plays a critical role in disease development of oilseed rape. A chemical treatment of the pathogen, like soil fumigation, is not allowed in Germany. Therefore, our aim was to develop a disease prognosis for oilseed rape on the basis of the MS concentration in soil. Various methods, e.g. direct soil assays or bioassays, have been developed for detection and quantification of the inoculum in soil (Evans et al., 1974; Harris et al., 1993). These classical techniques, however, are time consuming and require mycological expertise. An immunoassay for the assessment of *V. dahliae* in soil can be considered as a faster and easier detection method.

2. Materials and Methods

2.1. PRODUCTION OF ANTISERUM

Antiserum against soluble proteins form microsclerotia of *V. dahliae* (isolate from oilseed rape V10) was raised in a rabbit. The immunization was done in the BAZ by the Institute for Pathogen Diagnostics, Aschersleben. We isolated and purified the IgG as described by Clark and Adams (1977).

2.2. DOUBLE-ANTIBODY SANDWICH (DAS) ELISA

All working volumes were 100 µl per well, except blocking and washing (200 µl/well). Between incubation, all wells were washed three times with PBS containing 0.05 % Tween 20. The biotinylation of the IgG was done as described by Bieber (1985). The incubation of the microtiter plates was performed as described earlier (Heppner, 1995).

H.-W. Dehne et al. (eds.), Diagnosis and Identification of Plant Pathogens, 105–108.

2.3. PREPARATION OF THE SOIL SAMPLES FOR THE DAS-ELISA

The wet-sieving were based on the method of Harris et al. (1993). A soil suspension (50 g sieved soil/100 ml destilled water) was prepared by shaking 1 h at 270 oscillation/min. The slurry was wet sieved with tap water through nested 1000, 250, 125, and 20 µm sieves. The 20 to 125 µm soil fraction, corresponding in size to the MS, was recovered and dried at room temperature for 12 h. We sonicated the residues in the special extraction buffer (3.2.) for 30 sec. and used the supernatant for ELISA after centrifugation (20,000 x *g*, 30 min.).

3. Results and Discussion

3.1. SPECIFICITY OF THE ANTIBODY

Eight *V. dahliae* isolates, 5 from oilseed rape and 3 from other hosts, reacted highly positive (A_{405nm}:> 2.0) using DAS-ELISA, with the exception of an aggressive strain from cotton. As we have a polyclonal serum, we also observed less intense reactions with other *Verticillium* species, e.g. *V. albo-atrum*. Seventeen soil fungi of different genera, e.g. *Penicillium* spp. and *Trichoderma* spp., did not react with the antibodies. However, a weak reaction with *Rhizoctonia solani* could be observed. The low cross-reactivity with different *Verticillium* species and *R. solani* can be neglected, as structures of these fungi were eliminated by the wet-sieving procedure.

3.2. BUFFER FOR THE EXTRACTION OF *V. DAHLIAE* ANTIGEN FROM SOIL

As there is only little published information on the extraction of pathogens form soil before using ELISA (Harrison et al., 1993; Thornton et al., 1993), we tested various neutral (HEPES, MOPS, Tris-HCl, pH 7.4) and acidic (Na-citrate, Na-acetate, pH 5.0) buffers. Compared to all other buffers, Tris-HCl resulted in lowest background levels in noninfested soil. We also realized that high concentration of the buffer (2 M) resulted in increased positive reactions in soil artificially infested with *V. dahliae*. The same effect could be achieved with high concentrations of NaCl (2 M). The resulting buffer (0.1 M Tris, 2 M NaCl, pH 7.4) showed a false-positive reaction with an absorbance of A_{405nm}: 0.3, compared to the normal level of A_{405nm}: 0.1. For further optimization of the extraction buffer, we investigated a group of different proteins (BSA, low-fat milk, casein). Casein showed the best results, as on the one hand it strongly improved the positive reaction, on the other hand it highly reduced the false-positive reaction. Glycerine and Tween had a negative effect on both the infested and the non-infested soil. Triton-X-100 and EDTA clearly reduced the false-positive reaction and slightly improved the positive reaction. The summarized results of this experiment are demonstrated in FIGURE 1. The extraction of *V. dahliae* was finally performed by sonicating the soil fraction (20-125 µm) for 30 sec. in a buffer containing 0.1 M Tris-HCl, 2.0 M NaCl, 20.0 mM EDTA, 0.1 % Triton-X-100, 1.0 % casein, pH 7.4.

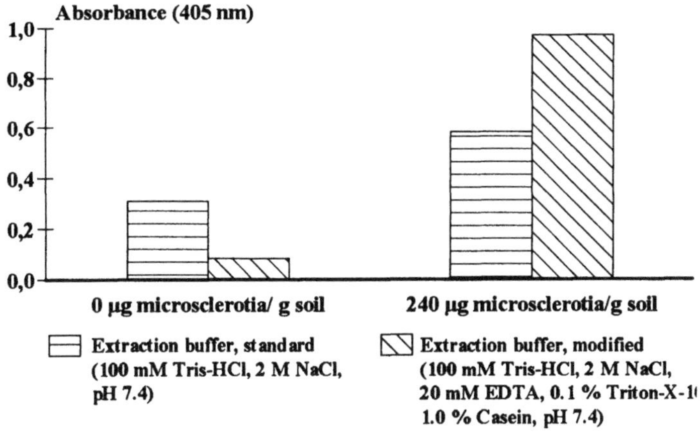

FIGURE 1. Effect of supplements to a buffer for the extraction of *V. dahliae* from soil

3.3. DETERMINATION OF *V. DAHLIAE* IN ARTIFICIALLY INFESTED SOIL USING DAS-ELISA

To quantitate the inoculum in soil, absorbance values have to be matched to inoculum density. Timmer et al. (1993) were not able to show a similar correlation for *Phytophthora* spp. For artificially infested soil, we found a linear positive correlation (FIGURE 2) after logarithmic transformation of both concentration of microsclerotia (0 to 5.6 MS/g soil) and the absorbance (A_{405nm}: 0.10 - 2.11). The detection limit of the ELISA was 2.4 µg MS/g soil (about 1-2 cfu/g soil). The detection limit can be further improved by ultrafiltration of the supernatant (30,000 MW) prior to ELISA.

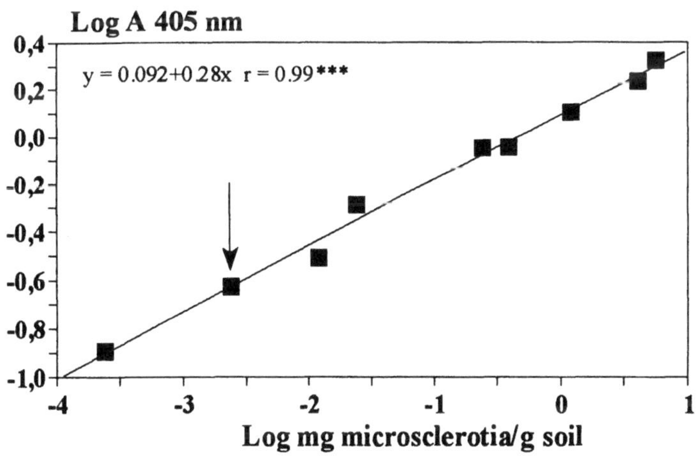

FIGURE 2. Relationship between absorbance and inoculum density of *V. dahliae* in artificially infested soil

In greenhouse and field experiments, up to 100-fold higher concentrations, corresponding to A_{405nm}: 0.62, led to significant yield losses for the susceptible oilseed rape cv. "Ceres" (FIGURE 3), indicating that this method is suitable for the estimation of *V. dahliae* in soil.

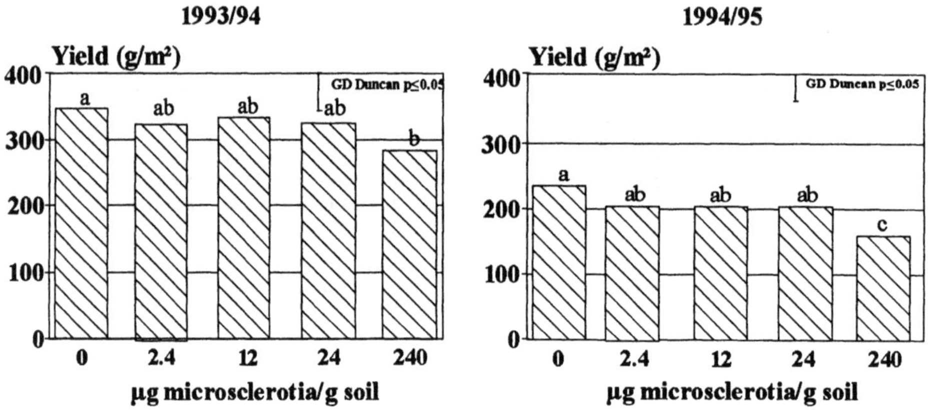

FIGURE 3. Effect of different inoculum densities of *V. dahliae* on the yield of oilseed rape cv. Ceres in 1993/94 and 1994/95

4. References

Amelung, D., Schulz, R.-R., and Daebeler, F. (1996) Einfluß der Fruchtfolge auf Rapskrankheiten, *Raps* **2**, 52-56.

Bieber, F. (1985) Biotinylierung monoklonaler Antikörper, in: J.H. Peters, H. Baumgarten, and M. Schulze (eds.), *Monoklonale Antikörper -Herstellung und Charakterisierung,* Springer Verlag Berlin, Heidelberg, New York, Tokio, pp. 209-212.

Clark, M.F., and Adams, A.N. (1977) Characteristics of the microplate method of enzyme-linked immunosorbent assay for the detection of plant viruses, *J. gen. Virol.* **34,** 475-483.

Evans, G., McKeen, C.D., and Gleeson, A.C. (1974) A quantitative bioassay for determining low numbers of microsclerotia of *Verticillium dahliae* in field soils, *Can. J. Microbiol.* **20,** 119-124.

Harris, D.C., Yang, J.R., and Ridout, M.S. (1993) The detection and estimation of *Verticillium dahliae* in naturally infested soil, *Plant Pathology* **42,** 238-250.

Harrison, J.G., Rees, E.A., Barker, H., and Lowe, R. (1993) Detection of spore balls of *Spongospora subterranea* on potato tubers by enzyme-linked immunosorbent assay, *Plant Pathology* **42,** 181-186.

Heppner, C. (1995) Nachweis von *Verticillium dahliae* Kleb. im Boden mit Plattengußverfahren und ELISA (enzyme-linked immunosorbent assay) sowie Untersuchungen zur Auswirkung des Inokulums auf den Befall von Winterraps (*Brassica napus* ssp. *oleifera* Metzg.), Dissertation Universität Göttingen.

Krüger, W. (1989) Untersuchungen zur Verbreitung von *Verticillium dahliae* Kleb. und anderen Krankheits- und Schaderregern bei Raps in der Bundesrepublik Deutschland, *Nachr.-Bl. Deut. PflSchutzd.* **41,** 49-56.

Thornton, C.R., Dewey, F.M., and Gilligan, C.A. (1993) Development of monoclonal antibody-based immunological assays for the detection of live propagules of *Rhizoctonia solani* in soil, *Plant Pathology* **42,** 763-773.

Timmer, L.W., Menge, J.A., Zitko, S.E., Pond, E., Miller, S.A., and Johnsonn, E.L.V. (1993) Comparison of ELISA techniques and standard isolation methods for *Phytophthora* detection in citrus orchards in Florida and California, *Plant Dis.* **77,** 791-796.

IMMUNOLOGICAL METHODS FOR THE DETECTION OF *VERTICILLIUM DAHLIAE* IN OILSEED RAPE

R. ČERNUŠKO, G.A. WOLF
Institut für Pflanzenpathologie und Pflanzenschutz der Georg-August-Universität, Grisebachstraße 6, D-37077 Göttingen, Germany

1. Introduction

Verticillium dahliae is an important plant pathogen causing vascular wilts in a wide range of dicotyledonous crops and wild plant species (Pegg, 1985). The fungus survives in soil as microsclerotia, initially embedded in fragments of plant tissue. After germination it penetrates directly into the vascular system of the plant. Verticillium wilt of oilseed rape (*Brassica napus* var. *oleifera* L.) is a serious problem in some regions with intensive cultivation (Krüger, 1989).
The current standard techniques used to screen oilseed rape for *Verticillium* incidence are based on isolation of the fungus and optical disease assessement. The quantitative analysis of fungal biomass of vascular wilt diseases has been particulary difficult using standard methods. These are laborious and time consuming. An alternative approach are serological procedures such as enzyme-linked immunosorbent assay (ELISA). This assay is routinely used for detection, identification and quantification purposes. Several polyclonal and monoclonal antibody-based immunoassays for *Verticillium* have been developed (Fitzel *et al.*, 1980; Strunnikova and Muromtsev, 1987). These antibodies were used to compare *Verticillium* isolates and to determine the antigenic similarities between different *Verticillium* species. Van de Koppel and Schots (1995) detected with monoclonal antibody Verticillium-wilt in ornamentals.

We developed a sensitive immunological assay to detect *V. dahliae* in infected oilseed rape plants from greenhouse and field experiments and in crop debris.

2. Materials and Methods

2.1. ANTISERUM PRODUCTION AND PREPARATION

Verticillium dahliae, isolated from oilseed rape, was cultivated on polysulphone membranes in petri dishes on Czapek Dox culture media. After incubation (4 days), the fungal mycelium was separated from the growth media. The tissue was ground in an ice-cooled mortar and pestle with phosphate-buffered saline (PBS). The suspension was

H.-W. Dehne et al. (eds.), Diagnosis and Identification of Plant Pathogens, 109–112.

centrifuged at 20,000 g for 20 min and at 75,000 g for 60 min at 4 °C. Protein extracts were stored at -20 °C until needed.
Soluble protein solutions from mycelium (1.0 mg) were injected 3 times in 2 week-intervals intramusculary into rabbits. The serum used in this study was collected after the third booster injection.
The antisera were prepared as described earlier (Clark and Adams, 1977). The immunoglobulin fraction was purified in ½ PBS using Fractogel TSK DEAE 650 (M) (Merck). The optical density at 280 nm was used to calculate the concentration of IgG. For the biotinylation the purified IgG fraction was conjugated with Biotin X-NHS (Sigma).

2.2. ELISA

The B-A ELISA (=DAS ELISA using biotinylated detection-antibodies and steptavidin-alkaline phosphatase) was conducted in microtitre plates using the following procedure: 100 µl of IgG in coating buffer (1 µg/ml) were incubated overnight. After every step the plates were rinsed 3 times with ½ PBST (PBST = PBS+ 0.05 % Tween 20). The plates were blocked by adding 200 µl coating buffer with 0.2 % BSA for 2 hours at room temperature. 100 µl of prepared samples (plant material were homogenized in PBST + 0.2 % PVP buffer, centrifuged, supernatant was used) were incubated overnight at 4 °C. 100 µl biotinylated IgG fraction (0.1 µg/ml) were incubated overnight at 4 °C. Streptavidin-alkaline-phosphatase conjugate (0.1 µg/ml) was added and incubated for 0.5 hour at 37 °C. 100 µl substrate solution (*p*-nitrophenyl phosphate, 1 mg/ml in substrate buffer) were added. After incubation for 2 hours at room temperature, the reaction were quantified by measuring the absorbance of each well at $\lambda = 405$ nm in a Spectra plate reader (SLT, Austria).

3. Results

The lowest level for the detection of *V. dahliae* could be detected was 20 ng fungal protein/ml. In the B-A ELISA, the antiserum did not cross-react with mycelial proteins from other fungal pathogens, e.g. *Phoma lingam*, *Sclerotinia sclerotiorum*, *Fusarium* spp., *Alternaria* spp., respectively. Cross reactivity was detected with *Verticillium albo-atrum*, *V. tricorpus*, *V. tenerum*, but not with *Verticillium nigrescens*.
The B-A ELISA was used to detect *Verticillium dahliae* in artificially inoculated oilseed plants. Roots, stems and leaves of susceptible and resistant cultivars were prepared to follow the spread of the fungus through the plant (*Fig. 1*). All roots and stems were highly infected. The differences between cultivars were found in the leaves.
We tested leaves of different lines with both ELISA and optical disease assessment (*Fig.* 2) to investigate disease severity of *V. dahliae* infected plants. The differences between oilseed rape lines found by optical disease assessment were small. In contrast to results found by optical estimation, the cultivars/lines investigated by ELISA resulted in different susceptibility succession.

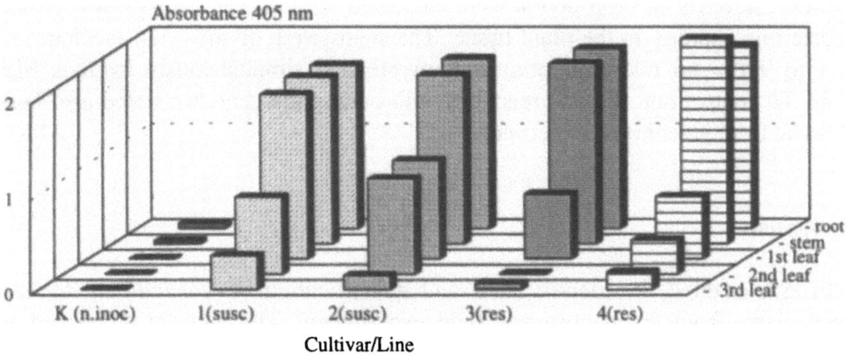

Figure 1. B-A ELISA with non inoculated and artificially inoculated susceptible and resistant oilseed rape cultivars, 35 dpi; (1st leaf= oldest leaf; K= non inoculated, susc.= susceptible, res.= resistant). Plant preparation with extraction buffer 1:5.

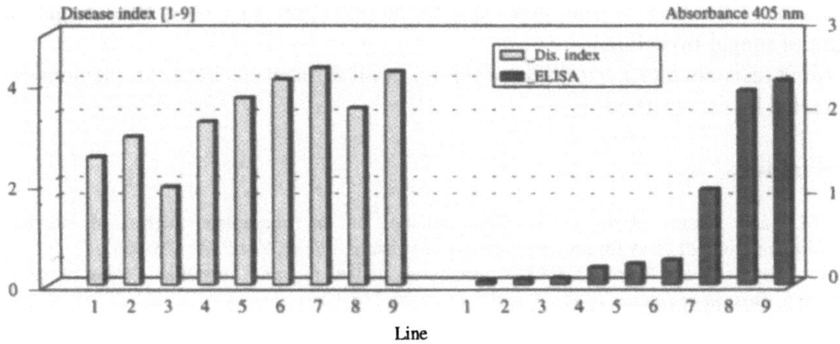

Figure 2. B-A ELISA and optical disease assessment of diferrent artificially inoculated oilseed rape lines. Plant preparation with extraction buffer 1:5.

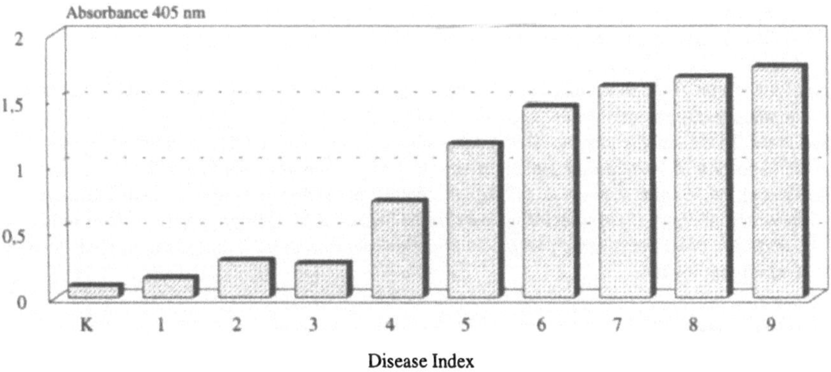

Figure 3. B-A ELISA and optical disease assessment with crop debris (K=non inoculated greenhouse plant, 1=microsclerotia were not observed; extraction with PBST buffer 1:320).

The disease severity of crop debris were evaluated with optical disease assessment by microsclerotia quantity in the plant tissue. The stems were divided into incidence class from 1 to 9 (1= no microsclerotia) and investigated simultaneously by B-A ELISA (*Fig. 3*). To control non specific reactions with other pathogens, we tested also healthy plant tissue from greenhouse experiment.

4. Discussion

B-A ELISA enabled the identification and quantification of *Verticillium dahliae* in infected plants from greenhouse and field experiments. The antisera developed from other authors (Fitzel *et al.*, 1980; Fortnagel and Schlösser, 1992) were used to compare *Verticillium* isolates or had cross-reactions with healthy plants (Van de Koppel and Schots, 1994). We did not get cross-reactions with oilssed rape fungal pathogens and healthy plant material. It is possible to detect *V. dahliae* already 7 days after inoculation. The highest antigen titer was found in roots and stems. In contrast to optical disease assessment on the base of plant reactions to the pathogen, ELISA is able to quantify the amount of fungal mycelium.

Regarding resistance to *Verticillium dahliae*, small differences between incidences could be detected by this method.

5. References

Clark, M.F and Adams, A.N. (1977): Characteristics of the microplate method of enzyme-linked immunosorbent assay for the detection of plant viruses. *J. gen. Virol.* **34**, 475-483.

Fitzel, R.; Fahy, P.C. and Evans, G. (1980): Serological studies on some Australian isolates of *Verticillium* spp. *Australian Journal of Biological Science* **26**, 115-124.

Fortnagel, B. and Schlösser, E. (1992): The ECL-western blotting system, a new tool in the production of monoclonal antibodies against *Verticillium dahliae*. *Med. Fac. Landbouww.* Univ. Gent **57/2a**.

Krüger, W. (1989): Untersuchungen zur Verbreitung von *Verticillium dahliae* Kleb. und anderen Krankheits- und Schaderregern bei Raps in der Bundesrepublik Deutschland. *Nachrichtenbl. Deut. Pflanzenschutzd.* **41**, 49-56.

Nachmias, A. and Krikun, J. (1984): Diagnosis of *Verticillium dahliae* in potato seed tubers with immunoassay techniques. Potato Research **27**, 423-426.

Pegg, G.F. (1985): Life in a black hole - the micro environment of the vascular pathogen. *Trans. Br. mycol. Soc.* **85**, 1-20.

Strunnikova, O.K. and G.S. Muromtsev (1987): Studying *Verticillium dahliae* and *V. albo-atrum* strains with immunoelectrophoresis. Mikologia i Fitopatologia **21**,155-160.

Van de Koppel, M. M. and Schots, A. (1995): Monoclonal antibody-based double-antibody sandwich-ELISA for detection of *Verticillium* spp. in ornamentals. Phytopathology **85**, 608-612.

Van de Koppel, M.M. and Schots, A. (1994): A double (monoclonal) antibody sandwich ELISA for the detection of *Verticillium* species in roses, in A. Schots, F.M. Dewey and R.P. Oliver (eds.), *Modern assays for plant pathogenic fungi: identification, detection and quantification*, CAB International, Oxford, pp. 99-104.

Acknowledgments

We thank Dr. F. Rabenstein, Institut für Pathogendiagnostik der Bundesanstalt für Züchtungsforschung, Aschersleben (Germany), for antisera production and for helpful suggestions. This work was supported by the Konrad-Adenauer-Stiftung, Bonn (Germany).

DEVELOPMENT OF A RAPID METHOD FOR THE DETECTION AND IDENTIFICATION OF *COLLETOTRICHUM ACUTATUM*.

HUGHES K.J.D., LANE C.R. and COOK R.T.A.
Central Science Laboratory, MAFF, Sand Hutton, York, YO4 1LZ, UK

1. Abstract.
The Central Science Laboratory has developed a rapid, easy to use "bio-amplification ELISA" method for the detection and identification of quiescent *Colletotrichum acutatum* infection in strawberry plant material. We believe this is the first use of an ELISA system for the routine detection of an EC quarantine fungus. The detection system has been extensively used on imported and UK derived material and is currently under evaluation in other EC laboratories.

2. Introduction.
The fungus, *Colletotrichum acutatum,* causes the economically important disease strawberry black spot (SBS). It is an EC listed quarantine organism, widely distributed in Europe, but not in the UK. It is predominantly a pathogen of ripe fruit, which develop sunken tan brown lesions on infection; more rarely black lesions may be observed on petioles and runners. However, a further problem faced by growers and plant health inspectors is that *C. acutatum* remains in a quiescent state in planting material. There is therefore, need for a rapid and accurate method for diagnosing the disease when no symptoms are apparent.

Cook (1993) developed a method, the "paraquat test," to induce sporulation in six to seven days by treating detached petioles with paraquat (a contact herbicide). However, following treatment, detection of *C. acutatum* relied on microscopic examination of petioles which was slow, laborious and required considerable experience in separating morphologically similar taxa. This identification problem was overcome by developing monoclonal antibodies (MAbs) which successfully identified conidia of *C. acutatum* in culture (Barker, *et* al 1994, Cook, *et al*, 1995). This paper describes the development of a rapid, simple serological assay for the diagnosis of SBS in strawberry planting material.

3. Materials.
Healthy plants (cv. Elsanta) were bought from a large UK propagator. Infected plants (cv. Elsanta) were obtained by the Plant Health and Seeds Inspectorate of the Ministry of Agriculture Fisheries and Food from a field where SBS had been detected previously. The healthy and infected plants were maintained in separate cubicles within a quarantine glasshouse. Cultures of *C. acutatum* were maintained on potato

H.-W. Dehne et al. (eds.), Diagnosis and Identification of Plant Pathogens, 113–116.

dextrose agar (PDA) at 22°C and a 12 h light/dark regime. Tissue culture supernatants of MAb MAFF 26 were produced (Barker, *et* al, 1994, Cook *et al,* 1995) and frozen in 1 ml aliquots.

4. Methods.

Petioles were detached from known infected and healthy plants; in each case these were divided into two batches. Each petiole of the first batch was placed individually in a heavy gauge plastic bag with 1 ml distilled water and either rinsed by gentle agitation or crushed using a wallpaper seam roller. Aliquots of 100 µl were challenged with MAFF 26 in an ELISA as described by Barker *et al* (1994). The second batch was treated in a similar way except the petioles were dipped in paraquat and incubated for 7 days as described by Cook, (1993), before being rinsed or crushed.

To determine the effect of sap on the MAb reaction, sap from two crushed, healthy petioles was diluted with distilled water to give 100%, 50%, 25%, 12.5% and 6.25% of the original concentration. The diluted sap was added in equal volume to spore suspensions (2×10^6/ml) in 1 ml Eppendorf tubes which were vortexed and 100 µl aliquots challenged with MAFF 26 in an ELISA. The spore suspension was obtained from a typical *C. acutatum* isolate (Herb IMI 351274) and its concentration determined using a haemocytometer.

To simplify the ELISA process, 100 µl aliquots of a *C. acutatum* spore suspension were loaded onto an ELISA plate (plate A) and incubated for 2 h at 35°C. The wells were then washed three times with phosphate buffered saline containing Tween 20 (PBSt), loaded with 100µl of MAFF 26 and incubated 2 h at 35°C. After another triple wash with PBSt, 100 µl of anti-rat conjugate (Sigma A6066) at 1/5000 dilution was added to each well and the plate incubated for another 2 h at 35°C. It was again triple washed with PBSt and 100 µl of substrate buffer added to each well. After incubation at room temperature (25°C) for 1 h the plate was read at 405 nm. Further plates were treated similarly except that the MAFF 26 was mixed with the conjugate and added in a single step ("cocktail ELISA") and the incubation steps were reduced to 1 h, apart from the loading step which remained at 2 h. Two dilutions of conjugate were tested: 1/5000 (Plate B) and 1/2500 (plate C).

To develop a "bio-amplification" method petioles from known healthy and infected plants were treated separately with paraquat, and incubated in damp chambers for three days. Eight healthy petioles and two tagged infected petioles were then placed in a polythene bag (20 x 20 cm) with 10 ml tap water. The ten petioles were rinsed, placed in fresh damp chambers and incubated for another four days, after which they were assessed using a dissecting microscope for the presence of *C. acutatum* acervuli.

5. Results.

ELISA tests on petioles gave low signals (1.1-1.2) for all the healthy petioles irrespective of treatment (Fig, 2). Similar results were obtained for untreated infected petioles, but paraquat treated petioles gave high signals. There was an indication that MAb activity was lower with crushing (4.2) compared with rinsing (5.5).

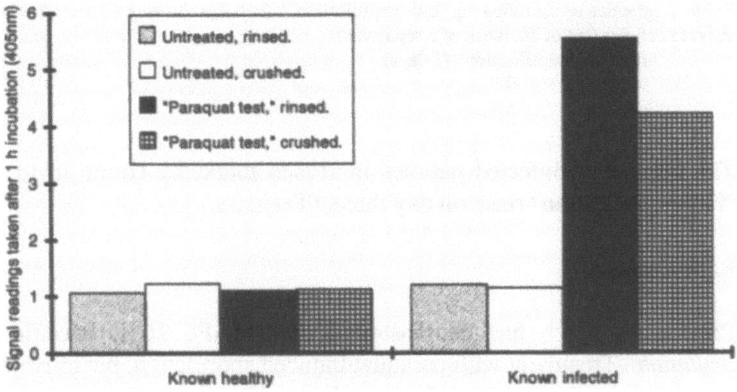

Figure 2. Detection of *C. acutatum* by ELISA in untreated and paraquated petioles, either rinsed or crushed.

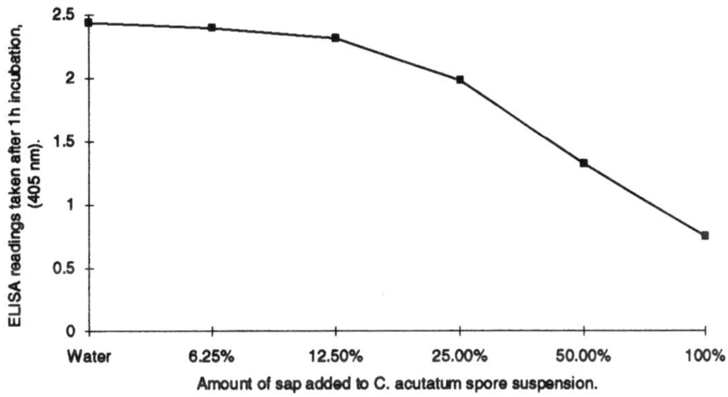

Figure 3. Effect of healthy petiole sap on ELISA values for a *C. acutatum* suspension (2 x 10^6 spores/ml).

With increased concentration of sap, there was a steady decrease in MAb activity from 2.4 (water only) to 0.7 (neat sap added) (Fig.3).

Table 1. Comparison of "standard ELISA" with "cocktail ELISA" (readings at 405 nm and 30 min incubation) using MAb MAFF 26 against a range of *C. acutatum* spore concentrations.

Spores/ml	"Standard ELISA" Conjugate 1/5000 (Plate A)	"Cocktail ELISA" Conjugate 1/5000 (Plate B)	"Cocktail ELISA" Conjugate 1/2500 (Plate C)
4 x 10^2	0.106	0.076	0.082
4 x 10^3	0.137	0.082	0.094
4 x 10^4	0.326	0.135	0.502
4 x 10^5	0.963	0.526	0.946
4 x 10^6	1.376	1.178	1.844

The "cocktail ELISA" gave reduced values compared to the "standard ELISA" using similar concentrations of conjugate (Table 1, plates A and B). Higher values were obtained for the "cocktail ELISA" by doubling the conjugate concentration (Plate C).

Table 2. Infection levels following "bio-amplification." (Petioles showing *C. acutatum* acervuli). Number of infected petioles (out of 10, mean of 3 replicates). At start (0 days)

After "bio-amplification" (7 days). 0

0 2

5.7

The number of infected petioles increased markedly (from 20 to 57%), following a "bio-amplification" rinse on day three. (Table 2).

6. Discussion.

Previous research has demonstrated that MAFF 26 is specific to conidia of *C. acutatum*. Treatment with paraquat induced sporulation, but crushing petioles released an inhibitor, which reduced MAb activity. This was overcome by rinsing petioles, this reduced sap inhibition and increased conidial numbers by spreading infection to healthy petioles. This "bio-amplification" enabled a serological assay to be developed for host material which could have potential for many other diagnostic protocols. Further improvements were made by combining ELISA steps thereby reducing the test time from 7 to 4 h. As a result of this work the procedure (fig. 1) has been extensively been used on imported and UK derived material and is under evaluation in other EC laboratories.

300 Petioles, bleached and treated with paraquat.

↓

Petioles incubated in a damp chamber in constant light at 25°C for 3 days.

↓

Petioles rinsed in 80 ml tap water.

↓ ↓

Petioles incubated for a further 4 days. Rinsing water stored at 4°C.

↓

Sample re-rinsed using the stored rinsing water.

↓

Aliquots tested by "cocktail ELISA" using MAb MAFF 26.

↓

Confirmation of *C. acutatum* viability by plating on PDA.

Figure 1. "Bio-amplification" ELISA for the detection of quiescent *C. acutatum* in strawberry plants.

7. Acknowledgements.

We wish to thank Plant Health Division of the Ministry of Agriculture, Fisheries and Food for financial support and their Plant Health and Seeds Inspectorate for the collection of samples. Also S. Freeman, R. Bowyer, S. Preston for technical assistance.

8. References.

Barker, I., Brewer, G., Cook, R.T.A, Crossley, S. and Freeman S. (1994) Strawberry black spot disease (*Colletotrichum acutatum*). In *Modern assays for plant pathogenic fungi: Identification, detection and Quantification.* (eds. A. Schots, F.M. Dewey and R.P. Oliver), pp 179-182. CABI, Oxford UK.

Cook, R.T.A. (1993) Strawberry black spot caused by *Colletotrichum acutatum. In Plant health and the European Single Market.* BCPC monograph no 54 (ed D. Ebbles), pp 301-304. BCPC, Farnham, UK.

Cook, R.T.A., Barker I., Brewer, G., Crossley, S., Freeman, S. & Lane, C. R. (1995). Detection of Strawberry black spot (*Colletotrichum acutatum*) with monoclonal antibodies. OEPP/EPPO 25 pp 57-63.

SELECTIVE ENRICHMENT IMPROVES ISOLATION, SEROLOGICAL AND MOLECULAR DETECTION OF PLANT PATHOGENIC BACTERIA

M.M. LÓPEZ, M.T. GORRIS, P. LLOP, J. CUBERO, B. VICEDO and M. CAMBRA
Instituto Valenciano de Investigaciones Agrarias (IVIA)
Apartado Oficial, 46113, Moncada, Valencia, Spain

1. Introduction

The detection of low numbers of plant pathogenic bacteria can be improved by using a combination of enrichment in selective media and specific monoclonal antibodies (mab) or specific primers. In this way, the enrichment increases the number of the target bacteria and the specific probes can accurately identify them in spite of some increase of the total bacterial population in the samples. The enrichment step must be optimised for each bacterium by defining the medium, time, temperature and incubation conditions in order to obtain high numbers of the target bacterium that will easily detected by the specific probes in a second step. Such procedures have been applied to the detection of *Agrobacterium tumefaciens* causal agent of crown gall, and to the detection of *Erwinia carotovora* subsp. *atroseptica* causal agent of potato black leg and *E. amylovora* causal agent of fire blight.

2. Materials and Methods.

2.1. SELECTIVE ENRICHMENT FOR *A. tumefaciens*

Samples of soil, tumors and different types of plant material were mixed or comminuted in sterile water and 0.5 ml of the sample was enriched in 5 ml of liquid selective media for *A.tumefacients* biovar 1 (Schroth *et al.*, 1965), biovar 2 (New and Kerr, 1971) and for biovar 3 (Roy and Sasser, 1983). After 3-5 days of incubation at 25°C in aerobiosis, aliquots were plated in the same selective media and processed by PCR using the primers *tmr* and *vir* defined by Nesme *et al.*, (1989). The efficiency of the detection of *A.tumefaciens* by the described methods was compared with direct plating on the same media or PCR without enrichment.

H.-W. Dehne et al. (eds.), Diagnosis and Identification of Plant Pathogens, 117–121.

2.2. SELECTIVE ENRICHMENT OF *E. carotovora* subsp. *atroseptica*

Samples of individual potato tubers and plants, or potato seed lots were enriched after being processed in different ways (Gorris *et al.*, 1994),. The enrichment was performed under anaerobic conditions for 40 h at 25° C after adding in every well of and ELISA-microplate 100 µl of the sample and 100 µl of D-PEM medium (Gorris *et al.*, 1994). The samples were processed by plating on CVP (Pérombelon and Burnett, 1991) and by ELISA-DASI using specific mab according to Gorris *et al.* (1994). The efficiency of the detection of *E. carotovora* subsp. *atroseptica* by the described methods was compared with direct plating on CVP and ELISA-DASI without the previous enrichment step.

2.3. SELECTIVE OR NON SELECTIVE ENRICHMENT FOR *E. amylovora*

Samples of buds, flowers, shoots, cankers or fruits of different fire blight hosts were processed by enrichment after being comminuted in an anti-oxidant buffer (Gorris *et al.*, 1996). The enrichment was performed for 40 h at 25°C in liquid King's B (King *et al.*, 1957) or CCT (Ishimaru and Klos, 1984) by adding 300 µl of the liquid media to 300 µl of the comminuted sample in an eppendorf tube. After enrichment, the samples were treated at 100°C for 10 min to be processed by ELISA-DASI using specific mab described by Gorris *et al.* (1996) or by PCR using the primers described by Bereswill *et al.* (1992). Unheated enriched samples, were plated on King's B or CCT. Non enriched samples were also plated and amplified using similar protocols. The sensitivity of the enrichment procedures described above for *A. tumefaciens, E. carotovora* subsp. *atroseptica* and *E. amylovora* was determined in spiked samples, prepared using plant extracts of the different hosts and suspensions of decreasing concentration of each plant pathogenic bacteria.

3. Results

3.1 DETECTION IN PLANT MATERIAL

The results of the comparison of several methods for their efficiency in the detection of the different pathogenic bacteria is shown in table 1. Results of detection of *A. tumefaciens* in plant material show that selective enrichment improved the efficiency of its diagnostic. Results of detection of *E. carotovora* subsp. *atroseptica* in potato show that the enrichment method set up improved the efficiency and sensitivity of the isolation and ELISA-DASI detection. Results of detection of *E. amylovora* in several hosts also show that the enrichment step improved its detection by isolation and ELISA-DASI.

TABLE 1. Detection of several plant pathogenic bacteria in healthy and naturally infected material by different techniques without and with a previous enrichment step. Data expressed as number of samples with the target bacteria divided by the analysed samples.

Species	Direct plating	Enrichment + plating	ELISA - DASI	Enrichment + ELISA - DASI	PCR	Enrichment + PCR
A. tumefaciens	57/85	83/85	--	--	38/39	39/39
E. carotovora subsp. *atroseptica*	77/144	90/144	65/144	93/144	--	--
E. amylovora	28/98	30/98	15/98	30/98	39/98	20/24

3.2 SENSITIVITY OF THE DETECTION OF THE DIFFERENT PATHOGENIC BACTERIA USING ENRICHMENT

The sensitivity of the different enrichment protocols assayed is shown in table 2. The enrichment procedure allowed the detection of 1 cfu/ml of *A. tumefaciens* in spiked peach seed samples by plating in selective medium and by PCR. Ten to 10^2 cfu/ml of *E. carotovora* subsp. *atroseptica* in inoculated potato tubers were detected by plating on CVP, ELISA-DASI and PCR after the enrichment step. Specific primers kindly supplied by Y. Bertheau (INA, Paris) were used for amplification. One hundred cfu/ml of *E. amylovora* were also detected in spiked samples by plating on King's B or CCT medium, ELISA-DASI and PCR after the enrichment.

Commercial kits for detection of *E. carotovora* subsp. *atroseptica* and *E. amylovora* based on ELISA-DASI with an enrichment step are available from Real-Durviz, Valencia (Spain) and Plantprint Diagnostics, Valencia (Spain) respectively, with license of the patent methods (Spanish patent P9302526; United Kingdom patent 9424387.0/1669004; The Netherlands patent 9402109).

In conclussion, appropiate enrichment on liquid selective medium for each plant pathogenic bacterium can improve the sensitivity of its isolation, and the serological or molecular detection. No interference of the saprophytic bacteria was observed in the analysed samples.

TABLE 2. Sensitivity of the different methods for detection of *A. tumefaciens*, *E. carotovora* subsp. *atroseptica* and *E. amylovora* in spiked samples.

Bacterial concen-trations	*A. tumefaciens*[a]				*E. carotovora* subsp. *atroseptica*[b]				*E. amylovora*[c]	
	D Pl[d]	E+Pl[e]	P[f]	E+P[g]	D Pl	E+Pl	P	E+P	El-D[h]	E+El-D[i]
10^6 cfu/ml	+	+	+	+	+	+	+	+	+	+
10^5 cfu/ml	+	+	+	+	+	+	+	+	-	+
10^4 cfu/ml	+	+	+	+	+	+	+	+	-	+
10^3 cfu/ml	+	+	-	+	+	+	+	+	-	+
10^2 cfu/ml	+	+	-	+	-	+	+	+	-	+
10 cfu/ml	+	+	-	+	-	+/-	-	+/-	-	+
1 cfu/ml	-	+	-	+	-	-	-	-	-	+/-
0 cfu/ml	-	-	-	-	-	-	-	-	-	-

[a]- Detection of *A. tumefaciens* in spiked peach seed samples.
[b]- Detection of *E. carotovora* subsp. *atroseptica* in artificially inoculated potato tubers.
[c]- Detection of *E. amylovora* in spiked pear samples.
[d]- DPl: Direct plating in the selective media appropiate for each bacterium.
[e]- E+Pl: Selective enrichment before plating in the selective media.
[f]- P: PCR using specific primers.
[g]- E+P: Selective enrichment before PCR.
[h]- El-D: ELISA-DASI (double-antibody sandwich indirect).
[i]- E+El-D: Selective enrichment before ELISA-DASI.

4. References

Bereswill, S., Pahl, A., Bellemann, P., Zeller, W., Geider, K.: (1992). Sensitive and species -specific detection of *Erwinia amylovora* by PCR- analysis. *Appl. Environ. Microbiol.* **58**, , 3522-3626.

Gorris, M.T., Alarcón, B., Lopez, M.M., Cambra, M.: (1994). Characterization of monoclonal antibodies specific for *Erwinia carotovora* subsp. *atroseptica* and comparison of serological methods for its sensitive detection on potato tubers. *Appl. Environ. Microbiol.* **60**, 2076-2085.

Gorris, M.T., Cambra, M., Lecomte, P., Llop, P., Chartier, R., Paulin, J.P., López M.M.: (1996). A sensitive and specific detection of *Erwinia amylovora* based on the ELISA-DASI enrichment method with monoclonal antibodies. *Acta Horticulturae* **411**, 41-46.

Ishimaru, C., Klos, E.J.: (1984). New medium for detecting *Erwinia amylovora* and its use in epidemiological studies. *Phytopathology* **74**, 1342-1345.

King, E.O., Ward, M., Raney, D.E.: (1954). Two simple media for the demonstration of phyocianin and fluorescein. *J. Laboratory Clinical Medicine,* **44**, 401-407.

Nesme, X., Leclerc, M.C., Bardin, R. (1989). PCR detection of an original endosymbiont: the Ti plasmid of *Agrobacterium tumefaciens* 4th International Colloquium on Endocytobiology and Symbiosis, Villeurbanne (France).

New, P.B., Kerr, A.: (1971). A selective medium for *Agrobacterium radiobacter* biotype 2., *J. Appl. Bacteriol.* **34,** 233-236.

Pérombelon, M.C.M., Burnett. (1991). Two modified crystal violet pectate (CVP) media for the detection, isolation and enumeration of soft rot *Erwinias*. *Potato Research.* **34**, 79-85

Roy, M.A., Sasser, M. (1983). A medium selective for *Agrobacterium tumefaciens* biotype 3 (abstr.) *Phytopathology* **73**, 810.

Schroth, M.N., Thompson, J.P., Hilderbrand, D.C.: (1965). Isolation of *Agrobacterium tumefaciens- A .radiobacter* group from the soil. *Phytopathology* **55,** 645-647.

APPLICATION OF DIFFERENT TECHNIQUES TO DETECT *PHYTOPHTHORA* SPP. IN ROOTS OF NURSERY PLANTS

SABINE WERRES
Federal Biological Research Centre for Agriculture and Forestry,
Institute for Plant Protection in Horticultural Crops,
Messeweg 11/12, D - 38104 Braunschweig

For the experiments *Chamaecyparis lawsoniana* plants originating from different nurseries and differing in age were used. Besides the rooted cuttings they were grown in 2 - 4 liter containers. Some of the *Chamaecyparis* showed symptoms such as wilting or needle or/and root discolouration, others looked healthy. For the experiments the roots were cut into pieces of about 1 cm in length.

For the detection of living propagules of *Phytophthora* spp., fresh roots were subjected to a bait test using *Rhododendron* leaves as baits. The rates of *Rhododendron* leaves showing discolouration and the isolation rate of *Phytophthora* propagules from leaf tissue pieces were estimated. Frozen root samples were taken for experiments with different serological techniques: A dot-immunobinding assay (DIBA) developed for *Phytophthora cinnamomi* detection in roots by a polyclonal antiserum to an isolate of this *Phytophthora* species (see HAHN and WERRES, present proceedings) and two commercial kits, the ALERT On-site Disease Diagnosis Kit and the AGRISCREEN Multiwell Laboratory Immunoassay from SAC ADGEN Diagnostic Systems for the identification of *Phytophthora* spp.

Phytophthora spp. could be detected with the *Rhododendron* leaf test in all 11 root samples of the symptomless and diseases *Chamaecyparis* plants. There was

H.-W. Dehne et al. (eds.), Diagnosis and Identification of Plant Pathogens, 123–124.

inconsistent correlation between the rate of discoloured *Rhododendron* leaves or the isolation rate and the symptomatology of the *Chamaecyparis* plants. This is probably due to a treatment of the *Chamaecyparis* plants with Aluminium-Fosetyl (Aliette) in the nurseries. DIBA also permitted *Phytophthora* detection in all 11 root samples. The ALERT *Phytophthora* kits and the AGRISCREEN *Phytophthora* Multiwell Laboratory Immunoassay gave clear reactions only in 4 and 9, respectively, of the 11 root.

SEROLOGICAL DETECTION OF *SPONGOSPORA SUBTERRANEA* F.SP. *SUBTERRANEA* IN SOIL

U. MERZ[1] and J. A. WALSH[2]
[1]*Institute of Plant Sciences, Pathology Group, Swiss Federal Institute of Technology, CH-8092 Zurich, Switzerland,* [2]*Horticulture Research International, Wellesbourne, Warwick, CV35 9EF, UK*

A bioassay with tomato as bait plants was compared with detection by ELISA using a dilution series of spore balls in uncontaminated soil. The relationship between A_{405} values and spore ball concentration/g soil appeared to be linear and raw antiserum detected down to 100 spore balls/g soil. Discrimination of spore ball levels improved with concentrations above 2000/g soil, whereas the bioassay showed better discrimination of spore ball levels below 2000/g soil. When a range of soils from Swiss potato fields with known bioassay scores were tested by ELISA, raw serum discriminated most of the soils with higher scores (≥ 3) from those with lower scores, but not between uncontaminated and infested soils with lower scores (<3). Of the three soils with scores higher than 3, the two where high tuber disease indexes were obtained also gave highest A_{405} values. The ability of the antiserum to discriminate infested field soils from uncontaminated ones was much improved by using the γ-globulin fraction, or serum which had been cross-absorbed with uncontaminated soil. The results are discussed and advantages and disadvantages of both detection methods compared.

1. Introduction

Spongospora subterranea f.sp. *subterranea* is the causal agent of powdery scab of potato. Control of the disease is difficult as the resting spores are able to survive in a dormant state in soils for a number of years and are difficult to kill. Disease avoidance through planting clean seed into uncontaminated land would be the only reliable form of control. The presence and contamination levels of resting spores in field soils are difficult to determine. Bioassays involving the baiting of soil samples with tomato (Merz, 1989) or potato (Wale *et al.*, 1993) seedlings have been used but they are very labour intensive and results may not correlate well with disease levels observed in the field. Rapid and sensitive detection of resting spores in field soils would be a great aid to the development of disease management strategies and the determination of disease thresholds.

H.-W. Dehne et al. (eds.), Diagnosis and Identification of Plant Pathogens, 125–128.

2. Materials and Methods

Preparation of spore ball material, antiserum production, ELISA technique, specificity of the antiserum and studies on the antigens detected by the antiserum are described by Walsh and Merz (these proceedings).

2.1. BIOASSAY DETERMINATION OF THE LEVEL OF *S. SUBTERRANEA* INFESTATION OF SOIL

A modified tomato bait plant test system as described by Merz (1989) was used. Soil samples were incubated for 10 days and baited for one day.

2.2. SENSITIVITY OF DETECTION OF RESTING SPORE BALLS IN SPIKED SOIL

A dilution series containing approximately 8000, 6000, 4000, 2000, 1000, 500, 100 and 0 spore balls per gram of soil was prepared from uninfested field soil and spore balls from potato cv. Bintje. A sample of each dilution (1g) was taken and ground with a pestle and mortar in 2 ml of 0.05M sodium carbonate buffer (pH 9.6), and tested by ELISA. The remainder of these spore dilutions in soil were tested by bioassay.

2.3. COMPARISON OF RESULTS FROM ELISA AND BIOASSAY TESTS ON FIELD SOILS AND DISEASE SEVERITY SCORES FOR TUBERS GROWING IN THE SOILS

A range of eleven soils collected one month before harvest from fields in Switzerland where potatoes were being cultivated and a known uncontaminated soil were tested by ELISA. Samples (1g) of these soils were ground with pestles and mortars in carbonate buffer (2 ml) and tested by ELISA. In addition, further samples of these soils were tested by bioassay. Tuber samples taken from these fields at the same time as the soil samples, were visually assessed for the severity of powdery scab, scored on a scale of 0-8 and disease indexes calculated as described by Merz (1993).

2.4. ATTEMPTS TO IMPROVE THE SENSITIVITY OF SEROLOGICAL DETECTION OF SPORE BALLS IN SOIL

The antiserum was also cross-absorbed with an uncontaminated field soil from Switzerland in order to reduce background absorbence levels. The resulting serum, the purified γ-globulin fraction and the original raw serum were tested by ELISA against a range of seven soils collected from Swiss potato fields and an uncontaminated soil.

3. Results

The raw serum detected spore balls mixed with uninfested field soil. The relationship between A_{405} values and spore ball concentration/g soil appeared to be linear. The antiserum detected 100 spore balls/g soil but discrimination of spore ball levels appeared to be better for concentrations greater than 2000/g soil. In contrast the bioassay discriminated spore ball levels better in soils containing less than 2000/g (Fig. 1).

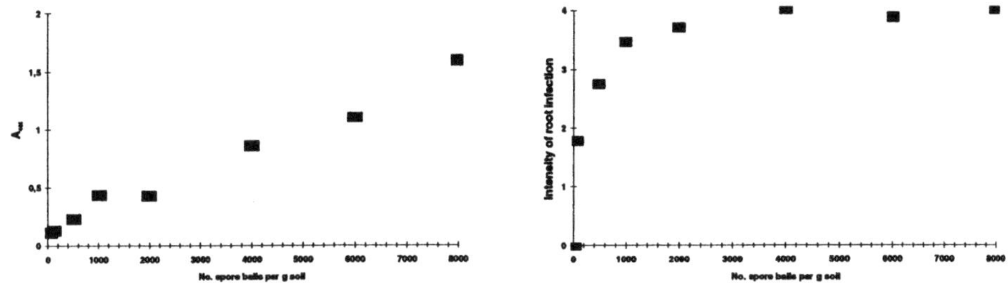

Fig. 1. The relationship between the concentration of spore balls of *Spongospora subterranea* in spiked soil samples and A_{405} in a PTA-ELISA (left) and intensitiy of root infection in a bioassay test (right).

The raw serum discriminated two of the three soils with high bioassay scores (>3) from those with lower scores in PTA-ELISA tests, however soils with bioassay scores ranging from zero to three were not differentiated from each other by ELISA (Table 1). The relationship between A_{405} values and bioassay scores for the field soils was similar to that observed with spiked soils and again demonstrated the different sensitivities of the two detection systems. Potatoes from the three soils which were discriminated from the others by ELISA also had the highest tuber disease indexes (Table 1).

When tested on other field soils the cross-absorbed serum and the purified γ-globulin frac-tion gave drastically reduced A_{405} values for the uncontaminated soil in PTA-ELISA compared to that of the raw serum. The

Soil sample number	A_{405}*	Bioassay score	Tuber score
12 (healthy)	0.55	0	0
14	0.46	0.4	0
19	0.49	3.0	0
15	0.51	0.9	0
16	0.52	1.4	0
17	0.34	2.7	0
18	0.44	2.7	26
28	0.43	2.0	31
30	0.49	3.6	60
20	0.63	3.0	84
21	1.41	4.0	228
29	1.65	3.2	281

*Data ranked for tuber disease index values

Table 1. Comparison of data from ELISA and bioassay tests on field soils and disease indexes for tubers from the soils

differentiation of the contaminated soils from the uncontaminated soil was also much improved although the overall response was reduced.

4. Discussion

The results described demonstrate for the first time the serological detection of an obligate pathogenic fungal virus vector in field soil. The detection of spore balls in spiked soil samples demonstrated the potential of the antiserum for identifying contaminated field soils. The reduced ability of the antiserum to discriminate spore ball concentrations of less than 2000/g soil in spiked soil may have been due in part to inaccuracies in the estimation of spore ball numbers in this range. However, it is also possible that detection of non-target antigens in the soil itself was masking spore ball antigen detection. Inconsistencies between ELISA and bioassay results may have been due to a large proportion of dead spore balls and/or the multiplication of the pathogen during the 8 days of the cultivation period in the bioassay.

ELISA identified the three soils where highest levels of tuber damage were observed (disease indexes of 80, 228 and 281) whereas the bioassay results suggested other soils where much lower tuber damage was observed (disease indexes of 0 and 60) were heavily infested. These limited observations suggest ELISA tests on soil using this particular antiserum may predict epidemiologically important field disease levels better than the bioassay. Swiss soils that are considered to be heavily infested are thought to contain >500 spore balls/g soil and such soils have given bioassay scores of >3 (Merz, 1993). However, as in the experiment described here, many soils that gave high bioassay scores in an earlier study, did not give rise to diseased tubers (Merz, 1993). It would be easy to conclude that in such soils the environmental conditions were unfavourable for infection. However, ELISA tests on some such soils suggest that they contain fewer spore balls, therefore bioassay scores may be anomalously high due to its "amplification" effect. Further detailed comparative studies are needed.

The improvement in sensitivity of detection of spore balls in soils after cross-absorption could make the serological identification of heavily infested soils more reliable. This was evident from tests on field soils where both the cross-absorbed serum and the γ-globulin fraction gave much better differentiation not only of contaminated soils from uncontaminated soil but also of low to medium contaminated soils from highly contaminated ones.

5. References

Merz, U. (1989) Infectivity, inoculum density and germination of *Spongospora subterranea* resting spores: a solution-culture test system, *EPPO Bulletin* **19**, pp. 585-592.

Merz, U. (1993) Epidemiological aspects of powdery scab of potatoes caused by *Spongospora subterranea*, *Proceedings of the Second Symposium of the International Working Group on Plant Viruses with Fungal Vectors*, 1993, McGill University, Montreal, Canada, pp. 103-106.

Wale, S. J.; Burgess, P. J., and Burnett, F. (1993) Bioassay for *Spongospora subterranea* detection in soil and its potential use in predicting powdery scab in the field, *Proceedings of 6th International Congress of Plant Pathology*, Montreal, Canada, Abstract 16.4.2, pp. 281.

PROSPECTS OF USING PCR FOR THE DETECTION OF BACTERIAL PATHOGENS IN POTATO

REIJO KARJALAINEN[1], AINO KANGASNIEMI[1] AND JUKKA TEGEL[2]
[1] Department of Plant Biology, Plant Pathology Section
00014 Helsinki, Finland
[2] Plant Protection Inspection Centre, Plant Quarantine Laboratory, Box 42, 00501 Helsinki, Finland

1. Abstract

Effective control of bacterial pathogens (*Clavibacter michiganensis* subsp. *sepedonicus*, *Pseudomonas solanacearum* and *Erwinia carotovora* subsp. *atroseptica*) of potato depends greatly on the specific, sensitive and accurate detection of the pathogen in infected tubers. Infected tubers are a major source of initial inoculum in the growing season, and therefore detection of low levels of bacterium in the seed plays a key role in disease control and eradication. Traditional diagnostic methods involves the use of bioassays, serological methods and culturing of the bacterium but these methods are often time consuming and cross-reacting saprophytic bacteria may cause problems. We have used PCR as an additional tool in the routine detection of potato ring rot pathogen. Generally PCR test have revealed the same positive and negative test results but in a few cases where cross-reacting problems prevented reliable immunofluorescence detection, PCR test gave a clear negative results. Our experiences suggest that when PCR test is carefully performed, it is a reliable test method for practical diagnosis of difficult samples. New developments of using more convenient test formats, such as immunocapture-PCR, will lead to the development of ELISA-like PCR for bacterial detection in potato.

2. Introduction

In Europe, three bacterial pathogens (*Clavibacter michiganensis* subsp. *sepedonicus, Pseudomonas solanacearum* and *Erwinia carotovora* subsp. *atroseptica* cause considerable losses to potato crops. In Europe (e.g. The Netherlands), on average 1 % of the potato seed lots is rejected or downgraded due to *Erwinia* contamination and therefore several millions of ECUs are lost in export business every year. In the Mediterranean area *E carotovora* subsp. *atroseptica* is considered a very serious

H.-W. Dehne et al. (eds.), Diagnosis and Identification of Plant Pathogens, 129–135.

disease causing significant emergent losses in potato because local culturing practices (seed cutting, flood irrigation) favours disease outbreaks. Seed certification has been the major means of disease control which aims to provide disease-free seeds for the potato farmers (De Boer et al. 1988, Bear et al. 1993). However, this is not an easy task because diagnostic methods used do not allow to detect seed stocks with a low level of contamination (De Boer and Ward 1995). Under favourable weather conditions, the blackleg incidence may develop and subsequently cause significant crop losses.

Currently two quarantine pathogens causing potato ring rot (*Clavibacter michiganensis* subsp. *sepedonicus*) and brown rot (*P. solanacearum*) pose a great risk to the European potato industry because of movement of potato material across countries. The recent outbreaks of the potato brown rot in several important potato farms of the Netherlands had serious economical consequences for the Dutch potato farmers which are the most important producers of seed potatoes in the world. This sudden outbreak of the brown rot suggest that the control of bacterial pathogens by seed inspection and eradication is currently ineffective because the available diagnostic methods are inadequate for the detection of both low level latent infections and pathogen spread and survival in the environment, and hence there is a risk of spreading these harmful bacterial pathogens within Europe via contaminated seed. Consequently, there is an increasing need for the improvement and further development of procedures for the rapid and reliable detection of potato bacterial pathogens.

Various diagnostic procedures, such as bioassay, gram stain test, serology (immunofluorescence, ELISA) and in some cases DNA-based assays have been used to detect bacterial contamination in seed stocks (Jansen et al. 1987). In biomedical laboratories, DNA-amplification (PCR) based diagnostic procedures has become popular, and in many cases they have replaced immunological test methods. Very high sensitivity and specificity makes PCR-based test methods attractive alternatives for other diagnostic systems. However, PCR-based test systems have been used only in limited cases in bacterial detection in potato despite many advantages over other detection methods (Henson and French 1993). Although, current PCR-methods have some drawbacks, such as sensitivity for contamination and they are not yet as robust test formats as immunological methods, advantages of PCR over other tests are, however, so great that these techniques will be an essential part of future disease detection of bacterial pathogens in potato.

3. Sources of PCR primers

Availability of specific primers is an important part of PCR based detection system. Various types of primers have been produced for bacterial detection in potato. Seal et al. (1993) construct PCR-primers based on 16S rRNA for sensitive detection of brown rot bacteria in potato. Using of the specific oligonucleotide with a non-specific oligonucleotide in the PCR, 1-10 cells of bacteria were detected after 50 rounds of amplification. Similar type of approach was reported by Li and De Boer (1995) for constructing PCR primers based on the intergenic spacer region, between 16S and 23S rRNA genes, for amplification of potato ring rot pathogen. PCR amplification of using these primers allowed the detection of Cms more sensitive than ELISA and immunofluorescence. PCR primers have also been constructed based on the DNA probes of the Cms plasmid region (Firrao and Locci 1994), but this type of primer sequence may have a problem that the PCR test system based on plasmid sequences may not be able to detect bacterial strains without plasmids. Genomic fragments isolated from ring rot pathogen provide also specific sequences for PCR based detection of Cms (Karjalainen et al. 1995). The PCR primers based on the sequence information of genomic fragments, enabled a detection limit less than 25 pq of Cms DNA using only of 25 cycle of amplification (Karjalainen et al. 1996). Recently a PCR based test for *Erwinia* detection was developed based on primers of consensus sequences of metalloprotease-coding genes present in soft rot erwiniae (Smid et al. 1995). The PCR assay in this work allowed a detection of less than 10^3 culturable erwiniae per tuber (Smid et al. 1995). Three approaches have been described for the isolation of genomic fragments of *E carotovora* subsp. *atroseptica* suitable for PCR amplification. Darrase et al. (1994) isolated Eca specific genomic fragments by genomic subtraction method and Maki-Valkama and Karjalainen (1994) isolated RAPD fragment directly from the gel and after cloning and sequencing of one fragment, Eca specific primers were successfully designed from sequence information of RAPD fragments. The detection level of 25 pq of Eca DNA after 25 cycle of amplification was achieved. Recently De Boer and Ward (1995) reported a sensitivity of 5-10 CFU per PCR reaction when they used PCR-primers designed from sequence information of specific genomic DNA probes. Thus available examples suggest that several types of primers can be successfully used for the specific amplification of target bacterial pathogens of potato.

4. Diagnostic applications

PCR based diagnostic methods have been used only in few cases for routine detection of bacterial pathogens in potato. We have used PCR for routine detection of potato ring rot pathogen during the last three years in collaboration with plant quarantine experts in Finland (Karjalainen et al. 1995, 1996). In order to improve PCR procedure, we have raised the annealing temperature up to 55°C which eliminated the nonspecific PCR products and finally a single, clear PCR product of 0,3 kb was detectable when tuber tissue was contaminated by Cms. We have used only simple sample preparation for PCR, and only one phenol-chloroform purification step has been added for normal IF- procedure.

For the evaluation of practical value of our PCR procedure, potato material for routine Cms detection were analyzed by both PCR and IF-methods. Composite samples consisting of 200-400 tubers from 20 000-25 000 kg of potatoes constitutes a single sample in a latent infection test (Karjalainen et al. 1996). Schematic scheme of the diagnostic procedure is described in figure 1 (Karjalainen et al. 1996).

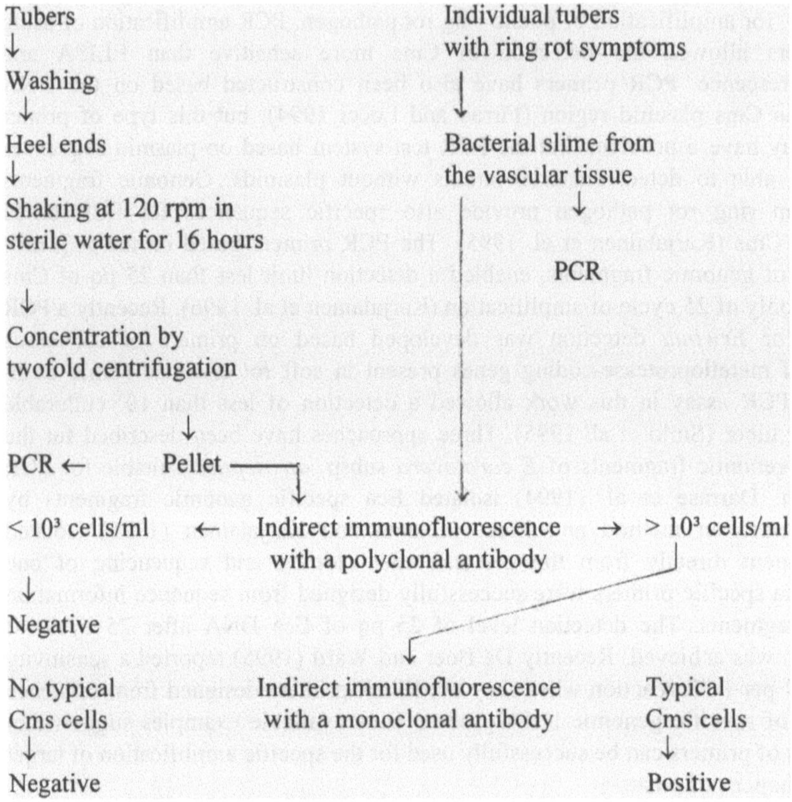

FIGURE l. Schematic scheme for the detection system used for analysing potato tubers by indirect immunofluorescence and by PCR.

Generally, our results have indicated that the same positive and negative results could be revealed in parallel tests with PCR and IF. Most of the IF positive samples tested in this work have contained a quite high amount of typically fluorescing cells/microscope field. However, in a few cases that were considered to be negative in the IF test had also fluorescing cells/window and some of the samples had a high amount of cross-reacting atypical fluorescent cells. These two type of cases were difficult to analyze by serological methods, but the PCR test gave a clear negative results (Karjalainen et al. 1996). Our experiences over three years of using PCR as an additional tool in routine diagnostics suggest that positive and negative samples are very easily detected on agarose gel after EtBr staining when PCR conditions have first optimized. As it is known that infected seed potatoes are the major source of initial inoculum each growing season, detection of low levels of Cms is an important step in the control and eradication of ring rot pathogen (Li and De Boer 1995). Sensitive PCR test will be therefore a crucial importance for the monitoring of successfulness of eradication programmes of ring rot pathogen and thereby help to prevent the further spread of disease from infested country into new areas. Reports from occasional outbreaks of ring rot from Canada and Denmark despite careful eradication schemes (Stead 1993) suggest that eradication is a difficult task. This is partly due to the fact that the epidemiology of Cms is insufficiently known. The survival of Cms in soils, potato waste and other host plants is not yet fully understood. Currently used serological methods are not sensitive enough for such studies, and PCR is a right tool for epidemiological analysis of ring rot pathogen.

Rapid and highly sensitive PCR-test is also needed for the detection of *P. solanacearum* in planting material as this quarantine pest can infect many important crops without causing visible symptoms. Seal et al. (1993) have developed a PCR-test which enables the detection of latent bacterial wilt infection of symptomless potato tubers. This test can be completed in less than 5 hours, and is much faster than the traditional methods and therefore the PCR-test is well suited for quarantine laboratories. Rapid eradication of potato brown rot pathogen in Europe is an urgent problem and PCR-tests will be an essential part of the successful eradication strategy. More information on the survival of the *P. solanacearum* in European conditions is also urgently needed for phytosanitary purposes and this information is difficult to obtain by conventional test methods and sensitive PCR test is therefore required. *P. solanacearum* is also a very devastating pathogen for a wide range of tropical crops and Seal (1995) have demonstrated that PCR- test system is also suitable for tropical and subtropical countries for revealing the presence of *P. solanacearum* in planting material. Application of PCR assay for the analysis of routine samples for the detection of potato brown rot and ring rot pathogen seems possible. Available information of using PCR primers for ring rot detection consistently suggest their greater sensitivity over the serological tests. Thus, PCR appears the most suitable for the verification of difficult samples that cannot be detected by other methods such as those of low levels of the bacterium in the seed. Further, successful eradication of the pathogen requires careful monitoring of the pathogen in various potential small pockets. Our experience of using PCR as a additional diagnostic tool in routine laboratories suggest that when PCR is carefully performed in the laboratory, it is very reliable and give reproducible results. However, in order to avoid contamination problems, high working standard is necessary in the laboratory.

More convenient PCR-test procedures is expected to be seen in the routine laboratories in the near future. Immunocapture-PCR is one alternative test format of considerable potential for bacterial detection in potato. In this system PCR tubes (or microtiter wells) are coated with an antibody that is specific or selective to the target pathogen. Thereafter, plant tissue is incubated within the tube to allow the antibody to bind to the pathogen. Compounds which may inhibit DNA-polymerase activity are subsequently washed away which eventually will improve the sensitivity of this PCR-test over direct PCR (Seal 1995). Immunocapture-PCR has been very succesful for the detection of plant viruses (Wetzel et al. 1992) and bacterial pathogens in foods (Fluit et al. 1993). About 250-fold increase in detection sensitivity over direct PCR has been reported in the plum box virus detection with immunocapture-PCR (Wetzel et al. 1992) apparently because interference of plant material in this test format cause less background problems. Development of immunocapture-PCR (or immunomagnetic separation PCR) seems realistic alternative because antibodies are available to be used as capture probes.

Even more convenient PCR-procedures for routine detection of bacterial pathogens may be seen from new developments of biosensor research. Automated closed and flow-through systems have been developed which allows to perform the amplification and detection steps within a single machine which already appears attractive procedure for the detection of bacterial contamination in food products (Strachan et al. 1995). Simultaneous detection of all potato bacterial pathogens in potato samples will be a realistic challenge for future diagnostic research.

5. References

Bear D k, Gudmestad N C (1993) Serological detection of non-mucoid strains of *Clavibacter michiganensis* subsp *sepedonicus* in potato. Phytopathology **83**, 157-163.

Darrasse A; Kotoujansky A & Bertheau Y (1994) Isolation by genomic subtraction of DNA probes specific of Erwinia carotovora subsp atroveptica Applications of Environmental Microbiology **60**, 298-306.

De Boer S H; Wieczorek A & Kummer A (1988) An ELISA test for bacterial ring rot of potato with a new monoclonal antibody. Plant Disease **72**, 874-878.

De Boer S H & Ward L J (1995) PCR detection of *Erwinia carotovora* ssp *atroseptica* associated with potato tissue. Phytopathology **85**, 854-858.

Firrao G A Locci R (1994) Identification of *Clavibacter michiganensis* ssp *Sepedonicus* using the polymerase chain reaction. Canadian Journal of Microbiology **40**, 148-151.

Fluit A C; Torensma R; Visser M J C; Aarsman C J M; Popellier M J J G; Keller B H I & Verhoeff J (1993) Detection of *Listeria monocytogenes* in cheese with magnetic immunopolymerase chain reaction assay. Applications of Environmental Microbiology **59**, 1289-1293.

Henson J M & French R (1993) The polymerase chain reaction and plant disease diagnosis. Annual Review of Phytopathology **31**, 81-109.

Janse J D & Van Vaerenbergh J (1987) Interpretation of the EC method for the detection of latent *Corynebacterium sepedonicum* infections in potato. European Plant Protection Organisation Bulletin **17**, 1-10.

Karjalainen R; Kangasniemi A; Hamalainen J & Tegel J (1995) Evaluation of PCR methods for the diagnosis of bacterial ring rot infections in potato. EPPO Bulletin **25**, 169-175.

Karjalainen R; Kangasniemi A; Hamalainen J & Tegel J (1996) Applications of PCR for the diagnosis of bacterial ring rot infections in potato. Diagnostics in Crop Production. BCPC Symposium Proceedings **65**, 133-138.

Li X & De Boer S H (1995) Selection of polymerase chain reaction primers for specific detection of *Clavibacter michiganenensis* subsp. *sepedonicus*. Phytopathology **85**, 837-842.

Mäki-Valkama T & Karjalainen R (1994) Differentiation of *Erwinia carotovora* subsp. *atroseptica* and *carotovara* by RAPD-PCR. Annual Applications of Biology **125**, 301-309.

Seal S E; Jackson L A; Young J P W & Daniels M J (1993) Differentation of *Pseudomonas solanacearum, Pseudomonas syzygii, Pseudomonas pickettii* and the blod disease bacterium by partial 16S rRNA sequencing: construction of oligonucleotide primers for sensitive detection by polymerase chain reaction. Journal of General Microbiology **139**, 1587-1594.

Seal S E (1995) PCR based detection and characterization of *Pseudomonas solanacearum* for use in less developed countries. EPPO Bulletin **25**, 227-231.

Smid E J; Jansen A H J & Gorris L G M (1995) Detection of *Erwinia carotovora* subsp. *atroseptica* and *Erwinia chrysanthemi* in potato tubers using polymerase chain reaction. Plant Pathology **44**, 1058-1069.

Stead D E (1993) Potato ring rot control through detection and certification. In Plant Health and the European Single Market, pp. 135-144. BCPC, Farnham (GB).

Strachan N J C & Gray D I (1995) A rapid general method for identification of PCR products using a fibre-optic biosensor and its application to the detection of *Listeria*. Letters in Applied Microbiology **21**, 5-9.

Wetzel T; Candresse T; Macquaire G; Ravelonandro M & Dunez J (1992) A highly sensitive immunocapture polymerase chain reaction method for plum pox potyvirus detection. Journal of Virological Methods **39**, 27-37.

DEVELOPMENT OF A SEMI-SELECTIVE MEDIUM AND USE OF SEROLOGY FOR FAST AND SENSITIVE DETECTION OF *XANTHOMONAS CAMPESTRIS* PV. *MANIHOTIS*, INCITANT OF BACTERIAL BLIGHT OF CASSAVA

FESSEHAIE[1], A., K. WYDRA[2] AND K. RUDOLPH[1]
1) Institut für Pflanzenpathol. & -schutz der Universität, Grisebachstr. 6, 37077 Göttingen, GERMANY
2) International Institute of Tropical Agriculture, Cotonou, BENIN, Africa

Cassava (*Manihot esculenta*, Crantz) is the natural host of *Xanthomonas campestris* pv. *manihotis* (*Xcm*), the causal agent of cassava bacterial blight (CBB). This crop is an important food for over 400 million people in Africa, Asia, and Latin America (El-Sharkawy, 1993), and one of the dominant starchy staples in the diet of the people in Sub-Saharan Africa. In Africa, CBB was reported in Zaire (Hahn and Williams, 1973), Benin, Ghana, Nigeria, and Cameroon (Wydra et al. 1995; Wydra and Msikita, 1995). In a few cases, high yield losses have been attributed to CBB, such as 75% in Nigeria (Ezelio, 1977) or 90-100% in Uganda (Otim-Nape, 1980). In Zaire, the epidemics led to severe starvation in the Kasai and Bandundu provinces between 1971 and 1973 (Maraite and Meyer, 1975).

DEVELOPMENT OF A SEMI-SELECTIVE MEDIUM

Xcm grows rather slowly *in vitro* and can easily be overgrown by saprophytes and accompanying bacteria during isolation from diseased plants. In order to develop a suitable semi-selective medium, the nutritional requirements of the bacterium were studied, especially its capability to utilize different carbon and nitrogen sources (Tables 1 and 2).

D(+)-Trehalose was found to be the best C-source, because all strains of *Xcm* grew equally well on it. Growth of *Xcm* on NH_4Cl was superior compared to other mineral N-sources. These tests led to a basal medium with 1% D(+)-trehalose as carbon source and 0.1% L-glutamic acid as a growth promoting substance, allowing good growth of the target organism whereas luxurious growth of some saprophytes was suppressed.
A further selective effect of the medium was achieved by incorporating three antibiotics in the basal medium.

H.-W. Dehne et al. (eds.), Diagnosis and Identification of Plant Pathogens, 129–135.

Tab. 1: Growth of *Xanthomonas campestris* pv. *manihotis* and two epiphytes on different C-sources

C-source	*Xanthomonas campestris* pv. *manihotis* ORST 36	CFBP 1860	*P. fluorescens* GSPB 1714	*E. herbicola* GSPB 450
Sol. starch	±	±	±	±
Alpha-D-Lactose	+	+	+	+
D(+)-Galactose	+	+	+	+
Succinic acid	+	+	+	+
D(+)-Cellobiose	+	+	+	+
D(+)Trehalose	+	+	+	+
Melibiose	+	+	+	+
Methylsuccinic acid	+	+	+	+
2-Oxoglutaric acid	+	+	+	+
Bromosuccinic acid	-	-	-	-
Citrate	-	-	+	+
L-Proline	+	+	+	+
Glycerol	+	+	+	+
Pepton	+	+	+	+

Tab. 2: Growth of *Xcm*-strains and two epiphytic bacteria on different N-sources

N-source	ORST 187	ORST 41	GSPB 1713	GSPB 450
TSA[1]	+++	+++	+++	+++
$(NH_4)_2SO_4$	+	+	+	+
NH_4Cl	++	++	++	++
$NH_4H_2PO_4$	+	+	+	+

[1)] TSA = 3% tryptic soy broth plus 1.4% agar (pH 7.0)

Testing the resistance of *Xcm* towards various antibiotics (Bauer *et al.*, 1966) revealed a natural resistance towards Cefazolin, Lincomycin, and Fosfomycin which were filter sterilized and added to the basal medium.

The newly developed Cefazolin-trehalose-agar (CTA) medium appeared to be well suited for a sensitive and reliable diagnosis of the pathogen (Fessehaie *et al.* 1996). It is suggested to use the medium for control strategies to eradicate infected planting material of cassava, as well as in epidemiological studies.

ELECTROPHORESIS AND IMMUNOBLOTTING

Polyclonal antisera raised against *Xcm* strains were obtained from ORSTOM (France) (Boher, 1994). These antisera directed against whole cells of *Xcm* were used for immunoblotting to select specific protein bands of *Xcm* (strains Xm 15, IITA 3c, Ug 21, I 1). Other bacterial strains used in these studies were the three *Xanthomonas campestris* pathovars *malvacearum*, *phaseoli* and *campestris* (GSPB 1247, 1241, 149, respectively) and *Erwinia carotovora* ssp. *carotovora* (GSPB 1405), *Erwinia herbicola* (GSPB 2314) and *Pseudomonas fluorescens* (GSPB 1713).

Bacterial proteins were prepared as described by Kersters (1990). Discontinuous gels were run in Dual Cooled Vertical Slab Electrophoresis Unit (18 X 16 cm gel; Hoeffer SE 600). 30 ml of a single concentration gel (makes for one 1.5 mm 18 X 16 cm Sandwich). Gels were usually 1.5 mm thick. The resolving gel was 12.6 cm long containing 10% total acrylamide, and the stacking gel composed of 4% total acrylamide was 1.5 cm long.

Bacterial proteins were separated electrophoretically using discontinuous sodium dodecylsulphate polyacrylamide gel electrophoresis (SDS-PAGE) according to the modified standard procedure after Laemmli (1970). Electrophoresis was initiated by applying a voltage of 80 V until proteins migrated (dye front) from stacking gel to the resolving gel followed by 120 V overnight. One gel was stained with Coomassie Blue and the other twin gel was electroblotted onto nitrocellulose membrane (NC) (Schleicher and Schuell, 0.20 μm).

The NC membrane containing the transferred proteins was blocked with 3% bovine serum albumine in TN overnight and incubated in *Xcm* polyclonal antisera diluted at 1:300 to 1:500 in TN. To avoid the growth of microorganisms, NaN_3 was added at a final concentration of 0.025% . The NC membrane was processed as described by Bdliya (1995). The colour development due to positive antibody reaction was demonstrated using the dye Fast Red and naphthol-AX-MX-phosphate or BCIP/NBT (Sigma Chemicals, USA). The staining reaction was stopped by washing the NC membrane with ddH_2O. The air-dried blot was stored at room temperature.

RESULTS

A *Xcm*-specific protein band with low molecular weight was detected (Figures 1 and 2). This protein band shall be used as a specific antigen to generate a monospecific antiserum.

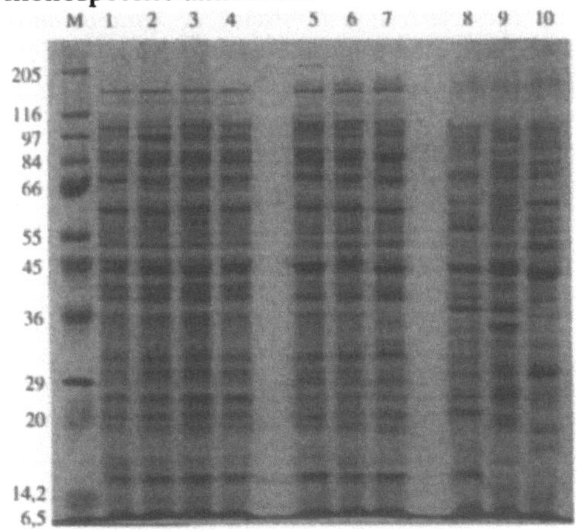

Fig. 1. Comparison between protein bands of *Xcm* strains and other bacterial species stained with Coomassie blue. Lane M: Mixture of molecular weight standard markers (Sigma, High Molecular Weight Range, Product No. M-3788 und Low Molecular Weight Range, Product No M-3913) and their respective values in Kilodalton (kD). Lanes 1-4: *Xcm* (strains IITA 3c, Xm 15, Ug 21, and I 1, respectively). Lanes 5-7: *Xanthomonas campestris* pathovars *malvacearum, phaseoli, campestris* (GSPB No.: 1247, 1241, and 149, respectively). Lanes 8-9: *Erwinia carotovora* ssp. *carotovora* (GSBP 1405), *Erwinia herbicola* (GSBP 2314). Lane 10: *Pseudomonas fluorescens* (GSPB1713).

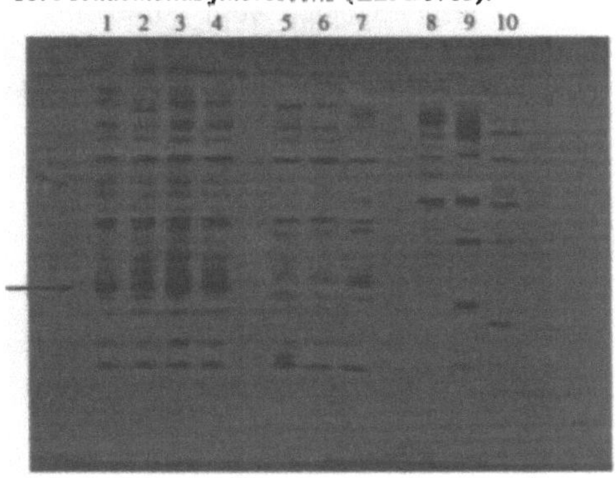

Fig. 2. Western blot corresponding to the Coomassie-stained gel shown in Fig. 1

REFERENCES

Bauer, A.W., Kirby, W.M.M., Sherris, J.C., and Turck, M. 1966. Susceptibility testing by a standardized single disk method. Am. J. Clin. Pathol. **45**: 493-496.

Bdliya, B.S. 1995. Studies on the detection and identification of soft rot causing *Erwinia* species (*Erwinia carotovora* ssp. *carotovora* and *Erwinia chrysanthemi*) on potato tubers. Dissertation, Institut für Pflanzenpathologie und -schutz, University of Göttingen.

El-Sharkawy, M. A. 1993. Drought tolerant cassava for Africa, Asia and Latin America. BioScience 43(7) 441-451.

Ezelio, W.N.O. 1977. Control of cassava bacterial blight. In: Report on a Workshop on Cassava Bacterial Blight, Nigeria, 1976 (eds. G. Persley, R.E. Terry, and R. MacIntyre). International-Development-Research-Centre, Ottawa, Canada, pp. 15-17.

Fessehaie, A., Wydra, K., and Rudolph, K. 1996. Nutritional characterization of *Xanthomonas campestris* pv. *manihotis* and development of a semi-selective medium. Phytopathology, (submitted).

Hahn, S. K., and Williams, R. J. 1973. Investigation on cassava in the Republic of Zaire. Report to the Commissaire d` Etat à l`Agriculture. Republic of Zaire, 12p.

Laemmli, U.K. 1970. Cleavage of structural proteins during the assembly of the head of bacteriophage T4. Nature (London) **227**: 680-685.

Kersters, K. 1990. Polyacrylamide gel electrophoresis of bacterial proteins. In: Klement, Z., K. Rudolph and D.C. Sands (eds.): Methods in Phytobacteriology, Akadémiai Kiado, Budapest, pp. 191-198.

Maraite, H., and Meyer, J.A. 1975. *Xanthomonas manihotis* (Arthaud-Berthet) Starr, causal agent of bacterial wilt, blight and leaf spots of cassava in Zaire. PANS, **21**, 27-37.

Otim-Nape, G.W. 1980. The effects of bacterial blight on germination, tuber yield and quality of cassava. In: Proceedings of the 10th International Congress of Plant Protection, Brighton 1983, British Crop Protection Council, London, p. 1205.

Terry, E., and Ezumah, H. C. 1974. Report on root crops disease survey and evaluation of production problems in Cameroon, 22-27th Aug. 1974. IITA, Ibadan, Nigeria, 4p.

Williams, R, J., Agboola, S. D., and Schneider, R. W. 1973. Bacterial wilt of cassava in Nigeria. Plant Dis. Rept. **57**, 824-827.

Wydra, K., Fessehaie, A., Assigbetse, K., Verdier, V., Janse, J.D., Fanou, A., and Rudolph, K. 1995. Physiological, genetic and pathological characterization of strains of *Xanthomonas campestris* pv. *manihotis*, incitant of cassava bacterial blight, from West and Central Africa. XIII Intern. Plant Protection Congress, The Hague, NETHERLANDS, July 2-7, 1995, Europ. J. Plant Pathology, Abstr. no. 463.

Wydra, K., and W. Msikita, 1995. An overview of the present situation of cassava diseases in West Africa. In: Proceedings of the 6th ISTRC-AB Symposium, Lilongwe, Malawi, October 22-28 1995 (in press).

CHARACTERISTICS OF LARGE-SCALE PRODUCTION OF PHYTO-SANITARY IMPROVED *IN-VITRO* FRUIT TREE CULTIVARS USING *IN-VITRO* ELIMINATION TREATMENTS AND SERODIAGNOSTICS

Knapp, E., Hanzer, V., Weiss, H., da Câmara Machado A., Katinger, H., and Laimer da Câmara Machado, M.
Institute of Applied Microbiology, University of Agriculture and Forestry
Nuszdorfer Lände 11, A-1190 Vienna, Austria

1. Abstract

We present results of five years of testing *in-vitro* cultures of over 30 apple and 20 stone fruit cultivars in a virus elimination programme. Our aim was to develop efficient large-scale screening designed for the special features of *in-vitro* plants to improve reliability in selection of truly negative *in-vitro* elite material following elimination treatment by thermotherapy and meristem preparation.
Commercial serodiagnostics were used in DAS or Two-Step ELISA. Screening for virus infection was generally carried out with 10-25 clones per cultivar and in accordance with the type of virus infections encountered, several cultivars were included in the investigations. Virus detection was undertaken at different growth phases during subculture, including short and long-term culture, a comparison of elimination treatments.
The following questions were considered:
1) Suitable sample sources for reliable large-scale screening
2) Influence of subculture intervals
3) Can virus elimination measures just reduce virus titres, but fail to eliminate the virus. If so how long does it take for virus to reaccumulate to detectable levels?
4) How many tests are necessary to confirm initial test results? Are serological methods suitable for selecting truly virus-free progeny plants?

Abbreviations: ASGV, apple stem grooving virus, ACLSV, apple chlorotic leafspot virus, PPV, plum pox virus.

Key words: Fruit tree virus, Virus elimination, ELISA, Tissue culture

H.-W. Dehne et al. (eds.), Diagnosis and Identification of Plant Pathogens, 141–147.

2. Introduction

In-vitro cultivation techniques are already widely used for the micropropagation and germplasm storage of rosaceous species (Boxus et al., 1977, Laimer et al., 1991). Moreover *in-vitro* culture techniques are the only effective tools for eliminating certain pathogens from desired germplasm.
Many fruit tree viruses are well identified and characterized (Németh, 1986, Fridlund, 1989). Conventional laboratory diagnostics based on serology have been developed and are routinely used for orchard surveys (Fuchs et al., 1988, Varveri and Bem, 1995, Myrta et al., 1995, Savino et al., 1995), but little is known from screening of *in-vitro* cultures (Knapp et al., 1995a).
Drawbacks in the certification of fruit plants are however encountered since many viruses are either latent, or irregularly distributed or highly localized. Thus the risk of false results is increased especially with field plants (Caspar and Meyer, 1981, Fridlund, 1983). In contrast sampling of *in-vitro* plants could be done systematically in accordance with detailed knowledge available on highest pathogen accumulation in tissues (Knapp et al., 1995 b, c).

3. Materials and methods

Propagation of fruit tree cultivars, thermotherapy and meristem preparation were done as described (Laimer et al., 1991, Knapp et al., 1995a,c). In a modified thermotherapy the temperature was set at 28°C for 4 days, then raised to 33°C for 3 days and finally to 38°C for a period of 29 days. DAS or Two-Step ELISA were essentially carried out according to Clark and Adams (1997) and Flegg and Clark (1979). Sampling for ELISA was done as described (Knapp et al., 1995a), unless otherwise stated. The threshold level was defined as mean value of 8 healthy replicates plus 2 times standard deviation. OD values slightly above the threshold were considered dubious positives.

4. Results

1) Suitable sample sources for reliable large-scale screening?
The localization of PPV, ASGV and ACLSV in host tissues was investigated by using the immuno-tissue printing (Knapp et al., 1995 b, c). These studies were undertaken complementary to the screening of cultures by ELISA and influenced choice of the sample (Knapp, 1996).
2) Influence of subculture intervals on outcome of diagnosis?
In-vitro plants undergo different growth phases (lag-phase, active growth phase, stationary phase) after transfer to fresh media. Preliminary results with four apple cultures indicated that diagnosis was more reliable with samples from short

subculture intervals (data not shown). To confirm these results the apple cultivar "Delbard Estival", infected with ASGV and ACLSV was chosen for a direct comparison of different sample sources. Leaf samples and the corresponding stem samples from single shoots, excluding the apical region, which was further micropropagated, were either investigated separately or mixed, when shoots were small. Samples were taken after a subculture period of 65 days (stationary growth phase) and in the succeding subculture interval after 23 days (active growth phase). Samples were stored at -20°C and the ELISA's were all performed at the same time. A number of 10 to 17 samples were tested per source. Table 1 shows that the earlier results were confirmed. Reliability in diagnosis of ASGV and ACLSV was either better or equal in all samples taken after 23 days than after 65 days.

TABLE 1: The percentage of false test results occurring at different growth stages

Sample	ASGV		ACLSV	
	23 days	65 days	23 days	65 days
Leaf	14.2	14.2	7.1	14.2
Stem	0	11.1	14.2	28.5
Mixed	0	10.5	0	6.6

Although reliability of diagnosis was higher in testing actively growing cultures (Table 1), the majority of samples from the stationary phase showed a higher virus titre, facilitating discrimination between positives and negatives (data not shown).

3) Can virus elimination measures just reduce virus titres, but fail to eliminate the virus. If so how long does it take for virus to reaccumulate to detectable levels?
It is known that elimination treatments may depress the pathogen titre below the threshold level of detection (Fridlund, 1989). In this respect the effects of thermotherapy on the titre of PPV, ASGV and ACLSV were investigated *in-vitro* with three apricot and nine apple cultivars. PPV was detectable in apricot tissues immediately after thermotherapy for 17 - 20 days in some but not in all of the samples tested. PPV levels in treated cultures were several times lower than in corresponding untreated cultures (data not shown). With apple cultures, irrespective of the length of thermotherapy, from 23 to 39 days, depending on heat sensitivity of the plant material, virus titre of both ASGV and ACLSV was decreased dramatically by thermotherapy. ELISA readings were either negative or just slightly above the threshold level (data not shown).
Thermotherapy was modified enhancing shoot growth in order to increase possibly the elimination success of the recalcitrant ASGV. The effects on virus titre of this modified regime were compared with the effects of the standard treatment using the apple cultivar "Rubinette" infected with both ASGV and ACLSV. The number of samples tested was 60 per treatment. Higher percentages

of samples tested positive with the modified temperature regime, but again titres were low (data not shown).
To study the speed of reaccumulation of virus in clones treated without success untreated apple cultures were included as positive control in ELISA. Treated plants showed a strong tendency for lower ACLSV level compared to the untreated cultures (Table 2).

TABLE 2: Comparison of ACLSVabsorbance values in non-treated control and treated apple cultures several month after treatment

Apple cultivar	Month	$\bar{x}$	$s\bar{x}$	n	Sample
Jonagored non-treated		1.240	0.338	8	a
treated	6	0.346	0.223	8	a
Landsberger Renette non-treated		1.754	0.453	9	a
treated	15	0.559	0.297	11	a
Landsberger Renette non-treated		0.769	0.330	10	a
treated	20	0.465	0.200	18	a
Gelber Bellefleur non-treated		0.207	0.134	10	a
treated	24	0.417	0.218	11	a
Gelber Bellefleur non-treated		2.190	0.231	8	b
treated	35	0.771	0.703	6	b

Month - tested after treatment, $\bar{x}$ - mean of OD values of samples tested, $s\bar{x}$ - standard deviation, n - number of samples tested, a) mixed leaf from shoot bundle, b) base of stem plus leaf of further propagated shoot.

Behaviour of ASGV was tested in four apple cultures as described for ACLSV. These cultures showed more variability than observed for ACLSV, as some of the treated cultures had higher virus levels compared to the untreated cultures (data not shown).

4) How many tests are necessary to confirm initial test results? Are serological methods suitable for selecting truly virus-free progeny plants?
Three ACLSV infected apple cultures were checked after treatment and no clear positives were found, but many dubious positives with OD values slightly above the threshold level. The second test was undertaken after 5 month, the third after 17 month and the results were again negative and dubious positive. In another ACLSV infected apple culture the majority of samples tested positive after treatment and although negatives were selected for multiplication, then turned into positives after 4 month (data not shown).
A higher variability in test results was encountered in apple cultures infected with both ASGV and ACLSV. In three of four apple cultures tested more than 75 percent ASGV positives were found after treatment. From these cultures only negative clones were multiplied, but again a high number of samples showed ASGV positive results after more than one year of *in-vitro* culture. Negative results for ACLSV were confirmed. The apple cultivar "Gelber Bellefleur" was

chosen for an extensive test as the two former tests did not contain ASGV positives after initial removal of unsuccessfully treated plants. Two types of samples from shoot bundle were used for comparison. Unsuccessfully treated ASGV positives were compared to selections with either twice negative or twice dubious positive results (Table 3).

TABLE 3: ASGV ELISA results of a third retest of the treated apple cultivar "Gelber Bellefleur".

Initial tests	Sample		Results as percentages		
Gelber Bellefleur	Source	Number	Positives	Dubious	Negatives
Negatives	a	46	26	39.1	34.9
	b	15	46.6	26.6	26.8
Dubious Positives	a	8	0	12.5	87.5
	b	8	0	12.5	87.5
Positives	a	6	100	0	0
	b	6	100	0	0

Sample source: a) base plus stem plus leaf of the further propagated main shoot, b) corresponding axillary shoots mixed

5. Discussion

Our investigations resulted in a better working system for obtaining phytosanitary improved plants. However there remain the limitations of current ELISA-based serological tests which might be not sensitive enough to detect very low levels of virus. Low levels of virus can be a virus specific trait, as is the case with ASGV, which occurs extremly limited and highly localized in host tissues (Fuchs et al., 1988, Knapp et al., 1995c). On the other hand generally high-titred viruses like ACLSV and PPV show an irregular distribution in host plants (Casper and Meyer, 1981, Fridlund, 1983). Detailed knowledge about virus distribution patterns - if localizable, as in the case of ACLSV or completely uneven distributed as in the case of PPV, helps to overcome sampling problems (Knapp et al., 1995c, Knapp, 1996). However plant physiology and culture growth conditions exert further influences on virus accumulation, as was shown in our experiments. The correct phase of *in-vitro* culture to take samples was determined in view of either maximum sensitivity or maximum reliability. Actively growing plants enhanced virus spread showing an increased reliability, whereas plants in the stationary phase accumulated higher virus levels thus allowing a better discrimination of infected plants, but being less reliable due to a higher variability of samples.

Treatment measures depressed virus titre, as has been shown for a wide range of cultivars. Thermotherapy altered or destroyed viral protein and in conclusion serodiagnostics are of little value for early and reliable screening. We furthermore investigated the question, if treatment measures exert prolonged virus depression, which was in fact a pronounced characteristic found in cultures where ACLSV elimination failed, but not in cultures remaining infected with ASGV. However ACLSV was readily detected after reaccumulation above the threshold level whereas the well-known problems in reliable diagnosis of ASGV were further encountered even after several tests of *in-vitro* cultures (Fuchs et al., 1988, Gilles and Verhoyen, 1992). Other but as yet not routine possibilities for laboratory diagnosis like detection by PCR or immuno-capture PCR will be introduced into the sanitation programme to possibly move towards a system of certifying *in-vitro* plants.

6. Acknowledgements

This research was supported by the Austrian Ministery for Agriculture and Forestry, project 853.

7. References

Boxus, P., Quoirin, M. and Laine, J. M. (1977). Large scale propagation of strawberry plants from tissue culture, p. 130-140. In J. Reinert and Bajaj, Y. P. S., editors, Applied and fundamental aspects of plant cell, tissue, and organ culture. Springer Verlag, New York, 803 p.

Casper, R. and Meyer, S. (1981). Die Anwendung des ELISA-Verfahrens zum Nachweis pflanzenpathogener Viren. Nachrichtenblatt Dt. Pflanzenschutzdienst (Braunschweig) 33 (2):49-54).

Clark, M. F. and Adams, A. N. (1977). Characteristics of the microplate method of enzyme-linked immunosorbent assay (ELISA) for the detection of plant viruses. J. Gen. Virol. 34:475-483.

Flegg, C. L. and Clark, M. F. (1979). The detection of apple chlorotic leaf spot virus by a modified procedure of enzyme-linked immunosorbent assay. Ann. Appl. Biol. 91: 61-65.

Fridlund, P. R. (1983). Distribution of chlorotic leafspot virus on various length on apple budsticks in sucessive years. Acta Hort. 130:85-87.

Fridlund, P. R., (1989). Virus and viruslike diseases of pome fruits and simulating noninfectious disorders. Cooperative Extension College of Agriculture and Home Economics, Washington State University, Pullman, Washington.

Fuchs, E., Grüntzig, M. and Al Kai, B. (1988). Der serologische Nachweis mechanisch übertragbarer Viren des Kern - und Steinobstes. Nachrichtenblatt für den Pflanzenschutz in der DDR, 42 (10):208-211.

Gilles, G. L. and Verhoyen, M. (1992). Viroses et maladies apparentees des abres fruitiers et ornamentaux assainissement et selection. I.R.S.I.A. - Institut pour l'Encouragement de la Recherche Scientifique dans l'Industrie et l'Agriculture.

Knapp E., Hanzer V., Weiss H., da Câmara Machado A., Wang Q., Weiss B., Katinger H. and Laimer da Câmara Machado M. (1995a). New aspects of virus elimination in fruit trees. XVI ISHS Symposium on Fruit Tree Viruses. Rome. Acta Hort 386: 409-418

Knapp E., Hanzer V., Weiss H., da Câmara Machado A., Wang Q., Weiss B., Katinger H. and Laimer da Câmara Machado M. (1995b). Distribution of ACLSV in apple shoots cultivated in vitro. XVI ISHS Symposium on Fruit Tree Viruses. Rome, 1994. Acta Hort 386: 187-194.

Knapp E., da Câmara Machado A., Pühringer H., Wang Q., Hanzer V., Weiss H., Weiss B., Katinger H. and Laimer da Câmara Machado M. (1995c). Localization of fruit tree viruses by immuno-tissue printing in infected shoots of *Malus* and *Prunus sp.*. J. Virol. Meth. 55(2): 157-173.

Knapp E. (1996). Diagnose und Eliminierung von Viruserkrankungen aus *in-vitro* Obstgehölzen. Dissertationsschrift, Universität für Bodenkultur, Wien.

Laimer da Câmara Machado M., da Câmara Machado A., Hanzer V., Kalthoff B., Weiss H., Mattanovich D., Regner F. and Katinger H.W.D. (1991). A new, efficient method for the initiation and establishment of tissue cultures of apple from adult material. PCTOC 27:155-160.

Myrta , A. Di Terlizzi, B. and Digiaro, M. (1995). A preliminary account of the sanitary status of stone fruit trees in Albania. Acta Hort.386:165-175.

Németh, M. (1986). Virus, Mycoplasma and Rickettsia Diseases of Fruit Trees. Martinus Nijhoff / Dr W. Junk Publishers. Dordrecht/Boston/Lancaster.

Savino, V. Di Terlizzi, B., Digiaro, M. Catalano, L. and Murolo, O. (1995). The sanitary status of stone fruit species in Apulia. Acta Hort.: 169-175.

Varveri, C. und Bem, F. (1995).Viruses of stone and pome fruit mother-tree plantations in Greece. Acta Hort. 386:431-438.

PRODUCTION OF ANTIBODIES AGAINST A SYNTHETIC PEPTIDE BASED ON A SEQUENCE COMMON TO ALL TOSPOVIRUS NON-STRUCTURAL Ns_S PROTEIN

P. Roggero*, M. Sohn#, Anna Maria Vaira*, Cornelia Heinze#, Vera Masenga* and G. Adam#
** Istituto di Fitovirologia Applicata, CNR, Torino, Italy*
 # Inst. of Applied Botany, Dept. of Plant Protection, University of Hamburg, Germany

1. Introduction

Only five years ago tomato spotted wilt tospovirus (TSWV) was considered the type and only member of the new genus *Tospovirus*, family *Bunyaviridae*. The worldwide presence and high economic impact of tospoviruses in many cultivated plants initiated intensive research on TSWV. This improved diagnosis and subsequently has led to identification of many new tospovirus species, whose unofficial number is now about 15 (German *et al.*, 1992; Adam and Kegler, 1994; Mumford *et al.*, 1996).The following features are common to all tospoviruses: enveloped quasispherical particles 80-100 nm in diameter; 3 ss RNA of negative (L) or ambisense (M, S) polarity which code for the structural proteins L, G_1, G_2 and N and the non-structural proteins NSM and NSs; circulative-propagative transmission by thrips; wide host range among angiosperms (Goldbach and Peters, 1995). Tospoviruses are separated into species by differences in the sequence of the N protein, and based on the serological properties of this protein they are grouped at present into four serogroups, with low or absent cross reactions among serogroups (Table 1).

Table 1. Tospovirus N protein sequences as available (Data from Heinze *et al.*, 1995).

Species	Acronym	Distribution	% of identities in amino acids of N protein[a]	Serogroup
tomato spotted wilt	TSWV	worldwide	100	I
groundnut ringspot	GRSV	S. Africa, S. America	79	II
tomato chlorotic spot	TCSV	S. America	78	II
impatiens necrotic spot	INSV	Europe, N. America	55	III
watermelon silver mottle,To	WSMV-To	Asia	33	IV
watermelon silver mottle	WSMV	Asia	33	IV

a- Homologies with respect to TSWV-L3.

Tospoviruses are usually detected by ELISA using monoclonal (Mab) and polyclonal (Pab) antibodies against the N protein, which necessitates several different ELISA kits and

H.-W. Dehne et al. (eds.), Diagnosis and Identification of Plant Pathogens, 149–154.

separate tests to detect each species. A genus-specific test, if available, would simplify the diagnosis and open the possibility of detecting new species within the genus.

Approaches towards a genus assay have been based on the presence of common epitopes within the glycoproteins of all species (see Roggero *et al.*, 1996). Efficient detection, however, has only been possible with Mab, and a Mab produced against TSWV glycoproteins is reported to detect members of serogroups I, II and III, but not serogroup IV (Adam *et al.*, 1995; Adam and Peters, 1996).

Another interesting target for serology is the NSs protein. NSs is present in the cytoplasm of both, infected plant and thrips cells; it has a Mw of about 50 kD and comprizes 482 amino acids (*aa*) (serogroups I and II), 453 *aa* (serogroup III), 439 *aa* (serogroup IV). NSs has a basic isoelectric point, does not contain hydrophobic regions, and it function is unknown. Polyclonal and monoclonal antibodies have been raised against TSWV NSs recombinant protein and used detect NSs in plant and thrips cells (Kormelink *et al.*, 1991; Wijkamp *et al.*, 1993). Detection of NSs in thrips was found to be highly correlated with their ability to transmit TSWV (Bandla *et al.*, 1994).

Heinze *et al*, (1995) described a remarkably conserved region of 24 *aa* (398-421 at the C-terminus of NSs. This is the longest sequence so far found common to all tospoviruses (Table 2).

Based on this sequence, the peptide Y-25-C (Table 2) has been chemically synthesized and used to produce polyclonal antibodies, with the aim of developing a general tospovirus assay. The sequence reflects those of the different viruses but also takes account of the even more similar hydrophobic profile of all the viruses. Homology of the peptide to the individual viral sequences is high and similar for all the viruses; 88% for serogroups I and II and 79% for serogroup III and IV.

Table 2. Amino acids sequences in the region *aa* 398-421 of NSs: comparison for serogroup I-IV and the synthetic peptide Y-25-C.

Serogroup	amino acid 398-421																							
I	Y	F	L	S	K	T	L	E	C	L	P	S	N	T	Q	T	M	S	Y	L	D	S	I	C
II	Y	F	L	S	K	T	L	E	C	L	P	S	N	V	Q	T	M	S	Y	L	D	S	I	C
III	Y	F	L	S	Q	T	L	K	S	L	P	R	N	S	Q	T	M	S	Y	L	D	S	I	C
IV	Y	F	L	S	K	T	L	E	V	L	P	K	S	L	S	T	L	S	Y	L	G	S	I	A
peptide Y-25-C	Y	F	L	S	K	T	L	E	V	L	P	K	N	L	Q	T	M	S	Y	L	D	S	I	C

2. Material and Methods

The peptide was synthesized by MedProbe (Sweden) at an immunological grade of purity (>70%) either free or conjugated to keyhole limpet haemocyanin (KLH) by MBS (3-maleiimidobenzoic acid *N*-hydroxysuccinimide ester); (4.05 mg/ml KLH and 1.9 mg/ml peptide). Two rabbits (A331 and A338) were injected twice with the peptide conjugated to KLH and received two further injections with the free peptide. A third rabbit (A341) was injected 4 times with free peptide only. Complete Freund's adjuvant was used for the first injection, and incomplete adjuvant for the rest.

Antisera were evaluated with an antigen coated plate (ACP) ELISA protocol. Plates were coated with the peptide at 1 μg/ml in carbonate buffer pH 9.6, followed by blocking with PBS, 0.05% Tween-20, 2% PVP, 2% skimmed milk. The antisera were diluted with the blocking solution and bound antibodies were detected using a goat anti-rabbit IgG-alkaline phosphatase conjugate.

3. Results and Discussion

The results obtained with the different bleedings from the three rabbits at a dilution of 1/10,000 in the same plate are shown in Fig. 1.

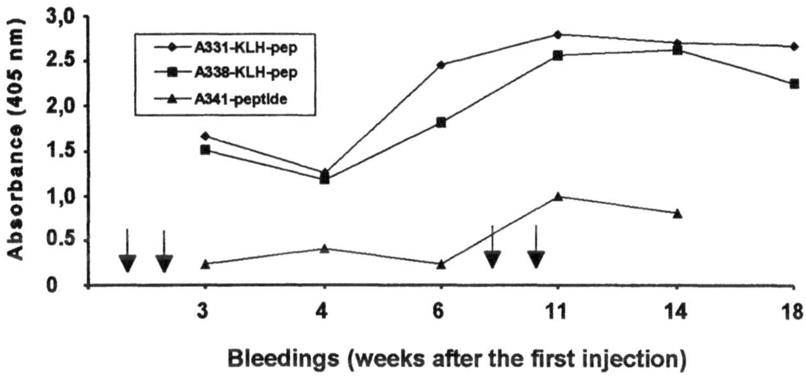

Fig. 1.Development of specific antibodies in three immunized rabbits against peptide Y-25-C. Arrows indicate injections of antigen. From each bleeding a 1/10.000 dilution was tested.

Bleedings from the rabbits injected with the KLH conjugated reacted similar, and high activities were obtained only after a considerable time from the first injection. The rabbit injected with the free peptide only gave lower activity. For further studies the two last bleedings from rabbits A331 and A338 were used.

The two antisera were tested in the same ACP ELISA format with the following tospoviruses: TSWV (2 isolates), GRSV (1 isolate), TCSV (2 isolates), INSV (2 isolates) and GBNV (2 isolates), all mantained in *Nicotiana benthamiana*. Plant sap from systemically infected and healthy *Nicotina benthamiana* plants was diluted 1/100, and twofold antiserum dilutions from 1/100 to 1/102,400 were tested. Results are summarized in Table 3.

Table 3. Titration of antisera A331 and A338 with peptide Y-25-C and different tospoviruses in plant sap.

Antiserum	Antigens					
	Y-25-C	TSWV (SG I)	GRSV (SG II)	TCSV (SG II)	INSV (SG III)	GBNV (SG IV)
A331	51,200†	1,600	<100	200	1,600	<100
A338	>102,400	51,200	12,800	102,400	12,800	<100

†The dilution giving an absorbance twice that of the healthy control at the same dilution was taken as the titre of the antiserum. Figures are reciprocals of the antiserum dilution.

Despite similar titres of both A331 and A338 with the peptide, only A338 had high titres with all tospoviruses, wich all reacted well, except GBNV. Taking the similarity of the common regions of the different Nss proteins into account, one explanation for this may be that the peptide contains more than one epitope against which antibodies were made.

The activity of A338 with NSs was confirmed in Western blots (Roggero *et al.*, 1996) and by immunogold labeling of thin section of infected plant tissues (Milne *et al.*, 1995), which could be expected for a continuous epitope.

Sap from TSWV-, INSV- and GBNV-infected and healthy *N. benthamiana*, the recombinant TSWV-NSs protein (Heinze *et al.*, 1995) and biotinylated MW markers were denaturated with SDS. After electrophoresis with Phast System (Pharmacia, Sweden) the proteins were transferred to nitrocellulose by diffusion and the membranes probed with antiserum diluted 1/1,000.

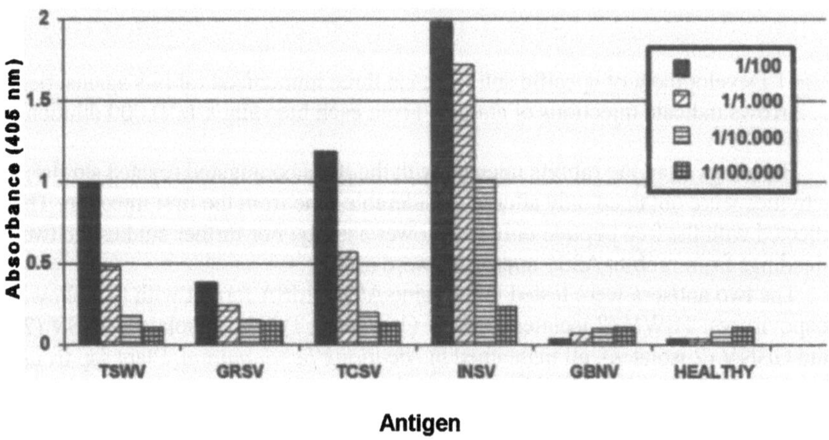

Fig. 2. Detection of tospovirus Nss in infected plant sap by ACP ELISA using the antiserum A338.

A band of the expected MW (about 50kD) was visible with the recombinant TSWV NSs protein and with TSWV and INSV infected *N. benthamiana* saps but not with GBNV (results not shown). Thin sections of *N. benthamiana* tissues from healthy or TSWV-infected plants were probed with A338 diluted 1/2,000, followed by 15 nm gold particles conjugated to goat anti-rabbit IgG. Clear labelling of TSWV infected tissues was found scattered in the cytoplasm only and no labelling was found with healthy controls (results not shown).

The antiserum A338 was also tested for its ability to detect different tospoviruses in *N. benthamiana* sap diluted in tenfold steps with ACP ELISA using an antiserum dilution of 1/1.000. The results are shown in Fig. 2.

Taking as detection limit the dilution giving an absorbance three times that of the healthy control, TSWV and INSV were detected up to 1/10,000, GRSV and TCSV up to 1/1,000 and there was no reaction with GBNV (Fig. 2). These values are in the same order of magnitude as described for the detection of tospoviruses by MABs (Adam *et al.*, 1995). Purified antibodies from this serum, however, failed to react in a double sandwich ELISA with any of the tospoviruses in plant sap when they were conjugated with alkaline phosphatase or biotin (results not shown). This could be due to the known modulation of the reactivity of PABs after conjugation (Koenig and Paul, 1982).

These preliminary results suggest the following conclusions.

1. The synthetic peptide based on the NSs common sequence was immunogenic, expecially when conjugated with KLH, but only one of the three antisera produced reacted well with NSs from tospovirus-infected tissues.
2. With ACP ELISA it was possible to detect viruses of serogroups I, II and III but not IV.
3. The reaction was specific for NSs, as confirmed in Western blots and immunogold localization in plant tissues.

Some questions are still open. In particular it is not clear why there was no reaction with GBNV (serogroup IV). Sequence data indicate similar homology levels of the peptide Y-25-C with NSs proteins of serogroups III and IV, and there was a strong reaction of the antiserum with INSV, serogroup III. Testing the recombinant GBNV NSs protein will probably clarify this point. Another possibility is a low concentration of GBNV NSs in *N. benthamiana* tissues. In fact serogroup IV viruses have been little studied, and the genome organization of their S segment appears different from that of the other tospoviruses (Heinze *et al.* 1995).

Acknowledgements

Michael Sohn was supported by a Short Term Scientific Mission grant from COST Action 823 "New technologies to improve phytodiagnosis".

References

Adam, G. and Kegler, H. (1994). Tomato spotted wilt virus and related tospoviruses, Archives of Phytopathology and Plant Protection **28**, 483-504.

Adam, G. and Peters, D. (1996). A European comparison of tospovirus isolates and related antibodies. Acta Horticulturae (in press)

Adam, G., Roggero P., Malavasi F., Milne R.G. and Papa G. (1995). Approach to a general tospovirus assay using antibodies to purified tomato spotted wilt tospovirus G proteins. Bulletin OEPP/EPPO Bulletin **25**, 247-257.

Bandla, M.D., Westcot, D.M., Chenault, K.D., Ullman, D.E., German, T.L. and Sherwood J.L. (1994). Use of monoclonal antibody to the nonstructural protein encoded by the small RNA of tomato spotted wilt tospovirus to identify viruliferous thrips. Phytopathology **84**, 1427-1431.

German, T.L., Ullman D.E. and Moyer, J. (1992). Tospoviruses: diagnosis, molecular biology, phylogeny, and vector relationship Annual Review of Phytopathology **30**, 315-348.

Heinze, C., Maiss, E., Adam, G. and Casper, R. (1995). The complete nucleotide sequence of the S-RNA of a new Tospovirus species representing serogroup IV. Phytopathology **85**, 683-690.

Goldbach, R.W. and Peters, D. (1995). Molecular and biological aspects of tospoviruses. In: H.H. Fraenkel Conrat (editor) , Vol. XX The Bunyaviridae, Plenum Press, New York, pp129-157.

Koenig, R. and Paul, H.L. (1982). Detection and differentiation of plant viruses by various ELISA procedure. Acta Horticulturae **127**, 147-158.

Kormelink, R., Kitajima, E.W., de Haan, P., Zuidema, D.; Peters., D.; and Goldbach, R.W. (1991). The nonstructural protein (NSs) encoded by the ambisense S RNA segment of tomato spotted wilt virus is associated with fibrous structures in infected plant cells. Virology **181**, 459-468.

Milne, R.G., Ramasso, E., Lenzi, R., Masenga, V., Sarindu, N. and Clark, M.F. (1995). Pre- and post-embedding immunogold labeling and electron microscopy in plant host tissues of three antigenically unrelated MLOs: primula yellos, tomato big bud and bermudagrass white leaf. European Journal of Plant Pathology **101**, 57-67.

Mumford, R.A., Barker, I. and Wood, K.R. (1996). The biology of tospoviruses. Annals of Applied Biology **128**, 159-183.

Roggero, P., Adam, G., Milne, R.G. and Lisa, V. (1996). Purification and serology of virions of impatiens necrotic spot tospovirus. European Journal of Plant Pathology **102**, 563-568.

Wijkamp, I., van Lent, J., Kormelink, R., Goldbach, R.W. and Peters, D. (1993). Multiplication of tomato spotted wilt virus in its insect vector, *Frankliniella occidentalis*. Journal of General Virology **74**, 341-349.

Session 4:

Nucleic Acid Based Hybridisation Techniques

THE USE OF THE POLYMERASE CHAIN REACTION FOR THE CHARACTERISATION AND DIAGNOSIS OF YAM POTYVIRUSES

R. A. MUMFORD, D. L. JAYARATNE & S. E. SEAL
Natural Resources Institute
Central Avenue, Chatham Maritime, Kent, ME4 4TB. UK

1. Introduction

Yams (*Dioscorea* spp.) are an important tuber crop in many tropical and sub-tropical regions (Degras, 1993), especially West Africa which accounts for over 90% of world yam production (FAO, 1994). Yams, in common with other vegetatively propagated crops, are susceptible to a wide variety of different viruses (Porth & Nienhaus, 1983). Of all these viruses, the potyviruses (family *Potyviridae*; genus *Potyvirus*) are the most diverse (Porth & Nienhaus, 1983; Porth *et al.*, 1987) and can cause significant reductions in crop yield (Thouvenel & Dumont, 1990). Of the yam-infecting potyviruses, yam mosaic virus (YMV) and yam mild mosaic virus (YMMV) are especially important, because of their high incidence and wide distribution amongst the two most widely cultivated yams; *Dioscorea rotundata-cayenensis* and *D. alata*. However, despite the importance of YMV and YMMV, very little information is available on the true level of variation found amongst these two viruses and the other potyviruses that have been reported (Porth & Nienhaus, 1983; Porth *et al.*, 1987). This lack of knowledge on variability, coupled with the recalcitrant nature of yam tissue, has resulted in a dearth of reliable diagnostic tests. This current situation is severely hampering efforts to improve the yam crop, especially by preventing the international exchange of breeding materials. In an effort to rectify this situation, an approach using PCR-based techniques has been employed and has resulted in an increased understanding of yam potyvirus variability and the development of reliable diagnostics.

2. Materials and Methods

2.1. VIRUS ISOLATES

Virus-infected yam tubers were obtained from a variety of yam-growing regions and grown in peat-based compost, in an insect-free glasshouse at a temperature of 23°C.

H.-W. Dehne et al. (eds.), Diagnosis and Identification of Plant Pathogens, 157–160.

2.2. IC-RT-PCR

Immunocapture-reverse transcription-polymerase chain reaction (IC-RT-PCR) was performed as described in Mumford & Seal (submitted), using 0.5 ml microfuge tubes coated with polyclonal antisera (raised against YMV or YMMV), that was kindly provided by Alan Brunt (Horticulture Research International, UK).

2.3. POTYVIRUS PCR, CLONING AND SEQUENCING

Total RNA was obtained from yam leaf tissue using a method adapted from Chang *et al.* (1993). RT-PCR was then performed using degenerate potyvirus primers, essentially as described by Pappu *et al.* (1993) and using Promega reagents. Resulting PCR products were purified using QIA-quick columns (Qiagen) before cloning using the pGEM-T vector system (Promega). Plasmids were obtained from selected clones using the Wizard Plus mini-prep system (Promega) and were sequenced automatically (SequiServe, Germany).

3. Results

3.1. COMPARISON OF YAM POTYVIRUS SEQUENCES

Sequences were obtained from 3'end of the viral RNA of thirteen different yam potyvirus isolates, from a variety of different host species and locations. The results indicate the existence of four distinct yam potyviruses groups (Figure 1). These include isolates known to be YMV (isolates Dr-Nig-N5/N20/8, Dr-Tog-3 and Dr-Cam-2) and YMMV (isolates Da-Van-L1, Da-SL-SLK and Da-Nig-316/2), in addition to two uncharacterised viruses identified from the Indian sub-continent from *D. esculenta* (DeV; isolate De-SL-SL15) and *D. dumentorum* (DdV; isolates Dd-SL-SL19/SL28 and Dd-Bang-Y60).

3.2. DETECTION BY IC-RT-PCR

Using sequences obtained for yam potyviruses, two virus-specific pairs of primers were designed. One pair, YMV CP F and YMV UTR R, were designed to be specific for YMV and the second, YMMV CP F and YMMV UTR R, were designed to detect YMMV. Both of these primer pairs, when used in IC-RT-PCR, were specific for their target virus, giving products of 586 and 259 base pairs, for YMV and YMMV respectively (Figure 2). Neither primer pair gave amplification with uninfected yam material, yam tissue infected with the alternative virus or buffer (no sap added) controls (Figure 2).

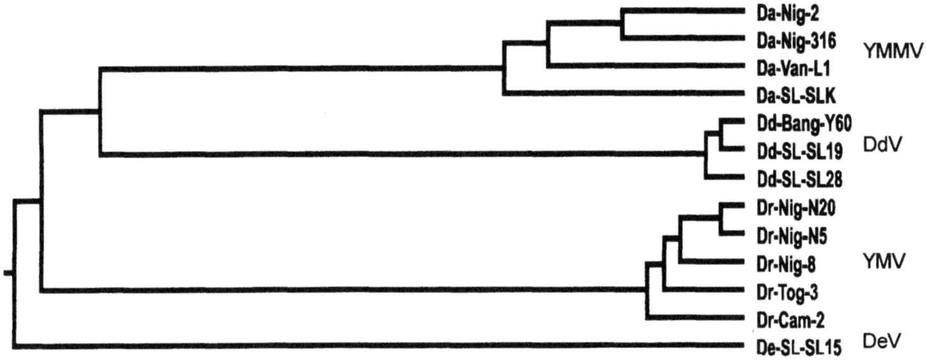

Figure 1. A basic phylogenetic tree generated from the similarities of the 3' RNA sequence for thirteen different yam potyvirus isolates. Figure shows four groupings representing YMV, YMMV, DeV and DdV.

4. Discussion

By using an approach based upon PCR techniques, it has been possible to rapidly generate sequence data from a number of different yam potyvirus isolates. As the results indicate, it has even been possible to obtain information on viruses that have not been previously characterised. The availability of sequence data has allowed the development of a preliminary phylogeny of yam potyviruses, against which future isolates can be compared. The sequence data obtained has also permitted the development of a number of primers, which can be used for the specific detection of YMV and YMMV. By utilising these primers with single-tube IC and combined RT-PCR, rapid, sensitive (1000 times more than ELISA) and reliable diagnosis of these viruses can be achieved.

5. References

Chang, S., Puryear, J. & Cairney, J. (1993). A simple and efficient method for isolating RNA from pine trees. *Plant Molecular Biology Reporter* **11**: 113-116.

Degras, L. (1993). *The Yam - a Tropical Root Crop.* London: Macmillan.

FAO. (1994). *Production Yearbook for 1993 (Vol. 47).* Rome: FAO.

Pappu, S.S., Brand, R., Pappu, H.R., Rybicki, E.P., Gough, K.H., Frenkel, M.J. & Niblett, C.L. (1993). A polymerase chain reaction method adapted for selective amplification and cloning of 3' sequences of potyviral genomes: applications to dasheen mosaic virus. *Journal of Virological Methods* **41**: 9-20.

Porth, A., Lesemann, D.E. & Vetten, H.J. (1987). Characterisation of potyvirus isolates from West African yams. *Journal of Phytopathology* **120**: 166-183.

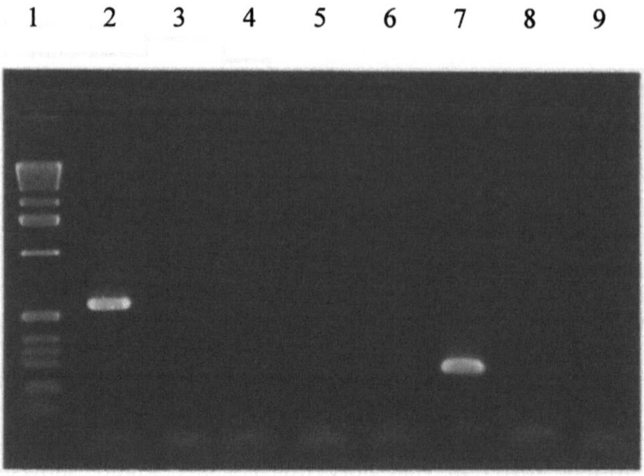

Figure 2. Detection of YMV and YMMV using IC-RT-PCR. Lanes 2-5, YMV antiserum coated tubes and primers; Lanes 6-9, YMMV antiserum coated tubes and primers; Test samples: Lanes 2+6, YMV isolate Dr-Nig-8; Lanes 3+7, YMMV isolate Da-Nig-2; Lanes 4+8, healthy yam; Lanes 5+9, buffer only. Lane 1, kb ladder (BRL).

Porth, A. & Nienhaus, F. (1983). Dioscorea alata ring mottle virus, a new potyvirus of yam in Togo. *Journal of Plant Diseases and Protection* **90**: 352-362.

Thouvenel, J.C. & Dumont, R. (1990). Yield decrease in yam infected with mosaic virus in Cote d'Ivoire. *L'Agronomie Tropicale* **45**: 125-129.

NOVEL METHODS BASED ON GENOMIC AMPLIFICATION FOR ROUTINE DETECTION OF POTATO LEAFROLL VIRUS IN POTATO TUBERS

G. LEONE AND C.D. SCHOEN
DLO-Research Institute for Plant Protection (IPO-DLO)
Binnenhaven 5, 6709 PD Wageningen, The Netherlands

1. Abstract

Two novel detection methods for potato leafroll virus (PLRV) in potato tubers based on genomic amplification have been developed: a one-tube, fluorogenic 5' nuclease RT-PCR assay and an isothermal NASBA~ amplification assay. The former method consists of an immunocapture of PLRV coupled to a one-tube RT-PCR using *Thermus thermophilus* (r*Tth*) instead of Taq DNA polymerase. Fluorescent detection in a microwell format was realized incorporating into the reaction a PLRV-specific probe (TaqMan™ probe) that carries a 5'- terminal 'reporter' fluorescein and a 3'-terminal rhodamine 'quencher'. During the extension phase of PCR, the probe was specifically degraded by the exonuclease activity of the DNA polymerase, resulting in an increase in reporter fluorescence at each cycle. The NASBA method allowed for amplification of PLRV RNA in a single enzymatic reaction at 41°C by the concurrent activity of AMV reverse transcriptase, T7 RNA polymerase and RNase H. A ten-fold dilution in water of extracts of both a healthy tuber spiked with purified PLRV and a primarily PLRV-infected tuber was sufficient to allow amplification and detection by agarose gel electrophoresis. Both systems offer the possibility to reduce the inspection time of seed-potatoes for infection of PLRV from the current five weeks to one day.

2. Introduction

The inspection of seed-potatoes for virus diseases by means of indexing programs requires extensive efforts from the Inspection Services worldwide. Difficulties are present, among others, because of: i) the massive amount of tubers that need to be inspected (for example, about 4 million tests per season are performed in the Netherlands) and; ii) the insufficient sensitivity of the current detection methods based on serology, that does not permit to detect directly latent infections in tubers. For these reasons the current seed-potato inspection program for PLRV and other potato viruses is routinely performed by ELISA tests on leaf extracts of sprouted tubers. This

H.-W. Dehne et al. (eds.), Diagnosis and Identification of Plant Pathogens, 161–167.

procedure is labor-intensive and time-consuming and necessitates at least five weeks for completion.

Recently, new detection methods based on genomic amplification of viral RNA have been introduced, that are sensitive, specific and have potential for diagnostic purposes. Reverse transcription polymerase chain reaction (RT-PCR) procedures have been described to detect PLRV in potato material (Hadidi et al., 1993; Robertson et al., 1991; Spiegel and Martin, 1993). However, these methods are impractical for routine applications mostly because of phenol or multistep RNA extractions and because detection of PCR products is heterogeneous, requiring an electrophoretic step.

The lack of possibility of automating the pre- and post-amplification steps is a major drawback for applications in routine laboratories, thus hindering wide adoption of these methods. Very recently, Schoen et al. (1996) showed that it is possible to significantly simplify the RT- PCR of PLRV in dormant potato tubers by coupling a pre-RT-PCR immunocapture of virions to a homogeneous, fluorogenic (TaqMan, Perkin Elmer) 5' nuclease RT-PCR. This latter assay allows for the detection of fluorescence specifically produced during the PCR process in microtiter plates, without the need of a post-PCR separation by e.g. electrophoresis. This new procedure not only reduces the inspection of seed-potatoes from five weeks to one day but is amenable to automation and therefore represents a step forward to high- throughput PCR.

An amplification method that could be performed in one-tube and/or does not necessitate thermocycling would require, however, even less manipulations, thus improving the applicability for very large-scale screenings. Here, we report preliminary experiments to develop a one-tube fluorogenic 5' nuclease RT-PCR and a NASBA assay for the detection of PLRV in potato tubers. The possibility of carrying out RT and PCR amplification in the same reaction mixture was investigated employing *Thermus thermophilus* (rTth) DNA polymerase, which has both reverse transcriptase and DNA polymerase activity. NASBA is a primer-dependent, isothermal enzymatic amplification technology particularly suited for the detection of RNA and can result in a billion-fold amplification of specific nucleic acid sequences in 90 minutes at 41'C (Kievits et al., 1991). Amplification is accomplished by the concurrent enzymatic activity of AMV reverse transcriptase, T7 RNA polymerase and RNase H. With this technology it has been possible to detect human RNA viruses and food pathogens that are very difficult to diagnose or occurring in very low concentrations (Sooknanan and Malek, 1995).

3. Materials and Methods

3.1 ONE-TUBE, 5' NUCLEASE RT-PCR ASSAY

The immunocapture step of PLRV virions, primers employed and the TaqMan probe (Perkin Elmer) were as described by Schoen et al. (1996). Intact PLRV and/or purified RNA were subjected to RT-PCR using a GeneAmp EZ r*Tth* RNA PCR Kit (Perkin Elmer). One reaction (50 pl) contained 5xEZ buffer, 0.3 mM each dNTP, 2.5 mM

Mn(OAc)z, 0.3 pM downstream and upstream primers (Hadidi et al., 1993), 40 nM TaqMan probe and 5.0 units of r*Tth* DNA polymerase. A thermocycler (PTC-200, MJ Research) was employed for amplification. Reaction were heated at 65°C for 60 min and at 94°C for 1 min, followed by 39 cycles at 94°C for 20 s, 65°C for 45 s and 72°C for 10 s.

3.2. POST-PCR ANALYSIS

The fluorescence spectra of each RT-PCR reaction were measured in a Perkin Elmer LS-50B luminescence spectrometer equipped with a microtiter plate reader. Calculation of the ARQ was as reported by Schoen et al. (1996).

The presence of a specific PCR product was also verified by electrophoresis of 10 pl of each reaction on a 1% agarose gel containing 0.5 µg of EtBr per ml. Bands were visualized by UV excitation and photographed with a DocuGel-V digital photodocumentation system (Scanalytics).

3.3. NASBA

Primer sequences were selected from the coat protein open reading frame of PLRV. Details about primers and detection probe design will be presented elsewhere. The downstream primer consisted of a 3' terminal, PLRV specific sequence and a 5' terminal T7 promoter sequence, which can be recognized by T7 RNA polymerase. The upstream primer was entirely PLRV specific. One reaction (20 µl) contained 2 pl of sterile water, 4 µl of 5xNASBA buffer and 4 µl of 5xprimer mix. The 5xNASBA buffer contained 200 mM Tris-HC1, pH 8.5, 60 mM MgClz, 350 mM KC1, 25 mM DTT, 5 mM each ATP, UTP and CTP and 7.5 mM GTP (Pharmacia) and 2.5 mM ITP (Boehringer). The 5xprimer mix contained 75% DMSO and 1 µM each downstream and upstream primers. To this reaction mix, 1 µl of PLRV material in 4 µl of water was added. The reaction mixture were then heated at 65'C for 5 min and put at 41'C for 5 min. Five µl of enzyme mix were added to the reactions, consisting of 375 mM sorbitol, 2.1 µg BSA, 0.08 units of RNase H, 32.0 units of T7 RNA polymerase (both Pharmacia) and 6.4 units of AMV-reverse transcriptase (Seigakaku). Reactions were then incubated at 41°C in a water bath for 90 min and stored at -20°C until use.

3.4. POST-NASBA ANALYSIS

The amplification products were processed by Northern blotting as follows: 5 µl of each reaction were run on a 2% agarose gel, blotted in 2xSSC (1xSSC: 0.15 M NaC1 and 15 mM Na-citrate) to a Zeta Probe nylon membrane (BioRad) and after UV-crosslinking hybridized to a P-labelled oligonucleotide probe in 5xSSC, 20 mM Na-phosphate, pH 6.7, 7% SDS and 10xDenhardt's at 50'C. After hybridization overnight the blot was washed in 3xSSC, 1% SDS at 50°C. The blots were exposed overnight to X-ray films at -80°C.

4. Results

4.1. DETECTION OF PLRV BY A ONE-TUBE, FLUOROGENIC 5' NUCLEASE RT-PCR ASSAY

A 10-fold serial dilution of PLRV RNA, in amounts ranging from 400 fg to 4 ng was amplified with the procedure described in Materials and Methods. Detection of the PCR products was performed both by measurements of fluorescence yields (ΔRQ) and by visual scoring after agarose gel electrophoresis. Figure 1 shows that 400 fg of input RNA still gave a clear signal with both assays used.

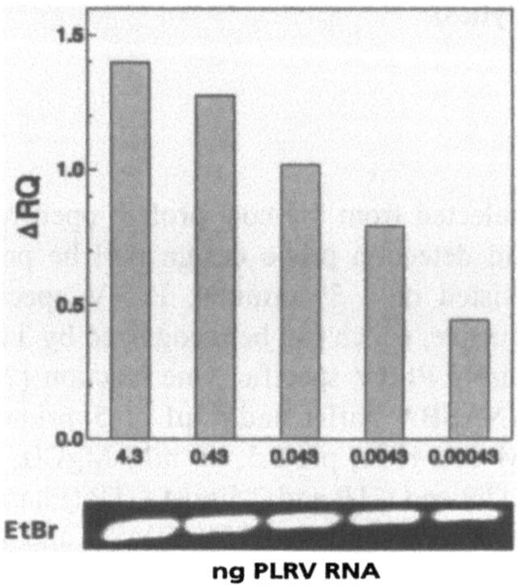

FIGURE 1. Comparison of fluorogenic 5' nuclease RT-PCR assay (expressed in ΔRQ) and gel electrophoresis (ethidium bromide staining) for detection of PLRV RNA. Figures represent ng amounts of pure input RNA. Amplification product is 486 bp.

A 10-fold serial dilution, ranging from 250 fg to 2.5 ng, was prepared using purified PLRV virions and used as input for the one- tube, fluorogenic 5' nuclease RT-PCR assay. Virions were immuno-captured to the inner wall of the tubes prior to RT-PCR. A 25 pg amount of PLRV gave a specific, detectable signal both after fluorescence measurements and agarose gel electrophoresis (Fig. 2). The limit of visual scoring of the specific PLRV amplicon band on agarose gel (2.5 pg) was in agreement with the limit of detection by fluorescence measurements.

4.2. DETECTION OF PLRV BY NASBA

NASBA amplification was tested using different PLRV-carrying material as input for the reaction. Presence of the expected RNA amplicon was checked by agarose gel electrophoresis and specificity was validated by Northern blot analysis. A specific, detectable signal was found starting with a 40 pg amount of purified PLRV, healthy tuber extracts spiked with 4.0 pg of PLRV or a PLRV-infected tuber only when the extracts were diluted at least 1:10 in water. Amplification directly from undiluted tuber extracts carrying PLRV was not possible.

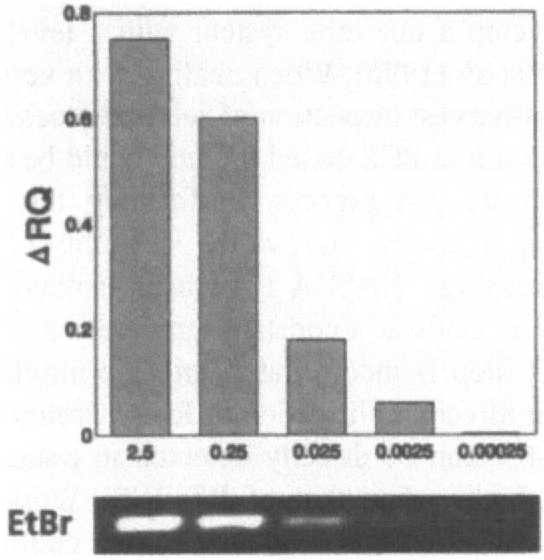

ng PLRV

FIGURE 2. Comparison of fluorogenic 5' nuclease RT-PCR assay (expressed in ΔRQ) and gel electrophoresis (ethidium bromide staining) for detection of PLRV. Figures represent ng amounts of input PLRV virions. Amplification product is 486 bp.

5. Discussion

The results presented in this report show that it is possible to develop new strategies for the inspection of seed-potatoes for PLRV, reducing the testing time from a minimum of five weeks for the current post- harvest ELISA, to a single work day. Schoen et al. (1996) demonstrated that PLRV could be reliably detected in seed-potatoes by a gel-free, fluorescent 5' nuclease RT-PCR assay. This method is amenable to automated sample analysis in a microwell format, allowing a significant simplication of the PCR technology for routine applications. The basis of the fluorogenic 5' nuclease RT-PCR assay is the digestion by Taq DNA polymerase of a probe (TaqMan probe), labelled with a reporter and a quencher fluorescent dye, which hybridizes specifically to the target PCR product during amplification (Livak et al., 1995). The cleavage of the probe during strand elongation, in the repeated cycles of the PCR process, generates a fluorescent signal that increases with the exponential synthesis of the amplicon.

However, the procedure described by Schoen et al. (1996) requires that the reverse transcription and the PCR reactions are carried out separately. Therefore, performing the two reactions in the same vessel would improve further the procedure. Our results show that this can be achieved employing *Thermus thermophilus* DNA polymerase which has both reverse transcriptase and DNA polymerase activity (Myers and Gelfand, 1991). In this way it has been possible to detect PLRV by coupling an immunocapture step to a one-tube, fluorogenic 5' nuclease RT-PCR assay. However, preliminary results (not shown) with primarily PLRV-infected tubers showed that the procedure was less sensitive compared with results obtained with purified virus. Work is in progress to develop a one-tube system with a level of sensitivity similar to the procedure of Schoen et al. (1996). When dealing with very high sample throughput, as in the case of the post-harvest inspection of seed-potatoes in the Netherlands, a limiting factor in the application of a PCR-based method could be represented by the number of reactions a thermocycler can process. Performing the amplification at one single temperature would significantly improve the flexibility of the system, facilitating very high- throughput screenings. NASBA is such an isothermal amplification technology able to amplify nucleic acids at a constant temperature of 41°C (Kievits et al., 1991). Moreover, as the RT step is incorporated into the amplification process, NASBA is particularly suited for direct amplification of RNA. Preliminary results presented in this report show that PLRV can be directly detected in potato tubers by NASBA, with a sensitivity that was higher than that of RT-PCR. Work is in progress to optimize NASBA for large-scale screenings of seed- potatoes coupling this technology to a gel-free detection method.

5. References

Hadidi, A., Montasser, M.S., Levy, L., Goth, R.W., Converse, R.H., Madkour, M.A., and Skrzeckowski, L.J. (1993): Detection of potato leafroll and strawberry mild yellow-edge luteoviruses by reverse transcription-polymerase chain reaction amplification. Plant Dis. **77**, 595-601.

Kievits, T., van Gemen, B., van Strijp, D., Schukkink, R., Dircks, M., Adriaanse, H., Malek, L., Sooknanan, R., and Lens, P. (1991): NASBA™ isothermal enzymatic in vitro nucleic acid amplification optimized for HIV-1 diagnosis. J. Virol. Methods **35**, 273-286.

Livak, K.J., Flood, S.J.A., Marmaro, J., Giusti, W., and Deetz, K. (1995): Oligonucleotides with fluorescent dyes at opposite ends provide a quenched probe system useful for detecting PCR product and nucleic acid hybridization. PCR Methods Appl. **4**, 357-362.

Myers, T.W. and Gelfand, D.H. (1991): Reverse transcription and DNA amplification by a *Thermus thermophilus* DNA polymerase. Biochem. **30**, 7661-7666.

Robertson, N.L., French, R., and Gray, S.M. (1991) Use of group-specific primers and the polymerase chain reaction for the detection and identification of luteoviruses. J. Gen. Virol. **72**, 1473-1477.

Schoen, C.D., Knorr, D., and Leone, G. (1996): Detection of potato leafroll virus in dormant potato tubers by immunocapture and a fluorogenic 5' nuclease assay. Phytopathology **86**, 993-999.

Sooknanan, R. and Malek, L.T. (1995): NASBA: A detection and amplification system uniquely suited for RNA. Bio/Technology **13**, 563-564.

Spiegel, S. and Martin, R.R. (1993): Improved detection of potato leafroll virus in dormant potato tubers and microtubers by the polymerase chain reaction and ELISA. Ann. Appl. Biol. **122**, 493-500.

6. Acknowledgements

The authors are indebted to Joan Lindner of IPO-DLO, Dave Knorr of Perkin Elmer's Applied Biosystem Division and to Harm van Schijndel and Bob van Gemen of Organon Teknika for their valuable support in this research.

This work was partly funded by contract PL921606 of the European Community specific program for Research, Technological Development and Demonstration in the field of Agriculture and Agro- Industry, including Fisheries.

ROUTINE DETECTION OF CHRYSANTHEMUM STUNT VIROID IN CHRYSANTHEMUM PLANTS BY REVERSE TRANSCRIPTION - POLYMERASE CHAIN REACTION

M.S.J. ARTS & J.W. ROENHORST
Plant Protection Service
P.O. Box 9102, 6700 HC Wageningen, The Netherlands

R. HOOFTMAN & A. VAN ZAAYEN
Inspection Service for Floriculture and Arboriculture (NAKB)
Johan de Wittlaan 12, 2517 JR Den Haag, The Netherlands

Summary

For detection of chrysanthemum stunt viroid (CSVd) in chrysanthemum plants, the reverse transcription-polymerase chain reaction (RT-PCR) method described by Hooftman *et al.* (1996) has been adjusted for practical use. This led to minor changes in the homogenization of leaf material, extraction method and conditions during PCR. The results obtained by the RT-PCR method were similar to those obtained by the current method, i.e. return-polyacrylamide gel electrophoresis (R-PAGE). With healthy chrysanthemum plants, also a slight reaction was obtained in the RT-PCR method. In this case however, the PCR product was some nucleotides smaller than in the case of infected plants. It appeared that the present RT-PCR method can be used for the detection of CSVd in chrysanthemum at practical scale. The reliability of the method still has to be proved in large-scale application, during which the results of this method are compared with those of the current method.

1. Introduction

The cultivation of chrysanthemum (*Dendranthema indicum*) cuttings and flowers is an economically important branche of horticulture in the Netherlands. The Dutch Inspection Service for Floriculture and Arboriculture (NAKB) executes a system of inspection, testing and certification to guarantee the quality level of the cultivation of chrysanthemum. One of the pathogens under testing is chrysanthemum stunt viroid (CSVd).

CSVd is first described as the cause of a serious disease in chrysanthemum in the United States of America in 1947 (Dimock, 1947). In the Netherlands, it is first reported by Noordam in 1952. Symptoms of this viroid are growth reduction and early flowering. In addition, flowers may be smaller in number and size and paler than flowers from

H.-W. Dehne et al. (eds.), Diagnosis and Identification of Plant Pathogens, 169–172.

healthy chrysanthemum plants. Symptoms often occur after a relative long incubation period (two to three months). The viroid is transmitted by contact and vegetative propagation (Lawson, 1987). CSVd consists of a circular, single-stranded RNA molecule of 354 to 356 nucleotides (Haseloff and Symons, 1981; Gross *et al.*, 1982). At room temperature it behaves like double-stranded RNA and is very stable (Keese and Symons, 1987; Sänger, 1988).

To control CSVd in the cultivation of chrysanthemum the NAKB performs a system of selection, visual inspection and laboratory tests. At the moment CSVd is detected by return-polyacrylamide gel electrophoresis (R-PAGE) (Schumacher *et al.*, 1983; Huttinga *et al.*, 1987). Because a more sensitive method is desirable and innovation is wanted, the suitability of the reverse transcription-polymerase chain reaction technique (RT-PCR) for this application was investigated. If the RT-PCR method indeed proves to be more sensitive than R-PAGE, the number of plants per sample can be increased, which will improve the efficiency of testing.

The purpose of this investigation was to develop a RT-PCR method for the detection of CSVd in chrysanthemum at practical scale. This method should be at least as sensitive and specific as the current method (R-PAGE) and should be quick and reliable in practical application.

2. Materials and methods

Both infected (21 cultivars) and healthy (6 cultivars) chrysanthemum plants were obtained from the NAKB. Leaf material from plants tested with R-PAGE was collected and sent for testing in RT-PCR.

The method for detection of CSVd in chrysanthemum by RT-PCR described by Hooftman *et al.* (1996), was used as starting point for this study. Different steps of the protocol were analysed.

Amount of leaf material per sample. In the original protocol only 30 mg were used per sample. Possibilities for increasing this amount were investigated.

Extraction of total nucleic acids. Phenol/chloroform extraction was compared with an alternative method that is quicker, cheaper and less toxic. After grinding the chrysanthemum leaves in TE-buffer and centrifuging the obtained sap, the supernatant was diluted ten times in TE-buffer and used directly for primer annealing in the reverse transcription process. In addition, the use of GeneReleaser was evaluated.

RT-PCR conditions. Conditions were evaluated by changing enzyme concentrations (per sample 22-44 units RNasin, 2-200 units M-MLV Reverse Transcriptase, 0.25-2.5 units Taq polymerase), primer concentrations (0.1-1.0 μg complementary primer in the reverse transcription process, 0-12 μM complementary and 6-12 μM homologous primer in the PCR process), omitting the 2-mercaptoethanol, and adding cDNA before and Taq polymerase after the 'hot start'.

By using the GeneAmp PCR System 9600 (Perkin Elmer) in stead of the DNA Thermal Cycler model 480, 96 samples could be processed at once without the use of mineral oil. The programme to be used was experimentally determined: 30 seconds at 94°C denaturation, 90 seconds at 64°C primer annealing and 45 seconds at 72°C extension, 30 cycles, ending up with 7 minutes at 72°C for final extension.

3. Results and discussion

Analysing the different steps of the protocol of Hooftman *et al.* (1996) for detection of CSVd by RT-PCR at a practical scale led to the following results.
Amount of leaf material per sample. The amount of leaf material per sample was increased from 30 mg to a maximum of 0.5 gram. The plant material was ground in a plastic bag with cambric by using a homogenizer (Bioreba). Results were satisfactory as grinding was sufficient and cross contamination was excluded. The method could be applied both manually and automatically.
Extraction of total nucleic acids. Extraction of total nucleic acids by dilution of the supernatant gave satisfactory results with young plants. However, from the appearance of flower buds, only phenol/chloroform extraction yielded reliable results (results not shown). Because very heterogeneous material is tested in practice, the phenol/chloroform extraction could not be replaced by the ten-fold diluted supernatant extraction. The only change which could be introduced without loss of sensitivity was the omission of the use of GeneReleaser. Since this time consuming step in the extraction protocol did not yield a clear improvement of the sensitivity of the RT-PCR method, it was abandoned.
RT-PCR conditions. In the protocol for practical use cDNA was added before and Taq polymerase after the 'hot start'. 6 μM complementary and 12 μM homologous primer were added in the PCR. Other adjustments of the protocol of Hooftman *et al.* (1996) yielded erratic results.

The 21 chrysanthemum cultivars that were found infected by R-PAGE at the NAKB, reacted positively with the CSVd primers in the RT-PCR by using the adjusted protocol. The six healthy chrysanthemum cultivars however, reacted in the RT-PCR with the CSVd primers as well, although to a very low extent (results not shown). In case of healthy plants the PCR product counted just a few nucleotides less than the product formed in the presence of CSVd. This sometimes hampered the interpretation of the results. In most cases the CSVd infected plants could be easily distinguished from the healthy plants because infected plants gave rise to much intenser bands on the gels.

In general, the tests at both experimental and practical scale yielded similar results for R-PAGE and RT-PCR. In four cases, discrepancies were encountered. At experimental scale, one sample that did not react in R-PAGE yielded a positive reaction in the RT-PCR. Since these results could be reproduced, the difference most likely reflects the different sensitivity of the two techniques. At practical scale, three samples reacted in R-PAGE, but not in RT-PCR. However, when the remaining of the nucleic acid extract used for RT-PCR was tested in R-PAGE, no reaction was obtained. It remained unclear why the original reaction in R-PAGE could not be reproduced. The infected material might have been unequally divided over the original samples or the amount used in the second R-PAGE was below the detection level of this technique.

The adjusted RT-PCR method has been succesfully introduced at the NAKB and the first tests at practical scale have been performed.

4. Conclusions and recommendations

From these findings it can be concluded that the present RT-PCR method for the detection of CSVd in chrysanthemum is suitable for application at practical scale. The method has been successfully implemented at the laboratory of the NAKB. Given the results so far, it is to be expected that cultivars of chrysanthemum, other than the 21 cultivars tested, react similar and can be tested in the same way. The reliability of the RT-PCR method still has to be proved in large-scale application, during which the results of this method are compared with those of the current method (R-PAGE). In the mean time a further decrease of the background reaction of healthy chrysanthemum plants is desirable, in order to increase the number of plants per sample and the efficiency of testing.

5. References

Dimock, A.W.(1947) Chrysanthemum stunt, *N.Y. State Flower Grow. Bull.* **26**, 2.

Gross, H.J., Krupp, G., Domdey, H., Raba, M., Jank, P., Lossow, C., Alberty, H., Ramm, K. and Sänger, H.L. (1982) Nucleotide sequence and secondary structure of citrus exocortis and chrysanthemum stunt viroid, *Eur. J. Biochem.* **121**, 249-257.

Haseloff, J. and Symons, R.H. (1981) Chrysanthemum stunt viroid: primary sequence and secondary structure, *Nucleic Acids Res.* **9**, 2741-2752.

Hooftman, R., Arts, M.-J., Shamloul, A.M., Zaayen, A. van and Hadidi, A. (1996) Detection of chrysanthemum stunt viroid by reverse transcription-polymerase chain reaction and by tissue blot hybridization, *Acta Horticulturae* (in press).

Huttinga, H., Mosch, W.H.M. and Treur, A. (1987) Comparison of bi-directional electrophoresis and molecular hybridization methods to detect potato spindle tuber viroid and chrysanthemum stunt viroid, *Bulletin OEPP/EPPO Bulletin* **17**, 37-43.

Keese P. and Symons R.H. (1987) The Structure of Viroids and Virusoids, in J.S. Semancik (ed.), *Viroids and Viroid-Like Pathogens*, CRC Press Inc., Boca Raton, Florida, pp 1-47.

Lawson, R.H. (1987) Chrysanthemum Stunt, in T.O Diener (ed.), *The Viroids*, Plenum Press, New York and London, pp. 247-259.

Noordam, D. (1952) Virusziekten bij chrysanten in Nederland, *Tijdschr. Plantenziekten* **58**, 121-189.

Sänger, H.L. (1988) Viroids and viroid diseases, *Acta Horticulturae* **234**, 79-87.

Schumacher, J., Randles, J.W. and Riesner, D. (1983) A two-dimensional technique for the detection of circular viroids and virusoids, *Analytical Biochemistry* **135**, 288-295.

This project was financed by the Ministry of Agriculture, Nature Management and Fisheries and the Product Board for Ornamental Horticulture.

IDENTIFICATION OF A TOBAMOVIRUS INFECTING *NEW GUINEA IMPATIENS*

J. MÖRBEL, T. WETZEL and GABRIELE KRCZAL
Landesanstalt für Pflanzenbau und Pflanzenschutz, Mainz, Germany

A putative tobamovirus was isolated from infected *New Guinea Impatiens* showing leaf symptoms. Virus particles were propagated in *Nicotiana benthamiana* and purified from infected plants, and the viral RNA was extracted. The ligation-anchored/PCR technique (LA/PCR), based on the ligation of the viral RNA to an EcoRI-linearised plasmid (Bluescript), was used to amplify the 3'end of the viral genome, using a degenerate primer designed from published tobamovirus sequences and the universal reverse primer on the plasmid. The LA/PCR-generated product, corresponding to part of the viral 3' non-coding region (3'NC), was cloned and sequenced. From this sequence information, we designed a specific primer, to be used in a RT/PCR reaction together with another degenerate primer, located upstream of the movement protein gene. The resulting PCR product, covering the putative movement protein gene, coat protein gene (CP), and part of the 3'NC, was cloned and sequenced.
The analysis and comparison of the 3' NC and CP of our virus with the corresponding sequences of other tobamoviruses revealed high homology (97% and 98% in the 3'NC and CP respectively) with the chinese rape mosaic virus (CRMV, Aguilar *et al.*, 1996), and ribgrass mosaic virus (85% in the CP)(cited in Oshima and Harrison, 1975).

References

Aguilar, I., Sanchez, F., Martin, A.M., Martinez-Herrera, D., and Ponz, F. (1996) Nucleotide sequence of chinese rape mosaic virus (oilseed rape mosaic virus), a crucifer tobamovirus infectious on Arabidopsis thaliana. Plant Molecular Biology 30, 191-197.

Oshima, O., and Harrison, B.D. (1975) Ribgrass mosaic virus. C.M.I./A.A.B. description of plant viruses No 152.

H.-W. Dehne et al. (eds.), Diagnosis and Identification of Plant Pathogens, 173.

IC-RT-PCR DETECTION OF BLACKCURRANT REVERSION ASSOCIATED NEPOVIRUS, AND PERSPECTIVES OF ITS USE FOR DIAGNOSIS OF THE REVERSION DISEASE

K. LEHTO[1], S. LATVALA[1,2], A.T. JONES[3], S. COX[3], P.E. SUSI[1,4], A. LEMMETTY[2]

[1]*Department of Biology, University of Turku, Finland;* [2]*Institute of Plant Protection, Agricultural Research Centre, Finland;* [3]*Scottish Crop Research Institute, Invergowrie, U.K;* [4]*Department of PlantProduction,University of Helsinki, Finland*

Recently we have succeeded to isolate a new nepo-type virus from a reverted blackcurrant. We have developed an immunocapture-reverse transcriptase-PCR (IC-RT-PCR) method to detect the virus, and have demonstrated that the virus is associated both with the E and R forms of reversion disease originating from Scandinavia, Scotland and New Zealand. Due to its origin and its consistent association with the different forms of reversion, but not with any other virus diseases of *Ribes* tested, we refer to the virus as Blackcurrant reversion associated virus (BRAV).

1. Introduction

Blackcurrant reversion is the most important of the about 14 virus or virus-like diseases that have been recognised in blackcurrant (*Ribes nigrum* L.) The causal agent of the disease is transmitted by the blackcurrant gall mite (*Cecidophyopsis ribis* Westw.). Experimentally it is transmitted between *Ribes* plants by grafting (1). Despite research over more than 50 years, the agent of reversion has remained unknown. At least two distinct forms of the disease are known. Symptoms of the common European form (E; 4) include change of the blackcurrant leaf form from the complex, wide leaf with deep-lobed basal sinus to narrow, straight-based or 'nettle-like' leaf shape. Reversion also causes decrease in the density of hairs on the flower buds and increased intensity of colour in the buds (1). In addition, the more

H.-W. Dehne et al. (eds.), Diagnosis and Identification of Plant Pathogens, 175–178.

severe form of the disease (R; 4), found in Eastern European and Scandinavian countries, induces formation of occasional and soon disappearing distinct yellow line pattern in some leaves and also occasional development of ten sepals instead of five and further increase of the pigmentation of flowers (1). However, the symptom severity and distictiveness greatly varies between different blackcurrant cultivars and may be easily overlooked due to the natural variation of *Ribes* leaf shapes. Still, observation of the leaf and floral symtoms has been the only easily accessible method for field diagnosis of the disease. The only test available for its diagnostics has been graft-inoculation to sensitive blackcurrant cultivars and waiting for development of symptoms for up to 2 y (1, 3).

Recently, an unknown virus was transmitted from reverted blackcurrant to *Chenopodium quinoa* Willd. After propagation in this host the virus was purified and partially characterized (5, Latvala *et al.*, this volume). The virus was shown to be a previously unknown Nepovirus, and was tentatively named 'Blackcurrant reversion associated virus' (BRAV). Based on the 3' terminal sequence of one of its genomic components oligonucleotide primers were designed for amplification of a 210 bp fragment by IC-RT-PCR-protocol (Fig. 1) (6,7). In amplification, the 'hot start' PCR-method (2) was used to prevent synthesis of unspecific PCR-products (5). BRAV was consistently detected from purified virus samples, from BRAV-infected *Chenopodium quinoa* host, from *Cecidophyopsis ribis* mites collected from galled buds of reverted blackcurrants, and from blackcurrant from which the BRAV had been isolated. Further on, the PCR-test has been used to study the relationship of BRAV with the reversion disease.

CGCTGGTGTCTCATTATATACTAGCGTTGCAGCGCTAGGACTACCTAAG
ACAACCGTATTATACGTTTGCTTTATTTATTTTTGTTACTCCTGTTTAGCA
GGTCGTGCCTTCAGTAAGCACACAAAAATATTTTCCATTTTTAGGTTTTG
GTTAAGAGTCATTTTATTGAGCACTTTTCTTTTAGAGAATAGAGTCTGAA
ATGTCCTTTCAAAAAAAAAAAAAAAAAAAA

Figure 1. Nucleotide sequence of the 210 3' terminal nucleotides of one of the genomic RNAs of BRAV. Sequences used for the oligonucleotide primers for the IC-RT-PCR are underlined.

2. Results

We tested 104 plant samples using IC-RT-PCR, including samples of healthy blackcurrant, blackcurrant infected with reversion or various other viral diseases, and *Chenopodium* samples infected with other Nepoviruses. BRAV was detected in each of the 20 tested blackcurrant plants with known positive infection of either the E or R form of reversion and originating either from Finland, Scotland or New Zealand. 41 unconfirmed field samples were also tested, showing putative symptoms of reversion. From these, the BRAV was detected in 34 samples. Lack of association of BRAV to observed symptoms of reversion in 7 samples with observed symptoms, may be due to slow spread and uneven and erratic distribution of the virus in *Ribes* plants: In systemic sampling of all buds of a reversion-infected bush an average of 50% of all buds were positive. Thus, association of BRAV to the positive samples of reversion disease was very high.

BRAV was not detected in any of the 24 controlled samples of either healthy blackcurrant or blackcurrant infected with other known virus or virus-like diseases (yellows disease, infectious variegation, gooseberry veinbanding, Arabis mosaic virus) or in 5 samples of *Chenopodium quinoa,* either healthy, or infected with Strawberry latent ringspot virus, Arabis mosaic virus (Scottish and Lilac isolates) or Raspberry ringspot virus. BRAV was not either detected in 13 field samples, showing no apparent symptoms of reversion. However, low level of BRAV was detected in one sample with no obvious symptoms of reversion at the time of testing. This was apparently due to slow development of reversion symptoms in infested plants, because in following summer (1996) several shoots of the same bush showed clear leaf symptoms of reversion. The slow development of reversion symptoms was demonstrated also in experiment where three plants of Ben Nevis blackcurrant, and two plants of each of the Ben Lomond and Ben More blackcurrants were mite-inoculated with reversion. The reversion symptoms developed in the plants of Ben Nevis variety, but not in the plants of the other two varieties within four years after inoculation. By IC-RT-PCR, BRAV was detected in all the inoculated plants.

4. Discussion

The concistent association of BRAV with reversion disease strongly suggests that it is the causal agent of the disease. Based on the apparent association of the virus and the disease, we believe that detection of BRAV using the IC-RT-PCR test, can be used at least for indicative diagnostics of the disease. The IC-RT-PCR-mediated detection offers, in comparison to the previously used indicator plant tests or the mere observation of symptoms, progress towards faster, more accessible, affordable and reliable diagnostics. Some essential problems in the detection of the BRAV by the IC-RT-PCR, for the diagnosis of the reversion disease, is that the virus is not detected from all of the leaves of virus-infested bushes. Occasionally this leads to very low level or negative detection of the virus from randomly collected samples from positive bushes. Therefore, to make the samples more representative of the whole bush an adequate amount of leaves must be combined to make a mixed sample. Detection of the positive PCR-amplified products may be improved, in comparison to electrophoresis in agarose gels, by their labelling with fluorescent nucleotides or primers and detection by an automated sequencer.

5. References

1. Adams, A. N., and Thresh, J. M. 1987. Reversion of blackcurrant. in: R. H. Converse, (ed.), *Virus Diseases of Small Fruits*. USDA Agriculture Handbook No. **631**. pp. 133-136.
2. Erlich, H. A., Gelfand, D., and Sninsky, J. J. 1991. Recent advances in the polymerase chain reaction. *Science* **252**, 1643-1651.
3. Jones, A. T. 1992. Appendix I. Recommended indexing procedures for detecting viruses, viroids, mycoplasma- and rickettsia-like organisms and virus-like diseases in small fruit crops: Currants and Gooseberries (*Ribes*). *Acta Hortic.* **308**, 51-154.
4. Jones, A. T. 1995. Recent developments affecting the control of two important diseases of small fruit crops in the UK. *Acta Hortic*. **385**, 64-69.
5. Lemmetty, A., Latvala S., Jones, A.T., McGavin, W.J., S., Susi, P.E., and Lehto, K. 1996. Purification and properties of a new virus from blackcurrant, its affinities with nepoviruses and its close association with blackcurrant reversion disease. *Phytopathology* (accepted for publication).
6. Nolasco, G.. de Blas, C., Torres, V., and Ponz, F. 1993. A method combining immunocapture and PCR amplification in a microtitre plate for detection of plant viruses and subviral pathogens. *J. Virol. Methods* **45,** 201-218.
7. Wetzel, T., Candresse, T., Ravelondandro, M., and Dunez, J. 1991. A polymerase chain reaction assay adapted to plum pox potyvirus detection. *J. Virol. Methods* **33,** 355-365.

DETECTION AND IDENTIFICATION BY PCR-BASED TECHNIQUES OF DIVERSE PHYTOPLASMAS INFECTING GRAPEVINE

P.A. BIANCO, P. CASATI AND G. BELLI
Centro CNR Miglioramento Sanitario Colture Agrarie, c/o Istituto di Patologia Vegetale, Università degli Studi di Milano, Italy

1. Introduction

Flavescence dorée (FD) is among the most serious diseases known in grapevine (*Vitis vinifera*). Unculturable wall-less prokaryotes termed phytoplasmas (previously named MLOs = mycoplasmalike organisms) were constantly found associated with the disease (Caudwell *et al.*, 1971, Quaroni *et al.*, 1988) and considered the causal pathogens.
In vineyard, the disease is transmitted by the leafhopper *Scaphoideus titanus* Ball. Since the first reports in France (Caudwell *et al.*, 1957) and in Italy (Belli *et al.*, 1973) severe damage and crop losses were observed in affected grapevines.
Similar diseases called "grapevine yellows" (GY) were then reported in other countries including Germany (Gartel, 1961), U.S.A. (Pearson *et al.*, 1985) and Australia (Magarey *et al.*, 1986) but until recently there is little evidence of the difference between the diverse phytoplasmas involved with the diseases. Molecular biology tools allowed to detect and identify at least three different phytoplasma groups associated with the diseases (Prince *et al.*, 1993, Daire *et al.*, 1993a, Bianco *et al.*, 1993, Maixner *et al.*, 1995). Particularly, in FD diseased grapevines, phytoplasmas belonging to 16SrV group were detected (Bianco et *al.*, 1993, Daire *et al.*, 1993a), while GY diseases were found associated with infections of phytoplasmas belonging to group 16SrI (aster yellows and related phytoplasmas) (Bianco *et al.*, 1993, Davis *et al.*, 1993, Daire *et al.*, 1993b) and group 16SrIII (X-disease and related phytoplasmas) (Prince *et al.*, 1993).
In the past 15 years, in Veneto region (northeastern Italy), a severe outbreak of Flavescence dorée has been observed (Belli *et al.*, 1983) and heavy crop losses were recorded. Therefore an extensive study was carry out in this and other areas of northen Italy in order to investigate on the presence and the distribution of the phytoplasmas involved with the diseases.

2. Materials and methods

Samples from sixty-one naturally diseased grapevines exhibiting symptoms typical of FD and from symptomless grapevines were collected, the beginning of September, in the provinces of Vicenza, Pavia, Bergamo and Siena. Aster yellows (AY) MLO strain AY1 was collected in a naturally infected plant of periwinkle (*Catharanthus roseus* (L.) G. Don in Beltsville, MD). EY 1 (from W.A. Sinclair) was the reference strain for 16SrV phytoplasma group. Diseased plants were maintained by grafting in a white flowering clone of periwinkle grown in a greenhouse. Healthy grapevine control plants were grown from seed in a greenhouse.
Total DNA from petioles and vein leaf samples were first amplified by PCR using oligonucleotide pairs r16SF2/r16SR2, which primes amplification of a 1.2 Kb ribosomal (r) DNA fragment from all known phytoplasmas (Lee *et al.*, 1993). Then 1μl of product was employed as template in a nested PCR using r RNA gene group-specific primer pairs designed (Lee *et al.*, 1994) for amplification of 16S rDNA from phytoplasmas affiliated

H.-W. Dehne et al. (eds.), Diagnosis and Identification of Plant Pathogens, 179–182.

with groups 16SrI (primer pair R16 [I] F1/R1) 16SrIII (primer pair R16 [III] F2/R1) and 16SrV (primer pair R16S [V] F1/R1) phytoplasmas.
RFLP analyses, using *Mse*I, were conducted by single enzyme digestions of 5 μl of products from PCR primed by R16[I]F1/R1 and R16[V]F1/R1 primer pairs. The fragments obtained were separated by electrophoresis through 5% polyacrylamide gel, stained in ethidium bromide and finally visualized with UV transilluminator.

3. Results

Polymerase chain reaction (PCR) tests. All the grapevine samples collected from symptomatic plants gave positive results when tested by use of PCR technique (TABLE 1), while no positive reaction was obtained from asymptomatic grapevines. In grapevine samples collected in Bergamo and Pavia provinces phytoplasmas belonging to group 16SrI were detected. Moreover the samples collected in Siena province were found to be infected by phytoplasmas in a 16SrI group and one sample was mixedly infected by phytoplasmas belonging to both 16SrI and 16SrV groups.
In 29 grapevine samples from vineyard located in Vicenza province phytoplasmas in a 16SrV group were detected; only 4 samples collected in this area, were infected by phytoplasmas belonging to 16SrI group. Negative results were obtained when primer pair R16 [III] F2/R1 were used with all cited grapevine samples.
RFLP analysis of 16S rDNA amplicons. RFLP profiles obtained from digestions by using of *Mse*I restriction enzyme of amplicons obtained by nested PCR with R16[I]F1/R1 primer pair, showed that phytoplasmas belonging to a group 16SrI subgroup G (Vibio *et al.*, 1995) were present in grapevine samples field collected in Bergamo, Pavia, Siena and Vicenza provinces (data not shown). In addition the digestions carried out on products of PCR with R16[V]F1/R1primer pair, indicate that tipical profiles of phytplasmas in a 16SrV group were obtained with *Alu*I and *Mse*I restriction enzymes (figure 1).

TABLE 1. Results of PCR tests on grapevine samples collected in four different areas.

Provinces.	n. of examined grapevines	AY	EY	AY/EY#
Bergamo	4	4	0	0
Pavia	20	20	0	0
Vicenza	33	4	27	2
Siena	4	3	0	1
total of samples	61	31	27	3

#: number of grapevine samples mixed infected by phytoplasmas belonging to groups 16SrI and 16SrV.

4. Discussion

In this communication we report the detection and the identification of two different phytoplasmas associated with grapevine yellows in northern Italy: the first one belonging to the 16SrV phytoplasma group, which contains the phytoplasma responsible of FD disease (Prince *et al.*, 1993,) and the second one belonging to the 16SrI-G phytoplasma subgroup, considered responsible of other yellows in Europe including Bois noir in France (Caudwell, 1993), GY in Italy (Bianco *et al.*, 1993, Davis *et al.*, 1993) and Vergilsbunkrankeit in Germany (Caudwell, 1993, Maixner *et al.*, 1995).
On the basis of our results FD disease is certanly present in Vicenza province; while it seems to be absent in Pavia province (the number of samples from Bergamo and Siena are too small for a conclusive remark). However it is not surprising the lack of FD diseased grapevine plants in Pavia and Bergamo provinces; in these areas in fact adequate chemical treatments are usually conducted in order to controle the presence of the vector *S. titanus* while in Vicenza province since 1985 (Belli *et al.*, 1985) a large presence of the cited insect were recorded due to the inadequacy of chemical controls. Moreover, in one of four grapevine samples from Siena province, we detected and identified a phytoplasma belonging to 16SrV group: in the same grapevine sample a 16SrI-G phytoplasma presence was thus revelead in mixed infection (TABLE 1).

5. References

Belli, G., Fortusini, A., Osler, R., and Amici, A. (1973) Presenza di una malattia del tipo "Flavescenza Dorée" in vigneti dell'Oltrepo' pavese, *Riv. Pat. Veg.* **Suppl. al vol. 9**, 51-56.

Belli, G., Fortusini, A., Rui, D., Pizzoli, L., and Torresin, G. (1983) Gravi danni da flavescenza dorata in vigneti di Pinot nel Veneto, *L'Informatore Agrario* **XXXIX** (7), 24231-24433.

Belli, G., Fortusini, A., Rui, D., Pizzoli, L., Torresin, G., Bianco, P.A., and Carraro, S. (1985) La flavescenza dorata della vite e le sue manifestazioni nel Veneto, Atti Convegno "Problemi attuali di patologia viticola", Vicenza, 24 maggio 1985.

Bianco, P.A., Davis, R.E., Prince, J.P., Lee, I.M., Gundersen, D.E., Fortusini, A., and Belli, G. (1993) Double and single infections by aster yellows and elm yellows MLOs in grapevines with symptoms characteristic of Flavescence dorée, *Riv Pat. Veg.* **Ser. V, 3**, 69-82.

Caudwell, A. (1957) Deux années d'études sur la flavescence dorée, nouvelle maladie grave de la vigne, *Ann. Amélior. Plant (Paris)* **4**, 359-393.

Caudwell, A., Giannotti, J., Kuszala, C., and Larrue, J. (1971) Étude du role de particules de type "mycoplasme " dans l'étiologie de la Flavescence Dorée de la vigne. Examen cytologique des plantes malades et des cicadelles infectieuses, *Ann. Phytopathol.* **3**, (1), 107-123

Caudwell, A. (1993) Advance in Grapevine Yellows research since 1990, Extended abstracts 11th Meeting ICVG, Montreux Switzerland, 6-9 September 1993 (Federal Agricultural Research Station of Changins, CH-1260 Nyon, Switzerland), 79-83.

Daire, X, Clair, D., Larrue, L., Boudon-Padieu, E., and Caudwell, A. (1993a) Diversity among mycoplasma-like organism inducing grapevine yellows in France, *Vitis* **32**, 159-163.

Daire, X., Clair, D., Larrue, J., Boudon-Padieu, E., Alma, A., Arzone, A., Carraro, L., Osler, R., Refatti, E., Granata, G., Credi, R., Tanne, E., Pearson, R., and Caudwell, A. (1993b) Occurrence of diverse MLOs in tissues of grapevine affected by grapevine yellows in different countries, *Vitis* **32**, 247-248.

Davis, R.E., Dally, E.L., Bertaccini, A., Lee, I.M., Credi, R., Osler, R., Savino, V., Carraro, L., Di Terlizzi, B., and Barba, M. (1993) Restriction Fragment Length Polymorphism analyses and dot hybridizations distinguish mycoplasmalike organisms (MLOs) associated with Flavescence dorèe and Southern European Grapevine Yellows disease in Italy, *Phytopathology* **83**, 772-776.

Gartel, W. (1965) Untersuchungen Äber das Auftreten und das Verhalten der Flavescence dorée in den Weinbaugebieten an Mosel und Rhein, *Weinberg Keller* **12**, 347-376.

Lee, I.M., Hammond, R.W., Davis, R.E., and Gundersen, D.E. (1993) Universal amplification and analysis of pathogen 16S rDNA for classification and identification of Mycoplasmalike Organisms, *Phytopathology* **83**, 834-842.

Lee, I.M., Gundersen, D.E., Hammond, R.W., and Davis, R.E (1994) Use of mycoplasmalike organisms (MLO) group-specific oligonucleotide primers for nested-PCR assay to detect mixed-MLO infection in a single host plant, *Phytopathology* **84** (**6**), 559-566.

Magarey, P.A., and Wachtel, M.F. (1986) Grapevine yellows, a widespread,apparently new disease in Australia, *Plant Dis.* **7 0**, 694.

Maixner, M., Ahrens, U., and Seemüller, E. (1995) Detection of the German grapevine yellows (Vergilbungskrankheit) MLO in grapevine, alternative hosts and a vector by specific PCR procedure, *European Journal of Phytopathology* **1 0 1**, 241-250.

Pearson, R.C., Pool, R.M., Gonsalves, D., and Goffinet, M.C. (1985) Occurrence of flavescence dorée-like symptoms on "White Riesling" grapevines in New York, U.S.A., *Phytopathol. Mediterr.* **2 4**, 82-87

Prince, J.P., Davis, R.E., Wolf, T. K., Lee, I.M., Mogen, B.D., Dally, E.L., Bertaccini, A., Credi, R., and Barba, M. (1993) Molecular detection of diverse mycoplasmalike organisms (MLOs) associated with grapevine yellows and their classification with aster yellows, X-disease, and elm yellows MLOs, *Phytopathology* **8 3**, 1130-1137.

Quaroni, S., Saracchi, M., Fortusini, A., and Belli, G. (1988) Osservazioni mediante microscopia elettronica a scansione su viti affette da Flavescenza dorata., *Riv. Pat. Veg.* **Ser. IV, 2 4**, 71-79.

Vibio, M., Bertaccini, A., Lee, I.M., Davis, R.E., and Clark, M.F. (1994) Characterization of aster yellows and related european mycoplasmalike organisms maintained in periwinkle plants and shoot-tip culture, *I.O.M. Letters* **3**, 297-298.

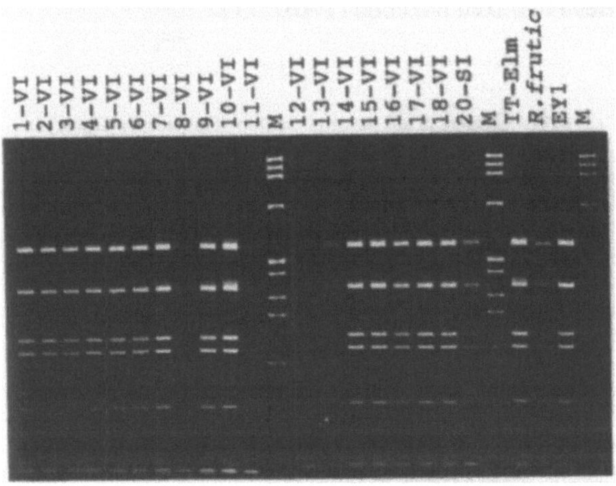

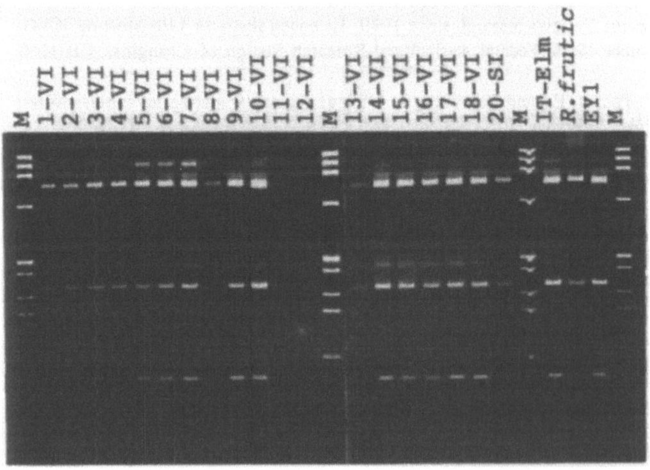

Figure 1. Restriction fragment lengh polymorphism (RFLP) analysis of 16S rDNA amplified from DNA templates of 19 different diseased grapevine plants (n. 1 to n.18 from Vicenza province and n. 20 from Siena province) by using 16S rRNA group V-specific primer pair R16 [V]F1/R1 in nested-PCR. DNA products were digested separately with restriction enzymes *Alu*I (lower) and *Mse*I (upper).

EY1, reference strains; *R. fruticosus*, field collected sample from spontaneous plants showing witches' broom symptoms in Pavia province; IT Elm, witches' broom diseased elm tree in Vicenza province; M, ΦX 174 RF1 DNA *Hae*III digest.

DIAGNOSIS OF VIROIDS IN FRUIT TREES: AN ITALIAN EXPERIENCE

F. FAGGIOLI, S. LORETI, M. BARBA
Istituto Sperimentale per la Patologia Vegetale - MiRAAF
Via C.G. Bertero 22, 00156 Rome, Italy

Abstract

In this paper, we report the results of comparison of different diagnostic techniques, such as bidirectional electrophoresis (dPAGE) and molecular hybridization, with radioactive and non radioactive probes, to detect the most important viroids in infected fruit tree matrices. The efficacy of the extraction method by chromatography column technique is compared with two other alternative ones.

The use of a rapid extraction method, in combination with digoxigenin labelled probes in dot or better, when it is possible (PLMVd), tissue blot molecular hybridization, is the best means in the viroids diagnosis in particular for ascertaining the sanitary status of fruit tree propagative material.

Introduction

Viroids are unencapsidated, single stranded circular RNA molecules with a size range between 246 and 345 nucleotide residues. Their diagnosis is mainly based on biological and molecular methods, as they are non-immunogenic pathogens.

In Europe, the most important are:

- apple scar skin viroid (ASSVd), present in the apple cultivars introduced from Japan;
- hop stunt viroid (HSVd), associated to different fruit tree species;
- peach latent mosaic viroid (PLMVd), occurring in high percentage in peach material coming from U.S.A. and Japan;
- pear blister canker viroid (PBCVd), infecting about 10% of European pear cultivars.

In order to minimize the risk of their spreading using infected plant germplasm, the protocols for the production of Certified Propagative Material foresee the absence, among other systemic pathogens, of viroids, too.

Yet, the use of reliable and rapid diagnostic methods is the prerequisite to guarantee the sanitary status of planting material and to improve knowledges on their molecular characterization, epidemiology and eradication.

In this paper, we report the results of comparison of different diagnostic techniques, such as bidirectional electrophoresis (dPAGE) and molecular hybridization, with radioactive and non radioactive probes, to detect viroids in different matrices of fruit trees using different extraction methods.

Moreover, the best diagnostic techniques were applied to screening pear and

H.-W. Dehne et al. (eds.), Diagnosis and Identification of Plant Pathogens, 183–187.

peach field material, for the presence of PLMVd and PBCVd.

Materials and methods

Source of material

Leaves and bark were collected from infected pear, apple and peach trees. Fruits were collected only from infected peaches.

ASSVd isolate PK13, PBCVd isolate P2098T, PLMVd isolate D-168 and HSVd isolate Tanyo were used as positive controls. Healthy 'A20' pear, GF305 peach and 'Virginia crab' apple were used as negative controls.

Nucleic acid extraction methods

Three different extraction methods were compared: **I)** partial purified viroids were obtained by cromatography column technique (CF11) according to the protocol of Flores and Llàcer (1988); **II)** total nucleic acids (TNA) extraction was performed by the method described by Ambròs *et. al* (1995a); **III)** TNA were extracted according to the method described by Loreti *et al* (1995).

Detection methods

Polyacrylamide gel electrophoresis (dPAGE)
Aliquots of TNA and purified viroid RNA were analyzed by bidirectional electrophoresis in 5% and 7% polyacrylamide slab gels, under non denaturing (first run) and denaturing (second run) conditions, respectively (Schumacher *et al.*, 1983; Flores *et al.*, 1985) and stained with silver nitrate.

Dot and Tissue blot hybridization.
Nucleic acid extracts were applied on nylon membranes pre-equilibrated with 20 X SSC (3M NaCl, 0.3 M tri-sodium citrate, pH 7) and fixed by UV irradiation. Alternatively, leaf and bark of peach and pear, freshly cut surface, were carefully pressed onto nylon membrane.

Hybridization assays were conducted as reported by Ambròs *et al.* (1995a), using four recombinant plasmids containing PBCVd, PLMVd, ASSVd and HSVd cDNA inserts.

Radioactive membranes were exposed at room temperature with intensifying screens, while digoxigenin labelled (DIG) ones were treated for chemioluminescent detection as reccommended by the supplier (Boehringer, DIG Luminescent Detection Kit for Nucleic Acids).

Reverse trascription and Polymerase chain reaction (RT- PCR).
TNA extracts were amplified (dilution 1:50) in RT-PCR, according to GeneAmp RNA PCR Kit instructions (Perkin Elmer Cetus) apporting some modifications in PCR cycles (Ambròs *et al.*, 1995b). PLMVd and PBCVd RNAs were amplified using the

following primers: - PLMVd, primer 3' complementary to nt 131 to 166 and primer 5' homologous to nt 167 to 202; - PBCVd, primer 3' complementary to nt 102 to 116 and primer 5' homologous to nt 117 to141.

Results

NUCLEIC ACID EXTRACTION METHODS

The two rapid extraction methods (II and III) as well as the cellulose chromathography extraction technique were effective in obtaining non-degraded viroid-RNA. The PBCVd, ASSVd and HSVd - RNAs quantity was similar no matter the starting infected matrix. Higher signal was observed when PLMVd RNA was extracted from fruits. Method II reduced the background signal when the viroid RNA was extracted from pear tissue.

DETECTION METHODS

No all infected samples of peach and pear were revealed in dPAGE (Tab. 1).

Probes, either radioactive (^{32}P) or chemically (digoxigenin) labelled, reacted with homologous viroid RNA in dot blot hybridization. Tissue blot hybridization was able to reveal the presence of PLMVd but not of PBCVd (Fig.1).

No reactions were observed in the healthy controls either in dot or tissue blot molecular hybridization assays.

Dot or tissue blot DIG hybridizations resulted reliable, respectively, for detecting PBCVd and PLMVd in field screening (Fig.2).

The primers were effective, in so far as products of expected size (336 nt for PLMVd, 315 nt for PBCVd) were amplified by RT-PCR from PLMVd and PBCVd TNA extracts. No bands were observed in the healthy and white controls (Fig.3).

Discussion

On the base of these results, it is possible to make the following considerations:

- the three compared extraction methods resulted effective. Among them, the two rapid ones have many advantages in comparison with cromatography column technique (CF11), such as the small amount of starting plant tissue, the rapidity and the possibility to run all extraction steps in two ml Eppendorf tubes. This is very important in massive diagnosis of viroids, because it is possible to analyze a large number of samples in few hours, starting from different matrices;
- diagnosis of viroids was achieved also using non radioactive labelled probes of PLMVd, PBCVd, HSVd and ASSVd in dot and, only for PLMVd and PBCVd, in tissue blot hybridizations. The positive results obtained during our activity strongly suggest the possibility to use digoxigenin labelled probe instead of radioactive labeled ones. In fact the former is more handy than the latter, without loosing sensivity;
- tissue blot hybridization resulted very effective in detecting, in short time, PLMVd in peach germplasm, avoiding the extraction step and the use of organic solvents;
- RT-PCR was confirmed as a highly sensitive and reliable method for diagnosis of viroids but seems less usefull in a massive screening work.

In conclusion, the use of a rapid extraction method, in combination with digoxigenin labelled probes in dot or better, when it is possible (PLMVd), tissue blot molecular hybridization, is the best means in the viroids diagnosis in particular for a preliminary screening for ascertaining the sanitary status of fruit tree propagative material.

Tab. 1.Comparison on the efficacy of different detection methods

	dPAGE	^{32}P Probe		Dig Probe		RT-PCR
		Dot	Tissue	Dot	Tissue	
PLMVd	15/20*	20/20	20/20	20/20	20/20	2/2
PBCVd	10/20	20/20	0/20	20/20	0/20	2/2
ASSVd	4/4	4/4	N.T.	4/4	N.T.	N.T.
HSVd	3/3	3/3	N.T.	3/3	N.T	N.T

* positive results/infected samples. N.T. = not tested

References

Ambròs S., Desvignes J.C., Llàcer G. and Flores R. (1995a.) Peach latent mosaic viroid and Pear blister canker viroid: detection by molecular hybridization and relationship with specific maladies affecting peach and pear trees, *Acta Horticolture* **386,** 515-521.

Ambròs S., Desvignes J.C., Llàcer G. and Flores R. (1995b) Pear blister canker viroid: sequence variability and casual role in pear blister canker disease, *Journal of General Virolology* **76**, 2625-2629.

Flores R., Duran-Vila N., Pallas V. and Semancik J.S. (1985) Detection of viroid and viroid-like RNAs from grapevines, *Journal of General Virolology* **66**, 2095-2102.

Flores R. and Llàcer G, (1988) Isolation of a viroid-like RNAs from grapevine, *Journal of General Virology* **66**, 2095-2102.

Loreti S., Faggioli F. and Barba M. (1995) A rapid extraction method to detect peach latent mosaic viroid by molecular hybridization, *Acta Horticulturae* **386**, 560-564.

Schumacher J., Randler J. W. and Riesner D. (1983) A two dimensional electrophoretic technique for the detection of circular viroids and virusoids, *Analytical Biochemestry* **135**, 288-295

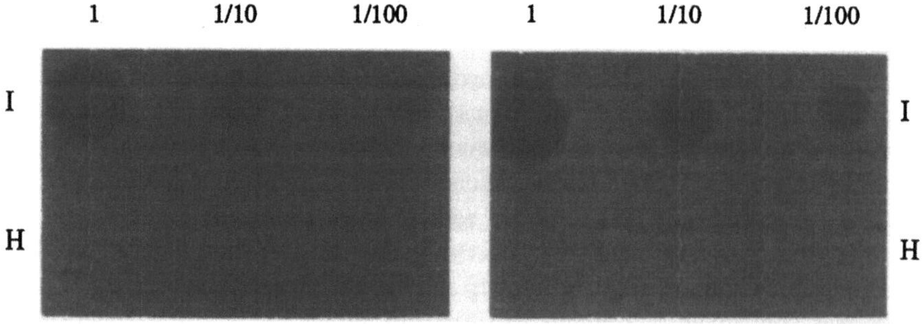

Fig. 1. Autoradiograph of dot blot hybridization analysis with digoxigenin (A) and ^{32}P (B) labelled cRNA probes of PLMVd at different diluitions. H = healthy control ; I = infected control

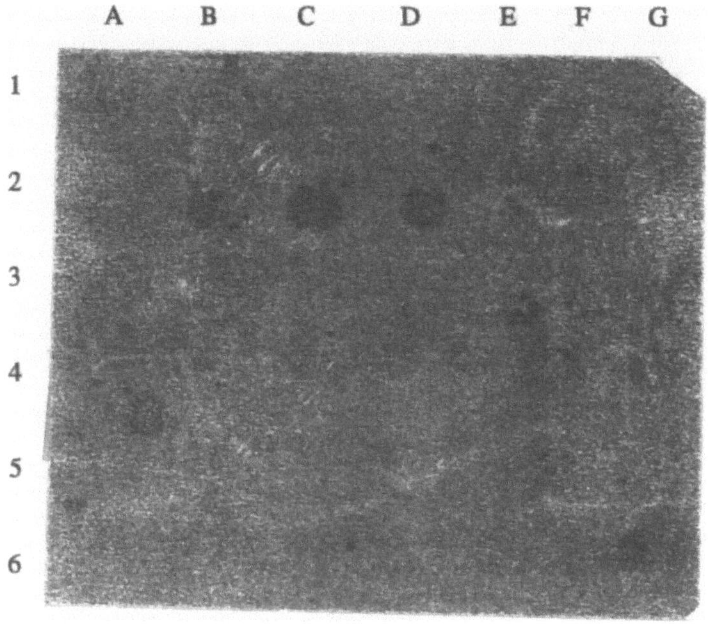

Fig. 2. Tissue blot of 42 peach leaves that were hybridazed with digoxigenin labelled cRNA of PLMVd. Position (G-6) and (G-5) were positive and negative control samples, respectively.

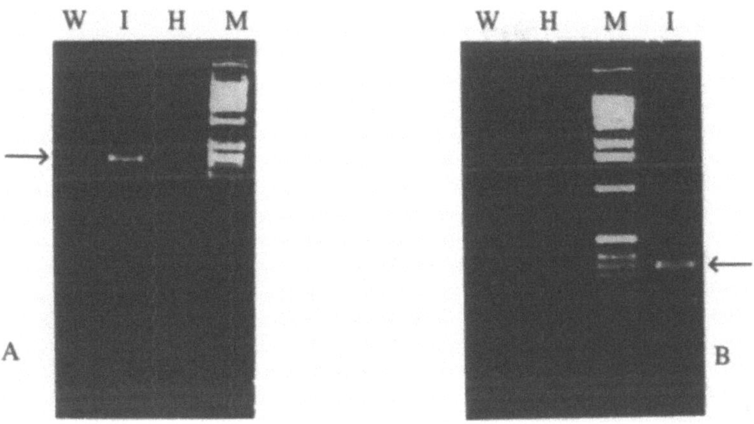

Fig. 3. Electrophoretic analysis of PBCVd (A) and PLMVd (B) RT-PCR amplified products from TNA extracts. Arrows indicate 315 bp (PBCVd) and 336 bp (PLMVd).bands respectively.
H = healthy control I = infected control W = water control M = 1Kb marker

DETECTION OF PLANT PATHOGENIC BACTERIA IN ORNAMENTALS BY THE PCR TECHNIQUE

MANULIS SHULAMIT, KOGAN NINA, VALINSKY LEA, LEVI MICHAL AND WIEMAN LIORA
Deprtment of Plant Pathology, ARO, The Volcani Center, Bet Dagan 50250, Israel

Introduction

In ornamentals the use of culture indexing to obtain disease-free propagation material is crucial. Therefore sensitive, rapid and reliable methods are necessary. Two examples of using the PCR method for detection of bacteria causing diseases in ornamentals will be presented.

The first is the disease caused by the bacterium *Erwinia herbicola* pv. *gypsophilae (Ehg).* This is the most destructive pathogen of *Gypsophila paniculata,* an ornametal used for cutflower production. In Israel it is considered as one of the main flowers for export. The pathogen induces gall formation at wound sites mainley in the crown region of the stem. The main demage occurs during the development of rooted cuttings in the nurseries. Neither resistant clones nor effective chemical treatment are available, and the only control measure is the production of pathogen-free cuttings through culture indexing.

Identification of the pathogen in Israel is based on the ELISA procedure. However, problems related to the specificity of the antiserum have been encountered due to the presence of more than one serotype. Moreover, saprophytic strains of *E. herbicola,* which are widespread on plants, can not be distinguished from pathogenic strains by the serological methods. Often pathogenicity tests are necessary to confirm the presence of pathogenic strains whenever a positive result is obtained by the ELISA procedure. Therefore it was necessary to develop a molecular procedure for identification of this pathogen. From previous work we found that all pathogenic strains of *E. herbicola* harbor a unique plasmid of about 150 kb which is absent in non-pathogenic strains (Manulis, *et al.* 1991). This plasmid may vary in size in different serotypes. Since this plasmid is associated with pathogenicity, genes residing on it may be used as candidates for molecular detection of the pathogenic strains. From this plasmid we cloned the genes necessary for the production of the phytohormones: indole acetic acid (IAA) and cytokinins (Clark, *et al.* 1993, Lichter, *et al.* 1995).

The genes for IAA and cytokinins biosynthesis form a cluster on the pathogenicity plasmid. A DNA fragment containing these genes was used by us as a DNA probe in colony hybridization (Manulis, 1992). This probe could distinguished between pathogenic and non pathogenic strains. However, in order to increase the sensitivity

H.-W. Dehne et al. (eds.), Diagnosis and Identification of Plant Pathogens, 189–192.

we used the PCR technique which is also more simple to performed than the hybridization.
The second example of using the PCR technique in ornamental is the bacterial wilt of geranium caused by *Xanthomonas campestries* pv. *pelargonii (Xcp)*. This is the most important disease of this ornamental. Symptomes associated with this disease include wilting of the plant, localized necrotic lesions on the leaves and stem rot. The disease is widespread in the USA, Europe and Israel and may cause heavy losses, particularly where geranium cuttings are propagated in a large scale.
Diagnosis of this pathogen is based on attempts to isolate the bacteria from plants in culture and by serological methods *e.g.* ELISA or immunofluorescence (IF). Although the diagnostic procedures are generally satisfactory, false positive results often occur. Since kowledge of the molecular biology of this pathogen is very scanty the RAPD-PCR technique was chosen as the method to evaluate variation among strains (Manulis, *et al.* 1994).

Results and Discussion

1. Detection of *Erwinia herbicola* pv. *gypsophila*

For detection of *Ehg* we designed primers according to the sequence of the IAA or the cytokinins genes. From the cytokinins genes two pairs of primers (one nested to the other) were chosen. The sizes of the fragments obtained in the PCR reactions were between 600 bp to 400 bp. The specificity of the primers for *Ehg* were examined with other gall forming bacteria e.g. *Agrobacterium tumefacience* and *Pseudomonas savastanoi* and with saprophytic strains of *E. herbicola*. Using these 3 pairs of primers no PCR product was obtained with the other gall forming bacteria or with the saprophytic strains.
We examined the sensitivity of the PCR reaction using the cytokinin drived primers with different concentrations of the pathogen. Ten-fold dilutions of the bacteria starting with 500 cells were made. Using the first pair of primers in the first round of PCR we could detect about 5 cells. Nested PCR carried out with the second pair of primers increased the sensitivity by 100 fold. Addition of plant extract reduced the sensitivity by 100 fold due to the presence of inhibitors in the extract. To overcome the inhibitory effect of plant material on the PCR reaction the bio-PCR method was employed. This method included an enrichment step by growing the bacteria on a semi-selective media (Manulis, *et al.* 1991) and then performing the first round of PCR. By this procedure it was possible to overcome the inhibitory effect of plant material however, it takes 2 days to grow the bacteria on plates. Detection of the pathogen by this procedure was slightly less sensitive than nested-PCR but still very sensitive for parctical purpose. After 7 days no disease symptomes are visible (Table 1). However, it is possible to detect the bacteria in the plants by either n-PCR or bio-PCR. After 14 or 21 days galls are formed at 10^4 cells/ml, but with PCR it is possible to detect pathogen concentrations that do not cause disease symptomes (Table 1).

The conclusions of this part are: 1) by using the PCR technique it is possible to detect the bacteria at a very low level and 2) bio-PCR exhibits the same sensitivity as n-PCR and it is the prefarable method since it is easier to perform.

2. Detection of *Xanthomonas campestris* pv. *pelargonii*

The random amplified polymorphic DNA (RAPD) was used for obtaining PCR primers and probes for diagnosis of the pathogen (Manulis, *et al.* 1994). We isolated a 1.2 kb fragment which was found to be specific for *Xcp*. This amplified fragment was cloned and sequenced. Based on the sequence 2 oligonucleotides were synthesized and used as specific primers for detection of *Xcp* by PCR. Amplification of *Xcp* DNA produced one major band. No amplification was observed with any other *X. campestris* strains tested.

Table 2 summarizes the results obtained with inoculated geranium plants by 4 methods: bio-PCR, n-PCR, IF and plating. The bio-PCR was slightly better than the n-PCR and the same as IF. The advantage of the PCR over IF is its specificity.

The PCR procedure for these two diseases are now being used routinely in Israel for confirmation of positive results obtained by the serological methods.

Table1 : Detection of *Erwinia herbicola pv. gypsophilae* in inoculated gypsophila cuttings by PCR

cells/ml	galls[a]	PCR I	n-PCR	bio-PCR	Days
10^5	–	– +	+	+	7
10^4	–	– +	+	+	
10^3	–	–	+	–	
10^2	–	–	–	–	
10^5	+	+	+	+	14
10^4	+	+	+	+	
10^3	–	–	+	+	
10^2	–	–	– +	– +	
10^5	+	+	+	+	21
10^4	+	+	+	+	
10^3	–	–	+	+	
10^2	–	–	+	+	

[a] Pathogenicity tests were conducted by dipping cuttings of *Gypsophila paniculata* into bacterial cell suspension.

+ positive result, – negative result, – + ambiguous result

Table 2: Detection of *Xanthomonas campestris pv. pelargonii* in inoculated geranium plants by 4 methods

Cells/Plant[a]	bio-PCR	n-PCR	IF	Plating	Days
1	–	–	–	–	
10	+	– +	+	–	7
100	+	+	+	+	
1	–	–	–	–	
1	+	+	+	–	14
100	+	+	+	+	
1	+	–	+	–	
10	+	+	+	– +	21
100	+	+	+	+	

[a] Inoculations were performed by puncturing the stem 6 times with a needle dipped in various concentrations of bacterial suspension.

+ positive result, – negative result, – + ambiguous result

References

Clark, E., Manulis, S., Ophir, Y., Barash, I. and Gafni, Y. (1993) Cloning and characterization of *iaa*M and *iaa*H from *Erwinia herbicola* pv. *gypsophilae*, *Phytopathology* **83**, 234-240.

Lichter, A., Barash, I., Valinsky, L. and Manulis, s. (1995) The genes involved in cytokinin biosynthesis in *Erwinia herbicola* pv. *gypsophilae*: characterization and role in gall formation, *J. Bacteriology* **77**, 4457-4465.

Manulis, S., Gafni, Y., Clark, E., Zutra, D. Ophir, Y. and Barash, I. (1991) Identification of a plasmid DNA probe for detection of strains of *Erwinia herbicola* paghogenic on *Gypsophila paniculata*, *Phytopathology* **81**, 54-57.

Manulis, S. (1992) Evaluation of a DNA probe for the detection of *Erwinia herbicola* strains pathogenic on *Gypsophila paniculata*, *Plant Pathology* **41**, 342-347.

Manulis, S., Valinsky, L., Lichter, A. and Gabriel, D.W. (1994) Sensitive and specific detection of *Xanthomonas campestris* pv. *pelargonii* with DNA primers and probes identified by Random Amplified Polymorphic DNA analysis, *Applied and Environmental Microbiology* **60**, 4094-4099.

DIFFERENTIATION AND DETECTION OF THE SUBSPECIES OF *CLAVIBACTER MICHIGANENSIS* BY PCR (POLYMERASE CHAIN REACTION) - TECHNIQUES

KARL-HEINZ PASTRIK [1]AND FRED RAINEY [2]
[1] *Pflanzenschutzamt Hannover, Germany,* [2] *DSM Braunschweig, Germany*

The five phytopathogenic subspecies of *Clavibacter michiganensis* (*C. m. sepedonicus, C. m. michiganensis, C. m. insidiosus, C. m. tessellarius, C. m. nebraskensis*) are causal agents of various diseases in agriculture. For the detection of each of these five subspecies a specific polymerase chain reaction (PCR)-assay was established. PCR primers were selected targeting genomic DNA. To obtain these primers, the spacer region between the 16S and 23S rRNA genes of these five subspecies was amplified and sequenced. On the basis of these sequence data primer pairs for each of the five subspecies were designed. The primer pairs (PSA 1-5) amplified a specific DNA fragment respectively only at that subspecies they were designed for and not at other related subspecies. The size of the amplified specific DNA fragment was for *C. m. sepedonicus* 502 bp (PSA-1), for *C. m. michiganensis* 270 bp (PSA-2), for *C. m. insidiosus* 393 bp (PSA-3), for *C. m. tessellarius* 587 bp (PSA-4) and for *C. m. nebraskensis* 393 bp (PSA-S).

To differentiate isolates of the five subspecies of *Clavibacter michiganensis* the RAPD-PCR (Random Amplified Polymorphic DNA -PCR) technique was used. In this technique, DNA fragments are amplified using single random primers of arbitrary sequence. Reaction conditions for the RAPD-PCR must be very carefully optimised to obtain informative and reproducible results. Once standardised, the technique is easily amenable to routine testing due to its speed and simplicity. Eighty different 10mer primers were tested for their ability to differentiate between the five subspecies. Due to the high GC-content of the *Clavibacter michiganensis* genom (>70 %) only primers with a GC-content of 80% produced reproducible informative patterns of amplified DNA fragments. The reproducibility of the RAPD patterns were examined using different template DNA concentrations in the PCR. Primers were found, which differentiate the five subspecies of *Clavibacter michiganensis* by RAPD patterns and subspecies specific DNA fragments. The usefulness of a PCR assay based on a cloned subspecies specific RAPD-DNA fragment is under investigation.

H.-W. Dehne et al. (eds.), Diagnosis and Identification of Plant Pathogens, 193.

Molecular diagnosis of rape seed pathogens

An experimental strategy for the development of taxon-specific genetic markers

K. VOIGT, S. SCHLEIER, J. WÖSTEMEYER
Friedrich-Schiller-Universität Jena, Institut für Mikrobiologie, Lehrstuhl für Allgemeine Mikrobiologie und Mikrobengenetik, Neugasse 24, D-07743 Jena, Germany

1. Detection of plant pathogens without the need for cultivation

Conventional methods for the precise determination of pathogenic fungi depend largely on morphological criteria and thus require axenic cultures. This process is time-consuming and needs appreciable mycological expertise. The analysis of mixed infections is especially complicated. The identification of fungi is highly facilitated by exploiting molecular markers directly at the DNA level. This approach does not depend on the development of micro-morphological characteristics of the fungi. For this purpose Restriction Fragment Length Polymorphism (RFLP)-analysis, various conventional Polymerase Chain Reaction (PCR)-approaches including Random Amplification of Polymorphic DNA (RAPD)-PCR has been proposed (Hassan *et al.*, 1991; Voigt *et al.*, 1995). All of these methods have their specific drawbacks. RFLP analysis depends on considerable amounts of purified, restriction grade DNA. Conventional PCR works only within highly conserved genomic regions, which may not offer enough variability for all differential purposes. RAPD-PCR has certainly enough resolution but needs an exact control of the reaction conditions for reproducible results. Here we propose a generally applicable experimental approach that makes use of RAPD-PCR for generating taxon-specific probes. In a second step, these probes are used as the basis for synthesis of highly specific PCR primer pairs that can be employed under conventional reaction conditions. This strategy combines the high resolution of RAPD with the ease of applicability and reproducibility of PCR diagnosis. Even mixed infections can be identified within a single working day with simple plant extracts as starting material. Plants can be analysed at early stages of infections, before manifestation of visible symptoms.

H.-W. Dehne et al. (eds.), Diagnosis and Identification of Plant Pathogens, 195–198.

2. Experimental procedure

The experimental layout for obtaining pathogen-specific PCR primer pairs can be summarized by the stepwise scheme outlined in *Figure 1*.

1. Generation of polymorphic DNA-markers by Random Amplification of Polymorphic DNA (RAPD)-PCR

↓

2. Specificity test of the purified RAPD-markers in Southern hybridization experiments

↓

3. Sequencing of the RAPD-fragments with diagnostic value

↓

4. Design of PCR primers of approximately 20 nucleotides length

↓

5. Conventional Polymerase Chain Reaction assay

Figure 1. Experimental procedure for obtaining pathogen-specific PCR primer pairs.

In a first step fungal DNAs are amplified by a collection of random primers. Bands that distinguish the fungus in question from related pathogens and from different pathogens that may occur in mixed infections are identified by gel electrophoresis and eluted for the further screening procedure.
In a second step the eluted polymorphic bands are labeled by multipriming (Feinberg and Vogelstein, 1983) and used as hybridization probes in Southern blot experiments against the unrelated fungi which may occur on the same plant. Molecular probes that hybridize with all fungal isolates but fail to do so with unrelated fungi are selected. RFLPs that may be observed during this second selection cycle are not important for the further procedure.
In the third step bands with the desired diagnostic properties are sequenced. Because the DNA-fragments were generated by short decamer primers, which are not generally suited as sequencing primers, this step is not necessarily trivial. Although cycle sequencing may be successful, we prefer to use a procedure of general applicability which is based on chemical sequencing (Voigt and Wöstemeyer, 1995). The DNA-fragments are end-labeled by re-amplification with the appropriate 5′-^{32}P-RAPD primers, and the termini are separated by digestion with restriction endonucleases cutting the particular RAPD-fragment into two sub-fragments of approximately equal molecular size. Conventional chemical sequencing (Maxam and Gilbert, 1977) yields enough sequence information for constructing longer primers. If more sequence data are desired, sequencing according to the chain termination method (Sanger *et al.*, 1977) is now considerably facilitated.
In step 4 longer PCR primers are synthesized based on the sequence information. These primers are now checked for reliability and specificity with an appropriate number of fungal isolates.

3. Molecular dissection of fungi growing on oilseed rape (*Brassica napus*)

The value of our experimental setup was demonstrated by the analysis of a selection of fungi parasitizing oilseed rape. Probably, the most important pathogen is the causative agent of blackleg, *Leptosphaeria maculans* (imperfect form: *Phoma lingam*). This pathogen has various pathotypes which can be ascribed to two major pathotype groups, designated aggressive (A) and non-aggressive (NA; Hoppe, personal communication; Koch *et al.*, 1989). As only A-isolates cause severe systemic infections on *Brassica napus*, the identification of A and NA strains is highly desirable. For both basic genotypes of *Leptosphaeria maculans* we have developed highly reliable PCR primer pairs using the sequence data of a 350 bp and a 615 bp RAPD-fragment specific for the NA and the A pathotype group, respectively (Schäfer and Wöstemeyer, 1992). The method was checked for 107 strains (A-, NA- as well as non-identified strains) of *Leptosphaeria maculans*. NA-strains are characterized by a 100 bp-fragment generated by the primer pair NA1-3 / NA2-1

5′-TCGAAACCTAGAGTGTTCCAAT-′3 / 5′-TGGCAGTTATTATGGTGGAT-3`, A-strains by 300 bp-fragment amplified with the primer pair A1 / A3 5′-AGGTGCCCAGCATGTGAAGA-3` / 5`-AATGGGATGGTGAGATCTCT-3′. The method was successfully employed for the detection of *Leptosphaeria maculans* in freshly infected plants and in stubbles. For a collection of other fungi associated with *Brassica napus* we have identified species-specific hybridization probes along the proposed experimental outline: *Alternaria brassicae*, *Alternaria brassicicola*, *Cylindrosporium concentricum*, *Fusarium moniliforme*, *Rhizoctonia solani*, *Pythium* sp., *Sclerotinia sclerotiorum* and *Verticillium dahliae* (Schleier *et al*, 1996).

4. References

Feinberg, A.P. and Vogelstein, B. (1983) A technique for radiolabeling DNA restriction endonuclease fragments to high specific activity, *Anal. Biochem.* **132**, 6-13.

Hassan, A.K., Schulz, C., Sacristán, M.D., and Wöstemeyer, J. (1991) Biochemical and molecular tools for the differentiation of aggressive and non-aggressive isolates of the oilseed rape pathogen, *Phoma lingam*, *J. Phytopathology* **131**, 120-136.

Koch, E., Badawy, H.M.A., and Hoppe, H.H. (1989) Differences between aggressive and non-aggressive single spore lines of *Leptosphaeria maculans* in cultural characteristics and phytotoxin production, *J. Phytopathology* **124**, 52-62.

Maxam, A.M. and Gilbert, W. (1977) A new method for sequencing DNA, *Proc. Natl. Acad. Sci. USA* **74**, 560-564.

Sanger, F., Nicklen, S., and Coulson, A.R. (1977) DNA sequencing with chain-terminating inhibitors, *Proc. Natl. Acad. Sci. USA* **74**, 5463-5467.

Schäfer, C. and Wöstemeyer, J. (1992) Random primer dependent PCR differentiates aggressive from non-aggressive isolates of the oilseed rape pathogen *Phoma lingam* (*Leptosphaeria maculans*), *J. Phytopathology* **136**, 124-136.

Schleier, S., Voigt, K., and Wöstemeyer, J. (1996) Molecular diagnosis of mixed fungal infections in the blackleg complex of oilseed rape (*Brassica napus*): evidence for genus- and species-specific sequences in the fungal genomes, *J. Phytopathology*: submitted.

Voigt, K., Schleier, S., and Brückner, B. (1995) Genetic variability in *Gibberella fujikuroi* and some related species of the genus *Fusarium* based on random amplification of polymorphic DNA (RAPD), *Curr. Genet.* **27**, 528-535.

Voigt, K. and Wöstemeyer, J. (1995) The combination of Gilbert/Maxam chemical sequencing and the dideoxynucleotide chain termination approach facilitates the construction of species-specific PCR-primers based on diagnostic RAPD-bands, *Microbiol. Res.* **150**, 373-377.

IDENTIFICATION OF SPACE-SPECIFIC MARKERS OF *PHYTOPHTHORA FRAGARIAE* WITH AFLP

P.J.M. Bonants[1]*, M. Hagenaar-de Weert[1], D.E.L. Cooke[2] and J.M. Duncan[2]

[1] Institute for Plant Protection (IPO-DLO), PO Box 9060, 6700 GW Wageningen, The Netherlands
[2] Scottish Crop Research Institute (SCRI), Invergowrie, Dundee DD2 5DA, United Kingdom
*to whom correspondence should be addressed

1. Introduction

AFLP stands for Amplified Fragment Lenght Polymorphism and has been developed by Keygene (Wageningen, The Netherlands) [1,2]. AFLP is a new DNA fingerprint technique which is based on the selective amplification of genomic restriction fragments and which gives more information than the conventional fingerprint techniques, like RFLP and RAPD.

The procedure of this new technique for fungi is as follows:

- DNA is isolated from mycelium of the fungus.
- DNA is digested with two restriction enzymes (a frekwent (Mse1) and a rare (EcoR1) cutter).
- At the ends of these restriction fragments adapters will be ligated.
- Next, these fragments will be amplified in a PCR with the so-called 0-primers, the non-selective amplification. This results in the amplification of all restriction fragments.
- The next step is a selective PCR with primers which have one or more extra bases at their 3'-end compared with the 0-primers. One of the selective primers is labeled radioactively with ^{33}P. This results in the selective amplification of a subset of all restriction fragments.
- After PCR the products will be seperated on a denaturing gel.
- After drying of the gel the DNA fingerprints will be recorded by a X-ray film.

The sequences of the two necessary AFLP-adapters is:

```
Mse1  : 5'-GACGATGAGTCCTGAG
                  TACTCAGGACTCAT-5'

EcoR1 : 5'-CTCGTAGACTGCGTACC
                  CTGACGCATGGTTAA-5'
```


H.-W. Dehne et al. (eds.), Diagnosis and Identification of Plant Pathogens, 199–201.

Some examples of AFLP-primers are shown below :

M00 :	5'-GAT GAG TCC TGA GTA A	0
M01 :	5'-GAT GAG TCC TGA GTA AA	+1
M12 :	5'-GAT GAG TCC TGA GTA AAC	+2
E00 :	5'-GAC TGC GTA CCA ATT C	0
E03 :	5'-GAC TGC GTA CCA ATT CG	+1
E17 :	5'-GAC TGC GTA CCA ATT CCG	+2

M00 and E00 are the 0-primers. A +1 primer has for example 1 selective base, e.g.. M01 and E00 an Adenine; a +2 primer has two and a +3 primer has three selective bases.
By varying the number of selective bases in the selective amplification, the AFLP-reaction, more or less PCR products obtained. In the +2/+2 combination there are far less bands than +1/+1 combination.

Phytophthora fragariae var. *fragaria* and *P. fragariae* var. *rubi* exist of several races, which are pathogenic to different cultivars of strawberry and raspberry, respectively [3,4,5,]. The AFLP fingerprinting technique is used to search for race-specific markers.

2. Results

DNA was isolated from a lot of *P. fragariae* var. *fragariae* and *P. fragariae* var. *rubi* isolates. AFLP DNA fingerprint patterns were obtained from ten races of *P. fragariae* var. *fragariae* with several +2/+2 primer combinations. Most of the fragments show up in all races. But some show up in one or more races. Many polymorphisms have been found. These specific bands can be cut of the gel and amplified again with the same primers.
Until now about some 150 bands were found showing polymorphism. Using the resistance data from van de Weg (CPRO-DLO, Wageningen, The Netherlands), who is breeding for resistance and has texted all fysio's against a set of different cultivars of strawberry, we found 10 bands showing some correlation with his data. van de Weg developed a gene for gene hypothesis and found in this way several resistance genes [6]. Those ten bands have been cloned and will be studied further to reveal if these fragments are indeed markers for race-specificity. Their sequence will be determined and race-specific primers will be developed.
The same approach has been used for the three races of *P. fragariae* var. *rubi*.

3. Conclusion

In conclusion we can say that AFLP is a powerful technique to make DNA fingerprints. Markers can be found for any feature, e.g. race, virulence, mating type, pathogenicity or hostrange. Because of the C/G content of a particular fungus, it may be necessary to use different restriction enzyme combinations and therefore also different adapters and primers have to be designed. At IPO-DLO different fungi, bacteria and nematodes are being studied with AFLP.

4. References

1. Zabeau, M. & Vos, P. (1993). Selective restriction fragment amplification : a general method for DNA fingerprinting. European Patent Office, publication number EP 0534858.
2. Vos, P., Hogers, R., Bleeker, M., Reijans, M., van de Lee, T., Hornes, M., Frijters, A., Pot, J., Peleman, J., Kuiper, M. Zabeau, M. (1995) AFLP : a new technique for DNA fingerprinting. *Nucleic Acids Research* **23**, 4407-4414.
3. Converse, R.H. & Scott, D.H. (1962) Physiologic specialization in *Phytophthora fragariae*. *Phytopathology* **52**, 802-807.
4. Converse, R.H. (1970) Occurrence of *Phytophthora fragariae* race A-10 in California. *Plant Disease Reptr.* **54**, 969-971.
5. Kennedy, D.M. & Duncan, J.M. (1993) Races in *Phytophthora fragariae* var. *rubi*. *Acta Horticulturae* **352**, 555-562.
6. van de Weg, W.E. (1996) A gene-for-gene model for the strawberry-*Phytophthora fragariae* var. *fragariae* pathosystem. Submitted for publication in *Theoretical and Applied Genetics.*

PHYSIOLOGICAL RACE DETERMINATION IN *FUSARIUM OXYSPORUM* F.SP. *DIANTHI* BY RAPD-PCR

Q. MIGHELI, E. BRIATORE, M. ANDRINA, and A. GARIBALDI
Di.Va.P.R.A. - Patologia vegetale, Università di Torino, Via P. Giuria 15, I-10126 Torino (Italy)

1. Introduction

Fusarium oxysporum f. sp. *dianthi* (*FOD*), incitant of vascular wilt, is the most important pathogen on carnation, where it causes severe losses in all growing areas. The use of resistant carnation cultivars still provides the most effective solution to this problem and indexing is routinely adopted for the obtainment of pathogen-free propagative material, (Baayen and Kleijn, 1989; Garibaldi and Gullino, 1987). Eight physiological races or pathotypes of *FOD* have been reported (Garibaldi, 1975), and knowledge of the distribution of pathotypes is a basic prerequisite for choosing of suitable resistant varieties in each growing area.

Distinction between saprophytic *Fusarium* spp. and *FOD* and among different pathotypes of *FOD* is based on pathogenicity tests, which are relatively long (30-40 days) and require extensive facilities. Moreover, susceptibility of differential cultivars can be strongly influenced by many environmental and technical variables as differences in the defence reaction appear to be largely quantitative (Ben-Yephet *et al.*, 1996).

DNA random amplification by the use of arbitrarily chosen primers (Welsh and McClelland, 1990; Williams *et al.*, 1990) proved very effective in the molecular characterization of several *formae speciales* of *F. oxysporum* (Grajal-Martìn *et al.*, 1993; Bentley *et al.*, 1994; Manulis *et al.*, 1994; Migheli and Cavallarin, 1994; Assigbetse *et al.* 1994). We report here the use of the RAPD analysis to distinguish among six physiological races of *FOD* within a collection of 76 isolates of different geographic origin.

2. Materials and methods

Seventy-six isolates of *FOD*, collected from different carnation-growing regions in Italy, Israel, The Netherlands, Japan, Spain, and France were maintained on potato dextrose agar (PDA, Merck) and carnation leaf agar (Fisher *et al.*, 1982).

DNA was purified with a miniprep method, adjusted to a final concentration of 100 ng per microliter in TE, pH 8.0 and stored at 4°C. PCR was carried out in 15 μl of reaction mix containing 10 mM Tris-HCl, pH 9.0, 1.5 mM $MgCl_2$, 50 mM KCl, 0.1% Triton X-100, 0.01% (w/v) gelatin, 60 μM each of dATP, dCTP, dGTP, and dTTP, 5 pM of primer, about 20 ng of template DNA, and 0.75 U of SuperTaq DNA polymerase (Stehelin). Eighty 10-mer oligonucleotide primers (series OPA, OPB, OPE, and OPM by Operon Technologies) were tested as primer sequences. Amplification was performed in a Perkin-Elmer (Norwalk) Cetus Gene Amp PCR System 9600

H.-W. Dehne et al. (eds.), Diagnosis and Identification of Plant Pathogens, 203–206.

programmed for one cycle of 2.5 min at 94°C, 45 cycles of 30 s at 94°C, 1 min at 36°C, 2 min at 72°C, with no ramping, followed by one cycle of 5 min at 72°C. Half of the reaction product was loaded in 1% SeaKem LE agarose (FMC BioProducts) gels containing 0.5 μg ml^{-1} ethidium bromide. Electrophoresis was performed for 15 h at 0.8 V cm^{-1} in 1 x TAE as running buffer and amplimers were directly viewed over a UV light source.

3. Results

RAPD patterns were obtained for the 76 isolates of *FOD* with 80 primers, 10 of which provided the most informative results, and a total of 74 polymorphic bands were chosen as markers. These bands were strongly amplified and reproducible by using different extraction protocols, different preparations of template DNA and different thermostable polymerases. The molecular size of amplimers ranged from 0.3 to 4 kb and all the primers tested revealed at least one polymorphic band which could be used to define homogeneous groups among the different isolates. An example of RAPD profile is given in Figure 1. Isolates of pathotype 1 and 8 shared identical RAPD profiles with all the 80 primers tested during this study. Isolate 4/325, previously assigned to pathotype 4, was grouped with members of pathotypes 1 and 8 based on RAPD profiles. Data obtained from a recent pathogenicity assay allowed us to reclassify this isolate as pathotype 8. Pathogenic isolates of race 2 formed an homogeneous group of 16 members, which comprised also isolates of pathotype 5 and 6. Representatives of pathotype 4 were grouped within a single group of 20 isolates according to RAPD profiles. Thirteen isolates of pathotypes 1, 2, 4, and 8, showing low or no pathogenicity on the tested carnation cultivars, shared very similar RAPD patterns and only isolate 2/1219 presented common RAPD markers with the nonpathogenic or slightly pathogenic isolates, although it was confirmed as a representative of pathotype 4 after pathogenicity test.

4. Discussion

The RAPD fingerprinting procedure was used in this study to characterize a representative collection of *FOD* isolates of different geographic origin. This technique offers several advantages over established analyses based on vegetative compatibility groups (VCG, Baayen and Kleijn, 1989; Katan *et al.*, 1989; Aloi and Baayen, 1993), DNA restriction fragment length polymorphisms (RFLP, Manicom and Baayen, 1993) and electrophoretic karyotype (EK) variability (Migheli *et al.*, 1995). Very little amounts of template DNA are required and results can be rapidly obtained, allowing a large number of isolates to be analysed at the same time.

The variability observed in RAPD patterns was largely correlated with pathotype specificity. Isolates assigned to pathotypes 1 and 8 shared the same RAPD patterns and were all clustered in the same group. This confirms that these two pathotypes are closely related, although differing from each other in virulence, and can be considered as near-isogenic (Waalwijk *et al.*, 1996). The same RAPD profiles were also shared by members of pathotypes 2, 5, and 6, in accordance to both VCG and RFLP grouping.

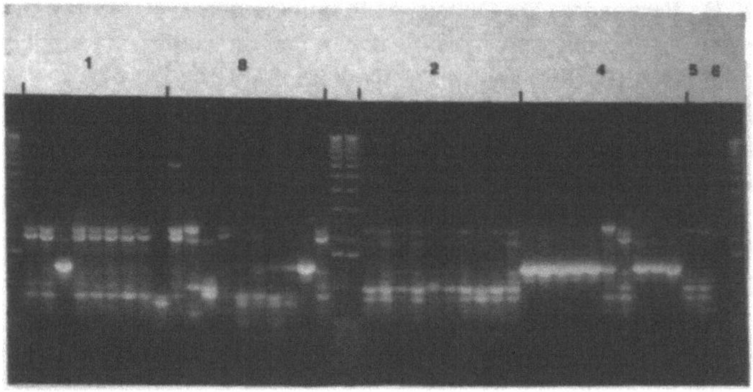

Figure 1. RAPD profiles of different *F. oxysporum* f.sp. *dianthi* isolates (from left: representatives of pathotypes 1, 8, 2, 4, 5 and 6) obtained with primer OPE-08 (5'-TCACCACGGT-3'). Molecular size marker: 1 kb ladder.

Pathogenic isolates of race 4 formed a very homogeneous group.
Interestingly, a correlation was found between the lack (or the low level) of pathogenicity presented by 13 isolates on the tested carnation cultivars and their RAPD profiles. This result appears even more striking as the correlation was maintained for representatives of different pathotypes isolated in Israel, Italy, Japan, Spain or The Netherlands.
No correlation was found between RAPD data and geographic origin of the tested isolates, as representatives of pathotype 2 which were isolated in Israel and Japan presented the same amplification profiles as other members of the same race isolated in Italy. This appears in contrast with results obtained in *F. oxysporum* f.sp. *vasinfectum*, where differentiation of isolates was directly related to both virulence and geographic origin (Assigbetse *et al.*, 1994).
The data presented here suggest that RAPD fingerprinting represents a powerful tool for the study of the population structure of *FOD*. The purification, cloning and sequencing of selected amplimers is now being performed and will allow the development of pathotype-specific probes and primers to be used in dot-blot or PCR-based diagnostic tools for use in a large-scale screening of pathogenic isolates.

Research supported by the Ministero delle Risorse Agricole, Alimentari e Forestali (M.I.R.A.A.F., Piano Nazionale "Biotecnologie vegetali", Area 10 - I diagnostici, Programma n° 451).

5. References

Aloi, C. and Baayen, R.P. (1993) Examination of the relationships between vegetative compatibility groups and races in *Fusarium oxysporum* f.sp. *dianthi*, *Plant Pathology* **42**, 839-850.

Assigbetse, K.B., Fernandez, D., Dubois, M.P. and Geiger, J.-P. (1994) Differentiation of *Fusarium oxysporum* f.sp. *vasinfectum* races on cotton by random amplified polymorphic DNA (RAPD) analysis, *Phytopathology* **84**, 622-626.

Baayen, R.P. and Kleijn, J. (1989) The *Elegans* fusaria causing wilt disease of carnation. II. Distinction of vegetative compatibility groups, *Netherlands Journal of Plant Pathology* **95**, 185-194.

Bentley, S., Pegg, K.G. and Dale, J.L. (1994) Optimization of RAPD-PCR fingerprinting to analyse genetic variation within populations of *Fusarium oxysporum* f.sp. *cubense*, *Journal of Phytopathology* **142**, 64-78.

Ben-Yephet, Y., Reuven, M., Zveibil, A. and Shtienberg, D. (1996) Effects of abiotic variables on the response of carnation cultivars to *Fusarium oxysporum* f.sp. *dianthi*, *Plant Pathology* **45**, 98-105.

Fisher, N.L., Burgess, L.W., Toussoun, T.A. and Nelson, P.E. (1982) Carnation leaves as a substrate and for preserving cultures of *Fusarium* species, *Phytopathology* **72**, 151-153.

Garibaldi, A. (1975) Race differentiation in *Fusarium oxysporum* f.sp. *dianthi* (Prill. *et* Del.) Snyd. *et* Hans., *Mededelingen van de Fakulteit der Landbouwwetenschappen, Rijksuniversiteit Gent* **40**, 531-537.

Garibaldi, A., Gullino, M.L. (1987) Fusarium wilt of carnation: present situation, problems and perspectives, *Acta Horticulturae* **216**, 45-54.

Grajal-Martìn, M.J., Simon, C.J. and Muehlbauer, F.J. (1993) Use of random amplified polymorphic DNA (RAPD) to characterize race 2 of *Fusarium oxysporum* f.sp. *pisi*, *Phytopathology* **83**, 612-614.

Katan, T., Hadar, E. and Katan, J. (1989) Vegetative compatibility of *Fusarium oxysporum* f.sp. *dianthi* from carnation in Israel, *Plant Pathology* **38**, 376-381.

Manicom, B.Q. and Baayen, R.P. (1993) Restriction fragment length polymorphisms in *Fusarium oxysporum* f.sp. *dianthi* and other fusaria from *Dianthus* species, *Plant Pathology* **42**, 851-857.

Manulis, S., Kogan, N., Reuven, M. and Ben-Yephet, Y. (1994) Use of RAPD technique for identification of *Fusarium oxysporum* f.sp. *dianthi* from carnation, *Phytopathology* **84**, 98-101.

Migheli, Q., Berio, T., Gullino, M.L. and Garibaldi, A. (1995) Electrophoretic karyotype variation among pathotypes of *Fusarium oxysporum* f.sp. *dianthi*, *Plant Pathology* **44**, 308-315.

Migheli, Q. and Cavallarin, L. (1994) Characterization of antagonistic and pathogenic *Fusarium oxysporum* isolates by random amplification of polymorphic DNA, *Molecular Biotechnology* **2**, 197-200.

Waalwijk, C., de Koning, J., Baayen, R. and van Pelt, H. (1996) Molecular analysis for virulence in *Fusarium oxysporum* f.sp. *dianthi*, *Proc. 3rd European Conference on Fungal Genetics*, March 27-30, Muenster, Germany, p.209 (abstract).

Welsh, J. and McClelland, M. (1990) Fingerprinting genomes using PCR with arbitrary primers, *Nucleic Acids Research* **18**, 7213-7218.

Williams, J.G.K., Kubelik, A.R., Livak, K.L., Rafalski, J.A. and Tingey, S.V. (1990) DNA polymorphisms amplified by arbitrary primers are useful as genetic markers, *Nucleic Acids Research* **18**: 6531-6535.

QUANTITATIVE ESTIMATION OF FUSARIUM GRAMINEARUM USING A NOVEL SOLID PHASE PCR-ASSAY

MARTIN L. NIESSEN, JÜRGEN KLUSMANN, AND RUDI F. VOGEL
Technische Universität München
Lehrstuhl für Technische Mikrobiologie
85350 Freising-Weihenstephan

1. Introduction

Fusarium graminearum (teleomoph: *Gibberella zeae*) is a widespread soil fungus isolated from temperate regions all over the world (Domsch *et al.*, 1980). It has a high phathogenic potential towards grasses (Cook, 1981) under which cereals and maize are of economical importance. Considerable losses occur due to reduced yields and a negative impact on the quality of grain crops infested by the fungus (Charmly et al., 1994). In addition to its pathogenicity towards plants, *F. graminearum* produces several potent mycotoxins under which nivalenol (NIV), deoxynivalenol (DON), and the estrogene analogous substance zearalenone (ZEA) are the ones which are most frequently isolated from cereals in different parts of the world (FAO/OPS, 1992; Gareis, 1989; Scott, 1989; Tanaka *et al.*, 1988). Moreover, *F. graminearum* has been suggested as one of the most important inducers of gushing in beer (Niessen *et al.*, 1992; Amaha and Kitabatake, 1981).
methods but chemical and immunochemical methods have also been described (Beyer *et al*, 1993, Notermans *et al.*, 1987). As these methodes are time consuming, unspecific or feasible only for fresh materials, there is need for alternative methods which can detect infested samples in a rapid and sensitive way. As a first step towards this direction, a PCR method for the specific detection of DNA isolated from strains of *F. graminearum* has previously been set up and optimized for high sensitivity (Niessen and Vogel, in press).
DIAPOPS (**D**etection of **I**mmobilized, **A**mplified **P**roduct in a **O**ne **P**hase **S**ystem) (Rasmussen *et al.*, 1994) makes use of the separation of a PCR primer pair between the solid phase to which one primer is covalently bound and the liquid phase which contains the second primer. The liquid phase contains an additional small amount of to he solid phase primer. Using this

H.-W. Dehne et al. (eds.), Diagnosis and Identification of Plant Pathogens, 207–211.

system, the DNA strand polymerized to the 3´terminus of the solid phase primer becomes trapped to the wells of a microwell module. All components are removed from the wells by washing. Detection of the solid phase bound product is done with a product specifc probe. All reactions are done in the same well. No pipetting of PCR products for detection is needed. DIAPOPS has been successfully used for the quantitative estimation of bacterial DNA from clinical and food samples (Rasmussen et al., 1994).

2. Materials and Methods

2.1 FUNGAL CULTURES, MEDIA, AND ISOLATION OF DNA

All fungal cultures used in this study were maintained on Oatmeal agar slants and kept at 4 °C. All cultures were kept in parallel at -75 °C in 80 % glycerole having grown on porous Leca® granules. Prior to DNA isolation, cultures were grown for 5 d at ambient temperature an a rotary shaker (120 rpm) in 250 ml Erlenmeyer flasks containing 100 ml of 3 % malt extract broth (0.3 % soy peptone).

Mycelia were vacuum filtered through filter discs (Schleicher & Schuell), washed with 2x50 ml of sterile tap water and freeze dried. Lyophilized mycelia were finely ground with a mortar and pestill adding a small amount of sterile seasand until no complete hyphae were seen under the microscope. Ground mycelium was subjected to DNA-extraction as described by Lee and Taylor (1990). DNA from cereals was isolated using the method described by Möller *et al.* (1992). DNA contents and purity of DNA-samples were estimated using a Genquant UV photometer at 260 nm (Pharmacia, Freiburg, Germany) and a model Biochrom 4060 spectrophotometer (Pharmacia, Freiburg, Germany) at 200-400 nm. DNA content was calculated from the peak highth at 258 nm. Purity was estimated from the peak width.

2.2 QUANTIFICATION OF *F. GRAMINEARUM* DNA BY DIAPOPS

Primers and probes used in this study were published elsewhere (Niessen and Vogel, in press). The assay was performed in NucleoLink® break appart modules (NUNC, Roskilde, Denmark). Primer P-polyT-GaoA-V2 was used as the solid phase primer which was covalently linked to the plastic support as described by Rasmussen *et al.* (1994). Dried modules with primer bound were stored at 4 °C until used. For DIAPOPS 25 μl of the PCR reaction mixture described by Niessen and Vogel (accepted for publication) were mixed in double concentration with the same volume of target DNA in sterile ddH_2O. A calibration curve was set up with a twofold serial dilution of the purified specific 889 bp PCR-product. DNA concentration of the standard target-solution was 0.29 pmol/μl. The cycle program used for DIAPOPS was pricipally the same as allready described (Niessen and Vogel, 1996). PCR-product not bound to the solid phase was eliminated by several washings with 0.4 M NaOH (0.25 % Tween 80).

2.3 DETECTION OF AMPLIFIED PRODUCT

For detection of ampliefied PCR-product bound to the solid phase, a product specific biotinylated probe was used in a concentration of 100 pmol in hybridization buffer. After hybridization, the wells were washed with 0.2 x SSC followed by incubation with steptavidin-alcaline-phosphatase conjugate for 1 h at ambient temperature. Wells were washed with TBS and inkubated with 200 μl 4-nitrophenylphosphate, 1 mg/ml (Boehringer, Mannheim, Germany) in 1 M diethanolamine (pH 9.8) as enyme substrate. OD of the reaction mixture at 405 nm was measured in an Emax microplate photometer (Molecular Devices, USA) after 30 min of enzyme reaction at 37 °C. Results were compared to the DNA content of the samples measured with a Biochrome 4060 spectral photometer (Pharmacia LKB Biochrom, England) and a GeneQuant DNA/RNA calculator (Pharmacia LKB Biochrom, England)

3. Results

3.1 OPTIMIZATION OF THE DIAPOPS PROCEDURE

Optimum coating concentration for the solid phase primer Gaoa-V1 was found to be 50 ng/well. Best ration between the liquid phase primer Gaoa-V2 and the amount of Gaoa-V1 in the liquid phase was 1:10. The liquid phase of the DIAPOPS reactions perfomed with a twofold serial dilution of purified PCR product as target was transferred to a 1% agarose gel and showed high specificity and performance.
A standard curve was processed by hybridization of the immobilized PCR product with the biotinylated specific DNA-probe which was incubated with streptavidin-alcaline phosphatase conjugate and pNPP as substrate. The standard curve was linear between 850 target molecules and $1,5*10^8$ target molecules corresponding to 425 to $0.75*10^8$ diploid genomes. Molar amounts of target molecules were recalculated to give μg/μl of fungal genomic DNA on the basis of the weight of the *S. cerevisiae* haploid

genome. Based on this, DNA contents of DNA preparations from several *F. graminearum* strains and strains of other *Fusarium*-species were compared. DNA contents of the preparations varied from 0.30 $\mu g/\mu l$ to 2.25 $\mu g/\mu l$ in *F. graminearum* preparations. Concentrations found with a spectral photometer and with a DNA calculator where in good accordance. Values found with DIAPOPS were at the same level as compared to the photometric values in those extracts showing high purities and narrow peaks at 258 nm in the UV-spectra. In extracts with low purity (<90 %) DIAPOPS values were below the photometric values in all cases. Extreme differences between the values were seen in two extracts with low purity (74 % and 67 %) where DNA contend found with the DIAPOPS method was only 10 % of that found with a photometer. Extracts of other *Fusarium*-species tested had high DNA contents but did not show a signal in DIAPOPS.

4. Discussion

The assay procedure set up here has proved to be a useful tool for the quantitative assessment of *Fusarium graminearum* DNA. Using this system, it will be possible to quantify the fungus not only in pure cultures but also in natually contaminated matrices, where DNA of other organisms and of the matrix itself has to be discriminated from the target DNA. The work presented proved, that DIAPOPS gives more accurate results with impure samples, in which DNA-peaks are often covered by protein peaks to give unrealistic measurements of the DNA content. The primer pair used here has shown to be highly specific for the DNA of *F. graminearum*. It did not react with DNA of any other organism tested. For qantification of the fungus in biological materials or in soil, the DIAPOPS signal should be calibrated to the specific biomass of the *F. graminearum* rather than to the amount of DNA. This is currently under development. An internal standard is beeing constructed to correct the values found in contaminated materials from influences of the matrix on the PCR reaction. Currently the DIAPOPS assay described requires 7 h of analysis time using purified DNA. Improvements in the extraction procedures and the assay itself will soon lead to a further reduction in assay time. This will make DIAPOPS an interesting new tool not only for quality control of cereals in the food industry but also for epidemiological studies.

5. References

Amaha, M. and Kitabatake, K.(1981) Gushing in beer, in J. R. A. Pollock (ed.), *Brewing Science* Vol. 2., Academic Press, Ney York, pp. 457-489.

Beyer, W., Höxter, H., Miedaner, T., Sander, E., and Geiger, H. H. (1993) Indirect ELISA for quantitative assessment of *Fusarium* spp. in rye, *Zeitschrift für Pflanzenkrankheiten und Pflanzenschutz* **100** (3), 278-284.

Charmley, L. L., Rosenberg, A., and Trenholm, H. L (1994) Factors responsible for economic losses due to *Fusarium* mycotoxin contamination of grain, foods, and feedstuffs, in J. D. Miller and H. L. Trenholm (eds.), *Mycotoxins in Grain - Compounds other than Aflatoxin*, Eagan Press, St. Paul, pp. 471-486.

Cook, R. J.(1981) *Fusarium* diseases of wheat and other small grains in north america, in P. E. Nelson, T. W. A. Toussoun, and R. J. Cook (eds.), Fusarium - *diseases, biology, and taxonomy*, The Pennsylvania State University Press, University Park.

Domsch, K. H., Gams, W., and Anderson, T.-H (1980) *Compendium of soil fungi*, Vol. 1, Academic Press, New York, pp. 303-341.

FAO/OPS (1992) *Taller conjunto FAO/OPS sobre prevención y control de micotoxinas en América Latina y el Caribe*, FAO: Rome, Costa Rica,San José, 23.2.1992.

Gareis, M., Bauer, J., Enders, C., and Gedek, B. (1889) Contamination of cereals and feed with *Fusarium* mycotoxins in European countries, in J. Chelkowski (ed.), Fusarium - *Mycotoxins, Taxonomy and Pathogenicity*, Elsevier, Amsterdam, pp. 441-472.

Lee, S. B. and Taylor, J. W. (1990) Isolation of DNA from fungal mycelia and single spores, in M. A. Innis, D. H. Gelsand, J. J. Sninsky, and T. J. White (eds.), *PCR Protocols: A Guide to Methods and Applications, Academic Press, New York,* pp. 282-287

Möller, E. M., Bahnweg, G, Sandermann, H., and Geiger, H. H. (1992) A simple and efficient protocol for isolation of high molecular weight DNA from filamentous fungi, fruit bodies, and infected tissues, *Nucleic Acid Research* **20**, 6115-6116.

Niessen, L. and Vogel, R. F. (accepted for publication) Detection of *Fusarium graminearum* by PCR using GaoA targeted primers, *Applied and Systematic Microbiology*.

Niessen, L., Donhauser, S., Weideneder, A., Geiger, E., and Vogel, H.(1992) Mykologische Untersuchungen an Cerealien und Malzen im Zusammenhang mit dem Wildwerden (Gushing) des Bieres, *Brauwelt* **132** (16/17), 702-714.

Notermans, S., Wieten, G., Engel, H . W. B., Rombouts, F. M., Hoogerhout, P., and van Boom, J. H.(1987) Purification and properties of extracellular polysaccharide (EPS) antigens produced by different mould species, *Journal of Applied Bacteriology* **62**, 157-166.

Rasmussen, S. R., Rasmussen, H. B., Larsen, M. R., Hoff-Jørgensen, R., and Cano, R. J. (1994) Combined polymerase chain reaction-hybridization microplate assay used to detect bovine leukemia virus and *Salmonella*, *Clinical Chemistry*, **40/2**, 200-205.

Scott, P. M. *(1989)* The natural occurrence of trichothecenes, in V. R. Beastley (ed.) *Tricho thecene Mycotoxicosis: Pathophysiologic Effects*, Vol. 1, CRC Press, Boca Raton, pp. 1-26.

Tanaka, T., Hasegawa, A., Yamamoto, S., Lee, U.-S., Siguira, Y., and Ueno, Y.(1988) Worldwide contamination of cereals by the *Fusarium* mycotxins nivalenol, deoxynivalenol, and zearalenone. 1. Survey of 19 countries, *J. Agric. Food Chem.* **36**, 979-983.

MOLECULAR DIAGNOSIS OF TOXIGENIC *FUSARIUM* SPECIES

GIUSEPPINA MULE' and ANTONIO LOGRIECO
Istituto Tossine e micotossine da parassiti vegetali, C.N.R.
Viale Einaudi 51, Bari, Italy.

1. Introduction

Correct identification of species in *Fusarium* genera is important because many species are plant pathogens and same isolates produce potent mycotoxins (Marasas *et al.*, 1984). Methods that have been used in *Fusarium* systematics include DNA-DNA hybridization (Ellis, 1988), RFLP analysis (Manicom & Baayen, 1993), RAPD markers (Schilling *et al.*, 1996), isozyme (Huss *et al*, 1993), sequence analysis (O'Donnell, 1996) and electrophoretic karyotyping (Fekete *et al.*, 1993). DNA-DNA hybridization may reflect relatedness at the level of isolates, but RFLP and karyotype analyses typically differentiate at level of genetically isolated groups. Problem with *F. oxysporum* in Boehm *et al.* (1994) where they found karyotype variation in strains belonging to the same VCG. RAPDs have been used to detect genetic differences between and within species. Variation in nucleotide sequence demonstrates relatedness at different levels depending on the sequence chosen for analysis. O'Donnell (1996) recently have developed a DNA sequence database using four loci: beta tubulin gene introns/exons, mitochondrial small subunit (mtSSU) rDNA, nuclear large subunit (LSU) rDNA, and nuclear rDNA internal transcribed spacer (ITS) region. We found positive correlation between phylogeny, secondary metabolites production and the sequence of nuclear LSU 28S rDNA .

2. Methods

All strains were initiated from single conidia prior to further study and stored on slants of special nutrient agar (SNA) (Niremberg, 1981). Chemical and genetical analyses were carried out using the methods described of Bottalico *et al.* (1989) and Mulè *et al.* (1995), respectively. Toxin reference standards were purchased from Sigma Chemical Co. (St. Louis, Missouri, USA).

Sequences of fragments were determined using an automated DNA sequencer and were aligned with the aid of computer alignment programs (e.g. GCG, Clustal). Phylogetic assessments were made using PHYLIP (Felsenstein, 1991) and the stationary Markov model, for more details see Saccone *et al.* (1990). This method

H.-W. Dehne et al. (eds.), Diagnosis and Identification of Plant Pathogens, 213–217.

allows one to calculate time of divergence ratios (t/t') for any pair of homologous sequences by analyzing any class of site, which is assumed to be subjected to the same evolutionary dynamics. Sequences are available from the authors.

3. Results and Discussion

Phylogenetic placement of toxigenic *Fusarium* strains based on the nucleotide sequence of partial LSU rDNA is discussed in relationship to morpho-physiological and chemotaxonomic information. Following preliminary studies of some variable rDNA regions, we focused our attention on one small region to obtain a fast, useful molecular method.

Logrieco *et al.* (1995) studied the phylogenetic relationship in the recent morphotaxonomic interpretation of *F. sambucinum* sensu lato (Nirenberg, 1995) which split the species into three distinct taxa: *F. sambucinum* sensu stricto, *F. torulosum* (Berk. & Curt.) Nirenberg, and *F. venenatum* Nirenberg. Interspecific comparisons revealed nucleotide sequence differences of 3-9 substitutions. Although these interspecific differences were not large in terms of percent nucleotide substitution, they were much larger than the observed intraspecific variation and supported the morphological interpretation distinguishing three taxa. When the data were analyzed using parsimony and bootstrapping, the three taxon tree was well supported and the phylogenetic arrangement of the strains was congruent with the secondary metabolite profiles (Altomare *et al.*, 1995). The trichothecene producing strains of *F. sambucinum* and *F. venenatum* clustered with *F. crookwellense,* another species know produce tricothecenes, *F. torulosum* which did not produce trichothecenes, was phylogenetically distinct.

An additional positive correlation between genetic data, obtained by comparison of LSU region nucleotide sequences, and secondary metabolites profiles was observed among species in *Fusarium* section *Sporotrichiella. Fusarium poae* (Peck) Wollenw. and *F. sporotrichiodes* Sherb. shared more common base sequences than with *F. tricinctum* (Corda) Sacc and *F. chlamydosporum* Wollenw & Reinking, in according with their common ability to produce trichothecenes (Logrieco *et al.*, 1991a).

On the basis of some morphological and ecological characteristics *F. acuminatum* Ell. & Ev. was recently split into two subspecies: *F. a.* subsp. *acuminatum* and *F. a.* subsp. *armeniacum* Forbes et al. (Burgess *et. al.*, 1993). These subspecies also differ in their ability to produce secondary metabolites. Specifically *F. a. armeniacum* produces high levels of T-2 toxin and neosolaniol (Logrieco *et al.*, 1992a). This pattern of secondary metabolites production has ben previously reported for some isolates of *F. sporotrichiodes* (Marasas *et al.*, 1984). *F. a. armeniacum* and *F. sporotrichioides* are close related based on LSU region partial nucleotide sequence (Logrieco *et al.,* 1992b) and electrophoretic karyotype (Nagy and Hornok, 1994).

Further studies on the *F. acuminatum* subspecies showed that only *F. a. armeniacum* strains could hybridize with a trichodiene synthase gene probe (Nagy and Hornok, 1994). More recently Hornok *et al.* (1996) showed that *F. poae* has a trichodiene synthase gene that is very similar to the *tox5* gene of *F. sporotrichioides*. In same cases the probe can hybridize to DNA from several species, if the toxin is not species specific.

Conidiogenesis and conidial form are critical characters in mycological taxonomy. The morphotaxonomic system regard the *Fusarium* conidiog6enous cell (e.g. mono- or poly-phialide) as a fundamental taxonomic character. However, mutants in this character may occur in culture making difficult or impossible to correctly identify the strain . The need for correct identification is particularly urgent when the atypical strain is also toxigenic. Phylogenetic reconstruction of degenerate, mutant or dubious toxigenic *Fusarium* strains have obtained by comparison LSU region sequences. Fn-2B, one of the most important trichothecene producing *Fusarium* strain, first reported as *F. nivale* but with a very controversial identification (Marasas *et al.* 1984), was compared to the LSU region from species. It was presumed to belong in order to assess its phylogenetic affinity. *Fusarium* strain FN-2B was most closely related to *F. poae* showing only one heteromorphic site. In comparison to other fungal strains, Fn-2B differed at 3, 11, and 34 bases from *F. sporotrichioides*, *F. tricinctum* and *Microdochium nivale* (Fries) Samuel & Hallett, respectively (Mulè *et al.*, 1995). This phylogenetic placement of Fn-2B correlates well with the production of trichothecene mycotoxins by the closely related species.

A second atypical *Fusarium* strain (ITEM-709), was degenerate in conidiogenesis but could synthesize trichothecenes (T-2 toxin and neosolaniol). The LSU of this strain was sequenced and compared with other from known toxigenic *Fusarium* strains. The nucleotide sequence obtained from ITEM-709 was identical to the homologous sequence from a morphologically typical and T-2 toxin producer of *F. sporotrichioides* Sherb. DNA/DNA hybridization confirmed their high genetic relatedness (90 % similarity) (Logrieco *et al.*, 1994).

Microdochium nivale (*ex Fusarium*) NRRL-3289 is third strain in a degenerate state and is reported to be only strain in this genus able to produce trichothecene mycotoxins. Its taxonomic disposition was investigated by sequence comparison with morphologically typical strains of *M. nivale* var. *major*. The sequences of all the isolates were the same and when thc mycotoxin analysis was repeated no trichothecene mycotoxins were detected (Logrieco *et al.*, 1991b). Thus, it was complitely excluded this specie as tricothecenes producer.

Phylogenetic alignment of toxigenic *Fusarium* species by sequence analysis of rDNA variable regions is undoubtedly useful for taxonomic and systematic studies. In addition, our data, obtained by using partial SLU region, suggest that phylogentic placement of strains, including those atypical and degenerate, in the *Fusarium* genus may also be indicative of potential mycotoxin production and other physiological characteristics.

Acknowledgements
This work was supported by the Italian Ministry of Agriculture and Forestry, D.M. 131/7240/94 del 14/02/1996, prog. 451 "Tecniche molecolari per l'identificazione di funghi di interesse in micologia e fitopatologia".

4. References

Altomare, C., Logrieco, A., Bottalico, A., Mulè, G. Moretti A., & Evidente A. (1995) Production of type A trichothecenes and enniatin B by *Fusarium sambucinum* Fuckel sensu lato, *Mycopathologia* **129**, 177-181.

Boehm, E.W.A., Ploetz R.C., & Kistler H.C. (1994) Statistical analysis of karyotipe variation among vegetative compatibility groups of *Fusarium oxysporum* f.sp. *cubense, Molec. Pl. Micr. Interact.* **7,** 196-207

Bottalico, A., Logrieco, A., & Visconti A. (1989) Fusarium species and their mycotoxins in infected cereals in the field and in stored grains, in Chelkowski J. (ed.), Fusarium: mycotoxins, taxonomy and pathogenicity. Elsevier Biomedical Press, Amsterdam, pp 85-119.

Burgess, L.W., Forbes, G.A., Windels C., Nelson P.E., Marasas W.F.O., & Gott K.P. (1993) Characterization and distribution of *Fusarium acuminatum* subsp. *armeniacum* subsp. nov., *Mycologia* **85**, 119-124.

Ellis, J.J. (1988) Section Liseola of *Fusarium, Mycologia* **80**, 225-258.

Fekete, C., Nagy, R., Debets, A.J.M., & Hornok, L. (1993) Electrophoretic karyotypes and gene mapping in eight species of the *Fusarium* sections *Arthrosporiella* and *Sporotrichiella, Curr. Genetic.* **24**, 500-504.

Felsestein , J. Phylip version 3.4.1991. Seattle: University of Washington.

Hornok, L., Fekete C., & Giezey G. (1996) Molecular characterization of *Fusarium poae, Sydowia* **48(1),** 23-31.

Hughes, S.J. (1953) Conidia, conidiophores and classification, *Canad. J. Bot.* **31,** 577-659.

Logrieco A., Peterson, S.W., & Bottalico, A. (1991a) Phylogenetic affinities of the species in *Fusarium* section *Sporotrichiella, Exp. mycol* **15**, 174-179.

Logrieco, A., Vesonder, R. F., Peterson, S.W., & Bottalico, A. (1991b) Reexamination of the taxonomic disposition and deoxynivalenol production by *Fusarium nivale* NRRL 3289, *Mycologia* **83**, 367-370.

Logrieco, A., Altomare, C., Moretti, A., & Bottalico, A. (1992a) Cultural and toxigenic variability in *Fusarium acuminatum., Mycol. Res.* **96**, 518-523.

Logrieco, A., Mulè, G., Peterson, S.W., & Bottalico, A. (1992b) Genetic relatedness in *Fusarium* section *Gibbosum* strains.- Third European Fusarium Seminar. IHAR, Radzikow, Poland. (Abstract.)

Logrieco, A., Bottalico, A., Peterson, S.W., Evidente, A., & Mulè, G. (1994) Molecular genetic criteria for the identification of atypical toxigenic *Fusarium* strains, in Llewellyn G.C., Dashek W.V. and O'Rear C.E. (eds.) *Biodeterioration Research 4,* Plenum Press, New York, pp. 177-188.

Logrieco, A., Peterson, S.W. & Bottalico, A. (1995) Phylogenetic relationship within *Fusarium sambucinum* Fuckel sensu lato, determined from ribosomal RNA sequences, *Mycopathologia* **129**, 153-158.

Marasas, W.F.O., Nelson, P.E., & Toussoun, T.A. (1984*) Toxigenic Fusarium species: Identity and Mycotoxicology*, Pennsylvania State University Press, University Park.

Manicom, B.Q., & Baayen, R.P. (1993) Restriction fragment length polymorphisms in *Fusarium oxysporum* f.sp. *dianthi* and other fusaria from *Dianthus* species, *Plant Pathol.* **42**, 851-857.

Mulè, G., Logrieco, A., Bottalico, A., (1995) *Fusarium* sp. FN-2B: a controversial strain genetically close to *Fusarium poae*, *Mycotoxin. Res.* **11**, 53-58.

Nagy, R. & Hornok, L. (1994) Electrophoretic karyotype differences between two subspecies of *Fusarium acuminatum*, *Mycologia* **86**, 203-208.

Niremberg H.I. (1981) A simplified method for identifying *Fusarium* spp. occurring on wheat, *Can. J Bot.* **59**: 1599-1609.

Nirenberg, H.I. (1995) Morphological differentation of *Fusarium sambucinum* Fuckel sensu stricto, *F. torulosum* (Berk & Curt.) Nirenberg comb. nov. and *F. venenatum* Nirenberg sp. nov. *Mycopathologia* **129**, 131-141.

Nelson, P.E., Toussoun, T.A., & Cook, R.J. (1981) *Fusarium: Disease, Biology and Toxonomy*, . Pennsylvania State University Press, University Park.

O'Donnell, K. (1996) Progress towards a phylogenetic classification of *Fusarium, Sidowia* **48(1)**, 57-70.

Schilling, A.G., Möller E.M., & Geiger H. (!996) Molecular differentiation and diagnosis of cereals pathogens *Fusarium culmorum* and *F. graminearum, Sidowia* **48(1)**, 71-82.

GENETIC VARIABILITY OF *ASCOCHYTA RABIEI*: POPULATION STRUCTURE REVEALED BY RAPD ANALYSIS AND CLASSIFICATION INTO PATHOTYPES

S.M. UDUPA[1], F. WEIGAND[1], J. GEISTLINGER[2] AND G. KAHL[2]
[1]*ICARDA, P.O. Box 5466, Aleppo, Syria;* [2]*Biocenter, University of Frankfurt, Marie-Curie Str. 9/N200, D-60439, Frankfurt/Main, Germany*

Abstract. The poor definition of variation in the ascochyta blight fungus has historically hindered strategies for reducing blight disease damage to the chickpea crop in the West Asia and North Africa. We applied RAPD and microsatellite markers to construct genotype-specific, DNA fragment profiles of Syrian field isolates of this fungus. By using conventional pathogenicity tests and genome analysis with DNA markers, we show that the DNA markers distinguished the major pathotypes in Syria, and defined the organization of clonal lineages within and among pathotype groups. Further, we demonstrate that these markers could be used for pathogenicity surveys to identify the most predominant genotype in the population. On the basis of the surveys, a general picture of *A. rabiei* evolution in Syria and its implications in the plant disease management are discussed.

1. Introduction

Ascochyta rabiei (Pass) Lab., the causal agent of ascochyta blight disease of chickpea (*Cicer arietinum* L.) is widespread in the West Asia, North Africa, and North America, where it represents a major constraint to production. This fungal disease occurs in chronic epidemic cycles despite the frequent introduction of new resistant chickpea cultivars (Singh and Reddy, 1993). The absence of durable resistance in the field has been attributed to the appearance of new pathotypes (Singh et al., 1992) and high levels of polymorphism in aggressiveness with in the pathogen population (Vir and Grewal, 1974; Reddy and Kabbabeh, 1985; Porta-Puglia, 1992). The nature of this polymorphism was determined by assaying its infection spectrum, with a pathogenicity test on a set of differential chickpea cultivars. However, the nature of this polymorphism is subject of much debate (Weising et al., 1991; Malik and Rahman, 1992; Porta-Puglia, 1992), marked by strongly contrasting views on the diversity and stability of pathotypes. For example, Reddy and Kabbabeh (1985) reported aggressiveness of the six isolates in the order of 6>5>4>3>2>1. For the same isolates, Weising et al., (1991) reported differently as 6>4>2>1>5>3. These results led to the view that many isolates of *A. rabiei* were pathogenically unstable, may be promiscuously polymorphic and continually changing. In contrast to this view, our extensive testing of single spore-derived cultures showed no changes in pathotypes, even after prolonged subculturing and storage. In addition, several researchers suggested that difficulty in standardizing assay procedures (Reddy et al., 1990; Porta-Puglia, 1992) and

H.-W. Dehne et al. (eds.), Diagnosis and Identification of Plant Pathogens, 219–222.

disease rating scales (Harrabi and Halila, 1992; Porta-Puglia, 1992,) lead to inflated estimates of pathotypic diversity and variability. The key to resolve these issues is to provide the phenotypic variation observed in pathogenicity assays with a genetic definition, using genetic markers for pathogenicity surveys.

2. Choice of DNA marker

Although a variety of genetic markers can be used to study fungal phytopathogen populations (for review see Michelmore and Hulbert, 1987), few have been definitive for the ascochyta blight fungus. For example, Bouznad et al. (1996) found that isozymes are of little use as genetic markers as they detected nearly uniform patterns for various isolates of Algeria, Tunisia and Syria. However, Weising et al. (1991) showed that oligonucleotide sequences complementary to microsatellites can be efficiently used as a probe in RFLP analysis to detect genetic diversity in *A. rabiei.* Similarly Udupa and Weigand (1996b) showed that PCR-based markers as e.g. RAPDs can be efficiently used for detecting genetic diversity in different pathotypic groupings of *A. rabiei.* Moreover, the RAPD technique is fast, less cumbersome, very little template DNA is required for the analysis, and a very large number of samples can be handled in a very short period of time. Therefore, RAPD markers were further used for pathogenicity surveys to identify the more predominant genotype of the pathogen.

3. Relationship between pathotype and genotype of the pathogen

To establish the relationship between pathotype and genotype, a set of 53 isolates sampled periodically in different chickpea growing regions of Syria were analyzed for pathogenicity and also for genetic polymorphism using a set of DNA markers (microsatellites and RAPDs; Udupa and Weigand, 1996a and b). The pathogenicity test (under growth chamber conditions) classified the isolates into three pathotypes, based on their reaction on three differentials, namely, ILC 1929 (susceptible), ILC 482 (moderately resistant) and ILC 3279 (resistant). The DNA marker analysis diagnosed the three pathotypes and resolved the pathotypes into 18 distinct genotypes; pathotype I (least aggressive) as resolved into 12 genotypes, pathotype II (aggressive) into three genotypes and pathotype III (most aggressive) into three genotypes. These genotypic groupings will provide a reference base for the detection of newly emerging genotypes as well as the appearance of new lineages.

4. Population structure of the pathogen in Syria

Knowledge on population structure of the pathogen in chickpea growing regions is important for deployment of suitable resistance genes and formulating and implementing suitable disease control strategies. Conventionally, the population structure of the pathogen is determined by pathogenicity surveys based on reactions on a set of differential cultivars. Initially we employed a set of microsatellite and RAPD markers for this purpose (Udupa and Weigand, 1996b). More recently (1994-95), we used only RAPD markers (unpublished results). All the surveys (1991-95) revealed the predominance of a single genotype (we designate it as genotype-H) in all the chickpea growing regions (Table 1), which was not detected during the earlier survey (1982).The genotype is increasing its frequency in all the chickpea growing regions of Syria in recent years. The genotype-H is highly aggressive (pathotype III) and can kill the most resistant cultivar (ILC 3279) released in Syria. This represents the breakdown of the resistance due to evolution of a new genotype (pathotype) of the pathogen.

TABLE 1. Frequency of genotype-H of *Ascochyta rabiei* sampled in Syria.

Year of sampling	No. of isolates sampled	Frequency of genotype-H (%)
1982	6	0
1991	6	33
1992	26	77
1993	15	73
1994	81	86
1995	82	83

5. Evolutionary forces acting on the population

On the basis of the surveys, we speculate about the evolutionary forces acting on the *A. rabiei* population of Syria. Increase in the aggressiveness observed in the population in response to a change in host resistance in the field, could be due to mutation and selection exerted by the host population. A recent study in a plant pathogenic fungus, *Rhynchosporium secalis*, by Goodwin et al. (1994) also showed that a high rate of mutation could occur for the pathogenicity trait under natural conditions in response to the selection pressure exerted by host resistance. This kind of selection-induced genetic changes in fungi is not solely due to random mutations but also to a process called "adaptive mutation" (Harris et al., 1994), recently described in bacteria and yeast. In addition to mutation and selection, migration/genetic drift affected genetic changes in the population. The most likely mechanism for migration is through infected seeds. Since a single genotype-H was periodically sampled in a very large area, we conclude that there exists only one clone. This observation further supports that under natural conditions the pathogen reproduces preferably asexually and sexual recombination plays a minor role in evolution of new pathotypes of *A. rabiei*. Furthermore, the increase in aggressiveness the pathogen population over the years in response to resistance of the host population, suggests that selection is a strong force in establishing and maintaining genetic changes in the population.

6. Implications in chickpea disease management

Our pathogenicity survey using DNA markers clearly demonstrated that there is a need to develop chickpea cultivars with better level of resistance to the genotype-H of the pathogen. This need was partly achieved with the identification of new sources of resistance to genotype-H in the germplasm of chickpea (unpublished results). We also found out that the level of resistance to genotype-H can be improved through pyramiding of resistance genes to ascochyta blight (unpublished results). Since the pathogen seems to be migrating through infected seeds, use of disease free seeds for planting and seed treatment with suitable fungicides can help in better management of the disease.

Acknowledgment: Research of the authors is supported by BMZ, Bonn (grant 95.7860.0-001.01) and UNDP (grant RAB/87/025).

7. References

Bouznad, Z., Corbiere, R., Elbiari, A. and Spire, D. (1996) Proteins and isozyme analysis in *Ascochyta* species of food legumes using isoelectric focusing method, in F. Weigand and S.M. Udupa (eds*), DNA markers and breeding for resistance to ascochyta blight in chickpea*, ICARDA, Aleppo, Syria, (in press).

Goodwin S.B., Webster, R.K. and Allard R.W. (1994) Evidence for mutation and migration as source of genetic variation in populations of *Rhynchosporium secalis, Phytopathology* **84**, 1047-1053.

Harrabi, M and Halila, H. (1992) Breeding for combined resistance to ascochyta blight and fusarium wilt in kabuli chickpea in Tunisia, in K.B. Singh and M.C. Saxena (eds), *Disease resistance breeding in chickpea*, ICARDA, Aleppo, Syria ,pp. 98-109.

Harris, R.S., Longerich, S. and Rosenberg, S.M. (1994) Recombination in adaptive mutation. *Science* **264**, 258-260.

Malik, B.A. and Rahman, M.S. (1992). Breeding for ascochyta blight-resistance desi chickpea in Pakistan, in K.B. Singh and M.C. Saxena (eds), *Disease resistance breeding in chickpea*, ICARDA, Aleppo, Syria ,pp. 77-85.

Michelmore, R.W. and Hulbert, S.H. (1987) Molecular markers for genetic analysis of phytopathogenic fungi, *Annual Review of Phytopathology*, **25**, 383-404.

Porta-Puglia, A. (1992) Variability in *Ascochyta rabiei*, in K.B. Singh and M.C. Saxena (eds), *Disease resistance breeding in chickpea*, ICARDA, Aleppo, Syria ,pp. 135-143.

Reddy, M.V. and Kabbabeh, S. (1985) Pathogenic variability in *Ascochyta rabiei* (Pass) lab. in Syria and Lebanon, *Phytopathologia Mediterranea* **24**, 265-266.

Reddy, M.V., Nene, Y.L., Singh, G. and Bashir, M. (1990) Strategies for management of foliar diseases of chickpea, in *Chickpea in the nineties*, ICRISAT Center, Hyderabad, India, pp. 117-127.

Singh, K.B. and Reddy, M.V. (1993) Resistance to six races of *Ascochyta rabiei* in the world germplasm collection of chickpea, *Crop Science* **33**, 186-189.

Singh, K.B., Reddy, M.V. and Haware, M.P. (1992) Breeding for resistance to ascochyta blight in chickpea, in K.B. Singh and M.C. Saxena (eds), *Disease resistance breeding in chickpea*, ICARDA, Aleppo, Syria ,pp. 23-54.

Udupa S.M. and Weigand, F. (1996a) Pathotyping of *Ascochyta rabiei* isolates of Syria, in F. Weigand and S.M. Udupa (eds), *DNA markers and breeding for resistance to ascochyta blight in chickpea*, ICARDA, Aleppo, Syria, (in press).

Udupa S.M. and Weigand F. (1996b) Genotyping with microsatellite and RAPD markers resolves pathotype diversity in the ascochyta blight pathogen of chickpea, in F. Weigand and S.M. Udupa (eds), *DNA markers and breeding for resistance to ascochyta blight in chickpea,* ICARDA, Aleppo, Syria, (in press).

Vir, S. and Grewal, J.S. (1974) Physiological specialization in *Ascochyta rabiei*, the causal organism of gram blight, *Indian Phytopathology* **27**, 265-266.

Weising, K., Kaemmer, D., Epplen, J.T., Weigand, F., Saxena, M.C. and Kahl, G. (1991) DNA fingerprinting of *Ascochyta rabiei* with synthetic oligonucleotides, *Current Genetics* **19**, 483-489.

DETECTION OF *LOPHODERMIUM SEDITIOSUM* FROM PINE NEEDLES USING PCR TECHNIQUES

E. STENSTRÖM AND K. IHRMARK
Department of Forest Mycology and Pathology
Swedish University of Agricultural Sciences,
P.O. Box 7026, S-750 07 Uppsala, Sweden

Introduction

Lophodermium seditiosum is a serious needle pathogen of pine. In Sweden it is particularly *Pinus sylvestris* in the south and middle of Sweden that is badly attacked. During certain years a heavy infection can result in serious needle loss in young plantations and in nurseries, which results in reduced growth or death of young seedlings. During recent years many nurseries in the south and middle of Sweden have had very serious attacks of the pathogen and the use of recommended fungicides has not had the expected results in many cases.

Although many different *Lophodermium* species have been found on pine needles (Minter 1981) only *Lophodermium seditiosum* is considered to be pathogenic (Diwani and Millar 1987). All other *Lophodermium* species are saprophytic and of these *L. pinastri* is the most common on pine in Sweden.

The infection of *L. seditiosum* takes place during late summer or autumn and later the same autumn it is possible to find signs of infection (Butin 1995). Small yellow spots can be seen on the infected needles. In spring after the snow melts the needles first become red- to brown-spotted and later the whole needles turn red. At the beginning of the summer small grey to black pycnidia are formed, resembling a small pencil line. Aphothesia are formed in summer or autumn. Identification of the species is based mainly on the morphology of the aphothesia. Nursery managers have therefore highlighted the need to detect and identify the pathogen during its latent phase.

In the present paper we compare three different PCR-based (polymerase chain reaction) methods to identify *L. seditiosum* and distinguish it from related fungi.

H.-W. Dehne et al. (eds.), Diagnosis and Identification of Plant Pathogens, 223–226.

Material and methods

Different isolates of *L. seditiosum* were isolated from pine needles showing infection symptoms. For the purpose of comparison we also used isolates of other related fungi from our own culture collection. Fungi were grown at room temperature on 2 % malt syrup agar amended with vitamins (folic acid, aneurin and biotin).

DNA was extracted both from fungal mycelium and from needles with and without symptoms from *Lophodermium* infection. Extraction from fungal mycelium was as described by Henrion et al. (1992) and Gardes and Bruns (1993) using the modification of Henrion et al. (1994). For DNA-extraction from needles, needles were cut into pieces a few mm long, freeze dried overnight and then ground in a mortar together with AlO. One ground needle was transferred to each Eppendorf tube before extraction. We used the same extraction buffer as above but for the final clean up steps we used a DNA clean up system from Promega SDS.

The ITS-region of the fungal ribosomal DNA was amplified using the PCR (polymerase chain reaction) followed by digestion with the restriction enzymes Hae III and Hinf I. The amplification was performed with approximately 0.3 ng DNA in a solution of 10 mM Tris-HCl, 3 mM $MgCl_2$, 50 mM KCl, 0,01% gelatine, 0.2 mM dNTP-mix, 0,025 u/µl Taq-polymerase and 2 µM of the universal primers ITS 1 and ITS 4 respectively (White et al 1990). In the thermal cycler (Perkin Elmer model TC-1) samples were first denaturated for 3 min at 95° C followed by 35 cycles of denaturation for 2 min at 95 °C, annealing for 25 sec at 50 °C and polymerisation for 2 min at 72 °C. The programme was finished by polymerisation for 10 min at 72 °C. The PCR products were separated by gel electrophoresis on 1.2 % agarose gels (Seakem ME agarose from FMC Bioproducts) 3 V/cm, 3 h in 0.5 x TBE buffer (45 mM Trisborat, 1 mM EDTA). The gels were stained in ethidium bromide and were checked and photographed in UV-light. Two different restriction enzymes (Hae III and Hinf I) were used for digestion. The mixture was incubated overnight at 37 °C whereafter the reaction was stopped with LB (0.25% bromphenolblue, 0,25% xylene cyanol FF, 30% glycerol) The fragments were separated with gel electrophoresis on 2% agarose gels (Agarose NA from Pharmacia Biotech), 3 V/cm, 3 h in 0.5 x TBE.

Fingerprints made by RAPD (random amplified polymorphic DNA) with the minisatellite M13 sequence as a primer were also made for the same isolates. The amplification was performed with 10 - 20 pg DNA in the same reaction cocktail as above (ITS) but with 150 nM of the minisatellite primer M13 (GAG GGT GGC GGT TCT) (Herrin and Smith 1988). The amplification was run for 3 min at 93 °C followed by 45 cycles with 1 min at 93 °C, 1 min at 48 °C, 1 min at 72 °C and the programme was finished with 10 min at 72 °C. The fragments were separated with gel electrophoresis as for the PCR products above.

To design primers specific to *L. seditiosum* the ITS regions from *L. seditiosum* and *L. pinastri* were sequenced and compared. At sites with differences in sequence between the two species we designed primers for *L. seditiosum* with as many base pair differences as possible, in the 3′ end, compared with *L. pinastri* .

Results and discussion

Amplification of the ITS region was successful for all isolates of the tested fungi. For all *Lophodermium* isolates, including *L. pinastri* and *L. piceae* the amplification product included 520 base pairs. For *Cenangium. ferruginosum* the amplification product was bigger, including 570 base pairs. Both Hae III and Hinf I cut the ITS amplification product into two fragments. The same band pattern was obtained for *L. seditiosum* and *L. pinastri* after digestion with Hinf I, resulting in two fragments of 130 bp and 270 bp. These fungi were easily distinguished from *L. piceae* (120, 160 and 190 bp) and *C. ferruginosum* (280 and 300 bp). Digestion with Hae III clearly distinguished all four tested species, *L. seditiosum* with two fragments of 140 and 360 bp, *L. pinastri* with one fragment of 330, *L. piceae* with two fragments of 250 bp each and *C. ferruginosum* with two fragments of 130 and 270 Bp.

Using the minisatellite M13 it was possible to distinguish between 22 different isolates of *L. seditiosum*. After amplification of their DNA they gave characteristic banding patterns with about 15 bands that were distinguishable from each others. These patterns were also different from the ones obtained after amplification of DNA from *L. pinastri, L. piceae* and *C. ferruginosum*. DNA from axenically incubated pine needles was also easily distinguished from the fungal DNA. These results were evaluated using principal components analysis. Using the first and second principal components it was possible to distinguish *L. seditiosum* from the other tested needle and shoot fungi. The different isolates of *L. seditiosum* can also be more or less divided into two populations originating from the two different nurseries from where the isolations were made.

After comparing the sequence of the ribosomal DNA between one isolate of *L. seditiosum* and one isolate of *L pinastri* we designed a few primers at each end of the ITS region. By using these newly designed primers instead of ITS 1 and ITS 4 for the PCR reaction we obtained an amplification product solely from isolates of *L. seditiosum* and not from related species. We also obtained amplification product from symptomless needles with a known infection of *L. seditiosum*. The preliminary results therefore indicate that this method should enable diagnosis of *L. seditiosum* in infected needles before symptoms of infection appears.

Conclusions

PCR-RFLP can be used for studying and identifying of *L. seditiosum* in pure culture.

PCR-RAPD can be used for studying intraspecific variation in *L. seditiosum* at the isolate level in pure culture.

It is possible to design specific primers against *L. seditiosum* after sequencing of the ITS region of the ribosomal DNA. In the near future it will be possible to study and identify *L. seditiosum* directly inside needles by using such specific primers.

References

Butin, H. (1995) *Tree Diseases and Disorders*, Ed; Lonsdale D, Oxford University press, pp 31-33.

Diwani, S. A. and Millar, C.S .(1987, Pathogenicity of three *Lophodermium* species on Pinus sylvestris L., *Eur. J. For. Path.* **17**, 53-58.

Gardes, M. and Bruns, T. (1993) Its primers with enhanced specificity for Basidiomycetes; application to the identification of mycorrhizae and rusts, *Mol Ecol.* **2**, 113-118.

Henrion, B., Le Tacon, F and Martin, F. (1992) Rapid identification of genetic variation of ectomycorrhizal fungi by amplification of ribosomal RNAgenes, *New Phyt* **12**, 289-298.

Henrion, B., Chevalier, G. and Martin, F. (1994) Typing truffle species by PCR amplification of the ribosomal DNA spacers. *Mycol. Res.* **98(1)** 37-43.

Herrin, D.L. and Schmidt, G.W. (1988)A rapid method for confirming cell line identity: DNA "fingerprinting" with a minisatelite probe from M13 Bacteriophage, *BioFeedback* **6**, 200-202

Minter, D.W. (1981) *Lophodermium* on Pines, *Mycological papers* No. 147, Commonwealth Mycological institute Kew, Surrey, England, 54p.

White, T.J., Bruns, T., Lee, S. and Taylor, J. (1990) Amplification and direct sequencing of fungal ribosomal RNA genes for phylogenetics. Eds. M.A, Innis, D.H. Gelfand, J.J. Snisky, and T.J. White, In; *PCR Protocols, A Guide to Methods and Applications*., Academic press, San Diego, pp 315-322.

Session 5:

Disease Assessment and Quantification

A COMMON BASIS OF GENETICALLY BASED AND INDUCED RESISTANCE IN CEREALS
NEW APPROACHES FOR THE ASSESSMENT OF RELIABLE ASSAYS FOR CHEMICAL INDUCERS

BIRGIT JAROSCH, ULI BECKHOVE, RUTH SCHIFFER,
SILKE HEISE, FRITZ KREUZALER, KARL-HEINZ KOGEL
Institut für Biologie I, RWTH Aachen, Worringer Weg 1, 52074 Aachen, Germany

1. Abstract

We have analysed the phenomenon of systemically activated resistance (SAR) in the interaction of the powdery mildew fungus (*Erysiphe graminis*) with the respective hosts, wheat (*Triticum aestivum* L.) and barley (*Hordeum vulgare* L.). One aspect of the work concerns the identification and assessment of molecular markers for SAR, i.e. gene products that accumulate with a certain kinetics after the onset of SAR. Such markers might be a valuable tool for the identification of new resistance inducing compounds.

As one prerequisite for the assessment of a molecular marker we have analysed the cytological events on barley leaves and coleoptiles following treatment with a resistance inducing compound. This study was aimed at the definition of cytological criteria for SAR.

Another rational way for identification and assessment of a potential molecular marker is the use of plant genotypes with altered sensitivity for chemical resistance inducers. We discuss here a preliminary study of the expression of PR genes in such genotypes.

2. Introduction

We have analysed the phenomenon of SAR in the interaction of the powdery mildew fungus (*Erysiphe graminis*) with its respective cereal hosts, wheat (*Triticum aestivum* L.) and barley (*Hordeum vulgare* L.).

SAR is a well known phenomenon in the plant kingdom (Chester, 1933; Sequeira, 1983; Métraux *et al.*, 1991; Kessmann *et al.*, 1994). First documented in 1901, this type of resistance is thought to play an important role in the preservation of plants in nature. SAR at its simplest meaning can be defined as a general activation of the

H.-W. Dehne et al. (eds.), Diagnosis and Identification of Plant Pathogens, 229–235.

plant's defense mechanisms by biotic or abiotic inducers, e.g. necrotising pathogens or chemicals. The acquisition of this resistance provides long lived protection against a broad spectrum of pathogens, thus performs compelling demands for modern disease management strategies. Among the few active compounds identified during for the last years, only one was promising enough to be envisaged for commercial use. This compound, the benzo(1,2,3)thiadiazole-7-carbothioic acid *S*-methyl ester (BTH), activates gene expression and disease resistance in wheat (Görlach *et al.*, 1996) and dicots (Friedrich *et al.*, 1996; Lawton *et al.*, 1996).

Obviously, there is a compelling need for the development of new resistance inducing compounds, if the SAR concept shall be broadly introduced into practice. However, the search for new chemicals is seriously encountered by several difficulties. Most importantly, SAR in barley and wheat typically is associated with an intermediate infection type (no complete resistance). This problem is even strengthened by the fact that our knowledge about the conditions controlling SAR is in its infancy. From the practical point of view, this often results in an inability to discriminate between a resistance inducing activity and direct toxic effects of a given compound. Because these assays suffer from the absence of clearly defined criteria for SAR, they usually are much more critical than tests for e.g. modern fungicides. Hence, it is a scientific demand to develop new criteria for SAR and subsequently provide improved alternatives i.e. more reliable assays for inducers.

We report here on a macroscopic and microscopic inspection of SAR in barley plants induced by BTH. Additionally, we emphasise the value of barley coleoptiles for the identification of cytological criteria for SAR. Additional progress may come from the use of new barley genotypes with modified sensitivity for chemical inducers. We report on an approach aimed at the identification of barley genotypes with altered sensitivity to the resistance inducing compounds 2,6-dichloroisonicotinic acid (DCINA) and BTH.

3. Results and Discussion

3.1. CYTOLOGICAL MARKERS FOR SAR

A prerequisite for the identification of new chemical inducers is the understanding of the mechanisms underlying SAR. This implies the analysis of the cytological events characterising this type of resistance. As an example, we briefly discuss here the criteria for SAR by selecting the interaction of barley with the powdery mildew fungus (*Erysiphe graminis* f.sp. *hordei*, *Egh*).

A protective infiltration of barley primary leaves with BTH resulted in a significant reduction of the disease. The number of powdery mildew colonies on treated leaves was reduced, the pustules were smaller and surrounded by a few necrotic cell layers. Thus, fungal pustules may produce fewer and less virulent conidia resulting in a reduced spread of the pathogen under field conditions. Typically, complete resistance was not achieved.

Figure 1 shows a microscopic analysis of the BTH action. Fungal growth was arrested at early developmental stages. A detailed inspection of single sites of attempted infection revealed formation of cell wall appositions (PAP(eff) = effective papillae) being the basis for penetration resistance. Papillae as well as cell walls close by were associated with a yellow autofluorescence in UV light.

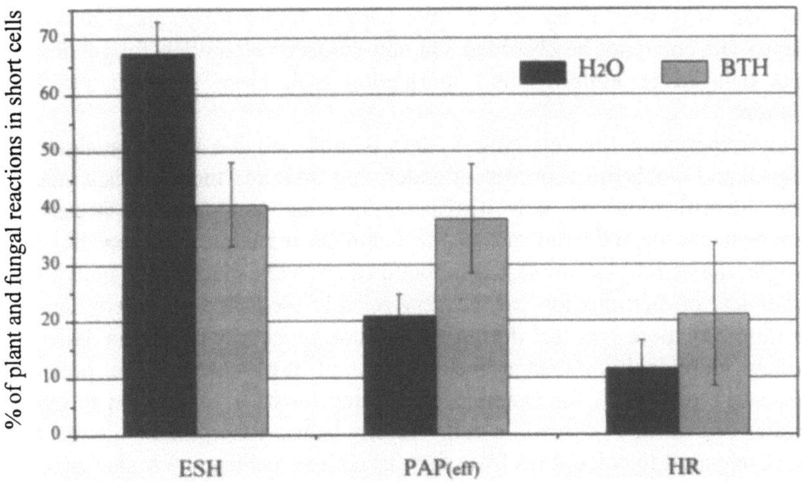

Figure 1. Occurrence of elongated secondary hyphae (ESH), effective papillae (PAP(eff)), and hypersensitive cell death (HR) after infiltration of barley primary leaves with water and BTH, respectively.
Seven-days-old leaves were infiltrated with the compound (1 mg/ml) or water and inoculated with *Erysiphe graminis* f.sp. *hordei* four days later. 48 hours after inoculation, 6 cm leaf segments were harvested and placed in a fixing solution (0.15 % trichloroacetic acid in ethyl alcohol:chloroform (4:1 [v/v]). For bright-field microscopy, fixed segments were stained with Coomassie blue (0.6 % Coomassie brilliant blue R-250, 15 % trichloroacetic acid, 1:1 [v/v]) for 15 s, washed in water and mounted in 50 % [v/v] glycerol. Whole cell autofluorescence was observed by fluorescence microscopy (wavelength = 310 nm). Each data point of the representative experiment shown is based on an inspection of 200 single-cell interaction sites on primary leaf segments from five individuals.

The autofluorescence may indicate the presence of polyphenols at the sites of attempted penetration, one reason for an enhanced toughness of the cell walls. As a consequence of the penetration resistance, the percentage of haustoria and secondary hyphae (ESH) was significantly reduced. In some cases, a hypersensitive response (HR) was seen in the epidermal cells attacked, but non-invaded by *Egh*, indicating that BTH modulates different defense responses.

There is now increasing evidence that papillae formation is sufficient for the expression of SAR. A strong argument is provided by the fact that SAR is expressed in tissues incapable to mount a HR. This is the case with the epidermal monolayers of barley coleoptiles. The infection process can be followed microscopically on the living tissue. On coleoptiles prepared from DCINA-induced barley seedlings, fungal development was arrested by an effective papilla. Importantly, the attacked cell still was alive - HR could not be observed. On non-induced coleoptiles, fungal development was not delayed or inhibited and sporulation took place at about six days after inoculation.

In our experience, the coleoptile system is most suitable for the evaluation of the cytological and biochemical processes underlying SAR and therefore is a valuable tool for the identification of new resistance-inducing compounds. Because fungal development can be followed during the complete infection cycle on the living cell monolayer, the activity of any test compound can be very specifically analysed. Effects of a chemical on germination can be very reliably identified on coleoptiles, whereas non-germinated spores get lost during the fixation procedure on leaves. Histochemical techniques additionally enhance the potency of this system. The potentiometric fluorescence dye $DiOC_6$, for example, was tested for its accumulation in cytoplasmic aggregates that occur in coleoptile cells directly beneath the primary germ tube of the fungus. Compared to coleoptiles from the susceptible seedlings, a higher percentage of attacked cells in those from either *Mlg* or chemically induced resistant seedlings showed an earlier accumulation of the fluorescent dye beneath the primary germ tube (Beckhove *et al.*, 1996).

3.2. ANALYSIS OF THE GENETIC BASIS FOR SAR

Previous work has shown that the microscopically visible defense reaction induced by DCINA is a phenocopy of the processes mediated by specific barley resistance genes governing penetration resistance. Examples for these genes are the *Mlg* gene (Görg *et al.*, 1993) and the recessively inherited *mlo* alleles (Aist *et al.*, 1988). The infection phenotype mediated by both traits is characterised by the formation of effective papillae resulting in fungal arrest.
Cytological investigation of processes leading to penetration resistance let to the hypothesis of a cross talk between the pathways of SAR and genetically based resistance. This consequently implies common genes required for these mechanisms. With regard to molecular markers, we therefore assume the existence of factors specifically accumulating in plants showing the phenotype of penetration resistance.

Moreover, there is increasing evidence that *mlo* genotypes are conditional SAR mutants. Table 1 displays the properties of resistance specified by *mlo* alleles and SAR.

There are great similarities between the two types of resistance mechanisms: *i. mlo* mediates fungal arrest solely by penetration resistance ($PAP_{(eff)}$); *ii. mlo* acts non-race-specifically i.e. almost all isolates of *Egh* are avirulent on *mlo* genotypes; *iii.* in *mlo* plants several PR genes like *PR1b* or chitinase are conditionally expressed, which, in SAR, typically accumulate regardless of the plant genotype.

Table 1. Comparison between *mlo*-specified and chemically induced resistance

	type of resistance	
characteristic features:	*mlo*-specified	induced
formation of effective papillae:	+	mainly
non-race-specific action:	+	+
accumulation of transcripts:	+	+

In an attempt to identify genes required for SAR, we have analysed barley genotypes with mutations in genes controlling *mlo*-mediated resistance against almost all known isolates of the powdery mildew fungus. We discuss here the utilisation of *Ror* mutants which were originally identified by the group of Schulze-Lefert in Norwich (Freialdenhoven *et al.*, 1996). *Ror* mutants derive from a mutagenization of *mlo*-resistant plants and are phenotypically characterised by a defect in the resistance response. The mutation does not lie in the *mlo* allele itself but in a second independent locus. According to the phenotype defined by the mutation, the mutated locus is called *Ror* for "required for *mlo*-specified resistance".

The obvious similarities of SAR and *mlo*-specified resistance posed the question whether the *Ror* locus is also required for the expression of SAR. Preliminary results indicate this is not the case. Most unexpectedly, compared to the *Mlo* wild-type plants, the *Ror-1* mutant tested so far showed a considerably higher sensitivity towards chemical inducers. Treatment with 0,1 ppm DCINA clearly retarded and partly inhibited fungal development, whereas wild-type plants were still completely susceptible at this concentration (Unpublished results: Jarosch, Freialdenhoven, Schulze-Lefert, Kogel). Because it is highly unlikely that enhanced sensitivity towards DCINA is due to the recessive mutation of the *Ror* locus, we speculate that the *mlo* allele itself controls SAR. Since *mlo*-resistant plants are incompatible with nearly all known isolates of *Egh*, the effect of *mlo* alleles on SAR is only detectable in these susceptible mutants. Thus, the use of *Ror* mutants in the elucidation of the SAR mechanism is a bright example for the potency of the mutational approach in plant science.

Other genotypes characterised by a mutation in a locus required for race-cultivar-specific resistance functions mediated by different alleles of the *Mla* locus (Jørgensen, 1996) were analysed for their competence to express SAR. These mutants proved to be less sensitive to DCINA and BTH (unpublished results of the authors). This observation provides further evidence for a cross talk between the pathways of SAR and genetically based resistance. Moreover, these mutants can serve as a tool to assess potential markers of SAR.

3.3. POTENTIAL MOLECULAR MARKERS FOR SAR

Genotypes with different sensitivity to resistance inducing compounds can be used for determining the value of a marker for SAR. This can be done by demonstrating a close correlation of accumulation of transcripts of marker genes and dose dependent occurrence of the resistance response.

The accumulation of classical PR proteins like PR1, chitinase or β-1,3-glucanase has been shown in barley, wheat and dicots when SAR is induced (Rebmann *et al.*, 1991; Ward *et al.*, 1991; Uknes *et al.*, 1992; Kogel *et al.*, 1994; Görlach *et al.*, 1996). These proteins also conditionally accumulate in *mlo*-resistant plants (Peterhänsel, personal communication). Further experiments will demonstrate whether the higher sensitivity of *mlo* genotypes towards resistance inducing compounds is reflected by the accumulation of these proteins upon chemical treatment.

In wheat, a group of genes called WCI (wheat chemically induced) were identified to be responsive to resistance inducing compounds (Görlach *et al.*, 1996). Because the kinetics of their expression is a characteristic feature only detectable after chemical treatment but not in e.g. a compatible interaction, these genes may be more specific molecular markers for SAR than genes encoding PR proteins.

Acknowledgement
This work is supported by the Bundesministerium für Bildung, Wissenschaft, Forschung und Technologie, Bonn, Germany.

4. References

Aist, J.R., Gold, R.E., Bayles, C.J., Morrison, G.H., Chandra, S., and Israel, H.W. (1988) Evidence that molecular components of papillae may be involved in *ml-o* resistance to barley powdery mildew, *Physiol. Mol. Plant Pathol.* **33,** 17-32.

Beckhove, U., Kogel, K.-H., Schmelzer, E. (1996) Acquired resistance in barley: Early accumulation of the potentiometric dye $DIOC_6$ in cytoplasmic aggregations beneath the primary germ tube of *Erysiphe graminis* f.sp. *hordei* of cells of barley coleoptiles induced by 2,6-dichloroisonicotinic acid, in H. Lyr, P.E. Russell, and H.D. Sisler (eds.), *Intercept of the 11th International Symposium of Modern Fungicides and Antifungal Compounds 1995*, Andover, pp. 475-481.

Chester, K.S. (1933) The problem of acquired physiological immunity in plants, *Q. Rev. of Biol.* **8,** 275-324.

Freialdenhoven, A., Peterhänsel, C., Kurth, J., Kreuzaler, F., and Schulze-Lefert, P. (1996) Identification of genes required for the function of non-race-specific *mlo* resistance to powdery mildew in barley, *Plant Cell* **8,** 5-14.

Friedrich, L., Lawton, K., Wilhelm, R., Masner, P., Specker, N., Gut Rella, M., Meier, B., Dincher, S., Staub, T., Uknes, S., Métraux, J.P., Kessmann, H., and Ryals, J. (1996) A Benzothiadiazole derivate induces systemic acquired resistance in tobacco, *Plant J.* **10**(1), 61-70.

Görg, R., Hollricher, K., and Schulze-Lefert, P. (1993) Functional analysis and RFLP-mediated mapping of the *Mlg* resistance locus in barley, *Plant J.* **3,** 857-866.

Görlach, J., Volrath, S., Knauf-Beiter, G., Hengy, G., Beckhove, U., Kogel, K.-H., Oostendorp, M., Staub, T., Ward, E., Kessmann, H., and Ryals, J. (1996) Benzothiadiazole, a novel class of inducers of systemic acquired resistance, activates gene expression and disease resistance in wheat, *Plant Cell* **8,** 629-643.

Jørgensen, J. H. (1995) Effect of three suppressors on the expression of powdery mildew resistance genes in barley, *Genome* **39**, 492-498.

Kessmann, H., Staub, T., Hofmann, C., Maetzke, T., Herzog, J., Ward, E.R., Uknes, S.J., and Ryals, J. (1994) Induction of systemic acquired resistance in plants by chemicals, *Ann. Rev. Plant Pathol.* **32,** 439-459.

Kogel,K.-H., Beckhove, U., Dreschers, J., Münch, S., and Rommé, Y. (1994) Acquired resistance in barley, *Plant Physiol.* **106,** 1269-1277.

Lawton, K.A., Friedrich, L., Hunt, M., Weymann, K., Delaney, T., Kessmann, H., Staub, T., and Ryals, J. (1996) Benzothiadiazole induces disease resistance in *Arabidopsis* by activation of the systemic acquired signal transduction pathway, *Plant J.* **10**(1), 71-82.

Métraux, J.P., Ahl Goy, P., Staub, T., Speich, J., Steinemann, A., Ryals, J., and Ward, E. (1991) Induced resistance in cucumber in response to 2,6-dichloroisonicotinic acid and pathogens, in H. Hennecke and D.P.S. Verma (eds.), *Advances in Molecular Genetics of Plant-Microbe Interactions*, Vol. 1, Kluwer Academic Publishers, Dordrecht, pp. 432-439.

Rebmann, G., Hertig, C., Bull, J., Mauch, F., Dudler, R. (1991) Cloning and sequencing of cDNAs encoding a pathogen-induced putative peroxidase of wheat (*Triticum aestivum L.*), *Plant Mol. Biol.* **16,** 329-331.

Sequeira, L. (1983) Mechanisms of induced resistance in plants, *Ann. Rev. Microbiol.* **37,** 51-79.

Uknes, S.J., Mauch-Mani, B., Moyer, M., Potter, S., Williams, S.C., Dincher, S.S., Chandler, D., Slusarenko, A., Ward, E.R., and Ryals, J. (1992) Acquired resistance in *Arabidopsis*, *Plant Cell* **4,** 645-656.

Ward, E.R., Uknes, S.J., Williams, S.C., Dincher, S.S., Wiederhold, L., Alexander, D.C., Goy, P.A., Métraux, J.P., and Ryals, J. (1991) Coordinate gene activity in response to agents that induce systemic acquired resistance, *Plant Cell* **3,** 1085-1094.

QUANTIFICATION OF FUNGAL BIOMASS OF A PATHOGEN AND AN ANTAGONIST IN NECROTIC LEAF TISSUE USING FLUORESCENCE MICROSCOPY AND IMAGE ANALYSIS

KESSEL, G.J.T.[1,2], LOMBAERS-VAN DER PLAS, C.H.[1], MEIJER, E.M.J.[1], KÖHL, J.[1] AND VAN DER WERF, W.[2]

[1] *DLO Research Institute for Plant Protection (IPO-DLO), Wageningen, the Netherlands*
[2] *Department of Theoretical Production Ecology (TPE-WAU), Wageningen Agricultural University, Wageningen, the Netherlands*

Weekly fungicide applications against Botrytis are common practice in Dutch flower bulb crops such as lilies and tulips. Governmental policy measures will restrict this pesticide use in order to reduce environmental side effects. Biological control might offer an alternative or supplement to present control options.
We study the possibilities of biological control of Botrytis elliptica with the saprophytic fungus Ulocladium atrum. This antagonist can colonize dead and senescent leaf tissue that would otherwise be colonized by Botrytis. Hence, the amount of sporulating Botrytis is reduced, which is expected to delay Botrytis epidemics.

The competition between Botrytis and Ulocladium atrum in dead and senescent leaves is studied microscopically. A dual-staining procedure was developed to detect and distinguish both fungi and to estimate their biomass within the senescent lily leaf. Samples are taken from leaves inoculated with both fungi, and sectioned using a freeze microtome. Botrytis hyphae in the sections are stained using the genus specific monoclonal antibody BC-KH4 (obtained from Dr F.M. Dewey, Oxford University), conjugated to the fluorochrome FITC (green fluorescent). After the Botrytis stain a general fungal stain, wheat germ agglutinin conjugated to the fluorochrome TRITC (red fluorescent), is applied. This general stain will attach to Botrytis as well as to the antagonist. Images of leaf sections, obtained from a fluorescence microscope, are interpreted using a dedicated image analysis program which calculates the area of both fungi relative to the area of the section. The area proportions of both fungi in the section are estimators for fungal biomass and can be used as indicators of fungal competition in the leaf .

Based upon the results of microscopic and macroscopic observations on the etiology of Botrytis lesions and the substrate competition process, a dynamic simulation model

H.-W. Dehne et al. (eds.), Diagnosis and Identification of Plant Pathogens, 237–238.

will be developed for the growth and interaction of both fungi in lily leaves as influenced by abiotic factors and leaf senescence. The model will be used as an integrative tool, to estimate the importance of environmental and other ecological factors on the competitive substrate colonization by *Botrytis* and *Ulocladium*. Later, the model may be used as an instrument to predict the outcome of the *Botrytis*-*Ulocladium* interaction under field conditions and to evaluate options for effective control strategies, using *Ulocladium*.

A NOVEL SPORE TRAP FOR MONITORING AIRBORNE *ALTERNARIA BRASSICAE* SPORES USING MONOCLONAL ANTIBODIES

D. SCHMECHEL[1], H.A. MCCARTNEY[1] & N. MAGAN[2]

[1]*IACR-Rothamsted*
Department of Crop and
Disease Management
Harpenden, Herts.
AL5 2JQ

[2]*Cranfield Biotechnology Centre*
Cranfield University
Cranfield, Bedfordshire
MK43 0AL

1. Introduction

For many crop diseases caused by fungi, the pathogen is well established before symptoms appear and control measures based on disease detection by direct observation may be less efficient than they would have been had the risk for disease been detected earlier. For air-dispersed fungal pathogens the potential for disease may be estimated by monitoring the spatial and temporal exposure of the crop to inoculum. If this information were readily available, it would allow the development of more rational, environmentally acceptable systems for disease control decision making.
The most commonly used spore samplers in plant pathology are Burkard spore traps and rotorod samplers (McCartney *et al.*, 1996). However sample assessment is time-consuming and the results may be biased according to the taxonomical skills of the personnel.
Immunometric techniques based on enzyme-linked immunosorbent assay (ELISA) have the potential to provide accurate, quantitative and specific data on target antigens and methods have been developed for the identification of fungal pathogens in soil samples and plant tissue (Dewey, 1996). However, due to the lack of an appropriate spore sampler, no attempts have been made to apply ELISA techniques for the detection of air-dispersed fungal propagules. In this paper we describe a spore trap referred to as strip rotorod trap (SRT) which provides samples compatible with direct ELISA processing.

2. Materials and Methods

2.1. DESCRIPTION OF THE WIND TUNNEL EXPERIMENTS

A detailed description of the wind tunnel has been published by Fitt *et al.* (1986).

H.-W. Dehne et al. (eds.), Diagnosis and Identification of Plant Pathogens, 239–242.

Briefly, air was drawn by an axial fan through an 11 meter long tunnel with an internal cross-section of 1 x 1 m. The wind speed range of the tunnel is between 0.5 and 8 ms^{-1}. Wind speed, temperature and relative humidity were continuously recorded.
Experiments were carried out with spores of *Lycopodium clavatum* L. (Pteridophyta) and *Alternaria brassicae* (Berk.)Sacc. The collection efficiency of the SRT was calibrated using a miniature suction trap with collection efficiencies comparable to that of a Burkard spore trap as reference trap. Test particles were released from a spore dispenser in case of *L. clavatum* or infected oilseed rape pods in case of *A. brassicae*. Spores were then simultaneously collected by the reference trap and the SRT about 5.5 m downwind from where the spores were released. Spores collected by each trap were counted and the relative spore collection factor as the ratio of the number of spores caught by the SRT divided by the number of spores collected by the reference trap was calculated. This factor is a measure for as how many times more spores are collected by the SRT in relation to the reference trap. In case of *A. brassicae*, spore numbers per well were also estimated by ELISA using monoclonal antibody 34/7/G (Schmechel *et al.*, 1996). The traps were tested at 1, 2 and 3 ms^{-1}.

2.2. CONSTRUCTION OF THE TRAP

A conventional rotorod trap was chosen as the starting design for the development of a prototype of spore sampler suitable for direct ELISA processing of samples. The main reason was the relatively high sample volume (150 l min^{-1} for traps used at Rothamsted) which results in a higher spore catch per unit sampling time. In order to facilitate direct ELISA processing, the adhesive-coated tape as collection material was replaced with 8 well ELISA strip modules (PolySorp Nunc Immuno Modules, Nunc InterMed, Cat. No. F8 469078). Modules were attached to the rotorod arms by two rubber bands for ease of removal so that the long axis of the module was parallel to the axis of rotation.The relative arrangement of the well orifice in relation to the direction of rotation was adjusted so that the majority of the spores impacted on the base of individual wells rather than the side wall of the wells (Schmechel, 1996).

3. Results

Experiments using smoke as simulator of air movement through the trap showed that air enters the trap from above and below the volume swept by the rotating arms during sampling. The nominal collection volume of the SRT was calculated to be 537 l min^{-1} when operated at a speed of rotation of 2800 rpm. The configuration of the SRT is shown in Plate 1. The two 6 mm wide rubber bands, which hold the strips in place when attached to the support arm, cover the third and forth top wells of each strip which are therefore unable to collect spores.
Figure 1 shows that the collection efficiency of the SRT was inversely correlated with

the wind speed, with collection factors ranging from 54 times at 1 ms^{-1} to 24 times at 3 ms^{-1} for *L. clavatum* and 24 times at 1 ms^{-1} to 15 times at 3 ms^{-1} for *A. brassicae*. The variability can probably be explained by the fact that particles travelling at higher wind velocities possess greater kinetic energies which make them less likely to be deflected enough to be collected by the trap. The collection efficiencies of the SRT when compared to those of a Burkard spore trap were found to be 0.74 for L. clavatum and 0.4 for *A. brassicae*. Figure 2 shows that individual wells collect similar proportions of the total amount of spores collected by the trap when averaged over wind speeds of 1 - 3 ms^{-1}. However, at 1 ms^{-1} inner wells seem to collect more spores than wells at the top or bottom of the ELISA module. The reverse was observed for 3 ms^{-1}. Thus, for ELISA-based calculation of air spore concentrations, it is suggested to use all wells of one module as test wells for the estimation of spore numbers collected by the trap and the wells of the second module as negative control wells in ELISA.

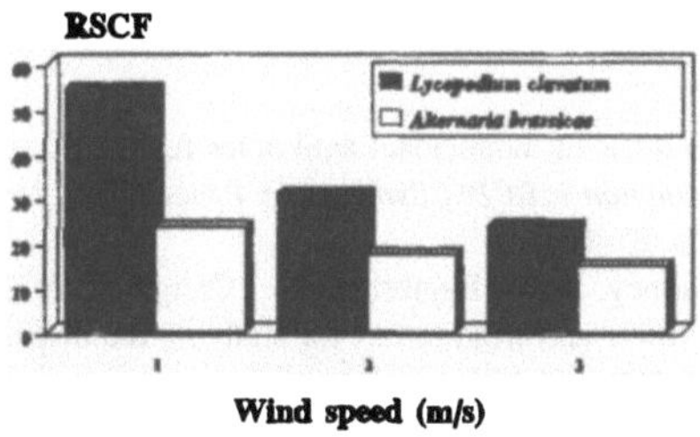

Figure 1 Relative spore collection factors (RSCF) at different wind speeds. The factor is a measure for as how many times more spores are collected by the SRT in relation to the reference trap which has the factor 1.

The suitability of the trap for direct ELISA-based sample analysis was tested by collecting spores of *A. brassicae* in wind tunnel experiments. Samples were then sequentially processed by ELISA using monoclonal antibody 34/7/G and the number of spores in each ELISA well was counted using a light microscope. A good correlation between the counted and estimated number of spores per well was obtained (Schmechel *et al.* 1996). Thus, the trap has the potential to provide the technical base for the development of ELISA-based immunomonitoring techniques.

4. Discussion

A prototype of a spore trap providing samples compatible with direct ELISA processing has been developed. Wind tunnel experiments showed that the collection efficiency is inversely correlated with the wind speed and particle size. The trap is portable, battery-driven and easy to install and maintain. For field operation the trap can be mounted on a pole and the height can be adjusted as required. Initial field experiments showed that the trap should be covered by a fine mesh (1 x 2 mm) in order to prevent the collection of insects. Future work should concentrate on: (a) field experiments; (b) the influence of vertical spore gradients in the field.

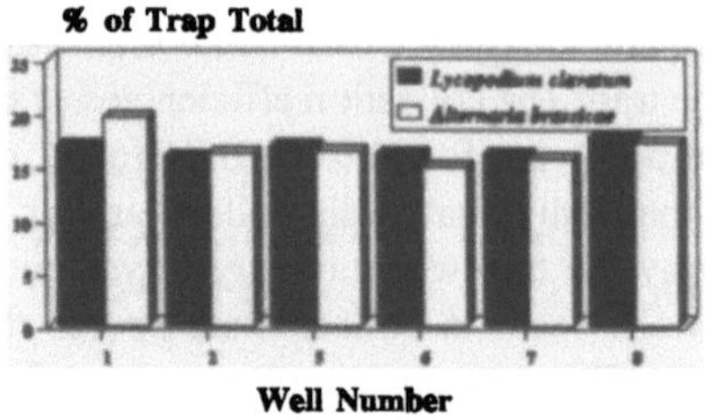

Figure 2. Distribution of test particles among wells as the average of 1 - 3 ms^{-1}. The values represent the proportion of spores collected by individual wells in relation to the total number of spores collected by the trap.

5. References

Dewey, F.M. (1996). Production and use of monoclonal antibodies for the detection of fungi. In: Marshall, G. *Diagnostics in Crop Production - BCPC Symposium Proceedings No. 65*, British Crop Protection Council, Farnham, UK, pp. 85-91.

Fitt, B.D.L.; Walklate, P.J.; McCartney, H.M.; Bainbridge, A.; Creighton, N.F.; Hirst, J.M.; Lacey, M.E. & Legg, B.J. (1986). A rain tower and wind tunnel for studying the dispersal of plant pathogens by rain and wind. *Ann. appl. Biol.* **109**, 661-671.

Schmechel, D. (1996). *Development of an immunological approach to monitoring airborne inoculum of fungal pathogens*. PhD thesis, Cranfield University.

Schmechel, D.; McCartney, H.A. & Magan, N. (1996). A novel approach for immunomonitoring airborne fungal pathogens. In: Marshall, G. *Diagnostics in Crop Production - BCPC Symposium Proceedings No. 65*, British Crop Protection Council, Farnham, UK, pp. 93-98.

McCartney, H.A.; Fitt, B.D.L. & Schmechel, D. (1996). Sampling bioaerosols in plant pathology. *Journal of Aerosol Science* (in press).

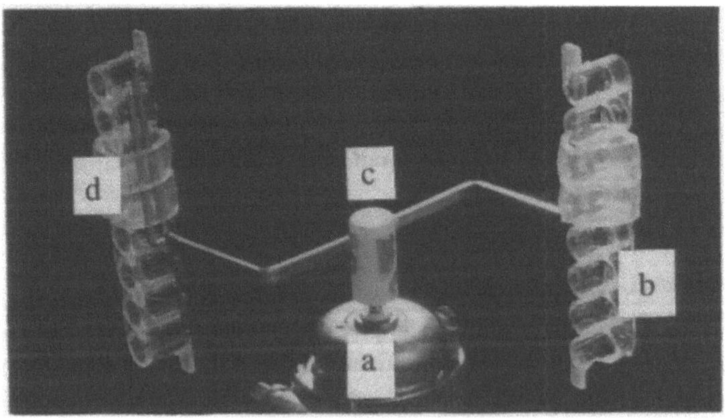

Plate 1. Configuration of the strip rotorod trap (SRT): (a) 12 V electric motor, (b) modified ELISA strip modules, (c) angular support arm, (d) 6 mm wide rubber bands.

FUSARIUM EAR BLIGHT ON WHEAT - EPIDEMIOLOGY AND CONTROL OF *GIBBERELLA ZEAE*, THE TELEOMORPH OF *FUSARIUM GRAMINEARUM*, WITH FOLICUR®

A. SUTY, A. MAULER-MACHNIK
Bayer AG
Agrochemicals Division, Research/Fungicides, D-51368 Leverkusen

1. Introduction

Ear fusariosis is caused mainly by *Fusarium graminearum, Fusarium culmorum* and *Microdochium nivale*. The significance and distribution of the fusarioses was investigated in various regions of Germany between 1987 and 1993 (Mauler-Machnik and Zahn, 1994). The most common species was *F. graminearum*. In the ear, this species could be found at numerous sites in Germany. In contrast to *F. culmorum* and *G. nivalis, F. graminearum* was only rarely isolated from wheat leaves in our investigations. It appears that the ear can be infected via a route which does not involve the leaves. Unlike the conidia of *F. graminearum*, which are carried over short distances in splash water droplets, spores of *Gibberella zeae* (teleomorph of *F. graminearum*) can be transported over long distances by wind (Reis, 1988a). It was therefore suspected that *G. zeae* could play an important role in the epidemiology of this disease. Results on epidemiology and control of *G. zeae* are presented in this paper.

2. Material and methods

Over 3 years (1993 to 1995) a total of 9 field trials (winter wheat) were carried out under natural infection conditions in various regions of Germany. Maize-crop debris - from previous crop maize - were checked for the formation of perithecia every month, starting from January. The quantity of perithecia formed was assessed and the degree of ripening was ascertained. At one site over two years, ascospore release was measured 10 cm above the ground and at ear height, with the aid of a spore trap.

Treatments with tebuconazole (as ®Folicur 250EW or ®Matador) were carried out once or twice starting at growth stage EC55, under natural infection conditions. The level of infection was evaluated on the basis of the per cent reduction in infected spikelets at EC85 (soft dough stage).

H.-W. Dehne et al. (eds.), Diagnosis and Identification of Plant Pathogens, 243–246.

In the laboratory, the *in-vitro* efficacy of tebuconazole against *G. zeae* was ascertained with the aid of a germ-tube test.

3. Results and discussion

3.1. PERITHECIA FORMATION IN THE FIELD

Perithecia on maize debris were formed in May between 1993 and 1995, in 6 of the 9 field trials (table 1). No perithecia were found before May. According to Ye (1980), the formation of perithecia was promoted by high temperatures and wet weather. Numbers of perithecia remained generally low, as did ear infection levels. However, at the Dingolfing (a) location perithecia formation was particularly high in 1994. More than 50% of the investigated maize-crop residues were affected by perithecia and more than 2000 perithecia could be counted on one fragment with a surface area of 50 cm^2. This represents an inoculum potential of about 10^8 ascospores, given that a perithecium can contain up to 45.000 ascospores (Kongha and Sutton, 1988).

TABLE 1: Formation of perithecia and dispersal of ascospores of *G. zeae* and infection of wheat ears at EC 85.

rural district	Year	% of maize debris with Perithecia	*Number of perithecia	Ascospore peak (number/cm2 of slide)	% Infected spikelets
Main-Kinzig	1993	19	+	-	0
Main-Kinzig	1994	0	0	-	0
Dingolfing (a)	1993	30	+	-	-
Dingolfing (a)	1994	52	++++	18.9	27
Dingolfing (a)	1995	20	+	3.2	0
Dingolfing (b)	1993	22	+	-	1
Herford	1993	0	0	-	0
Herford	1994	0	0	-	0
Unstrut	1994	10	+	-	-

*+ = few; ++ = moderately abundant; +++ = abundant;
++++ = very abundant; - = not determined

3.2. RELEASE OF ASCOSPORES

At the Dingolfing location, ascospore dispersal was investigated in 1994 and 1995. In 1994 ascospores were caught from the end of May to the end of June. In general the mean number of ascospores caught at ground level was approximately twice that at the wheat ear level. Two peaks of ascospore release at ear level could be observed (Figure 1): at heading and at flowering. In each case, it rained shortly before or during ascospore release. Consequently, ear infection was high at EC 85 and reached 27% infected spikelets in the control plot.

The number of ascospores caught during the same period in 1995 was very small and no clear ascospore peak was detected.

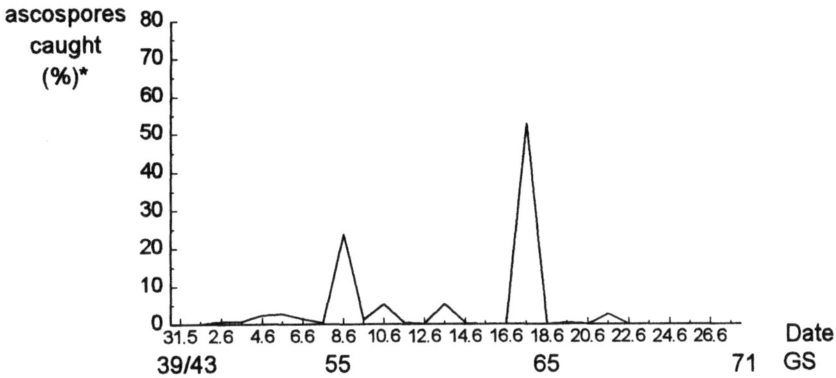

*as a % of the total ascospores caught during GS39-GS71

FIGURE 1: Ascospores caught at ear level from EC39 to EC71

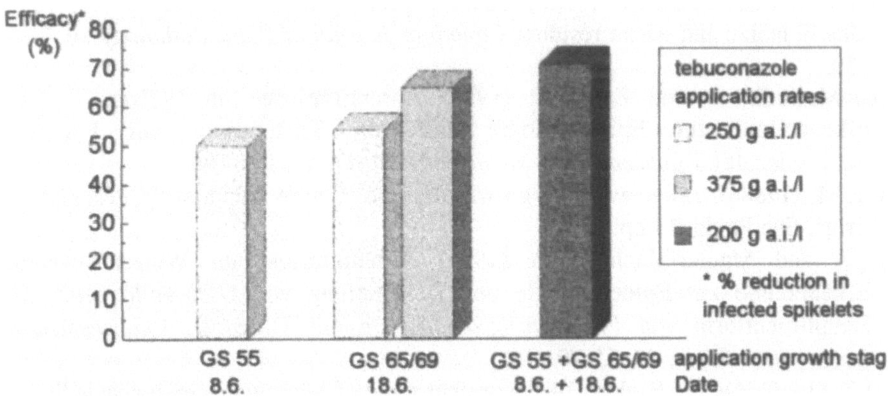

FIGURE 2: Efficacy of products containing tebuconazole in relation to ascospore peaks

3.3. EFFICACY OF TEBUCONAZOLE

In vitro studies with the aid of a germ tube test demonstrated that the active ingredient tebuconazole was similarly very effective against ascospores of *G. zeae* and conidia of *F. graminearum* (Suty and Mauler-Machnik, 1996). At the Dingolfing site, treatments with products containing tebuconazole were carried out on the basis of the ascospore peaks observed (Figure 2). The two ascospore peaks at ear level show clearly the difficulty of correctly timing fungicide applications against ear fusarioses. The efficacy achieved after a single treatment (250 g/l a.i.) at EC55 or EC65 was about 50%. The increase of active ingredient (375 g/l a.i.) at EC 65 improved the efficacy level to 65%. On the other hand, a double application using reduced application rates (200 mg/l a.i.) increased efficacy to more than 70%.

4. Conclusion

Field studies have demonstrated for the first time the efficacy of products containing tebuconazole against *G. zeae*. The efficacy level achieved (50-70%), corresponds to the known efficacy against the imperfect stage of *F. graminearum*. Since the treatment of ear fusarioses is difficult to time correctly, it should be supported by cultivation measures in accordance with the principles of integrated crop protection.

5. References

Khonga, E.B. and Sutton, J.C. (1988):Inoculum production and survival of *Gibberella zeae* in maize and wheat residues.*Canadian Journal of Plant Pathology* **10**, 232-239.

Mauler-Machnik, A and Zahn, K (1994) Ährenfusariosen an Weizen - Neue Erkenntnisse zur Epidemiologie und zur Bekämpfung mit Folicur® (tebuconazole). *Pflanzenschutz-Nachrichten Bayer,* **47**, 133-160.

Reis, E.M. (1988a): *Doencas do Trigo III Giberela.* Centro nacional de Pesquisa de Trigo. Sao Paulo, 13 pp.

Suty, A and Mauler-Machnik, A (1996) Ährenfusariose an Weizen - Neue Erkenntnisse zur Epidemiologie und Bekämpfung von *Gibberella zeae*, der Hauptfruchtform von *Fusarium graminearum* mit Folicur®. *Pflanzenschutz-Nachrichten Bayer,* **49**, 55-70.

Ye, H.Z. (1980): On the Biology of the perfect stage of *Fusarium graminearum* Schw. *Acta Phytophylacica Sinica* 7, 35-42.

RISK ASSESSMENT THROUGH QUANTITATIVE DETECTION OF SOIL SUPPRESSIVENESS AGAINST PLANT DISEASES CAUSED BY *Rhizoctonia solani* AG 2-1, AG 2-t and AG 4, *Fusarium solani* f.sp. pisi, *Thielaviopsis basicola, Aphanomyces euteiches*, trichodorid nematodes and *Meloidogyne hapla*

G. Dijst, P.J. Oyarzun[#], F.C. Zoon, C.J. Kok, M. Gerlagh, P.W.Th. Maas, and N.J. Fokkema. *DLO-Research Institute for Plant Protection (IPO-DLO), P.O. Box 9060, 6700 GW Wageningen, the Netherlands. E-mail: name@IPO.DLO.NL ([#]CIP, P.O.17-21-1977 Quito, Ecuador)*

1. Introduction

To develop integrated pest management (IPM) against soil-borne diseases in field grown crops, we seek new options for disease control. Crops, crop residues, soil amendments and cultural methods are known to affect the soil-ecosystem and thus the natural inoculum potential of the soil (IPS) as well as the disease suppressive capacity of the soil-ecosystem. Many publications reported on such effects in qualitative terms. We aim to develop quantitative measurements and risk analyses to allow conscious management based upon choices between fields, crop rotation etc. in order to avoid uneconomic control inputs, yield loss and loss of 'soil quality' in a sustainable agriculture.

A soil-ecosystem approach is used in our studies to develop risk assessment through quantitative measurements and analyses of soil suppressiveness. In risk analyses for soil-borne pathogens, an estimation of the presence of inoculum or the density of inoculum can be meaningless, as shown by e.g. Amir and Alabouvette (1993). They demonstrated that *F. oxysporum* f.sp. *lini* survived much better in a clay than in a sandy soil. Yet, the sand was highly conducive for disease development whereas the clay strongly suppressed the disease. Oyarzun et al. (1996) demonstrated the pathogen-specificity in soil suppressiveness. So, as a qualitative measure we use the term '*soil suppressiveness*' defined as the (pathogen specific) capacity of a soil-ecosystem to reduce disease in a susceptible host. As a quantitative measure we use the term '*soil receptivity*' defined as the (pathogen specific) extend of the effect of a soil-ecosystem on the inoculum potential of the soil, ranging from conducive to suppressive.

Perspectives created by our soil suppressiveness studies are in the first place the development of techniques (tests and test conditions) and (statistical) procedures to measure, indicate and analyse the soil receptivity to disease development and, if necessary, to the population dynamics of the pathogen as affected by soil and cropping factors. Measurements of soil receptivity open new options for IPM: to select land for a crop and to protect and improve the natural resource of disease suppressiveness of the field soil through risk analysis of the effects of crop residues, crops, compost and other cultural methods. The assays are also used for analysis of the mechanisms of suppression. This adds to the existing options for IPM by isolation of new biocontrol

H.-W. Dehne et al. (eds.), Diagnosis and Identification of Plant Pathogens, 247–251.

agents and integration of these with existing control measures. Further analyses may reveal new options to fool or fight the pathogens and prevent disease outbreaks. Integration of these studies leads to the development of 'site-specific' crop & soil management and prediction strategies for small scale cropping areas. At a larger scale, options for 'field-specific' crop & soil management may be developed, upon extension of the knowledge on distribution patterns of suppressiveness and infestation, development of field sampling protocols, studies on economic threshold levels and integration with expert systems.

2. Our accomplishments

At IPO-DLO, bioassays have been developed to estimate the natural disease suppressiveness of field soils and their infestation. A special system was developed to regulate the soil water retention and soil temperature during these bioassays (Oyarzun et al., 1994a, Oyarzun and Dijst, 1991). This is essential for comparisons between different soils because these factors influence the dynamics of both the pathogens and of its natural enemies. For the eight pathosystems studied, the differences in soil receptivity measured between (still small sets of) samples corresponded with growers' reports on differences in disease incidence between the sampled sites. The state of the art differs between the pathosystems studied.

Proceedings [#)]	Oyarzun Pea root rot			Kok *Meloidogyne hapla*	Zoon Trichodorid nematodes	Dijst *R. solani* AG		
	Fsp	Tb	Ae[##)]			2-1	2-t	4
1. Techniques : Assay, quantification	p	p	p	+	+	p	+	+
2. Risks assessments: crops, residues, etc.	p	p	p		+		+	+
3. Analysis: - indicators for suppression	p	p	p				+	+
- new options: ecology & damage					+		+	+
- new antagonists for biocontrol	p						+	+
4. Site-specific Crop & Soil Management: - Integrate 1.,2. and 3: Prediction - Expert systems								
5. Field-specific Crop & Soil Management - economic threshold levels - field sampling protocols	+	+	+					

#): + = under investigation, p = published (Oyarzun 1993, 1994; Oyarzun et al., 1991, '94a, '94b, '95, '96)
##): Fsp = *Fusarium solani* f.sp. *pisi*, Tb = *Thielaviopsis basicola*, Ae = *Aphanomyces euteiches*

Oyarzun et al. (1994b, 1996) determined the receptivity separately for each of the three most dominant fungal pea root rot pathogens *Fusarium solani* f.sp. *pisi* (Fsp), *Aphanomyces euteiches* (Ae), *Thielaviopsis basicola* (Tb),. Bulk samples were collected from large fields and the measured IPS correlated with the disease incidence in these fields (Oyarzun 1993, 1994). For a set of 30 of these soils with a low natural IPS, the disease severity was determined per pathogen at an artificially created range of inoculum levels. Multivariate analyses of the disease responses in the bioassay enabled a distinct classification of the tested soils for their receptivity, but these procedures only yield a ranking depending on the set of soils. An independent characterization of the receptivity per soil was found in the scale parameter B obtained by Weibull model fitting of each individual disease response curve. The Weibull shape-parameter C further differentiated the type of response. The receptivity of each soil appeared specific for each pathogen and different mechanisms seemed to be involved. Soil receptivity to Fsp largely differed between the tested soils. To Tb, most soils responded more or less suppressive. To Ae, most soils were more or less conducive, which indicates that this may be a risk pathogen for Dutch soils and that determination of its presence may be more important than the determination of the soils' receptivity. For both Tb and Fsp, correlations were found between the receptivity of soils and certain abiotic soil factors (Oyarzun, 1994). Heat sterilization eliminated any suppressive capacity of the soils, indicating that the soil biota effectuate the suppression. New antagonists were isolated from suppressive soils (Castejon-Munoz and Oyarzun, 1995).

Kok et al. developed a bioassay to measure the effect of soils on the root knot nematode *Meloidogyne hapla* (Maas et al., 1996). In the bioassay a fixed number of nematodes is artificially inoculated near a lettuce plantlet. Because the effect of the distance of the inoculum from the plantlet was the same in different soils, the shortest distance was chosen. In the bioassay not the number of galls in the roots, but the number of second generation juveniles from the galls corresponded with the difference in population densities between the sampled field sites. Heat treatment of the soils reduced a general antagonism to gall formation to an equal extent in all soils.

Zoon et al. developed a bioassay for trichodorid nematodes. Only light soils are known to be conducive to these nematodes (Maas et al., 1996). The estimation of the soil receptivity is based on the host-finding activity (activation, attraction, migration). The test system consists of three soil layers separated by sieves that don't permit root penetration but allow the migration of nematodes: a bottom-layer of soil with inactive natural inoculum, a middle-layer with a soil-to-be-tested and an upper-layer with sandy soil planted with sugarbeet plantlets. The percentage of nematodes that arrive in the upper rooted soil layer is determined. A soil sample with a higher organic matter content appeared more suppressive than other sandy soil samples. Amending a conducive sandy soil with different composts or manure significantly increased the soil suppressiveness. Aspecific antagonism seems responsible for these effects. Effects of pre-crops, composts and other soil treatments are under further investigation.

For three anastomosis groups (AG's) of *Rhizoctonia solani* Dijst et al. developed bioassays based on the spread of disease in a point inoculated row of plants (Dijst and Schneider, 1996; Oyarzun et al., 1994a; Oyarzun and Dijst, 1991). For AG-2-1 in cauliflower, disease progress curves were obtained by counting the number of diseased seedlings (damping off) at weekly intervals. For AG-2-t (at 12 °C) and for AG-4 (at 18 °C) disease progress curves were obtained by measuring the severity of bulb rot in a row of iris after five weeks. Soil samples had been collected from small sites within different commercial fields which differed in soil type or disease incidence. For two sites within one field, which differ in disease incidence, a difference in soil suppressiveness was determined in three successive years. This confirms previous reports in that also stable differences in soil receptivity for *R. solani* exist in practise (Huber and Sumner, 1996). Multivariate analyses of the disease response curves yielded a clear ranking of the soils for receptivity. An independent receptivity characterization for each soil is sought now, according to the procedures developed by Oyarzun et al. (1996, 1994). Current investigations look for possible correlations between soil receptivity and abiotic characteristics in the soil samples of the 40 sampled field sites. The soil biota appear to effectuate the suppressiveness, since sterilization (gamma radiation) eliminated any suppressiveness in each sample tested. Amending the soil with wheat straw some months before the test, induced a strong disease suppression. The results suggest that different mechanisms are involved in the disease suppression and that competition for food may play an important role.

Each of these results are in line with the many previous reports on disease suppression in other rhizoctonia-pathosystems as reviewed by Huber and Sumner (1996). According to these authors soil suppressiveness against rhizoctonia-diseases has been related to both aspecific and specific microbial activity, can occur due to stable soil characteristics and can be introduced by organic and inorganic amendments. Not every type of disease suppression goes with a reduction in population density. Introduced suppression may be as much from modifying the abiotic or biologic soil environment to affect virulence, pathogenicity or density of the pathogen or, to induce plant resistance and, may also be based on alterations in the availability of nutrients to the pathogen or the plant.

Our further analyses of the iris-pathosystems showed that the food supply to the inoculum influenced the inoculum aggressiveness and that the presence of inoculum did not necessarily cause disease. The presence of iris host plants helped the pathogens to partly overcome the soils' suppression against inoculum spread and density. The effects of amending soil with crop residues (both on the soil inoculum and on the soil suppressiveness) is under further investigation. In the bioassay at 18 °C for AG-4 all soils performed a stronger disease suppression than in the test at 12 °C for AG-2-t, for which an effect of the difference in temperature on the soil biota may account. Results suggest that the natural suppression of these field soils against *Rhizoctonia*-disease in iris and tulip is mainly based on aspecific mechanisms of the soil-ecosystem and that it can be improved.

3. Discussion

Our results clearly emphasise that, with regard to soil-borne pathogens, the benefits (and risks) of soil treatments, crops etc. should be assessed per pathogen and both with regard to the soil inoculum and the soil suppressiveness. Because different suppressive factors and mechanisms can act interactively (Huber and Sumner, 1996), our bioassay stay as comparable to the original pathosystem as possible and take the effect of different soil types into account. For the investigated pathosystems the obtained techniques and information offer the possibility for an intelligent management of the soils' suppressiveness to disease. In line with Alabouvette (1986) our results indicate that soil treatments with a negative effect on the soil biota may hold a risk.

For the development of site-specific or even field-specific crop & soil management, the long term scenario may seem discouraging. Yet, realization is feasible because of the opportunity to recognize sub-projects in the fields of risk assessment, biocontrol, cultural and integrated control, if necessary detection and identification, population dynamics, epidemiology and soil-ecosystem analyses. There is a lack of essential knowledge of spatial distribution and temporal dynamics of the inoculum and of the soil suppressiveness. Also, relations between inoculum density, IPS, damage and economic threshold levels need clarification. As for *R. solani,* such studies in field soils are hampered by a lack of easy and adequate techniques for isolation, identification and detection at the AG-sub-group levels. Furthermore, expert systems are needed to integrate with and optimize multiple goals. We welcome any initiative to develop a joint EU approach either per pathosystem or across pathosystems.

4. References

Amir, H., and Alabouvette, C., 1993. Involvement of soil abiotic factors in the mechanisms of soil suppressiveness to fusarium wilt. *Soil Biology and Biochemistry* **25**: 157-164

Alabouvette, C., 1986. Fusarium-wilt suppressive soils from the Chateaurenard region: review of a ten-year study. Agronomy **6**: 273-284

Castejon-Munoz, M., and Oyarzun, P.J., 1995. Soil receptivity to *Fusarium solani* f.sp. *pisi* and biological control of root rot of pea. *European Journal of Plant Pathology* **101**: 35-49

Dijst, G., and Schneider J.H.M., 1996. *Rhizoctonia* species in flower bulbs. Chapter V-4 in: Sneh, B., Jabaji-Hare, S., Neate, S., and Dijst, G. (eds), 1996. *Rhizoctonia species: Taxonomy, Molecular Biology, Ecology, Pathology and Control.* Kluwer, Dordrecht. pp. 279-288.

Huber, D.M., and Sumner, D.R.,1996. Suppressive soil amendments for the control of *Rhizoctonia* species. Chapter VI-A3 in: Sneh, B., Jabaji-Hare, S., Neate, S., and Dijst, G. (eds), 1996. *Rhizoctonia species: Taxonomy, Molecular Biology, Ecology, Pathology and Control.* Kluwer, Dordrecht. pp. 433-443.

Maas, P.W.Th., Kok, C.J., and Zoon, F.C., 1996. Measuring and description to soil-ecosystem resistance to nematode damage. In: *Proc. Third Nematology Congress,* July 7-12 1996 Guadeloupe. p. 60.

Oyarzun, P.J., 1994. *Root rot of peas in The Netherlands; Fungal pathogens, inoculum potential and soil receptivity.* PhD Thesis, Agricultural University Wageningen, the Netherlands.

Oyarzun, P.J., 1993. Bioassay to assess root rot in pea and effect of root rot on yield. *Netherlands Journal of Plant Pathology* **99**: 61-75.

Oyarzun, P.J., Maas, P.W.Th., Zoon, F.C., and Dijst, G., 1996. Comparison of soil receptivity to *Thielaviopsis basicola, Aphanomyces euteiches* and *Fusarium solani* f.sp. *pisi* causing root rot in pea. *Phytopathology*: submitted.

Oyarzun, P.J., Dijst, G., and Maas, P.W.Th., 1994b. Determination and analysis of soil receptivity to *Fusarium solani* f. sp. *pisi* causing dry root rot of peas. *Phytopathology* **84**: 834-842

Oyarzun, P.J., Zalm, A.J. van der, Roelofsen, H.W. and Dijst, G., 1994a. a computerized system to control soil water potential and soil temperature in soil inoculum potential assays of soil-borne pathogens. *Plant and Soil,* **166**: 125-135.

Oyarzun, P.J., and Dijst, G., 1991. Assessment of site-specific receptivity of soils to soil-borne diseases of pea and cauliflower. in: A.B.R. Beemster, G.J. Bollen, M. Gerlagh, M.A. Ruissen, B. Schippers and a. Tempel (eds), 1991. Biotic Interactions and Soil-borne Diseases. Elsevier, Amsterdam. pp. 322-328.

EFFECT OF CROP ON THE DIVERSITY AND THE POTENTIAL OF FUNGAL EGG PATHOGENS OF CYST NEMATODES

RALF-PETER SCHUSTER
Institute for Plant Diseases
Phytopathology in soil ecosystems
University of Bonn, Germany

1. Abstract

Fungal egg pathogens are the most important factor for the natural biological control and decline of cyst nematode populations. Their occurrence and potential is dependent on different factors like crops, organic matter etc. which can be managed in agricultural systems. The effect of different monocultures as well as the effect of intercrops on the occurrence of egg pathogenic fungi and the degree of nematode egg infection was studied. The fungal egg pathogens found were isolated directly from the nematode eggs, identified by microscopic techniques and indices for species richness (R), diversity (H') and eveness (E) were calculated. Isolates were furthermore identified by fatty-acid-methyl-ester analysis (FAME). Different cluster analysis systems were compared for correspondence of results.

2. Materials and methods

2.1. MEASUREMENT OF FUNGAL EGG INFECTION AN ISOLATION OF FUNGI

For the measurement of the degree of cyst nematode egg infection, *Globodera pallida* cysts were inoculated into soil using the slide-frame method described by Sikora *et al.* (1994). After 2 weeks of incubation in a soil sample, one half of the cysts was removed and stained with Bengal-rose-solution (Sikora *et al.*, 1994) for measurement of the fungal egg infection. The second half of the cysts was plated on water agar at a density of 500 eggs/plate. 2 days later, hyphae growing from infected eggs were isolated and transferred to petri dishes containing potato-dextrose agar. After 7-14 days, the fungal egg pathogens were identified microscopically, and groups of identical species were set up. For the determination of the biodiversity of the isolated egg pathogens, the following indices were calculated for each soil sample and each treatment: number of species **NO**; species richness **R**: S-1/ ln/n; diversity **H'**: [(ni/n)ln(ni/n)]; eveness **E**: H' / ln(S) with S = number of species, n = number of isolates.

H.-W. Dehne et al. (eds.), Diagnosis and Identification of Plant Pathogens, 253–255.

2.2. FATTY-ACID-METHYL-ESTER-ANALYSIS (FAME) AND CLUSTER ANALYSIS METHODS

Isolated fungi were multiplicated on Saboraud dextrose agar (SBA) and then fermented in Saboraud dextrose broth (SBB) on a rotary shaker for 3-4 days. Fungal mycelium was separated from the culture by filtration. Extraction of the fatty acid methyl esters (FAME) from the fungal mycelium was performed according to the MIDI-manual. Different cluster analysis were performed in the following way: 1. **MIDI:** Cluster method used by the MIDI-Programme was "unweighted pair grouped" using the percentage of occurrence of all fatty acids found in each analysis, with the data standardized and different squared euclidian distances (4, 6, 8 and 10) used as the border for different species. 1.**WARD**: Cluster method was the method of WARD using recalculated percentages of the occurrence of 5 selected fatty acids (16:1, 16:0, 18:2, sum in feature 8, 18:0) with standardized data and squared euclidian distances. Number of clusters was set according to the number of species found (WARD-Cluster) or euclidian distances of 2, 4, 6 or 8 (WARD-Distance) were used to differentiate between species.

3. Results

The potential of egg pathogenic fungi measured as the degree of fungal infection of *G. pallida* -eggs in the two monocultures during 4 month was consistently higher (1.1-1.5 fold) in the bean compared to the potato field. The degree of egg infection was varying greatly between the different dates of investigation, but the variation was the same for both fields. Additionally, the investigations concerning the fungal diversity of the egg pathogens showed, that the indices for number of species, species richness, diversity and eveness were higher in the bean monoculture field during all 4 months. This proved not only the higher potential in the bean field but also a greater diversity, which means a better potential for buffering during unsuitable conditions for the egg pathogenic fungi, which is important for their effect on the nematode population.

A big difference in the degree of egg infection between the two sugarbeet fields investigated, could be also be detected. Sugarbeet field number 5 consistently showing better values concerning the fungal egg pathogenic diversity for the number of species, species richness, diversity and eveness. The growth and incorporation of the intercrops mustard and oil radish led to a different spectrum of egg pathogenic fungi in the two sugarbeet field soils compared with the untreated soil. Furthermore, the intercrop treatments had a strong effect on the fungal species diversity by reducing the number of species isolated as well as all diversity indices (R, H', E) calculated for both field soils. The intercrop mustard, in both cases had a stronger effect, reducing the number of species from 13 to 7 in field number 5 and from 9 to 4 in field 20. Sugarbeet field 5 showing the higher values for all diversity indices calculated for egg pathogenic fungi was not influenced as strong as field 20 by the intercrop treatments, again showing the importance of fungal egg pathogenic species diversity for nematode control.

The comparison of the different cluster analysis used showed, that the percentage of correspondence between FAME-analysis and microscopically identification varied greatly between the methods used. Overall, the cluster method performed by MIDI was not as good as the WARD-method. In a number of cases, the percentage of correspondence was below 40% for all methods. The number of isolates incorporated in the cluster analysis was very important for a good correspondence of the identification systems used. Best results were obtained when the ratio between isolates and species was low. The shifting of the euclidian distance from 10 to 4 for MIDI or 8 to 2 for the WARD-method gave better results concerning the different species identification methods, FAME and microscopically, with up to 80% of correspondence.

4. Conclusion

With the method used, it was consistently possible to measure differences in the antagonistic potential of egg pathogenic fungi towards cyst nematodes in different monoculture fields as well as in fields from sugar beet rotations. It could clearly be seen, that not only the potential but also the diversity of the fungi is of great importance for the natural control of cyst nematodes. Due to a higher diversity, unsuitable conditions for the fungi could be buffered. The use of FAME-analysis for identifying and comparing fungi is useful, but still needs further investigation.

5. References

Sikora, R.A., Schuster, R.-P. and Kiewnick, S. (1994) Indexing Biodiversity and antagonistic potential in agricultural soil of Madagascar. *Med. Fac. Landbouww. Univ. Gent,* **59/2b**, 781-790.

MODIFIED SPORE GERMINATION INHIBITION TECHNIQUE FOR EVALUATION OF CANDIDATE FUNGITOXICANT (*EUCALYPTUS* SPP.)

Sushil Shahi, A.C Shukla, Shalini Dikshit* and Anupam Dikshit
Biological Product Laboratory,
Department of Botany; Allahabad University,
Allahabad - 211002, India

1. Abstract

The technique of spore germination inhibition has been modified by introducing the mathematical model for standardization of inoculum, reducing the cost as well as observation period thereby making the technique simple and no chance of contamination for the evaluation of candidate fungitoxicant(s). The present technique is compared with prevalent poisoned food technique and we found that quantitative results are almost the same. Amongst the candidate essential oils extracted from *Eucalyptus citriodora, E. dalrympleana, E. laveopinea, E. pauciflora*, only the oil of *E. citriodora* at 1000 mg/ml showed complete inhibition of spore germination of the storage spoilage test fungus *Penicillium italicum*. The oil of E. *dalrympleana* and *E. laveopinea* showed fungistatic toxicity against the test fungus at 3000 mg/ml but the oil of *E. laveopinea* showed partial inhibition at 3000 mg/ml. The oil of *E. citriodora* at 1000 mg/ml exhibited fungicidal nature and withstood heavy inoculum. Since the oil at sublethal concentration exhibited no abnormalities at morphological level, the total picture can be drawn only after undertaking the work on cytological and at enzymatic level against the test fungus.

1. Introduction

Several techniques such as Slide germination method (Anon, 1943), Agar plate spore germination technique (Gattani, 1954), Cell- volume assay method (Mandels and Darby, 1953), Modified paper disc method (Sharvelle and Pelletier, 1956), Poisoned food technique (Grover and Moore. 1962) adopted by Asthana et al. (1982), Dikshit et al. (1986), Pandey et al. (1996) for evaluation of fungitoxicant have been described. The present paper reports the modification in the spore germination inhibition technique for getting quick results at low cost and without any contamination as well as screening the candidate essential oils extracted from *Eucalyptus citriodora* Hooks, *E. dalrympleana* Maiden, *E. laveopinea* Baker and *E. pauciflora* Sieb, for their fungitox-

H.-W. Dehne et al. (eds.), Diagnosis and Identification of Plant Pathogens, 257–263.

icity against the test fungus *Penicillium italicum* Wehmer, causing deterioration of food commodities in storage.

2. Materials and Methods

Isolation of active constituent: The essential oils were extracted from the fresh leaves of the plant by hydrodistillation using Clevenger's apparatus, (Clevenger, 1928) and tested at 3000, 2000, 1000 mg/ml concentration against the test fungus *Penicillium italicum*, using spore germination inhibition technique with necessary modification.

Preparation of medium: The medium (broth) was prepared in the following composition with slight modification :Peptone 10 g, Dextrose 40 g, Agar 1 g, Distilled water 1000 ml, pH 5.60. Penicillin G (5 mg) and streptomycin sulphate (5 mg) were thoroughly mixed in the medium at 400C in order to prevent bacterial growth, as suggested by Gupta and Banerjee (1970).

Preparation of inoculum suspension: Spore discs (5 mm diameter) cut out from the central region of 7 day old cultures, grown in petri dishes were transferred aseptically in flasks containing medium, and then shaked the flask thoroughly for even distribution of spores.

Standardization of inoculum suspension: The number of spores were counted per microscopic field using `Modified Cytometer Technique' (MCT) given below. The diameter of Microscopic field was measured , by using micrometer and then calculated the area and volume of microscopic field by the following formula: AMF= πr^2, VMF= (AMF) h, where AMF= area of microscopic field; VMF= volume of microscopic field; r= radius of microscopic field; h = thickness of medium (in between slide and cover glass) 0.1mm. After determining the volume, the number of spores, average value of 5 different sites (microscopic field) were counted eliminating the overlapping spores.

The number of spores in the volume of microscopic field (NSV) was calculated by the following formula: NSV = ANS / VMF, where ANS = average number of spores in microscopic field. The volume of liquid medium (VLM) per microscopic field was calculated by the formula VLM = 2 r h. The total inoculum density (TID) was calculated in the initial volume of medium as per formula TID = (NSV / VLM) IVM, where IVM = initial volume of medium.

To test the desired concentration, the candidate fungitoxicants were dissolved in acetone (2% of the required quantity of seeded medium) and mixed in standardized inoculum suspension. In controls, sterile water was used in place of the candidate fungitoxicant. Culture tubes thus prepared were incubated at 27°±1°C and the observation were recorded at the interval of 24 hrs up to 96 hrs, by counting the number of germinated spores. Percentage of spore germination inhibition (SGI) was calculated as per formula: SGI [%] = (100 (Gc-Gt)) / Gc, where Gc = Number of spore germination in control; Gt = Number of spore germination in treatment.

To see the toxic effect of the oil at morphological level the partially inhibited spores at sub lethal concentration were examined carefully by measuring the length (mm) of germ tube in control as well as in treatment. Along with the length of germ tube, the shape of spores was also examined. The percentage of germ tube inhibition (GTI) was calculated as per the formula GTI [%] = (100(Lc-Lt)) / Lc, where Lc = Germ tube length in control; Lt = Germ tube length in treatment

The nature of toxicity (fungicidal/fungistatic) of the oils were tested by reinoculating the inhibited spore suspension (1 ml) to the presterilised culture tube containing 5 ml medium and incubated at 27°±1°C for 24 hrs. Observation recorded as usual under the microscope on 2nd day, germination of the spore indicated fungistatic activity while its absence denoted fungicidal action.

The effect of inoculum density using the most effective oil from *E.citriodora* (1000 mg/ml) was determined following the procedure outlined by Dikshit and Dixit (1982). Spore discs of 5 mm diameter cut out from the centre of 7 day old culture of the test fungus were inoculated in culture tubes containing the oil dissolved in acetone (2% of the required medium) and mixed in requisite amount of modified Sabouraud's dextrose agar medium. Controls, using sterile water and acetone mixed in the medium were run simultaneously. The number of fungal discs in the treatment as well as in controls, were increased progressively up to 30 in multiple of five. Observations recorded as usual under the microscope, absence of germination of the spore on the second day exhibited the oil's potential to withstand heavy inoculum of the test fungus.

3. Results

Out of the four species tested viz., *Eucalyptus citriodora, E. dalrympleana, E. laveopinea, E. pauciflora* tested at 3000,2000, 1000 mg/ml, only the oil of *E. citriodora* showed complete inhibition of spore germination of the test fungus *Penicillium italicum* at 1000 mg/ml while the oil of *E. dalrympleana* and *E. pauciflora* showed 100 per cent inhibition against the test fungus at 3000 mg/ml and these sample were found to be fungistatic in nature. The oil of *E. laveopinea* showed partial inhibition of spore germination at 3000 mg/ml (Table 1) The oil of *E. citriodora* at 1000 mg/ml exhibited fungicidal nature and withstood heavy inoculum density (Table 3) showing minimum inhibitory concentration (MIC) at 900 mg/ml by modified spore germination inhibition technique as well as by poisoned food technique (Table 4)

TABLE 1. Effect of *Eucalyptus* oil on the germination of spores of *Penicillium italicum.*

Incubation period(hrs)	Percent inhibition of spore germination (SGI) at various conc. (μg/ml)											
	E.citriodora			*E. dalrympleana*			*E. laveopinea*			*E. pauciflora*		
	3000	2000	1000	3000	2000	1000	3000	2000	1000	3000	2000	1000
48	100^-	100^-	100^-	100^+	80	60	80	60	40	100^+	80	60
72	100^-	100^-	100^-	100^+	80	60	80	60	40	100^+	80	60
96	100^-	100^-	100^-	100^+	80	60	80	60	40	100^+	80	60

- = denote fungicidal; + = denote fungistatic

The oil at sublethal concentration exhibited no abnormalities in the shape of the spore as well as of germ tube. However, the toxic effect of the tested oil could be noticed in the form of per cent inhibition of germ tube length (mm). The oil of *E. pauciflora* at 2000 mg/ml exhibited maximum inhibition of germ tube, 80.5% while that of *E. laveopinea* minimum of 14.5% at 1000 mg/ml (Table 2).

TABLE 2. Effect of *Eucalyptus* oil on the length of germ tube of *Penicillium italicum.*

Incubation period(hrs)	Percent inhibition of spore germination (SGI) at various conc. (μg/ml)											
	E.citriodora			*E. dalrympleana*			*E. laveopinea*			*E. pauciflora*		
	3000	2000	1000	3000	2000	1000	3000	2000	1000	3000	2000	1000
48	100^-	100^-	100^-	100^+	78	44	78	45	33	100^+	78	28
72	100^-	100^-	100^-	100^+	77	57	67	60	15	100^+	81	34
96	100^-	100^-	100^-	100^+	67	60	71	71	42	100^+	80	53

- = denote fungicidal; + = denote fungistatic

TABLE 3. Effect of inoculum density on SGI by *E. citriodora*[a] oil.

Inoculum density at MIC (No. of fungal disc)	SGI % against *Penicillium italicum*
1	100
5	100
10	100
15	100
20	100
25	100
30	100

MIC = Minimum inhibitory concentration; SGI = Spore germination inhibition;
[a] = concentration 1000 mg/ml

4. Discussion

The 27 essential oils of *Eucalyptus* spp. have been tested by Egwa et al. (1976) by inhibition zone technique against two fungi and a bacterium and 11 species of *Eucalyptus* have been tested against dermatophytes using poisoned food technique by Dikshit (1988). Yadav and Dubey (1994) tested *Eucalyptus citriodora* oil against two animal pathogens- *Microsporum audouinii* Gurby and *Trichophyton mentagrophytes* (Robin) Blanchard, and found effective at 1000 mg/ml. However, in the present investigataion four oils extracted from *E. citriodora, E. dalrympleana, E. laveopinea* and *E.*

pauciflora have been tested against storage spoilage test pathogen *Penicillium italicum* Wehmer.

TABLE 4. Comparison of two techniques for fungitoxic evaluation.

Oil concentration [μg/ml]	SGI [%[by MSGIT	MGI [%] by PFT
1000	100	100
900	100	100
800	76	67
700	65	60
600	50	53
500	25	33

SGI = spore germination inhibition; MGI = mycelial growth inhibition; MSGIT = modified spore germination inhibition technique; PFT = Poisoned food technique

Earlier the techniques used for such evaluation were either spore germination or mycelial inhibition, qualitative as well as quantitative. However, with our past experience we have modified the technique for fungitoxic evaluation and to our knowledge the mathematical model for standardization of inoculum suspension has been introduced for the first time. Prevalent poisoned food technique adopted by various workers has also been used by the authors in the present communication for comparison, and found that quantitative results are almost the same. (Table 4). The comparative parameters of presently modified spore germination inhibition technique (MSGIT) with that of prevalent poisoned food technique (PFT) have been tabulated. (Table 5) and concluded that the technique is having edge over poisoned food technique. To see the nature of toxicity i.e. static or cidal, the technique of Garber and Houston (1959) and that of Thompson (1989) are prevalent and used by the authors in the past as well as by other worker (Dikshit et al. 1986, Pandey et al. 1996). However, in the present communication, we have also made necessary modifications as mentioned in the material and methods. Since the oil at sublethal concentration exhibited no abnormalities at morphological level, the total picture can be drawn only after undertaking the work on cytological and enzymatic level against the test fungus.

Thus, the essential oil of *E. citriodora* due to strong fungitoxicity, fungicidal nature, with standing heavy doses of inoculum, can be exploited as a potential source for post harvest protectant against mould infestation after undergoing thorough in vivo investigation with different food commodities.

TABLE 5. Comparative characteristic of the present technique with prevalent poisoned food technique.

Parameter	MSGIT	PFT
Cost	* cheaper culture tubes * low quantity of medium * low quantity of the agar in the medium	* Costlier petri plate * large quantity of medium * large quantity of the agar in the medium
Observation period	* after 2nd day up to 4th day of inoculation	* after 7th day of inoculation
Contamination	* lesser chance due to less exposed area	* greater chance due to larger exposed area
Technique	* simple	* complicated
Standardization of inoculum	* by counting the number of spore by using `Modified Cytometer Technique' instead of Heamocytometer	* by physical parameter i.e. area of disc, days and site
Quantitative estimation	* MIC Similar to PFT	* ---

MSGIT = Modified spore germination inhibition technique; PFT = Poisoned food technique; MIC = Minimum inhibitory concentration

5. References

Anon, 1943: The slide germination method of evaluating protectant fungicides. Phytopathology **33**: 627-632.

Asthana A., Chandra H., Dikshit A. and Dixit S.N., 1982: Volatile fungitoxicant from leaves of some higher plants against *Helminthosporium oryzae*. Journal of Plant Disease and Protection **89**, 475-479.

Clevenger J.F., 1928: Apparatus for the determination of volatile oil. J. Am. Pharm. Assocn. **17**: 346.

Dikshit A., Singh A.K., Tripathi R.D. and Dixit S.N., 1979: Fungitoxic and phytotoxic studies of some essential oils. Biol. Bull. India **1**, 45-51.

Dixit S.N., Tripathi S.C. and Upadhyay R.R., 1976: The antifungal substance of rose flower (*Rosa indica*), Econ. Bot **30**, 371-374.

Dikshit A., 1988: Antidermatophytic action of essential oils from *Eucalyptus* spp. J. Indian Bot. Soc. **66** (Supplement): 4 (abs).

Dikshit A. and Dixit S.N., 1982: *Cedrus* oil- a promising antifungal agent. Indian Perfumer **26,** 216-227.

Dikshit A., Naqvi A.A. and Husain A., 1986: *Schinus molle*: a new source of natural fungitoxicant. Applied and Enviromental Microbiology **51**, 1085-1088.

Egwa H., Tsutsui O., Tatasuyama and Halta. T., 1977: Antifungal substances found in leaves of *Eucalyptus* species. Experientia **33**, 889-890.

Garber R.H. and Houston B.R., 1959: An inhibitor of *Verticillium albo-atrum* in cotton seed. Phytopathology **49**, 449-450.

Grover R.K. and Moore J.D., 1962: Toxicometric studies of fungicides against brown rot organism *Sclerotinia fructicola* and *S. laxa*. Phytopathology **52**, 876-880.

Gupta S. and Banerjee A.B., 1970: A rapid method of screening antifungal antibiotic producing plants. Ind. J. Exp. Biol. **8**, 148-149.

Gattani M.L., 1954: Agar plate spore germination technique. Phytopathology **44**, 113-115.

Mandels G.R. and Darby R., 1953: A rapid cell volume assay for fungitoxicity using fungus spore. J. Bacteriol. **65**, 16-26.

Pandey M.C., Sharma J.R. and Dikshit A., 1996: An evaluation of the essential oil of *Cymbopogon pendulus* (Nees ex Steud) Wats var. Parman. Flavour and Fragrance Journal. **11**, 257-260.

Sharvelle E.G. and Pelletier E.N., 1956: Modified paper disc method for laboratory fungicidal bioassay. Phytopathology **46**, 26 (abst.).

Thompson D.P., 1989: Fungitoxic activity of essential oil component on food storage fungi. Mycologia **81**, 151-153.

Yadav P. and Dubey N.K., 1994: Screening of some essential oils against Ringworm fungi. Indian Journal of Pharmaceutical Sciences **6**, 227.

Acknowledgments

Thanks are due to Head, Department of Botany, University of Allahabad providing necessary facilities; to Dr. G.L. Tiwari for reviewing the manuscript and to Council of Science and Technology Lucknow (U.P.) for financial assistance.

Session 6:

Disease Prognosis and Expert Systems

INTRODUCING DECISION SUPPORT SYSTEMS AVAILABLE IN GERMANY

H. HINDORF
Institut für Pflanzenkrankheiten, University
Nussallee 9, D-53115 Bonn

Since WAGGONER & HORSFALL (1969) developed the prognosis model "EPIDEM" in the USA an intensive period of modelling started in Germany by KRANZ (1974), SCHRÖDTER (1983), GUTSCHE & KLUGE (1983) and others. During the last decade a number of decision systems have been established for farmer's practical use forecasting the occurrence of diseases and pests and guiding an optimal input of pesticides.

1. Systems for individual disease and pest organisms

1.1 PHYTPROG, West-Germany (potato late blight prognosis)

The oldest effective warning system on late blight of potatoes, *Phytophthora infestans*, was developed by SCHRÖDTER & ULLRICH (1967) and conducted by the German Weather Service (DWD). The goal is a negative prognosis forecasting disease free periods and to predict the start of the disease in different localities of Germany. That system includes a number of meteorological parameters influencing the epidemiology of the pathogen, crop emergence and disease begin. A mathematical function calculates with these parameters certain values. For farmer's use two numbers (Grenzbewertungsziffern, GBZ) are of importance:

GBZ 150: 0.1 % disease frequency = fungicide application during the next fortnight
GBZ 270: 1.0 % disease frequency = disease has been established in the stand and spraying programmes are to late

More recently the system is not well accepted due to a number of wrong decisions. A change in epidemiological behaviour of the pathogen like stem infestation via the tuber is not considered in that programme.

1.2 PHYTEB, East- Germany (potato late blight prognosis)

A basic model for future prognosis systems was developed in the former DDR forecasting the first appearance of the disease *Phytophthora infestans* and the disease progress curve incorporating meteorological parameters, potato varieties and fungicide effectiveness. This practical system was widely used in Eastern Europe and is now replaced by a more precise system called SIMPHYT.

1.3 HALMBRUCH-WARNDIENST (eyespot prognosis)

H.-W. Dehne et al. (eds.), Diagnosis and Identification of Plant Pathogens, 267–270.

SCHRÖDTER & FEHRMANN (1971) implicated a system announcing infection probabilities of the eyespot disease, *Pseudocercosporella herpotrichoides*, based on meteorological parameters only. Weekly presentations of the probable infection rates in graphs for at time 29 localities are distributed in the publication organ "Agrarmeteorologischer Wochenhinweis" of the German Weather Service (see fig. 1 as example for the season 1995/96 at Bonn). The Plant Protection Services more and more use recently developed diagnostic methods during growth stages EC 29-31 for deciding of treatments like the ink staining method of MAULER-MACHNIK or the field ELISA test Advisor of DU PONT.

Fig.1 : Infection rates for eyespot at Bonn during season 1995/96 (DWD)

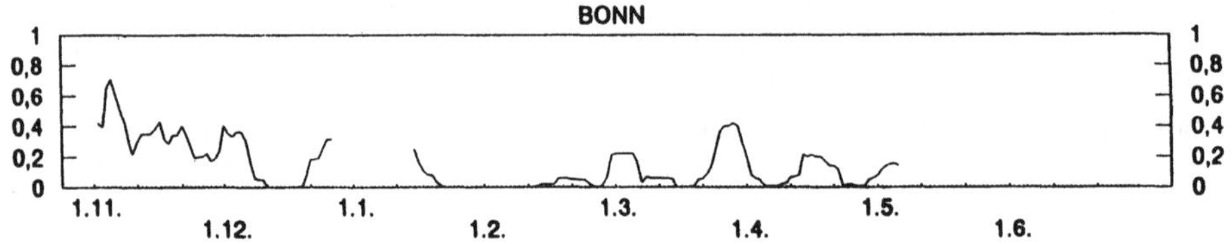

1.4 HOPFEN-PERONOSPORA-WARNDIENST (hop downy mildew prognosis)

During 1976 and 1982 a warning system was evaluated in Bavaria for forecasting infection periods of downy mildew in hop, *Pseudoperonospora humuli*, based on leaf wetness duration during daylight and the number of sporangia being caught in intervals of 4 days. In addition during early growth development young shoots with primary inoculum are assessed (KREMHELLER 1983). It was proved in a number of phytotron and field experiments that temperature, humidity, rainfall and wind speed are regulating the sporangia production of the fungus and being represented in the sporangia catches. Farmers will be encouraged to spray fungicides only when sporangia catches exceed the threshold of 30 sporangia before flowering and 10 sporangia after flowering, and when leaf wetness occurs during the last day of the 4-days period. The prognosis system is available by telephone from the Bavarian Plant Production Service, Section Hop and is well accepted by a high number of farmers in locally restricted areas of Southern and Eastern Germany.

1.5 PASO (forecasting of noxious organisms of importance for agricultural production)

Recently in a pilot project conducted by the "Biologische Bundesanstalt (BBA)" 9 models, elaborated in different institutions during 1989-1993, have been evaluated for practical use in advisory services.

1.5.1 SIMCERC (eyespot prognosis)
During EC 32 disease intensities of *Pseudocercosporella heroptrichoides* for ripening stages (EC 75) will be predicted for 5 different classes of emergence of winter wheat and winter rye. Included to the system are longterm disease severities, climatic data in general and a regional factor as a constant value for each locality (GUTSCHE 1988). Only the very early sown fields were proposed for fungicidal application on April, 30th after exceeding the threshold of 25%.

1.5.2 SIMERY (powdery mildew prognosis)

Monitoring powdery mildew, *Erysiphe graminis*, can be achieved with parameters like temperature, relative humidity, age of mildew colonies and host plant development. The SIMERY system results in disease severities on the three upper leaves for three differnet classes of susceptibility of winter wheat and winter barley (ROSSBERG & KLUGE 1989). Fig.2 shows mildew intensities on leaves F to F-3 at Cologne in 1996. The system recommends to assess disease progress when leaf F-2 has exceeded the threshold.

fig. 2 : Mildew intensities on wheat leaves F to F-3 at Cologne in 1996

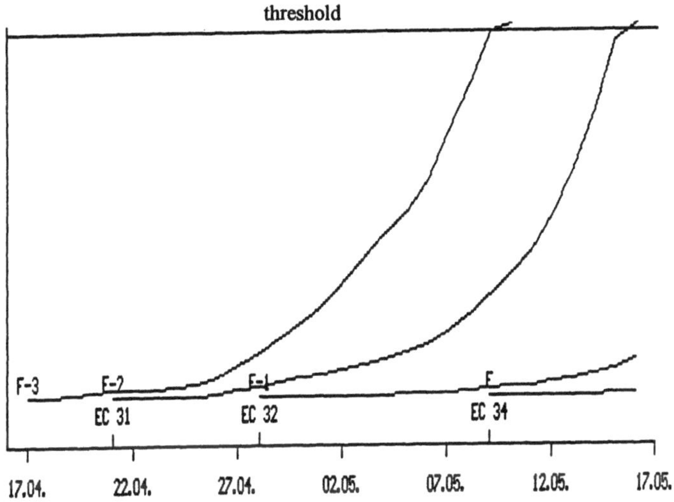

1.5.3 RHYNCHO-OPT (scald prognosis)

This decision system predicts fungicide applications after observing disease occurrence of *Rhynchosporium secalis* in rye on the 5th leaf and measuring leaf wetness duration (WERRES & HINDORF 1991). Both parameters are too labour extensive and difficult to score, therefore the practice does not accept such a model.

1.5.4 and 5 BONN-LAUS and SIMSIT (cereal aphids as vectors for BYDV)

1.5.6 SIMPHYT (potato late blight prognosis)

This system predicts the first appearance of the disease when fungicide applications should start. The model works with an initial value depending on climatic conditions, susceptibility of the host and actual precipitation (KLUGE & GUTSCHE 1990).

1.5.7 SIMLEP (potato Colorado beetle prognosis)

!.5.8 BBA-DELIA and BBA-PSILA (vegetable flies prognosis)

1.5.9 EURO-BUGOFF (apple codling moth prognosis)

2. Integrated systems for multiple disease and pest organisms

More effective economically and therefore demanded by farmers are decision systems for a certain crop integrating disease and pest organisms, meteorological parameters, phenology of hosts and production data.

2.1 EXPERTENSYSTEM-BAYERN

Observing threshold of different fungal pathogens and predicting spraying intervals was first developed for wheat diseases, but now extented to barley and incorporating climatological data.

2.2 PRO_PLANT

First evaluated for wheat diseases, now being extented for oilseed rape pests and sugarbeet diseases.

Gutsche, V. (1988) *Die Entwicklung und Nutzung von Schaderregermodellen in Forschung und Praxis des Pflanzenschutzes*. Diss. Berlin.

Gutsche, V. und Kluge, E. (1983) Phyteb-Prognose, ein neues Verfahren zur Prognose des Krautfäule-Auftretens (Phytophthora infestans (Mont.) de Bary). *Nachrichtenbl. Pflanzenschutzd. DDR*. **37**, 45-48.

Kluge, E. und Gutsche, V. (1990) Krautfäuleprognose mittels Simulationsmodell - Ergebnisse der Anwendungen 1982 - 1988. *Arch. Phytopath. Pflanzensch.* **26**, 265-281.

Kranz, J. (1974) *Epidemics of plant diseases. Mathematical analysis and modeling*. Springer-Verlag, Berlin.

Kremheller, H.T. (1983) Peronospora-Warndienst: Hinweise und erste Erfahrungen. *Hopfen-Rundschau*, 257-262.

Rossberg, D. und Kluge, E. (1989) PC- Programm zur Überwachung und Bekämpfung von Mehltau an Wintergetreide. *Nachrichrenbl. Pflanzenschutzd. DDR* **43**, 37-39.

Schrödter, H. (1983) Meteorological problems in the practical use of disease-forecasting models. *EPPO- Bull.***13**, 307-310.

Schrödter, H. und Fehrmann, H. (1971) Ökologische Untersuchungen zur Epidemiologie von *Cercosporella herpotrichoides*. *Phytopath. Z.* **71**, 97-112, 203-222.

Schrödter, H. und Ullrich, J. (1967) Eine mathematisch-statistische Lösung des Problems der Prognose von Epidemien mit Hilfe meteorologischer Parameter, dargestellt am Beispiel der Kartoffelkrautfäule (Phytophthora infestans). *Agric. Meteorol.* **4**, 65-68.

Waggoner, P.E. and Horsfall, J.G. (1969) EPIDEM, a simulator of plant disease written for a computer. *Conn. Agric. Exp. Sta. Bull.* **698.**

Werres, G. and Hindorf, H. (1991) Development of a forecasting model for scald, *Rhynchosporium secalis*, *EPPO Bull.* **21**, 485-493.

ESTIMATION OF DISEASE SEVERITY USING REDEFINED DISEASE INCIDENCE

B. HAU and P. SCHULD
Institut für Pflanzenkrankheiten und Pflanzenschutz
Universität Hannover
Herrenhäuser Str. 2
D-30419 Hannover

1. Introduction

The disease situation in a field can be characterized in terms of disease incidence (*DI*) or of disease severity (*DS*). If quantifiable relationships exist between the two disease measures, it is recommended to estimate severity based upon incidence data. Seem (1984) reviewed various aspects of *DI-DS* relationships and Campbell & Madden (1990) listed the application in various pathosystems.

Under the assumption that the lesions are randomly distributed on leaves, the relation between *DI* and *DS*, both given in percentage, can be derived (Seem, 1984):

$$DS = 100\ \kappa\ [-\ln(1 - DI/100)]$$

The term 100 κ can be interpreted as the disease severity which one lesion can cause on one leaf (Hau & Krüger, 1994). The graph of the *DI-DS* relationship is linear for lower *DI* and turns sharply upward for higher *DI* with an increase in the variance of the data. For small values of κ, *DS* remains low even when *DI* is in the order of 60 to 70 %, which is usually the upper limit of the linear range. For instance with κ equal to 0.005, *DI* of 70 % corresponds to *DS* of 0.6 %. If in this case the economic threshold is 1.0 % *DS*, a decision to control the disease cannot be based on the information about *DI*. In general, *DI-DS* relationships are not applicable to estimate *DS* in pathosystems expressing only small lesions as disease symptoms, for instance with κ-values < 0.005. In this paper a modification of the *DI-DS* relationships is proposed for these diseases by redefining disease incidence.

2. DI-DS relationships with redefined disease incidence

In the calculation of *DI* all leaves showing any disease symptoms are taken into consideration. In a study on the correlation between *DI* and *DS*, Analytis & Kranz (1972) defined a leaf as diseased only if the severity of the leaf was above a certain threshold, which was 0.67 % severity for apple scab. Following this approach, the

H.-W. Dehne et al. (eds.), Diagnosis and Identification of Plant Pathogens, 271–274.

incidence is now calculated from leaves which have more than one lesion, i.e. at least two lesions. Accordingly DI_1 means the disease incidence of those leaves which have more than one lesion.

For the relationship between DI_1 and DS no closed form can be given but the graph can be shown (Fig. 1). Using the redefined incidence results in a higher severity value compared to the original incidence. For $\kappa = 0.005$, an incidence DI_1 of 70 % is related to DS of 1.2 % which is above the assumed economic threshold. It would therefore possible to apply the DI_1-DS relationship in decision making.

DS as a function of DI_1 increases monotonically with higher slopes for low and high incidence values and a linear part in between (Fig. 1). Consequently, the curve must have a turning point which can exactly be determined. For the practical application the tangent through the turning point can be used as an approximation for the linear part of the relationship. The tangent is given by the following formula:

$$DS = \kappa\,[100(3 - e) + e\,DI_1] = \kappa\,[28.17 + 2.72\,DI_1]$$

As κ is not known exactly, the parameters of the regression line have to be estimated by means of linear regression analysis.

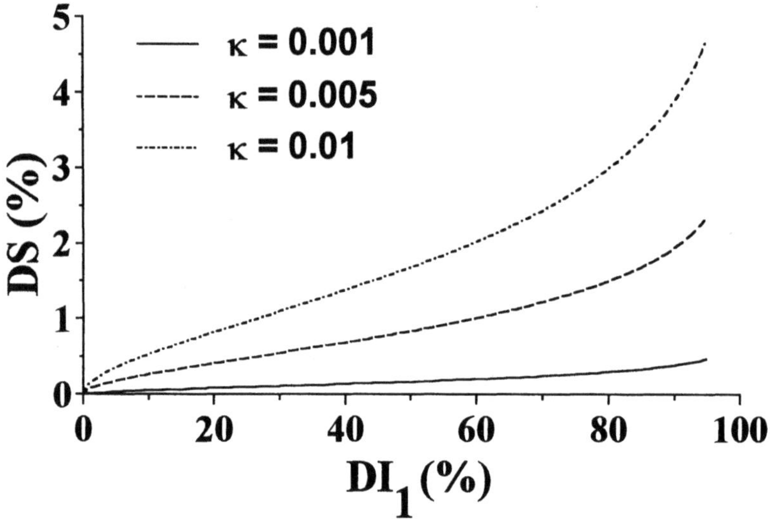

Figure 1. Relationships between DI_1 and DS for different κ-values.

In the first modification the leaves with more than 1 lesion have been used to determine the disease incidence. However, disease incidence can also be based on leaves which have more than any defined number of lesions. Thus in a general definition, DI_i means the incidence of leaves with more than i lesions. In Fig. 2 the relationships between DI_i and DS are shown for $i = 1$ to 4 for a value of $\kappa = 0.005$.

Again for each DI_i-DS relationship, a linear part around the turning points (TP) exists and the tangent can be used to calculate the severity (Tab. 1).

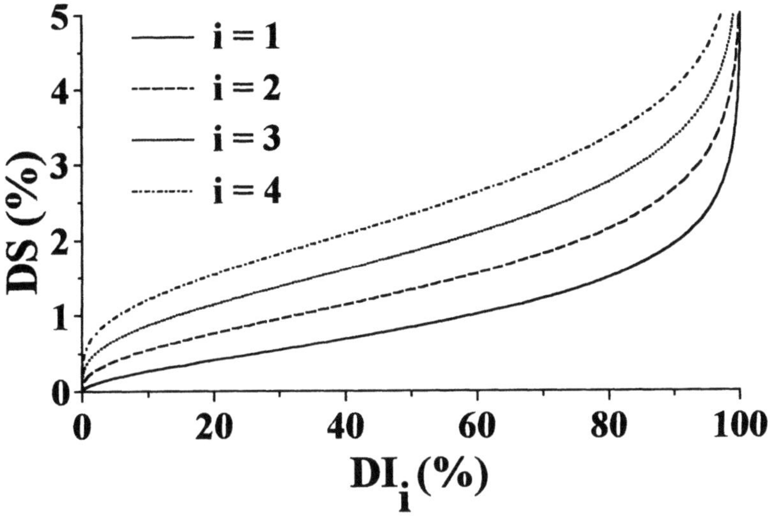

Figure 2. Relationship between DI_i and DS for different values of i ($\kappa = 0.005$).

TABLE 1. DI_i and DS at the turning point TP of the DI_i-DS relationship as well as the parameters of the tangent $DS = a + b\ DI_i$ at the turning point for $i = 1$ to 4

i	DI_i (%) at TP	DS (%) at TP	intercept a	slope b
1	26.42	100 κ	28.17 κ	2.72 κ
2	32.33	200 κ	80.55 κ	3.69 κ
3	35.28	300 κ	142.54 κ	4.46 κ
4	37.12	400 κ	210.02 κ	5.12 κ

The intercept and slope of the lines increase with higher values of i. Therefore, it is possible to reach much higher severity values with the modified incidence than with the original disease incidence. The parameter values for a and b in Tab. 1 are theoretically derived from the assumption of a random distribution of lesions on leaves and have to be estimated by regression analysis for real disease data.

As an example, bean rust (*Uromyces appendiculatus*) on Phaseolus beans is simulated. It is assumed that the leaf size of trifoliate leaves is normally distributed with a mean of 75 cm^2 and a standard deviation of 20 cm^2. For the pustule size including the halo an area of 3 mm^2 ($\pm$ 0.5 mm^2) is a valid assumption. The κ-value is then 0.03/75 = 0.0004, i.e. on average one pustule means 0.04 % disease severity on a single leaf. In the study 100 samples of 100 leaves each with up to 1200 pustules were simulated. Most of the original incidences DI are nearly 100 % (Fig. 3 A) so that DS cannot be determined. However, if the redefined incidence DI_5 is used, a linear relationship for incidences between 10 and 80 % can be achieved (Fig. 3 B) and DS can be assessed.

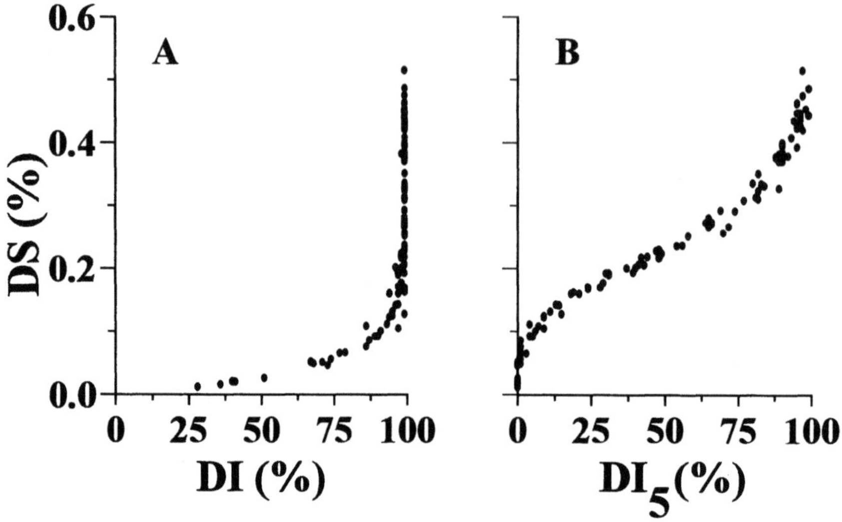

Figure 3. Relationship between *DS* and *DI* (A) and DI_5 (B), respectively.

3. Conclusion

The proposed redefinition of disease incidence can lead to useful relationships to assess disease severity. The application of this concept in field research has to be tested and the thresholds necessary for the modified incidences have to be determined. Alternative approaches have also to be checked, for instance to define the incidence below and above a threshold and to estimate *DS* by means of multiple linear regression equations as suggested by Schuld (1996).

4. References

Analytis, S. and Kranz, J. (1972) Über die Korrelation zwischen Befallshäufigkeit und Befallsstärke bei Pflanzenkrankheiten. *Phytopath. Z.* **73**, 201-207.

Campbell, C. L. and Madden, L. V. (1990) *Introduction to Plant Disease Epidemiology.* John Wiley and Sons, New York.

Hau, B. and Krüger, U. (1994) Zur Interpretation von Befallshäufigkeit-Befallsstärke-Relationen bei Blattfleckenkrankheiten. *Mitt. a. d. Biol. Bundesanst.* **301**, 446.

Schuld, P. (1996) Auswirkungen des Befalls durch drei Blattpathogene (*Uromyces appendiculatus* f. sp. *phaseoli*, *Phaeoisariopsis griseola* und *Colletotrichum lindemuthianum*) auf das Wachstum und den Ertrag von *Phaseolus* - Bohnen. Dissertation, Univ. Hannover.

Seem, R. C. (1984) Disease incidence and severity relationships. *Ann. Rev. Phytopathol.* **22**, 133-150.

A DECISION SUPPORT SYSTEM FOR CERCOSPORA LEAF SPOT ON SUGARBEET

P. BATTILANI, S. GIOSUÈ, P. RACCA, V. ROSSI
Istituto di Entomologia e Patologia vegetale
Università Cattolica S. Cuore
Via E. Parmense, 84 - 29100 Piacenza - Italy

1. Introduction

In the last years, decision support systems (DSS) have had a great development due to the increasing possibility given by computers to store and manage data. The DSS have interesting applications in crop protection, as shown by Pc-Plant protection in Denmark (Murali, 1991; Secher, 1991), Pro-Plant in Germany (Frahm *et al*, 1991), Epi-pre in The Netherlands (Daamen, 1991) and Nor-pre in Norway (Magnus and Munthe, 1991). The studies carried out on Cercospora leaf spot (caused by *Cercospora beticola* Sacc.) on sugarbeet in the last few years give a base of knowledge large enough to develop a DSS able to give all the information necessary to obtain the best integrated disease control. So CERCOESY (CERCOspora Expert SYstem) was elaborated, a computerized DSS for Cercospora leaf spot control able to emulate the reasoning of a human expert in solving the problems of this area of knowledge. It is a great help for technicians, farmers and researchers in managing the pathogen. The user can communicate with the system in a very simple way using windows and buttons. In fact, the system was written in Visual Basic Professional 3 and runs under Windows operating system ver 3.1 or Windows 95 on PC with 4 MB RAM (recommended 8) and S-VGA screen with 1024 x 768, 256 colors resolution. In the future the system will be inserted in a regional information system usable *via* internet.

2. The structure of the DSS

The program uses, by means of a management program, four data bases and two simulation models.

2.1. THE DATA-BASES

The first data-base contains information about the sugarbeet growing areas of the Emilia Romagna region, such as crop prevalence, crop rotation and disease risk,

H.-W. Dehne et al. (eds.), Diagnosis and Identification of Plant Pathogens, 275–279.

predicted with geophytopathological studies (Battilani *et al.*, 1996). In the second data-base are stored hourly or daily meteorological data (air temperature and relative humidity, rainfall) collected in the regional network of standard meteorological stations. In the third data-base information about cultivars are collected with regard to both quantitative and qualitative yield characteristics (Rosso *et al.*, 1995), rhizomania and Cercospora leaf spot resistance. The fourth data-base has information about fungicides used against Cercospora leaf spot, their cost and efficacy in reducing the infection rate.

2.2. THE SIMULATION MODELS

CERCOPRI (CERCOspora PRimary Infections) is an empirical-statistical model able to define the appearance of primary disease symptoms in the field which gives, for each day, the incidence of affected crops in the sugarbeet-growing area. It uses data from both the land and meteo data bases; daily data of useful air temperature and relative humidity (respectively greater than 5°C and 60%) are summarized starting from January 1 (Rossi and Battilani, 1986). CERCODEP (CERCOspora Development of EPidemics) is a simulation model based on the analysis of the plant-pathogen-environment system; it is able to simulate, day by day, disease progress in the field under the form of affected leaf area. It uses data from all the data bases (Rossi *et al.*, 1994; Giosuè *et al.*, 1995).

2.3. THE MANAGEMENT PROGRAM

The program manages data bases and simulation models to give information useful for strategical and tactical Cercospora leaf spot control. Strategical actions are taken before crop seeding and they can involve several years or one year only. Tactical actions are taken during crop growth and they depend on what happens during the growing season; they involve a period of a few months or a few days (Rabbinge *et al.*, 1993).

2.3.1. *Strategical decisions*

The program gives information to define the influence of the cropping system and cultivar on disease development and control. The first step to run the program is to define: the sugarbeet growing-area, cropping system (crop rotation, amount of nitrogen manuring and irrigation), presence of rhizomania, harvest time and disease threshold (amount of disease intensity to reach at harvest), by means of video helps like maps, drawings, tables and diagrams; otherwise the program automatically chooses ordinary conditions for the considered crop area and 10% of affected leaf area at harvesting (Canova *et al.*, 1996). Then the program gives, in a graphic shape, a prediction of disease progress without sprays, for each cultivar useful in the defined situation, and suggests the most suitable number of fungicidal sprays necessary to contain disease development within the fixed threshold.

2.3.2. *Tactical decisions*

The program gives information to define the time of fungicidal sprays, the active ingredients to use, their dosage and costs. Once the conditions previously cited and the actual disease severity, if available, have been defined, the program shows several different sprays schedules in decreasing order of cost. This response is obtained by simulating disease development, using CERCODEP and data from the meteo data base. Actual disease severity can be directly assessed in the field or in the pilot-fields, such as those inserted in the network realized by Eridania, Isi and Anb (Canova *et al.*, 1996); the lack of this data is simulated by the program.

3. A practical example

This example shows, in a simplified way, the kind of support that CERCOESY can give in the definition of cropping area, in the choice of cultivar and in defining the chemical control of Cercospora leaf spot. The example regards the plain of Piacenza province (North Italy).

3.1. THE CROPPING AREA

On the map which appears on the PC monitor, putting the slider on a desired point, all the information becomes available in a graphic form. In this example, CERCOESY defines an area in the plain, without limitations for the crops, except for climate. More prevalent crops are forage crops (40-50% of the utilized agricultural area), cereals (30-35%) and arable crops (10%), especially tomatoes and sugarbeet. The prevalent crop rotation is sugarbeet, wheat, tomatoes and sometimes forage crops. The risk for Cercospora leaf spot is mean-low: yield losses caused by the disease are about 10-11% of gross sucrose, with a range between 4-5% and 17-18%. Rhizomania has a low prevalence. Yearly mean temperature is 6-7°C. During *C. beticola* development mean temperature is 19-20°C in June, 22-23°C in July, 21-22°C in August and 17-19°C in September. Yearly rainfall is 800-900 mm, which decreases to 200 mm during the June-September period. The meteorological station of Paderna is the nearest to the considered point. Meteo data starting from 1986 are available; they can be seen graphically on video.

3.2. THE CULTIVAR TO SOW

An intermediate time of harvesting (first 10 days of September) is presumed to verify the effect of cultivar susceptibility on disease development. The date of harvesting can be chosen putting the slider on the video-calendar. The system automatically uses the cultivar data base to simulate, for each cultivar, disease onset and progress, number of sprays necessary to limit to 10% the final value of affected leaf area, using a compound with propiconazole (12.5%) + TPTA (18%), with a dosage of 1.2 kg/ha, 18 days between two successive sprays and a safety range of 45 days. The result is graphically shown: on video the list of cultivars and the number of sprays are visible. It is also possible to visualize in detail each simulated case with a graph which shows disease progress over time, from disease appearance to harvest. For instance, cv.Adrienne and cv. Formula need respectively 3 and 1 sprays.

3.3. THE SPRAY SCHEDULE

Support in deciding the spray schedule is requested for the cv. Adrienne with a forecasted harvesting date in the first ten days of September; other conditions are chosen by the system, except for the date of disease appearance which is observed in the field. The system simulates disease development and the effect of different spray schedules, chosing those able to keep disease severity below 10% of affected leaf area. The schedules are then visualized as dates for sprays, active ingredients (and commercial compounds), dosage and costs. In the example, to obtain the desired conditions, 3 sprays are necessary (on 29 June, 17 July and 4 August) using 2 kg/ha of cyproconazole + TPTA (commercial compound: Alto BS Pepite) with a mean cost/ha for each spray of 84.000 Lit. (Meriggi and Paganini, 1995).

4. Conclusion

The decision support system for the control of Cercospora leaf spot described is a modern tool of information management. It allows a rapid diffusion of recent results obtained by scientific research; it is a useful support for technicians because it gives them, in a very simple and rapid way, all the information useful in defining crop management but leaving them the possibility of defining the criteria of choice; it makes it possible to diffuse the techniques of integrated production without increasing assistance costs.

5. References

Battilani, P., Giosuè, S., Nassisi, A., Racca, P. and Rossi, V. (1996) Use of geostatistics in geophytopathology: *Cercospora* leaf spot on sugarbeet in Emilia

Romagna (Northern Italy) as a case study, *Phytopathologia mediterranea*, **35**, in press.

Canova, A., Meriggi, P., Ponti, I. and Rossi, V. (1996) La conoscenza della sensibilità varietale per definire le linee di intervento contro cercosporiosi e rizomania nella barbabietola da zucchero. *Sementi elette*, in press.

Daamen, R.A. (1991) Experiences with the cereal pest and disease management system EPIPRE in the Netherlands. *Danish Journal of Plant and Soil Science*, **85** (S2161), 77-87.

Frahm, J., Volk, T. and Streit, U. (1991) PRO-PLANT-A knowledge based advisory system for cereal diseases control. *Danish Journal of Plant and Soil Science*, **85** (S2161), 101-109.

Giosuè, S., Racca, P. and Rossi, V. (1995) Use of stochastic processes in simulating Cercospora leaf spot epidemics on sugarbeet. *Phytopathologia mediterranea*, **34**, 204-206.

Magnus, H.A. and Munthe, K. (1991) Performance of "NORPRE" cereal warning system. *Danish Journal of Plant and Soil Science*, **85** (S2161), 111-117.

Meriggi, P.L. and Paganini, U. (1995) L'importanza della lotta anticercosporica nella barbabietola da zucchero. *Agronomica*, **3**, 14-23.

Murali, N.S. (1991) An information system for plant protection: I. Development and testing the system. *Annales Anpp*, **2**, 143-148.

Rabbinge, R., Rossing, W.A.H. and Van der Werf, W. (1993) System approaches in epidemiology and plant disease management. *Netherland Journal of Plant Pathology* **99**, supp. 3, 161-171.

Rossi, V. and Battilani, P. (1986) Impiego di variabili climatiche nella previsione delle infezioni primarie di *Cercospora beticola* Sacc. *Informatore fitopatologico*, **36**, 29-39.

Rossi, V., Racca, P. and Battilani, P.(1994) A simulation model for Cercospora leaf spot epidemics on sugarbeet. *Phytopathologia mediterranea*, **33**, 105-112.

Rosso, F., Bimbatti, M., Bartolucci, G., Tugnoli, V. and Zavanella, M. (1995) Metodologie per la valutazione agronomica ed industriale delle varietà di barbabietola. *Agronomica* **5**, 8-18.

Secher, B.J.M. (1991) An information system for plant protection: II. Recommendation models structure and performance. *Annales Anpp* **2**, 153-160.

AN ADVISORY SYSTEM FOR THE CONTROL OF BROWN RUST ON WINTER WHEAT IN NORTHERN ITALY

V. ROSSI[1], P. RACCA[2], S. GIOSUE'[1], D. PANCALDI[3], R. BOTTAZZI[4]
[1]*Istituto di Entomologia e Patologia vegetale, Università Cattolica S. Cuore, Via Parmense 84, 29100 Piacenza, Italy.* [2]*Consorzio Emiliano Romagnolo Aziende Sperimentali (CERAS), Via Levante 18, 40026 Imola (Italy).* [3]*Dipartimento Protezione e Valorizzazione Agroalimentare, Via Re 8, 40126 Bologna, Italy.* [4]*Consorzio Provinciale Fitosanitario, Via Mazzini 40, 29100 Piacenza, Italy*

1. Introduction

The aim of this work is to develop an advisory system for the control of brown rust (caused by *Puccinia recondita* Rob. & Desm. f. sp. *tritici* Eriks. & Henn.) on winter wheat by combining three simulation models previously elaborated and validated with the networks of weather station, spore collection and field monitoring which act in the Emilia Romagna region (Bugiani *et al.*, 1996). It is a part of research promoted and supported by the Emilia-Romagna region by means of the co-ordination of CERAS.

2. Material and methods

Three mechanistic simulation models based on system analysis were used; they make it possible to simulate wheat growth, rust appearance and progress, using hourly data of temperature, relative humidity, leaf wetness, and rain.

Simulation of winter wheat growth was made by the WHEGROSIM model (WHEat GROwth SIMulation). It simulates both the phenology and the dynamic of the green area of each leaf from its appearance to complete senescence, for both main stems and tillers (Rossi *et al.*, 1996b). The model uses thermal time summations (phyllochrone) to predict both the beginning and the ending dates for the phenological phases. The growth of each leaf is divided into: appearance, expansion, full photosynthetic activity, and senescence. The length of each phase depends on the phyllochrone. Following leaf appearance, the model assumes that each leaf expands during a fixed thermal time at a rate determined by air temperature. Thermal time is computed using daylength (in hours), photoperiod-effective hours, and daily effective temperature.

The day when rust symptoms could appear for the first time was determined by the RUSTPRI (RUST PRimary Infection) model, which simulates the first infection cycle of *P. recondita* (Rossi *et al.*, 1996a).

The number of rust infections established on day i (GPL = germ tubes penetrated into leaves) depends on the number of uredospores deposited on leaves and on their

H.-W. Dehne et al. (eds.), Diagnosis and Identification of Plant Pathogens, 281–284.

infection efficiency (INF). INF is the proportion of available uredospores which cause infection; it depends on temperature and leaf wetness (Clifford and Harris, 1981).

The number of uredia that erupt on leaves on each day after i (UEL) is:

$$UEL_y = \sum_{i+LP-2}^{i+LP+2} (GPL_i \cdot ERU_y \cdot FAI_y) \qquad (1)$$

where: y = day counter. LP is the latent period or the number of days from inoculation until 50% of the uredia erupt; it depends on temperature (Johnson, 1980). ERU is the eruption rate of uredia on each yth day around LP (Shaner *et al.*, 1978). FAI is the failure rate of latent infections or the proportion of penetrated hyphae that cannot invade the leaf tissues till the eruption of an uredium; it depends on the leaf area that dies physiologically, as simulated by WHEGROSIM.

The progress of rust epidemics was obtained from the RUSTDEP (RUST Development of EPidemics) model, which simulates disease severity, expressed as a percentage of rusted leaf area (RLA) (both as average and its 95% confidence interval), on individual leaves, with a time step of one day (Rossi *et al.*, 1996b).

The model uses the basic structure of the logistic model: the daily increase of rusted leaf area (DRLA) on each jth day is calculated as:

$$DRLA_j = ILA_i \cdot INF_i \cdot ALA_i \cdot FAI_j \cdot CV \qquad (2)$$

where: i = j-LP. ILA is the infectious leaf area or the amount of leaf area that is occupied by infectious uredia; it varies according to the length of infectious period (that is the period of uredospore production by a uredium), which depends on temperature (Tomerlin *et al.*, 1983). ALA is the affectable leaf area or the total amount of a leaf area that can be infected; it is a part of the leaf area vulnerable to infection (simulated by WHEGROSIM), because the leaf area having either latent or visible infections is no longer vulnerable to infection. CV is the host resistance level; it considers that in many wheat cultivars (the slow-rusting cultivars) resistance towards *P.recondita* is a rate-reducing resistance: therefore its values are less than 1.

Validation was effected over three years (1994 to 1996) at several sites in the Emilia Romagna region. In this work the results obtained at Gariga (district of Piacenza, western Po Valley) are shown as an example. Hourly data of air temperature, relative humidity and rain were collected by an automatic meteorological station, while leaf wetness was estimated using a numerical model. Uredospores in the air were trapped by volumetric samplers which operate continuously at the canopy level. Silicone grease-coated tape on drums was changed weekly; then they were examined through a microscope to determine the number of uredospores per m^3 of air per day. To record disease onset and progress a pilot-field was arranged sowing several winter wheat cultivars (*Triticum aestivum* L. and *T. durum* Desf.) in big plots every year. The plots were carefully observed to note out the appearance of uredia; afterwards rust severity was assessed every 7-10 days on five 100-culm-samples per plot, as a percentage of the rusted leaf area on each leaf.

3. Results

In 1994, the uredospores were trapped for the first time on April 13, with a very low value (3 spores per m^3 of air). Other uredospores were trapped on April 15 ($3/m^3$) and continuously from April 23 to 27 (maximum of $6.1/m^3$); a greater number of spores were caught starting from April 29. Rust was not observed until the end of April. The first symptoms were recorded on May 10 on the flag leaf of 'Eridano', when plants were at the end of heading; mean disease severity was 0.7% of rusted leaf area. Afterwards the disease progressed till a mean of 14% of RLA on June 14. According to the RUSTPRI model, infection was established on April 25, when uredospores were washed onto the ground by a rain of 28.2 mm. Uredospores deposited on the wheat leaves had a high infection efficiency (72% were able to establish infection). Disease onset was simulated between May 1 and May 5, after a latency of 8 days. Simulation of rust progress made by the RUSTDEP model was similar to disease severity assessed in the field.

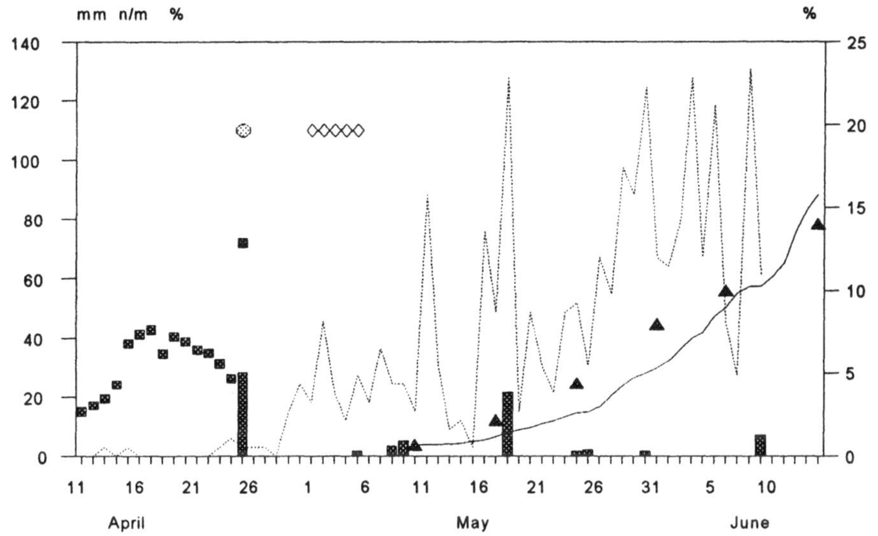

Figure 1. Information obtained daily by the run of the advisory system for brown rust on winter wheat at Gariga, in 1994: rainfall (bars in mm), uredospores/m^3 of air (- -), infection efficiency of uredospores % (□), estimation of the day when rust infection is established (O), estimation of the days when uredia erupt (◊), simulation of rusted leaf area % (---), actual rusted leaf area % (Δ).

In 1995, uredospores were trapped occasionally between April 12 and May 24 (3-12 uredospores/m^3). At the end of May a bigger number of uredospores was found, with a maximum of 33.4 on May 29. Rust was observed on June 6 on the two top leaves of 'Simeto', when the plants were at milk ripening. Disease severity started from 0.1% of rusted leaf area (on June 6) and progressed to 28% on the flag leaf (on June 27). RUSTPRI simulated disease establishment on May 31; it was a rainy day (34.2 mm) with an infection efficiency of uredospores of 50%. Rust onset was predicted between June 6 and 10, after a latency of 8 days. RUSTDEP simulated disease progress well.

In 1996, a few of uredospores were found on April 16, 22 and May 4. Between May 7 and May 14 spores were trapped frequently, with a maximum of 9/m^3 of air.

Uredospores were then trapped continuously after May 19. In the experimental plots rust was noticed at the end of May on the flag leaf of 'Spada' (0.5% of rusted leaf area), but in contiguous wheat crops uredia were observed starting from May 22 during heading. Afterwards the disease increased slowly, reaching a maximum severity of 3.8%. Model simulations showed that infection was established on May 13, when the inoculum deposited by a rain of 5.6 mm had an infection efficiency of 54.5%. Simulation of rust progress resulted in a very slow epidemic.

4. Discussion

The combination of meteorological data, uredospore trapping, simulation models and surveys in pilot-fields made it possible to obtain satisfactory simulations of both onset and progress of brown rust on winter wheat. In fact, in all the validations the detection of rust in the field was in accordance with model outputs. Furthermore simulations of the progress of rusted leaf area were similar to the actual values assessed in the field (they always clashed in the 95% confidence interval for assessed values).

The advisory system for brown rust on winter wheat can be organized according to the following scheme. The system is set going when uredospores are present in the wheat-growing area, that is when spores are trapped in a significant number for a few days. The presence of weather conditions conducive to infection establishment is then checked by RUSTPRI, especially when uredospores could be washed onto the ground by rainfall from the cloud of spores in the air. When the infection might be established, the days when uredia could appear in the field are determined and the technicians take action: they intensify surveys in the pilot-fields to note rust appearance. With disease onset they assess disease severity, as an input for RUSTDEP together with the meteorological data and the simulation from WHEGROSIM. The outputs of epidemic simulation are then used in order to advise growers.

5. References

Bugiani , R., Tiso, R., Butturini, R., Govoni, P. , and Ponti, I. (1996) Forecasting models and warning services in Emilia Romagna (Italy), *EPPO Bull.*, in press.

Clifford, B.C., and Harris, R.G. (1981) Controlled environment studies of the epidemic potential of *Puccinia recondita* f. sp. *tritici* on wheat in Britain, *Trans. Br. Mycol. Soc.* 77, 351-358.

Johnson, D.A. (1980) Effect of low temperature on the latent period of slow and fast rusting winter wheat genotypes, *Plant Dis.* **64**, 1006-1008.

Rossi, V., Racca, P., Pancaldi, D., and Alberti, I. (1996a) Appearance of *Puccinia recondita* f. sp. *tritici* on winter wheat: a simulation model, *EPPO Bull.*, in press.

Rossi, V., Racca, P., Giosuè, S., Pancaldi, D., and Alberti, I. (1996b) A simulation model for the development of brown rust epidemics on winter wheat, in press.

Shaner, G., Ohm, H.W., and Finney, R.E. (1978) Response of susceptible and slow leaf-rusting wheats to infection by *Puccinia recondita*, *Phytopathology* **68**, 471-475.

Tomerlin, J.R., Eversmeyer, M.G., Kramer, C.L., and Browder, L.E. (1983) Temperature and host effects on latent and infectious periods and on urediniospores production of *Puccinia recondita* f. sp. *tritici*, *Phytopathology* **73**, 414-419.

Taxonomy and Differentiation

VEGETATIVE COMPATIBILITY GROUPS IN *ACREMONIUM CUCURBITACEARUM*

ABAD P.[1], HACK T.[1], VICENTE M.J.[1], BRUTON B.D.[2] AND GARCÍA-JIMÉNEZ J.[1].
[1]. *Dpto.Producción Vegetal. Universidad Politécnica de Valencia, Spain.*
[2] *Agricultural Research Service, U.S.D.A. Lane OK 74555. USA.*

1. INTRODUCTION

Over the last fifteen years, a new disease of muskmelon (*Cucumis melo* L.) has produced serious losses to this crop in Spain. The causal agent of this disease is a Deuteromycete fungus named *Acremonium cucurbitacearum* A.Alfaro-García, W.Gams et J.García-Jiménez (1). The disease has been called acremonium melon collapse and the most obvious symptom is a nonvascular wilt of the plant as the first fruits approach maturity. Root symptoms appear soon after planting and include yellow discoloration and corking of the upper root, generalized death of adventious roots, and continous production of superficial roots. Aerial part of the plant does not show symptoms until wilting occurs (5). Application of molecular techniques to study genetic variations in field isolates of this fungus (8) provided the necessary information to resolve its taxonomic position, being described as a new species. Recently, the same disease has been observed in melon production areas of California (2) and Texas (3).
Vegetative compatibility group (VCG) analysis has been shown to be a useful technique for examining genetic diversity among populations of plant pathogenic fungi. Two physiologically distinct individuals of the same species belong to the same VCG if they can fuse asexually, forming an stable heterokaryon. Puhalla found that mutants resistant to $KClO_3$ were usually nitrate-nonutilizing (*nit*) mutants. He proposed a model using *nit* mutants to subdivide fungal populations into different VCGs. VCGs are well-suited for measuring the frequency of different genotypes within a population and for determining if two strains are identical to one another (6).The purpose of this study was to ascertain the relationships among field isolates of *A.cucurbitacearum* from Spain and USA (California and Texas states) by VCG analysis.

2. MATERIAL AND METHODS

2.1. Origin of fungal isolates. *A.cucurbitacearum* strains were isolated from diseased muskmelon plants. Twenty-nine isolates came from nine different Spanish provinces,

H.-W. Dehne et al. (eds.), Diagnosis and Identification of Plant Pathogens, 287–289.

twenty-two isolates were obtained from distinct counties of Sacramento Valley and northern area of San Joaquim Valley (California) and six isolates came from two counties placed in the southern region of Texas. Monosporic cultures were obtained from each isolate to ensure genetic purity.

2.2. Media. PDA and PDC were used for isolation of *A.cucurbitacearum* strains and to mantain cultures, respectively. Two chlorate-containig media, PDC and MMC (4), were used to recover *nit* mutants. These media were supplemented with increasing concentrations of $KClO_3$ ranging from 1.5 to 4.5%. Minimal medium (MM) contains 2 g/l $NaNO_3$ as unique nitrogen source (7). Phenotype of *nit* mutants (*nit1, nit3* or NitM) was determined by their growth on MM amended with sodium nitrate, sodium nitrite or hypoxanthine as the sole nitrogen source (4).

2.3. Vegetative compatibility tests. Procedures were based on Puhalla's technique (7). Small mycelial blocks of PDA cultures of each strain were placed in Petri dishes containing chlorate medium and were incubated at 28°C and 12 h light/dark intervals during 3 weeks. Then, plugs of chlorate-resistant sectors were selected and transferred to MM, where *nit* mutants showed thin but expanding growth. To test the vegetative self-compatibility, four *nit1* and two NitM of each strain were paired on MM plates. Vegetative compatibility between them was revealed by the presence of dense aerial mycelium where the colonies came in contact. Complementary mutants selected from each VCG previously determined with Spanish *A.cucurbitacearum* strains (8) served as tester pairs for screening the remaining strains. The *nit* mutants testers from all American *A.cucurbitacearum* strains were paired in all combinations to determine whether they belonged to the established VCGs or new ones.

3. RESULTS AND DISCUSSION

All of the strains, with one exception, produced spontaneous chlorate-resistant sectors when were cultivated on PDC and MMC media. Strain A-470 grew poorly on these media, perhaps due to potassium chlorate toxicity. The majority of chlorate-resistant sectors recovered were *nit* mutants and their colonies showed a thin, expansive non-aerial mycelium on MM. But, all the chlorate-resistant sectors obtained from strain CA-951096 grew as wild-type on MM, indicating that they were not *nit* mutants. Phenotype identification revealed that all *nit* mutants recovered from *A.cucurbitacearum* were *nit1* or NitM mutants. *Nit* mutant yield was higher on PDC than MMC and the majority of them were *nit1* mutants, but NitM mutants were obtained with more frequency on MMC medium.

Heterokaryon formation between complementary mutants (*nit1* and NitM) derived from the same strain were not detected in only one strain, A-499. Ten distinct VCGs were identified among this collection of *A.cucurbitacearum* strains. VCGs 2 and 3 are the most prevalent. Strains in VCG2 (32%) came from numerous locations of Spain, three of the six californian counties and the two Texas counties. VCG 3 (41.5%) included strains recovered from sites of three spanish provinces and five California counties. VCG4 contained few strains (5.7%), but also from both continents. VCG1 (7.5%) and VCG7 (3.8%) included only spanish strains. The rest of the VCGs identified (5,6,8,9 and 10) included only one strain each one. Nit mutants of strains A-473 and A-527 formed heterokaryons with complementary *nit* mutant testers belonging to three (1, 2 and 3) or two (1 and 2) different VCGs respectively. This "bridge" behaviour between distinct VCGs has been described previously by other authors suggesting that these VCGs must be closely related.

VCG analysis reveals that there is not much genetic diversity among the strains of *A.cucurbitacearum* studied, although in Spain we have found more heterogenicity (nine different VCGs). In California, the majority of strains belongs to VCG3. The number of strains from Texas must be increase but, at this time, all except one are included in VCG2.

4. ACKNOWLEDGEMENTS

We acknowledge financial support from the Spanish Comisión Interministerial de Ciencia y Tecnología (AGF94-0520). We also thank Erasmus Grant to T. Hack from R.F.W. University of Bonn.

5. REFERENCES

1. Alfaro-García A., Armengol J., Bruton B.D., Gams W., García- Jiménez J. and Martínez-Ferrer G. 1996. The taxonomic position of the causal agent of acremonium collapse of muskmelon. Mycologia 88(5):804-808.
2. Bruton B.D., Davies R.M. and Gordon T.R. 1995. Occurrence of *Acremonium* sp. and *Monosporascus cannonballus* in the major cantaloupe and watermelon growing areas of California. Plant Disease 79(7):754.
3. Bruton B.D., Miller M.E. and García-Jiménez J. 1996. Comparison of *Acremonium* sp. from the Lower Rio Grande Valley of Texas with *Acremonium* sp. from Spain. (Abstr.) Phytopathology 86: (in press).
4. Correll J.C., Klittich C.J.R. and Leslie J.F. 1987. Nitrate nonutilizing mutants of *Fusarium oxysporum* and their use in vegetative compatibility tests. Phytopathology 77:1640-1646.
5. García-Jiménez J., Velázquez M.T., Jordá C. and Alfaro-García A. 1994. *Acremonium* species as the causal agent of muskmelon collapse in Spain. Plant Disease 78:416-419.
6. Leslie, J.F. 1996. Fungal Vegetative compatibility - Promises and Prospects. Phytoparasitica 24(1):3-6.
7. Puhalla J.E. 1985. Classification of strains of *Fusarium oxysporum* on the basis of vegetative compatibility. Canadian Journal of Botany 63:179-183.
8. Vicente M.J., Abad P. and Cenis J.L. 1994. RFLPs del ADN mitocondrial, RAPDs y VCGs de *Acremonium* sp.: marcadores genéticos para su identificación taxonómica. In: La Protección Vegetal en España. J.M. García Baudín, A. Garrido Vivas and R. Jiménez Díaz (ed.). Instituto Nacional de Investigación y Tecnología Agraria y Alimentaria. Madrid. pp.401-415.

CAUSAL AGENTS OF HOP-DYING IN SPALT

W. AHRENS
Fachhochschule Weihenstephan, Abteilung Triesdorf, Germany

1. Introduction

In the hop-growing area of Spalt near Nuremberg, where hop is grown since the 15th century, hop yields are decreasing year by year. Yields dropped from about 32 q/ha in 1980 to 18 q/ha nowadays, without any external disease-symptoms to be seen at the hop plants.

2. Methods

As causal agents remained unknown for the present, the damage was called "soil exhaustion", according to Otto & Winkler (1985). This naming means heavy and lasting culture damages, which also occur after longer crop intervalls.

The test-principle of detecting cultivation damages is a vegetational trial with cuttings in pots. The test-soil is treated with different measures for controlling the cultivation damages. Reference quantitiy is the cuttings' growth in steamed soil (100 °C). By applying a nematicide, it can be detected, if the decrease in growth is only caused by nematodes.

The extent of growth-differences gives information about the degree of cultivation damages. The increase of plant-growth in steamed soil, compared with the untreated control, is relevant for estimating the damage.

Considering a positive steaming effect, the following threshold values are provided:
up to 140 % - no or slight damages
from 141 to 170 % - medium damages
from 171 to 200 % - heavy damages
more than 200 % - very heavy damages.

When growth in the nematicidal controlled soil (Temik 5 G, a.i. aldicarb) and in the soil, steamed at a temperature of 60 °C (thermic nematode control) does not differ significantly from the growth in the untreated soil, nematodes can be excluded as the only cause of damage. When there are no differences between growth in soil, treated chemically and thermicly at 60 °C against nematodes, to growth in soil steamed at 100 °C, nematodes are the main causal agents of damage.

Nematode-analysis was done by Dr. J. Schlang, Biologische Bundesanstalt, Elsdorf and, in combination with *Actinomycetes*-analysis, by Dr. H. Winkler, Biologische Bundesanstalt, Dresden-Pillnitz.

H.-W. Dehne et al. (eds.), Diagnosis and Identification of Plant Pathogens, 291-293.

3. Results

Old hop-stocks were attacked by the vessels-invading fungus *Ceratocystis paradoxa*. The fungi *Fusarium oxysporum* and *F. equiseti* could be isolated from the soil. The culture filtrates of both *Fusarium*-species caused severe hop-wilting within 24 hours. By microscopic examination, fungi of the genera *Pythium* and *Rhizoctonia* were found, which are known as root-rot fungi.

The filamentous roots were severely attacked by the hop-cyst-nematode *Heterodera humili*. Eight nematodes/cm root length were counted. Besides this, an infestation of the hairy roots by *Actinomycetes* was detected.

The results of the nematode analysis and vegetation test with hop cuttings in pots confirmed a high soil contamination with cyst-nematodes and a severe damage through it (TABLE 1and 2):

TABLE 1. Soil contamination with *Heterodera humili* in a hundred years old hop-garden of the farmer Weid in Wassermungenau, municipality Abenberg, Middle Frankonia; hop garden replanted five years before

Depth (cm)	cysts/100 ml soil	eggs and larvae/100 ml soil	eggs and larvae/cyst
0-30	598	4 700	7.9
60-90	313	6 413	20.5
120-150	27	45	1.7
150-180	9	19	2.1

TABLE 2. Vine length of two hop-varieties in 1995 (planting year) and 1996 in a vegetation pot trial with cuttings

Treatment of the soil	vine length (%)			
	var. Hallertauer Mittelfrüh		var. Perle	
	1995	1996	1995	1996
untreated	162 cm = 100 a*)	176 cm = 100 a	140 cm = 100 a	223 cm= 100 a
Temik 5 G	120 b	167 b	128 a	113 a
steam, 60 °C	111 a	160 b	127 a	108 a
steam, 100 °C	127 b	180 b	122 a	118 a

*) Values with the same notation do not differ significantly according Duncan's multiple range test.

In a pre-trial it became obvious, that searching the effect of the causal agent, it is sufficient to measure the vine length.

4. Discussion

A severe damage by nematodes became obvious in the second year of the trial in var. Hallertauer Mittelfrühe whereas var. Perle does not react in vine length significantly. The secondary invaders (root-rot fungi) however seem to be more important; they are injected by the nematode larvae, especially *Actinomycetes*.

In wet years, hop plants obviously can tolerate a high infestation rate with nematodes, because the non-infected deep roots compensate the water intake of the damaged year roots in the first 50 cm soil depth.

5. References

Otto, G. and H. Winkler (1985) Prognose von Nachbauschäden in Obstanlagen, *Akademie der Landwirtschaftswissenschaften der DDR*, 1-11

6. Summary

The filamentous roots of hop from a hundred year old hop garden were severely attacked by *Heterodera humili*. Besides this, an infection of the hairy roots by *Actinomycetes* and other root-rot fungi was detected.

It is discussed, which of the found invaders is the first causal agent of hop-dying and which is a following mycose.

FUSARIUM SECTIONS *ELEGANS* AND *LISEOLA*: TAXONOMY, rDNA INTERNAL TRANSCRIBED SPACER SEQUENCES AND DIAGNOSTICS

R.P. BAAYEN[1], C. WAALWIJK[1] and W. GAMS[2]
[1] *DLO Research Institute for Plant Protection, P.O. Box 9060, 6700 GW Wageningen, The Netherlands*
[2] *Centraalbureau voor Schimmelcultures, P.O. Box 273, 3740 AG Baarn, The Netherlands*

1. Introduction

The genus *Fusarium* harbours important plant pathogens and mycotoxin producers. One of the economically most important species is *F. oxysporum*, that comprises over one hundred formae speciales as well as saprophytic and antagonistic strains. *F. oxysporum* belongs to section *Elegans*, together with *F. redolens*. The questionable morphological distinction between the latter two species has been the subject of many studies [1, 3, 4, 7, 11]. Sequence analysis of part of the 28S ribosomal RNA gene indicates that *Elegans* is most closely related to *Liseola* [5]. The section *Liseola* comprises many mycotoxin-producing species such as *F. fujikuroi*, *F. proliferatum*, *F. sacchari*, *F. subglutinans* and *F. verticillioides*. The distinction between section *Liseola*, without chlamydospores, and *Elegans*, with chlamydospores, is increasingly questioned and a new section, *Dlaminia*, has even been proposed to accommodate species such as *F. dlaminii* that combine characteristics of both sections [6].

Ribosomal DNA (rDNA) analyses have been successfully employed to study major evolutionary events. Relationships at species level generally are reflected in the sequences of the internal transcribed spacer (ITS) region of the rDNA. We have previously shown that isolates of *F. oxysporum* f.sp. *dianthi* and *F. redolens* f.sp. *dianthi* can be distinguished unambiguously from one another on basis of ITS restriction fragment length polymorphisms [2, 8]. Apparent similarities between the ITS RFLPs of the species in sections *Elegans* and *Liseola* prompted us to determine the ITS sequences in these sections as well as section *Dlaminia*. The present paper summarizes results of these studies published elsewhere [8, 9, 10] and discusses the possible use of ITS sequences for diagnostic purposes.

2. Analysis of ITS restriction fragment length polymorphisms

The ITS region of the rDNA was investigated for 25 isolates belonging to eight formae speciales of *F. oxysporum*, 11 non-pathogenic isolates of *F. oxysporum*, seven

H.-W. Dehne et al. (eds.), Diagnosis and Identification of Plant Pathogens, 295-297.

isolates of *F. redolens* f.sp. *dianthi*, and ten isolates of *F. redolens* that do not belong to f.sp. *dianthi*. Amplified ITS fragments were larger in *F. redolens* than in *F. oxysporum*. Apart from this, eight out of twelve restriction enzymes tested produced RFLP patterns that further discriminated between the two size class groups. Results were independent of pathogenicity or host specificity: all isolates belonging to *F. oxysporum* had ITS type I, while all belonging to *F. redolens* had ITS type II [9].

Studies of the ITS region of the rDNA repeat were subsequently extended to 13 species and varieties of sections *Liseola*, *Dlaminia* as well as the related species *F. polyphialidicum*. Species belonging to sections *Fusarium*, *Sporotrichiella*, *Gibbosum* and *Martiella* were included as an outgroup. Two ITS types were encountered in *F. oxysporum* and *F. redolens* (section *Elegans*). But the other species examined from sections *Liseola* and *Dlaminia* also belonged to one of these types, while in species of other sections six other ITS types were encountered (Table 1). This supports the close relationship between sections *Elegans*, *Liseola* and *Dlaminia*. However, species groupings based on ITS types did not coincide with section boundaries [9].

TABLE 1. ITS types encountered among 27 species and varieties in seven sections of the genus *Fusarium*

ITS type	Species
I	*F. oxysporum* (section *Elegans*); *F. anthophilum*, *F. sacchari* var. *sacchari*, *F. sacchari* var. *elongatum*, *F. subglutinans*, *F. succisae*, *F. verticillioides* (section *Liseola*); *F. napiforme* (section *Dlaminii*); *F. polyphialidicum*
II	*F. redolens* (section *Elegans*); *F. annulatum*, *F. fujikuroi*, *F. proliferatum* (section *Liseola*); *F. beomiforme*, *F. dlaminii*, *F. nygamai* (section *Dlaminii*)
III	*F. culmorum*, *F. graminearum* (section *Fusarium*)
IV	*F. sambucinum* (section *Fusarium*); *F. poae* (section *Sporotrichiella*)
V	*F. sporotrichioides* (section *Sporotrichiella*)
VI	*F. tricinctum* (section *Sporotrichiella*); *F. acuminatum*, *F. longipes* (section *Gibbosum*)
VII	*F. equiseti*, *F. scirpi* (section *Gibbosum*)
VIII	*F. solani* (section *Martiella*)

3. ITS sequences and the development of diagnostics

Sequence analysis revealed that differences between ITS types I and II were mainly due to differences in the ITS2 region. UPGMA similarity dendrograms based on the sequences of the ITS1 and ITS2 regions were discordant with each other (not illustrated). The dendrogram for ITS1 revealed a close similarity for most species except *F. beomiforme* and *F. polyphialidicum*. In contrast, species groupings based on the ITS2 sequences formed two main clusters representing ITS types I and II, respectively [10].

The two ITS2 groupings do not seem to accurately reflect the phylogeny of these fungi (K. O'Donnell, personal communication). Differences in the sequences can nevertheless be used to develop diagnostics at species level. In section *Elegans*, such differences allowed unambiguous discrimination of *F. redolens* from *F. oxysporum*. Pathogenic specialization on, among others, carnation has thus developed independently in both species. This has important consequences for the development of diagnostics for wilt-inducing isolates of *Fusarium* spp. from carnation. In sections *Liseola* and *Dlaminia*, most species can be discriminated using ITS sequences, although this will not be possible for very closely related species such as *F. fujukuroi* and *F. proliferatum*.

For the development of diagnostics at infraspecific level (formae speciales, races) one should consider the mode of evolution within species. While mating populations are the natural genetic entities for some species in section *Liseola*, in the asexual fungus *F. oxysporum* vegetative compatibility groups (VCGs) rather than formae speciales or races should be the target of diagnostics development. Diagnostics at VCG level are particularly promising for those formae speciales where VCGs correspond to (groups of) races.

4. References

1. Baayen, R.P. and Gams, W. (1988) The *Elegans* fusaria causing wilt disease of carnation. I. Taxonomy. *Neth. J. Plant Pathol.* **94**, 273-288.
2. Baayen, R.P., Van Dreven, F., Krijger, M.C., and Waalwijk, C. (1997) Revision of the classification of races and vegetative compatibility groups in *Fusarium oxysporum* f.sp. *dianthi* and *F. redolens* f.sp. *dianthi*. *Eur. J. Plant Pathol.* **103**, submitted.
3. Booth, C. (1971) The genus *Fusarium*. Commonwealth Mycological Institute, Kew, U.K.
4. Gerlach, W. and Nirenberg, H.I. (1982) The genus *Fusarium* - a pictorial atlas. *Mitt. Biol. Bundesanst. Land.- Forstw. Berlin-Dahlem* **209**, 1-406.
5. Guadet, J., Julien, J., Lafay, J.f., and Brygoo, Y. (1989) Phylogeny of some *Fusarium* species, as determined by large-subunit rRNA sequence comparison. *Mol. Biol. Evol.* **6**, 227-242.
6. Kwasna, H., Chelkowski, J., and Zajkowski, P. (1991) *Grzyby (Mycota), tom XXII. Sierpik (Fusarium)*, Polska Akademia Nauk., Flora Polska, Warszawa-Kraków, Poland.
7. Nelson, P.E., Toussoun, T.A., and Marasas, W.F.O. (1983) *Fusarium species. An illustrated manual for identification*, Pennsylvania State University Press, University Park, PA, USA.
8. Waalwijk, C. and Baayen, R.P. (1995) Identification of races of *Fusarium oxysporum* f.sp. *dianthi* by DNA fingerprinting and vegetative compatibility. *Bull. O.E.P.P.* **25**, 117-124.
9. Waalwijk, C., Baayen, R.P., De Koning, J.R.A., and Gams, W. (1996) Ribosomal DNA analyses challenge the status of *Fusarium* sections *Elegans* and *Liseola*. *Sydowia* **48**, 90-104.
10. Waalwijk, C., De Koning, J.R.A., Baayen, R.P., and Gams, W. (1996) Discordant groupings of *Fusarium* spop. from sections *Elegans*, *Liseola* and *Dlaminia* based on ribosomal ITS1 and ITS2 sequences. *Mycologia* **88**, 361-368.
11. Wollenweber, H.W. and Reinking, O.A. (1935) *Die Fusarien, ihre Beschreibung, Schadwirkung und Bekämpfung*, Paul Parey, Berlin, Germany.

INVESTIGATION OF *CLAVICEPS PURPUREA /FR./ TUL* IN THE *POACEA* FAMILY PLANTS IN LITHUANIA

Z. DABKEVIČIUS
Lithuanian Institute of Agriculture
5051 Dotnuva-Akademija, Kėdainiai, Lithuania

INTRODUCTION

Ergot is one of the most harmful diseases of seed grasses, rye and triticale in Lithuania. In separate seed lots as much as 0.01 - 2.00 % of ergot sclerotia are found after pre-cleaning in the cereal and grass seed production farms. Since 1992 the incidence of ergot has increased considerably as a result of a higher number of small farms and natural pastures and as a result of a reduction in the application of chemical plant protection measures. In order to have an adequate control of ergot spread it is necessary to know the sources of the disease spread, biology of the causal agent and efficiency of the control measures.

MATERIALS AND METHODS

This research work was carried out in the Lithuania Institute of Agriculture over the period 1977-1996 with a view to establishing the whole range of *Poaceae* family plants attacked by ergot (*Claviceps purpurea Fr/Tul*). The plants were collected in different regions of Lithuania as well as in the Lithuanian Institute of Agriculture's collection and nurseries. In the tests of ergot causal organism specialization, 6 grass species and winter rye were cross-infected with ergot conidiaspores collected from these plants. In the experiments on the influence of the environmental factors on ergot sclerotia germination, differently-sized sclerotia were incorporated at various depths in the heavy loam, sandy and organic soil. With a view to establishing the influence of high temperatures on ergot sclerotia viability, dry and wet seeds contaminated with sclerotia were heated at temperatures of 40-80°C for a period from 10 minutes to 168 hours. The effect of different seed treaters and fungicides on ergot sclerotia viability was established under laboratory and field conditions. The seeds contaminated with ergot sclerotia were treated with various seed treaters at a rate of 1.5-8.0 g for 1 kg of seeds and afterwards were germinated in the laboratory in wet sand and outdoors in the soil.

H.-W. Dehne et al. (eds.), Diagnosis and Identification of Plant Pathogens, 299-301.

RESULTS

In Lithuania 38 ergot-affected plant species had been identified before we started our research. Over the period 1979-1996, 74 species and forms of cultivated and wild plants affected by ergot were collected in different regions of Lithuania as well as in the Lithuanian Institute of Agriculture's grass nurseries. Out of these 74 species, 39 plant species and forms were identified by us in Lithuania for the first time. Two grass species affected by ergot and described by Hb. dr. M.Strukčinskas were included in the ergot host list [1,2]. Thus, a total of 79 ergot-affected cereal plant species are known in Lithuania today. The disease was found in: 5 cereal species; 14 cultivated and 33 wild grasses; 4 major grass weeds; 17 introduction plants, grown in collections and ornamental plants, 3 interspecific grass hybrids and 3 tetraploid plants.

It was established that the following plant species are more infected by ergot: *Bromis inermis, Calamagrostis arundinacea, Elitrigija repens, Festuca arundinacea, F. gigantea, F. mairei, F. pratensis, F. rubra, Gliceria fluitans, Milium effussum, Molinia caerulea, Phalaris arundinacea, Poa palustris, P. pratensis, Secale cereale, Triticosecale,* tetraploid *Lolium multiflorum* as well as interspecific hybrids of grasses.

The majority of cereal plants may catch infection from one another; however, a certain specialization of ergot causal organism has been ascertained. Rye and meadow grass ergot causal organisms were found to be more specialized. Ergot sclerotia chemical composition, especially the content of amino acid depends on the origin of infections.

The size of ergot sclerotia depends on the species of the host plant and on the number of sclerotia in a spike or in a panicle. The smallest sclerotia were found in *Agrostis gigantea* (length 0.9-1.9 mm; width - 0.5-0.6 mm; thickness 0.3-0.4 mm and weight 0.1 mg) and the largest sclerotia were found in *Secale cereale* (6.8-47.0; 2.3-7.5; 1.9-5.5 mm and 16-817 mg respectively), which exceeded the grain size and weight by several times. The more the plants are affected by ergot, the smaller the sclerotia are and more complicated it is to separate sclerotia from seeds. The largest sclerotia were found in meadow feacue where there were 1-5 sclerotia in a panicle, while the smallest sclerotia was observed - in the case of 26 and more sclerotia per panicle. A strong negative correlation was established between the number of sclerotia in a spike and their length (r = -0.86), thickness (r = -0.70), weight (r = -0.54). A possitive correlation was established between the sclerotia length and their thickness (r = 0.56) and weight (r = 0.82). It was recommended to clean fescue seed from ergot sclerotia first of all according to weight and length and later according to width.

It was established that the most favourable conditions for ergot sclerotia germination and for stroma formation was in organic soil, where the number of stromas was significantly higher than that in heavy loam or sandy loam. A major factor determining sclerotia germination was the ability of the soil to accumulate moisture.

Larger sclerotia were the best at germinating and produced the highest number of stromas. Sclerotia germinated only from the soil surface and from the depth of 0.5-4 cm; however, the best conditions for their germination were at a depth of 0-1 cm. Although sclerotia placed at a depth of 6, 12 and 25 cm germinated; but, their stromata did not reach the soil surface. Thus we could reduce the risk of the spread of ergot by avoiding to grow seed grasses in organic soils and wet places and if the soil is deeply ploughed after seed harvesting. Infection level would be reduced if sclerotia especially large ones with the high infection potency were removed from the seed. It was established in our tests that when ergot infected seeds were stored for a long period the ergot viability was reduced but not so quickly as other authors indicate. Only after 2-3 years sclerotia lose their viability completely. Mechanically - damaged sclerotia or their debris form stromas are able to spread infection similarly to healthy ones.

It was known earlier that ergot sclerotia lost germinating power when they were kept for a longer time in warm storages. However, no findings were found in literature on how higher temperatures (40-80°C) affect sclerotia germinating power, and if it is possible to clean the infected seed from infection using a thermal treatment.

It was proved in tests that sclerotia viability is very much dependent on their moisture content, heating temperature and exposure , whereas grass seed is less sensitive to higher temperatures. It was established that ergot sclerotia - infested seed with a moisture content of 23.1-29.4% may be heated at 45°C temperature for not more than 24 hours with no damage for seed germinating power; however seed heating at 50° C should be not longer than 6 hours. Although the biggest part of ergot sclerotia is killed when temperatures are increased and when exposure time is extended, the seed germinating power, and especially seed vigour goes down.

Sclerotia and seed with a moisture content of 10.6-13.5% were less sensitive to temperature increasing than the sclerotia and seed with a moisture content of 23.1-29.4%. However, a significant reduction in ergot sclerotia viability was observed at 50° C temperature at a 24 hour exposure. Sclerotia germination power was lost completely when they were kept at 80°C temperature for one hour. Such regime did not damage the seed germinating power; however, it is risky to dry bigger seed lots using this method. It should be noted that, when seeds are dried at temperature of 40-45°C, part of sclerotia lose their viability and the other part start germinating several days later.

It is difficult to apply a thermal seed treatment in practice and therefore simpler, safer and cheaper seed treatment methods are being looked for. During 1980-1983, having investigated the efficiency of various seed treaters and their rates against meadow fescue ergot in laboratory and pot experiments, it was established that all seed treaters inhibited ergot sclerotia germination. The most effective turned out to be benlate and thiram. Therefore seed which cannot be cleaned from ergot sclerotia should be applied with these seed treaters at a rate of 3 - 4 g/kg.

REFERENCES

1. Dabkevičius, Z. (1995) The main Fungous diseases at Poacea family plants, their development and control. The work of doctor habilitatus.-Dotnuva-Akademija,.

2. Дабкявичюс З. В., Струкчинскас М. (1985) Поражаемые грибами рода *Claviceps Tul.*злаковых растений в Литовской ССР // Труды А.Н. Литовской ССР.-СерияВ.-Т.4 (92).-С.3-11.

FIRST REPORT ON ANTHRACNOSE DIAGNOSIS AND *COLLETOTRICHUM* SPP. IDENTIFICATION IN WHITE LUPIN (*LUPINUS ALBUS* L.) INFECTION IN POLAND

Anthracnose on lupins in Poland

I.M. FRENCEL,[1] E. LEWARTOWSKA[1] AND A. CZERWIŃSKA[2]
[1]*Institute of Plant Genetics of the Polish Academy of Sciences, Strzeszyńska 34, 60-479 Poznań*
[2]*Plant Breeding Station Wiatrowo of the Poznań Plant Breeders, 62-100 Wągrowiec, Poznań*

Key words: anthracnose, *Colletotrichum* ssp., *Lupinus albus* L.

Abstract. In Poland the occurrence of anthracnose on white lupin was first encountered in one of the breeders' experimental plots in 1995. The disease was apparently not known here before. Observation during the consecutive vegetation season pointed out to its rapid widespread, embracing also yellow and narrow-leafed lupin plots. Also, disease incidences developed mostly on top pods were found in some white and yellow commercial lupin fields.

1. Introduction

In the early summer of 1995, at the very beginning stage of flowering, some of white lupin (*Lupinus albus* L.) plants growing in the occasionally inspected breeders' experimental fields of the plant breeding station in the mid-western part of Poland showed typical symptoms of anthracnose. Diseased plants were easily visible by their characteristic "sheperd's crook" top-benting, with elliptic, pinkish brown, elongated lesions on leaves. Infection foci occurred in local patches and rather infrequently were widespread throughout the growing plots. The weather course records during spring months and before field observation were conductive to the fungal infection according to the biology of the pathogen and the anthracnose development described elsewhere.
Until now the disease has not been encountered on white lupin or other lupines in Poland. One of the author's of this report (I.M.F.) observed anthracnose on growing white lupin fields in Chile, where the disease is widespread and commonly known. To our knowledge, there has been no but one announcement on the occurrence of anthracnose on white lupin and/or lupin spp. growing in Europe, and the only one has been reported in France

H.-W. Dehne et al. (eds.), Diagnosis and Identification of Plant Pathogens, 303–306.

(Gondran, 1990). As it has appeared recently, the disease is spreading rapidly in European countries producing lupin crops. This conclusion may be conceived from the last International Lupin Conference, which was held in USA (Abstract Book, 1996).
The aim of our work was to initiate detailed studies on lupin(s) anthracnose developing in Poland: 1) to identify the pathogen(s) - causal agent(s) of anthracnose in local isolates, 2) to keep at observation of the disease development under local environmental conditions, 3) to contribute to the eventual coordinated European programmes in the aspect of disease threat evaluation and potential epidemic, and 4) to study on genetical variability, pathogenicity and races evolution within pathogen(s) populations for comparison in range of Europe and worldwide.

2. Material and Methods

In June 1995, diseased white lupin plants were collected from the breeding plots and fungal pathogen(s) were isolated; single-spore cultures on PDA were selected and subcultured, and diagnosed routinely by microscopical examination. Fungal pathogenicity was proved by artificial inoculation of lupin seeds in laboratory tests and/or stems in preliminary glasshouse tests.
In 1996, observations of disease incidence were carried out on lupin breeding plots and some seed production fields of commercial varieties, embracing also yellow and narrow-leafed lupines within several mid-western localities. Main records concerning daily temperature and rainfall levels were gathered for the vegetation season of 1996.

3. Results and Discussion

3.1. PATHOGEN IDENTIFICATION AND PATHOGENICITY

All of the several fungal isolates collected so far were identified as *Colletotrichum gloeosporioides* Penz. (Arx, 1987). Conidia obtained from re-isolation of artificially inoculated white lupin seeds or stems were approx. 16-20 μm long and their morphology was typical of that species, according to the literature. The host-pathogen interaction in preliminary glasshouse pathogenicity biotests on white lupin cultivars was positive, showing lesions on infection sites, similar to the diseased plants in natural infection.

3.2. LUPIN NATURAL INFECTION IN 1996

In the consecutive vegetation season, i.e. in 1996, the first disease symptoms were found (June, 16-18) on some multiplication plots of white lupin breeding lines (WTD 395, WTD-493), previously selected throughout full-sib family, which was found to have incidents of plant infection in 1995. Later in the season (June, 24-26), diseased plants

occurred also among other breeding lines, as well as on other experimental plots. The disease developed progressively and their incidences were widespread up to 100 per cent (Table 1). Infection appeared also in yellow lupin, and to smaller extent, on narrow-leafed lupin plots. Disease symptoms occurred mainly on pods and young branches.
The occurrence of disease was found on several, but not on all, fields of white and yellow lupin commercial varieties grown for seed production, located more or less in the vicinity of anthracnose-attacked fields. On the contrary, anthracnose has not been found on white or yellow lupin plantations of commercial varieties located in remoted private plant productive fields.

4. Conclusions

In conclusion, from our first observation in 1995 and 1996 it may be inferred, that the disease once appeared is widespread dramatically. The seed transmission of *Colletotrichum* spp. should be considered rather as the most important primary source of infection. The potential threat of epidemic should be focused.
Further investigations are underway.

References

Abstract Book (1996) Eight International Lupin Conference. 11-16 May 1996, Pacific Grove, California, USA.
Arx von JA (1987) Plant Pathogenic Fungi. Beihefte zur Nova Hedwigia, Heft 87, Cramer J (ed) Berlin, Stuttgart, pp.218-220.
Gondran J. (1990) The disease of the white lupin crops in France prevention possibilities. Proceedings 6th International Lupin Conference, Nov 25-30, 1990 Temuco-Pucon, Chile, pp. 277-279.

TABLE 1. Anthracnose on white lupin in experimental plots (Plant Breeding Station - Wiatrowo), 1996

Breeding strains/ cvs	B-103	B-107	B-108	B-109	B-85	B-92 A	B-74	B-76	B-9	B-63	B-13 E	B-14 E	WAT	BARDO	BAC
Disease score (0-9)	3.0	7.0	3.0	3.0	3.0	5.5	2.5	2.5	6.5	5.5	5.5	8.0	2.0	4.0	5.5
Diseased plants/m^2 (%)	97	94	83	85	91	87	92	89	90	100	99	100	98	93	79

Disease score: 0 = healthy, 9 = plant severe attacked; mean values from three repetitions, scoring date: June 28, 1996.

Diseased plants per square meter in per cent; mean values from four repetitions, date of observation: Aug 1, 1996.

SUITABILITY OF ELECTRON MICROSCOPY FOR ROUTINE DIAGNOSIS OF PLANT VIRUS DISEASES IN EXTENSION SERVICE

JOACHIM HAMACHER, RAINER GIERSIEPEN and MONIKA HEUPEL*
Institut für Pflanzenkrankheiten, Rheinische Friedrich-Wilhelms-Universität Bonn, Nussallee 9, 53115 Bonn
**Landwirtschaftskammer Rheinland, Pflanzenschutzamt Bonn, Siebengebirgsstr. 200, 53229 Bonn*

1. Introduction

A collaboration between Plant Protection Service and University has been installed in 1993 to support the diagnosis service. By using the electron microscope (EM)-facilities of the Institute of Plant Pathology, the efficiency of plant virus diagnosis could greatly be improved. Advantages and difficulties of electron microscopic standard methods in routine service are examplified by several virus diseases.

2. Materials and Methods

Incoming samples are at first inspected in the central diagnosis lab of the Plant Protection Service. Virus suspicious samples, that can not be identified by ELISA-kits at hand, are transferred by courier to the EM-lab of the Institute of Plant Pathology. Here, incoming samples are again inspected and virus etiology is confirmed by symptomatology. Leaves or other suspectedly virus-infected tissues are homogenised in PO_4- or other suitable extraction buffers with optionally added fixatives. Incubation of carbon fronted pioloform films on nickel or copper grids (400 mesh) is done for 2-20 min on a drop (40-100 ml) of crude plant sap placed on laboratory plastic film. The coated side of the grid is then rinsed with 40 falling drops of deionized, particle free water, drained on moist filter paper and negatively stained by just touching the surface of a drop of 2 % uranyl acetate with the coated face of the grid. The grid then is immediately drained on moist filtèr paper, filmed face upside, and allowed to dry for 20 min on an adhesive rubber plate in a petri dish. If no particles are visible, but symptoms suggest a viral etiology of the disease, immunosorbent electron microscopy (ISEM) may be performed using suitable antisera from the institute's antiserum collection. Antiserum dilutions applied for ISEM range from 1:200 to 1:1000. Incubation times are 5 - 30 min. For precise serological identification of the particles, decoration is performed with antiserum dilutions of 1:50 up to 1:200 with 15 min

H.-W. Dehne et al. (eds.), Diagnosis and Identification of Plant Pathogens, 307–309.

incubation time (Milne and Lesemann 1984). Crude antisera work equally well, if not better, than purified IgG. On request mechanical virus transmission is done on a set of the most usual test plants.

3. Results

In 1993 39 samples were tested, 14 proved virus-positive, in 1994 75 samples were tested, 37 were positive, in 1995 48 of 108 samples tested positive and in 1996 (January - August) 8 of 28 samples were positive. Most samples were ornamentals (184), whereas legumes (42), field crops (18) and fruit (6) had to be tested less frequently.

In 1993-95 most of the samples were vegetatively propagated *Petunia*-hybrids. They tested positive for tobamoviruses (either tobacco mosaic- or tomato mosaic virus), for potato Y potyvirus or alfalfa mosaic alfamovirus. Most infections occurred with tobamoviruses. A disease of *Impatiens* New Guinea-hybrids was caused by a tobamovirus and serologically identified as ribgrass mosaic virus by Lesemann in 1993 (pers. comm.). In carnation (*Dianthus caryophyllus*), carnation latent carlavirus and carnation yellow fleck closterovirus could easily be detected in mixed infections. Lilies grown in the field often revealed infections with filamentous viruses, presumably lily symptomless carlavirus and tulip breaking potyvirus. In *Helleborus* spp., a disease causing veinal necrosis, malformation and black spotting of the flowers was associated with helleborus carlavirus. The observation of virus particles in symptomless plants, suggests the presence of another disease-causing factor. Occurrence of cucumber mosaic cucumovirus in spinach and lettuce (in 1996) did not pose any problems. EM-detection of viruses in *Pelargonium*- hybrids (e.g. pelargonium flower break carmovirus) was unsatisfactory because of low virus particle numbers and substances in crude plant sap, reducing a clear contrast of particles on the support film.

Labile viruses, for which fixation is recommended, could be recognized efficiently without any special treatment (alfalfa mosaic alfamovirus from *Petunia*-hybrids and tomato spotted wilt tospovirus, acronym TSWV, from tomato and *Tagetes patula*) when incubation times were kept short. Mechanical transmission to *Nicotiana tabacum* cv. Xanthi nc, however, proved to be a more reliable method for TSWV-detection than absorption preparates for electron microscopy. Rod shaped particles of lettuce big vein virus (LBVV) could be observed in one case without the recommended fixation and trapping by antiserum (Vetten *et al.*, 1987). This result, however, could not be repeated. A potassium phoshate buffer containing sodium disulphite, sodium diethyl-dithiocarbamate and activated charcoal was used for mechanical transmission of this labile virus to *Chenopodium quinoa* (Huijberts *et al.*, 1990). Cherry leaf roll nepovirus particles in infected of wild cherry or birch trees could only be detected when trapped with a suitable antiserum.

No reliable results could be obtained with luteovirus-infections (BYDV, BMYV) or when samples from suspectedly infected woody ornamentals (*Paulownia tomentosa*) and shrubs (*Rhododendron japonicum*) had to be examined. Symptomatology or ELISA were more reliable in some of these cases to confirm viral etiology.

4. Discussion

The electron microscope proved to be a valuable tool for the routine diagnosis of virus diseases. The routine preparation of samples is easy and requires a minimum of time. Most viruses occurring in extension routine testing could be identified by simple absorbance preparates and negative staining. Usually the confirmation of viral etiology is sufficient for the customer, therefore a specification of the virus is not obligatory. In most cases, however, virus genus or species can be related from morphological or host range data. Sometimes the use of antisera is necessary for determination of viral etiology. The application of antioxidants or other additives like charcoal may further increase the success of EM-diagnosis as was shown by Blystad *et al.* (1995) for pelargonium flower break carmovirus in *Pelargonium* spp..

Requirements for a successful EM-routine diagnosis are: The use of carbon fronted grids (preferably 400 mesh) fixed on parafilm, antimagnetic tweezers with a fine tip, the use of ultra pure rinsing water, and the availability of antisera for trapping and decoration. An electron microscope with a turbomolecular pumping unit proved to be very useful. Because of the high pumping capacity and relative insusceptibility to remaining humidity on the grids, a high throughput of samples can be achieved. A quick courier service helps to prevent decay of samples.

In combination with biotest, electron microscopy-based diagnosis ensures a reliable system for most occurring virus diseases.

5. References

Blystad, D.-R., Naess, V., Haugslien, S. (1995) Optimizing immunosorbent electron microscopy for detection of pelargonium flower-break carmovirus in pelargonium, *EPPO-Bulletin*, **25**, 239-245

Huijberts, N., Blystad, D.-R., Bos, L. (1990) Lettuce big-vein virus: mechanical transmission and relationships to tobacco stunt virus, *Ann. appl. Biol.* **116**, 463-475.

Milne, R.G., Lesemann, D.E. (1984) Immunosorbent electron microscopy in plant virus studies, *Meth. Virol.*, **8**, 85-101

Vetten, H.J., Lesemann, D.E, Dalchow, J. (1987) Electron microscopical and serological detection of virus-like particles associated with lettuce big vein disease, *J. Phytopathology*, **120**, 53-59

USE OF THE SCANNING ELECTRON MICROSCOPE IN THE IDENTIFICATION OF POWDERY MILDEWS

A.J. INMAN, R.T.A. COOK & C. BILLINGS
Central Science Laboratory, MAFF, Sand Hutton, York, YO6 1LZ, UK

1. Introduction

The taxonomy and identification of powdery mildews is based largely on characteristics of the teleomorph. However, often only the conidial stage is present and it can be impossible to identify the mildew to species or even genus. Furthermore, dependence on the cleistothecium in identification often causes problems when a powdery mildew increases its host or geographical range, because the teleomorph may not be seen for many years, or may not be produced at all. The inability to conclusively identify the powdery mildews affecting *Rhododendron* and tomato, recorded in the UK in 1980 and 1987 respectively, and more recently one on *Fuchsia*, illustrate this point well. In plant quarantine, action against a potentially non-indigenous mildew must be taken quickly because air-borne conidia spread rapidly; a fast and accurate identification is essential.

2. Methods

Fresh powdery mildew specimens were examined by cryomicroscopy in the SEM. Small leaf pieces were frozen in a sputter-cryo in the absence of a vacuum, sputter coated with gold and transferred to the SEM for examination at relatively high magnification (x7–12,000). Dried herbarium material was used for a few genera unavailable locally. In total, 59 powdery mildew taxa on 84 host species were studied.

3. Results

3.1. OBSERVATIONS OF POWDERY MILDEW CONIDIA UNDER THE SEM

The SEM revealed previously undescribed patterns on the surfaces of powdery mildew conidia. Distinct patterns on conidial septa (end walls) combined with patterns on the outer wall enabled anamorphs to be separated into eleven major groups.

The anamorphs of *Leveillula*, *Phyllactinia* and *Pleochaeta* all had smooth or warty septa, but each had a different distribution of warts on the outer wall. A warty, annular septum and an echinulate outer wall was unique for the *Oidium* anamorph of *Blumeria*. The remaining genera, all presently covered by *Oidium*, had septa which were either 'whorled' (concentrically ridged) or 'fibrillar' (patchwork of rods arranged in

H.-W. Dehne et al. (eds.), Diagnosis and Identification of Plant Pathogens, 311–313.

fascicles). Those with 'whorled' septa encompassed anamorphs of *Podosphaera*, *Sphaerotheca*, *Sawadaea* and *Cystotheca*. Those with fibrillar septa included anamorphs of *Erysiphe*, *Microsphaera*, *Uncinula* and *Arthrocladiella*. (See Figs 1-12.)

3.2. PROPOSED REVISION OF THE FAMILY ERYSIPHACEAE *

Observations under the SEM prompted a proposed major revision of the family Erysiphaceae, based on both teleomorphic and anamorphic characters. In particular, subgenera were proposed for the anamorphic genus *Oidium*, enabling more refined diagnosis of genera with these conidial stages by the plant pathologist.

Oidium anamorphs with 'whorled' septa were divided into three newly proposed subgenera: '*Fibroidium*' for anamorphs of *Podosphaera* and *Sphaerotheca* having conidia with fibrosin bodies and smooth outer walls; '*Octagoidium*' for the dendritic patterned, eight-sided conidia of *Sawadaea*; and '*Setoidium*' for the anamorph of *Cystotheca*, reflecting its setae on the mycelium. Three new subgenera were also proposed for those anamorphs with 'fibrillar' septa: '*Reticuloidium*' for conidia with roughcast outer walls of *E.* sect. *Golovinomyces*; '*Graciloidium*' for the unswollen conidia of *Arthrocladiella*; and '*Striatoidium*' for the conidia of *E.* sect. *Galeopsidis* with striate outer walls. The emended term '*Pseudoidium*' was retained for the anamorphs of *E.* sect. *Erysiphe*, *Microsphaera* and *Uncinula* having conidia with scaly, fibrous outer walls.

3.3. CONSTRUCTION OF KEYS *

SEM, light microscope and host range data were used to construct both dichotomous and synoptic keys for the prediction of all powdery mildew anamorphs found in the UK. SEM and other morphological characters could not distinguish anamorphs of *Sphaerotheca* and *Podosphaera*, or anamorphs of *E.* sect. *Erysiphe*, *Microsphaera* and *Uncinula*, although host range data could predict the mildew genus in many cases.

4. Conclusion

Patterns on powdery mildew conidia under the SEM have been reported previously, but only for a few genera. Therefore, the significance of such patterns in the identification and taxonomy of the Erysiphaceae has not been recognised. The characters revealed here by the SEM enable more accurate identifications of powdery mildew anamorphs in the absence of the teleomorph. Moreover, they have prompted a taxonomic revision which overcomes anomalies caused by over emphasis of the cleistocarp appendage.

*(Cook, Inman & Billings (1996), in press.)

5. Acknowledgements

We acknowledge taxonomic advice from Dr U Braun and Dr B Sutton, and funding by Plant Health Division of the Ministry of Agriculture, Fisheries and Food.

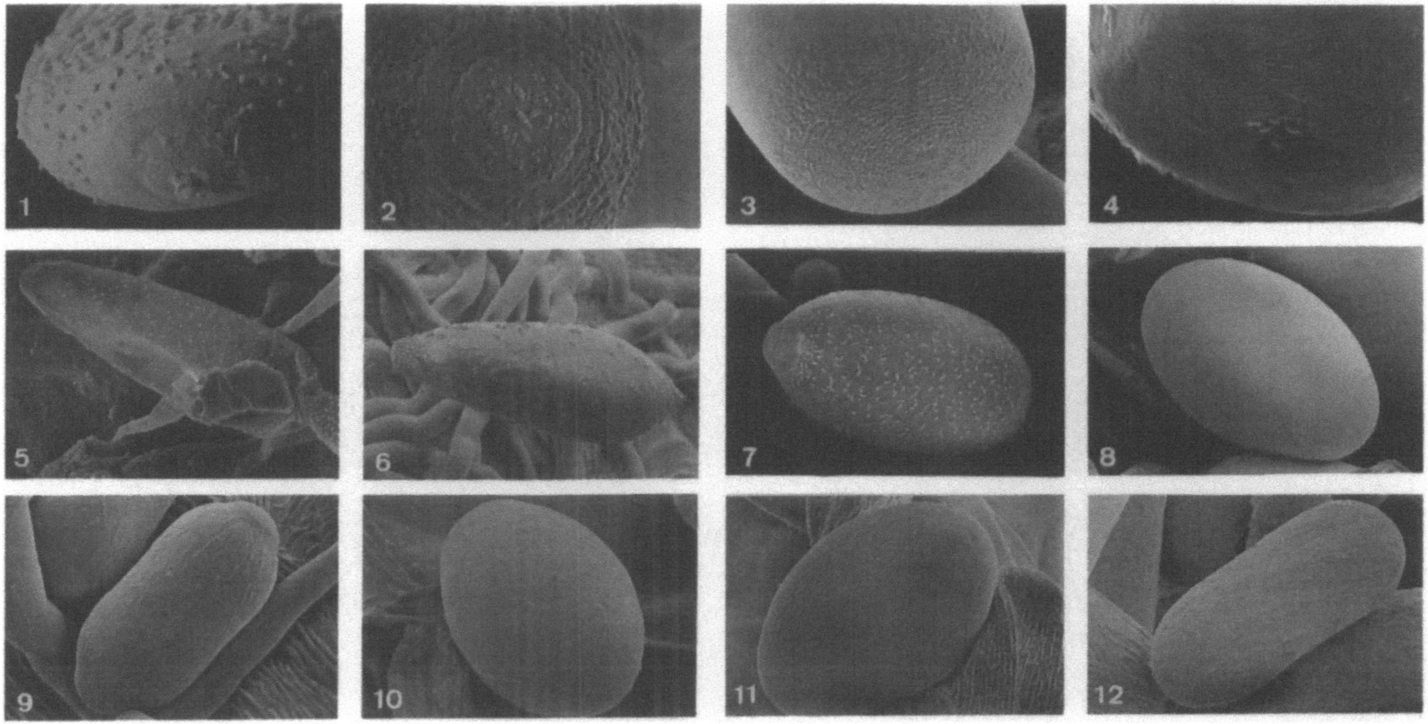

Figures 1-4. **Scanning electron micrographs of powdery mildew conidia showing septal patterns.** ***Fig. 1*. Smooth, surrounded by warts (*Leveillula*).** ***Fig. 2*. Warty and annular (*Blumeria*).** *Fig. 3*. Concentric 'whorled' (*Sphaerotheca*, *Podosphaera*, *Sawadaea* and *Cystotheca*). *Fig. 4*. 'Fibrillar' (*Erysiphe*, *Microsphaera*, *Uncinula* and *Arthrocladiella*). *Figures 5-12*. Patterns of outer walls. *Fig. 5*. Evenly warty (*Leveillula*). *Fig. 6*. Warts in groups, arranged helically (*Phyllactinia*). *Fig. 7*. Echinulate (*Blumeria*). *Fig. 8*. Smooth (*Sphaerotheca*, *Podosphaera*, *Cystotheca*). *Fig. 9*. Vein-like, embossed strips (*Sawadaea*). *Fig. 10*. 'Roughcast' (*E.* sect. *Golovinomyces*). *Fig. 11*. Finely ribbed (*E.* sect. *Galeopsidis*). *Fig. 12*. Scaly ridges (*E.* sect. *Erysiphe*, *Microsphaera*, *Uncinula*).

ANALYSIS OF *PYRENOPHORA TERES* (DIED.) DRECHS. POPULATION FROM SLOVAKIA.

M. JÁNOŠOVÁ AND J. KRAIC
Research Institute of Plant Production
Bratislavská cesta 122, 92168 Piešťany, Slovakia

Pyrenophora teres (Died.) Drechs. is the causal agent of net blotch of barley that has become in the last ten years one of the most widespread barley leaf mycoses in Slovakia due to the absence of resistent cultivars in present local assortment. From the viewpoint of new biological material creation with resistance to this facultative parasite the analysis of the Slovak population aimed at symptoms production, culture morphology, virulence, molecular differentiation has been carried out. All these parametres studied showed high level of inter- and intrapopulation variability.

Within the Slovak P. teres population three kinds of symptoms has been observed : *net type lesion* which is characterized by dark brown blotchies crisscrossed with a netlike venation and accompanied by chlorosis; *spots* of various shapes and sizes encircled by varying width of chlorosis and finally *stripe like pattern*.

When studying in vitro characteristics of individual monoconidial isolates from various localities the considerable variability of morphological expression has been recorded and between the dark and light-coloured isolates the highly significant difference in sporulation ability has been found. P. teres populations collected in 1995 at seven and in 1996 at four localities from diverse

H.-W. Dehne et al. (eds.), Diagnosis and Identification of Plant Pathogens, 315–316.

geographical regions of Slovakia have been undergone an virulence analysis. Based on 153 monoconidial isolates obtained, the set of differentials and the laboratory method of testing on detached leaves in benzimidazole solution the significant differences of aggressiveness among the local populations have been revealed. The most aggressive population was from the Čadca locality with a mean of 51,6% of virulent clones. Based on responses of the set of differentials to 153 individual isolates 73 pathotypes has been identiefied. Genotypes of the Ethiopian origin CI. 5791, CI. 9819, CI. 9820, and CI. 9825 have been attacked by the lowest percentage of virulent clones and confirmed their permanent high resistance not only to the local populations but to the Slovak P. teres population as a whole. On the other hand the Manchurian genotype CI. 739 which exhibited the low frequency of virulent clones to the population collected in 1995 was highly defeated by the Malý Šariš and Nové Zámky populations from 1996.

High level of inter- and intrapopulation variability of two populations has been revealed by amplification of simple sequence repeats (SSR). The (GATA)4 and (GACAGATA)2 sequences used as primers generated multiband polymorphic patterns. Populations isolated from six localities of Slovakia were distinguished each of others.

As a result of the analysis of P. teres populations from Slovakia the genotypes with the stable resistance and wide effectiveness against populations have been found out. For this reason they may have been used for barley resistance breeding to net blotch.

DETECTION OF PLUM POX VIRUS BY DIFFERENT METHODS

M. KÖLBER[1], M. NÉMETH[1], L. KRIZBAI[1],
E. KISS-TÓTH[2], M. KÁLMÁN[2]

[1]*Plant Health and Soil Conservation Station of Ministry of Agriculture, H-1118 Budapest, Budaörsi út 141-145 ,Hungary*

[2]*Bay Zoltan Institute of Biotechnology, H-6726 Szeged, Derkovits fasor 2. Hungary*

1. Introduction

During the production of virus-free propagating material the exact detection of plum pox virus (PPV) is of primer importance for the practice.
In order to determine reliability of different detection methods, field indexing and laboratory methods: variations of ELISA, nucleic acid sandwich hybridization (NASH) and RT-PCR were compared.

2. Material and Methods

2.1. PLANT MATERIALS

In the *first part* of the experiment the above methods were compared by testing eighty, previously identified PPV isolates collected from eight species and maintained on GF 305 indicators in the Hungarian virus collection.
In the *second part* of the experiment bearing trees of unknown health state were tested using the same methods. For this purpose forty-four trees of different stone fruit species were selected in orchards where PPV infection has been previously found.

2.2. WOODY INDEXING

PPV isolates, originating from eight species, and bearing trees were budded (chip budding) onto GF 305 peach seedlings and onto GF 31 Myrobalan hybrids under field conditions.

2.3. LABORATORY METHODS

Double antibody sandwich ELISA (DAS ELISA) and biotin-avidin ELISA (B-A ELISA) were simultaneously used with nucleic acid sandwich hybridization (NASH) method developed by Palkovics *et* al. (1994). Results were compared to RT-PCR assay, adapted by Krizbai *et* al. (1996) modifying the method designed by Levy *et* al. (1993).
RNA extraction. Total RNA extraction from leaf tissues, applying a lithium chloride-based method, was used for RT-PCR reaction. The RNA suspension was electrophoresed in 1% agarose gel stained with ethidium bromide.

H.-W. Dehne et al. (eds.), Diagnosis and Identification of Plant Pathogens, 317–319.

Primers. The following oligonucleotide primers were used:
sPPV: 5' GTAGTGGTCTCGGTATCTATCATA 3'
asPPV: 5' GTCTCTTGCACAAGACTATAACC 3'
In positive samples the PCR reaction generated 220 bp length PPV fragments.
PCR reaction. The reverse transcriptase (RT) and PCR reactions were done in one, non-interrupted thermal cycling programme. RT-PCR conditions: 30 min. at 42°C, 3 min. at 95°C, followed by 40 cycles of 30 sec. at 92°C, 30 sec. at 54°C, 1 min. at 72°C, with a final extension of 5 min. at 72°C. PCR products were electrophoresed in a 2 % agarose gel and stained with ethidium bromide.

3. Results

3.1. FIELD INDEXING AND LABORATORY TESTING OF PPV ISOLATES

3.1.1. Woody Indexing

In this experiment 91,25% of the 80 PPV isolates, originating from 8 plant species, were detected on GF 305 indicators and 95,00% of these isolates provoked symptoms on GF 31 indicators (Table 1).

3.1.2. Laboratory Tests

All the 80 isolates were revealed by RT-PCR and 95,00% by NASH method, while 93,75% and 83,7% of the isolates were detected by B-A ELISA and DAS ELISA, respectively (Table 1.).

3.2. FIELD INDEXING AND LABORATORY TESTING OF BEARING TREES OF UNKNOWN HEALTH STATE

3.2.1. Woody Indexing

PPV was detected in 75,00% and 72,73% of the 44 bearing trees of 5 species on GF 305 and GF 31 indicators, respectively (Table 2.).

3.2.2. Laboratory Tests

88.64% of the 44 trees were found PPV-infected by RT-PCR. NASH method was as sensitive as B-A ELISA (75,00 %), whereas only 68,18% of these trees proved to be infected in DAS ELISA (Table 2).

4. Conclusions

* Data suggest that reliability of woody indexing can be increased by using both GF 305 and GF 31 indicators.
* B-A ELISA and NASH method were more sensitive than DAS ELISA.
* RT-PCR proved to be the most sensitive of all the compared method.
* PCR, being labour-consuming and expensive,at present it is recommended for testing only of the most valuable basic mother plants and breeding sources, and not for large scale screening.

This research was supported under Grant No. C12-127, U.S. - Israel Cooperative Development Research Program, Human Capacity Development, Bureau for Global Programs, Field Support and Research, U.S. A.I.D.

TABLE 1. Result of field indexing and laboratory testing of PPV isolates

Source species of PPV isolates	No. of trees tested	Woody indexing				Laboratory tests							
		GF 305		GF 31		ELISA				NASH		PCR	
						DAS		B-A					
		Positive	Negative	Positive	Negative	Positive	Negative	Positive	Negative	Positive	Negative	Positive	Negative
Almond	1	1	0	1	0	1	0	1	0	1	0	1	0
Apricot	8	7	1	7	1	5	2	7	1	7	1	8	0
Myrobalan	1	1	0	1	0	1	0	1	0	1	0	1	0
Peach	44	40	4	42	2	37	7	42	2	43	1	44	0
Plum	22	20	2	21	1	19	3	21	1	21	1	22	0
Prunus glandulosa	1	1	0	1	0	0	1	1	0	1	0	1	0
Prunus amygdalopersica	2	2	0	2	0	2	0	2	0	1	1	2	0
Prunus sp.	1	1	0	1	0	1	0	0	1	1	0	1	0
Total	*80*	*73*	*7*	*76*	*4*	*67*	*13*	*75*	*5*	*76*	*4*	*80*	*0*
in %:		*91,25*	*8,75*	*95,00*	*5,00*	*83,75*	*16,25*	*93,75*	*6,25*	*95,00*	*5,00*	*100,0*	*00,00*

TABLE 2. Results of field indexing and testing of bearing trees with unknown health state

Species	No. of trees tested	Woody indexing				Laboratory tests							
		GF 305		GF 31		ELISA				NASH		PCR	
						DAS		B-A					
		Positive	Negative	Positive	Negative	Positive	Negative	Positive	Negative	Positive	Negative	Positive	Negative
Almond	9	7	2	7	2	6	3	7	2	6	3	8	1
Apricot	2	1	1	1	1	1	1	2	0	1	1	2	0
Amygdalopersica	4	1	3	2	2	1	3	1	3	1	3	2	2
P. myrobalan	2	1	1	1	1	1	1	1	1	1	1	1	1
Peach	27	22	5	22	5	21	6	22	5	24	3	26	1
Total	*44*	*32*	*12*	*33*	*11*	*30*	*14*	*33*	*11*	*33*	*11*	*39*	*5*
in %:		*72,73*	*27,27*	*75,00*	*25,00*	*68,18*	*31,82*	*75,00*	*25,00*	*75,00*	*25,00*	*88,64*	*11,36*

IMAGE ANALYSIS AND ARCHIVING AS AN AID TO FUNGAL IDENTIFICATION

C. R. LANE
Central Science Laboratory, MAFF, Sand Hutton, York, YO4 1LZ, UK.

1. Introduction

Image analysis has many applications in a wide range of disciplines, including medicine and industry. However, its potential use in plant pathology has not been investigated widely. The diagnosis of fungal plant diseases involves symptom recognition and a detailed knowledge of fungal taxonomy. Traditionally help can be obtained by reference to preserved material, slide and culture collections, and illustrated texts. However, this information is not always easily accessible and specimens deteriorate with time. Identification of fungi requires measurement and description of many structures in order to separate morphologically similar taxa. Imaging software may help by simply and reliably recording, analysing and retrieving information on specimens This paper describes the procedures involved in image capture, analysis and archiving and how this equipment can help in diagnosing fungal plant diseases.

2. Image capture

Images for analysis or archiving can be obtained using either a video camera or a flat-bed scanner. The camera can be mounted onto dissecting or compound microscopes fitted with a suitable photo-tube or a photographic copy stand to view larger objects. A frame grabber acquires the image and a file is created. The flat-bed scanner can be used to acquire images of line drawings, photographs, etc. Files can be saved in a number of formats (e.g. TIF, BMP) and although quite large (0.5-1.5 Mb) can be compressed. Images can either be acquired in black and white or colour, although, it is usual to convert the latter to black and white before analysis.

3. Image analysis

The first step in image analysis is to select desired objects from the acquired image - a process called 'thresholding'. The software measures the optical density of the entire image and this is expressed as a 'grey scale' range. By choosing an appropriate range, objects can be discriminated from the background. This can either be done manually or

H.-W. Dehne et al. (eds.), Diagnosis and Identification of Plant Pathogens, 321–323.

the software has an 'autothreshold' facility to select either light or dark objects. Many fungal structures are hyaline and their grey scale values are very similar to background values, thus a threshold range to select only desired objects may not be achieved. However, a suitable threshold may still be obtained if it is processed using image averaging, spatial filters, etc., but, these may change the object measurements.

Once a threshold has been chosen all objects within the grey scale range are highlighted. It is then necessary to check if all the desired objects have been selected, the similarity between their actual and highlighted boundary, and the number of highlighted but unwanted objects. Either a more suitable threshold value is chosen or unwanted highlighted objects are deleted.

The analysis software is capable of extracting data about points, lengths and areas which may be simple (e.g. number, breadth, length) or complex (e.g. perimeter, feret diameter and circularity). Data are extracted very quickly and can be displayed in dialogue boxes alongside the image and simple statistics calculated (e.g. mean, range) or exported to Microsoft Excel for further analysis.

4. Image archiving

Images can be exported directly to a database for archiving, rapid viewing or searching. Each imported image creates a new record where text and data can be added. Records are displayed individually with accompanying text and data, in a 'Gallery' of 28 small images, or as a table listing certain text fields. The database can help to identify an unknown image - either displaying all records with matching text or by making a direct visual comparison. The unknown image can be copied to make a 'Reference' image that can then be superimposed on archived records.

5. A practical application of image analysis

C. acutatum, the causal agent of strawberry black spot, which is an EC listed quarantine organism is separated from *C. gloeosporioides* in part by examination of conidia which are morphologically similar. Images of conidia for analysis were either scanned from line drawings representative of the species (Sutton, 1980) or acquired from slides prepared from cultures. Thresholding of the latter was difficult as the conidia are hyaline and a number of positive stains and a negative stain were tried to improve contrast. Of these cotton blue in lactoglycerol (0.1% v/v) gave the best results, uneven staining of the cell contents being avoided by gently warming the slides. Images were captured, the autothreshold facility used and unwanted objects deleted. The number, area, breadth and length of conidia were measured and data exported to Microsoft Excel where the area bounding the spore (length x breadth) and the area/bounding area were calculated. Results are shown in Table 1.

TABLE 1. Comparison of the area/bounding area of conidia of *C. acutatum* and *C. gloeosporioides*.

Identity	Source	Number of spores	Mean	S.E.
C.acutatum	Line drawing	4	0.7420	0.0115
C.gloeosporioides	Line drawing	5	0.7910	0.0039
C.acutatum	Culture	56	0.7095	0.0081
C.acutatum	Culture	46	0.7392	0.0058
C.acutatum	Culture	23	0.7570	0.0087
C.acutatum	Culture	69	0.7661	0.0043
C.acutatum	Culture	54	0.7687	0.0069
C.acutatum	Culture	27	0.7716	0.0092
C.acutatum	Culture	38	0.7822	0.0092
C.gloeosporioides	Culture	17	0.7668	0.0078
C.gloeosporioides	Culture	68	0.7986	0.0030

There was a statistically significant difference between *C. acutatum* and *C. gloeosporioides* for the line drawings. However, this difference was not observed when cultures were compared because conidia of *C. acutatum* are very variable and their shapes overlap those of *C. gloeosporioides*.

5. Discussion

There is great potential for the use of image capture, analysis and archiving as an aid in routine diagnosis. The equipment allows rapid and permanent capture of images from a wide range of sources that are readily accessible by searching the image database. This is extremely useful for recording symptoms on arrival and after incubation, for recording prior to destructive steps in sample and slide preparation, for reviewing diagnoses with colleagues and helping to identify unknown images.

Image analysis offers great potential for identification of fungi due to the rapid measurement of simple characters and quantification of shapes. The former is easily achievable and is already of great benefit. However, development of protocols to separate morphologically similar taxa requires considerable investigation and their usefulness may become apparent as image analysis becomes more widely used.

6. Acknowledgements

The author wishes to acknowledge the financial support of Plant Health Division of the Ministry of Agriculture, Fisheries and Food and the kind assistance of his colleagues Dr Roger Cook, Mr Richard Thomas, Mr Kelvin Hughes and Mr Steve Freeman.

7. References

Sutton, B.C. (1980). The Coelomycetes. Commonwealth Mycological Institute, Kew.

ANASTOMOSIS GROUPS AND PATHOGENICITY OF *RHIZOCTONIA SOLANI* ISOLATES FROM TOBACCO IN ITALY

R. NICOLETTI, E. LAHOZ, R. CONTILLO and A. OLIVA
Istituto Sperimentale per il Tabacco
Via P. Vitiello 66, 84018 Scafati (SA), Italy

1. Introduction

Within *Rhizoctonia solani* the identification of anastomosis groups (AGs) as taxonomic infraspecific groups has become prevalent in recent years since they must be considered biological species. This differentiation has been repeatedly confirmed by biochemical and molecular assays, such as isozyme analysis, DNA-DNA hybridization and RFLPs.
As a whole *Rhizoctonia solani* AGs prove to be polyphagous. Nevertheless the characterization of anastomosis groups that are pathogenic towards every single plant species often show the prevalence of one or few AGs. In this sense Solanaceous crops are probably the most important exception; for example potato is known to be the host of at least seven AGs: AG-1, AG-2-1, AG-2-2, AG-3, AG-4, AG-5 and AG-9.
Tobacco has not been extensively studied with reference to the subject. Shew and Lucas (1991) reported three anastomosis groups (AG-1, AG 2-2, AG-4) to be able to damage such crop; in addition AG-3 was found to cause target spot on tobacco leaves in both Southern Africa (Meyer *et al.*, 1990) and the United States (Stevens Johnk *et al.*, 1993).
In a survey carried out to determine the composition of the *R. solani* complex on tobacco in Italy, AG-4 and AG-2 appeared to be the most widespread, while both AG-1 and AG-3 were absent (Nicoletti and Lahoz, 1995b). Isolates belonging to the previously undetected AG-5 were also recovered (Nicoletti and Lahoz, 1995a).
In this paper we report the results of pathogenicity tests and isozyme analysis on isolates belonging to the several AGs recovered from tobacco. For each isolate these studies confirmed that virulence is a feature unrelated to their AG belonging.

2. Materials and Methods

Isolates of *R. solani* treated below were recovered from infected tissues of tobacco plants with sore shin or damping-off symptoms, or from other crop species from fields previously cultivated with tobacco (Tab. 1). Isolations procedures and identification of *R. solani* isolates were made as described in a previous paper (Nicoletti and Lahoz, 1995b).

H.-W. Dehne et al. (eds.), Diagnosis and Identification of Plant Pathogens, 325–327.

was carried out by pairing isolates on 2% WA. Tester isolates of AGs 1 through 6 and AG-BI were used. About 15 microscopic fields were examined for every pairing to determine the fusion frequency (FF).
Thiamine requirement of AG-2 and AG-5 was checked by evaluation of growth on Czapek-Dox broth. Pathogenicity tests were carried out in greenhouse at 24 ± 2°C by inoculating 4-week old plants grown in 13 cm-diameter pots containing a mixture of sandy soil and corn meal previously colonized by the fungus (4%). Polygalacturonase isozymes electrophoresis was perfomed on pectin-polyacrylamide gels according to Cruickshank and Wade (1980), with slight modifications.
Electrophoretic patterns of both tobacco and tester isolates were compared by multivariate analysis. The data set was submitted to cluster analysis using euclidean distances and complete linkage method.

3. Results and Discussion

The AG determination showed that at least four anastomosis groups are involved in causing disease at the crown level. AG-4 and AG-2-1 were prevalent (Tab. 1).

TABLE 1. *Rhizoctonia solani* isolates tested, and their aggressiveness on tobacco

Isolates	RT9	RT13	RT16	RT23	RT27	RP6	RF4	R92.1	R93.1	RT3	RT4	RT5	RT10	RT17	RT18
Symptoms	S.S.	D.O.	D.O.	S.S.	S.S.	/	/	S.S.	S.S.	D.O.	D.O.	D.O.	S.S.	S.S.	S.S.
AGs	AG2-1	AG2-1	AG2-1	AG2-1	AG2-1	AG2-1	AG2-1	AG2-1	AG2-1	AG-4	AG-4	AG-4	AG-4	AG-4	AG-4
D.S.I.	47.2	43.7	45	56.2	82.5	57.5	43.0	87.0	91.2	33.0	33.0	57.0	63	55.0	55.0
Isolates	RT20	RT21	RT22	RT25	RT30	RT31	RT32	RTL7	RTL8	RG1	RF1	RT11	RT12	RT7	RT19
Symptoms	S.S.	S.S.	S.S.	S.S.	S.S.	S.S.	S.S.	D.O.	D.O.	/	/	S.S	D.O.	S.S.	S.S.
AGs	AG-4	AG-4	AG-4	AG-4	AG-4	AG-4	AG-4	AG-4	AG-4	AG-4	AG-4	AG2-2	AG2-2	AG-5	AG-5
D.S.I.	57.0	71.0	50.0	53.0	86.0	30.0	54.0	87.0	71.0	35.0	43.0	47.2	39.3	25.0	27.0

AGs = Anastomosis groups; S.S. = Sore shin; D.O. = Damping off.
RT and RTL = Tobacco; RF = Bean; RG = Carnation; RP = Pepper; R. 92.1 and R 93.1 = French isolates from tobacco.
D.S.I. = Disease Severity Index (Mc Kinney)

Although only AG-2-2 had been previously reported on tobacco, cultural appearance of the colonies on PDA, thiamine requirement, and electrophoretic patterns of polygalacturonase isozymes (Fig. 1) showed that both AG-2-1 and AG-2-2 are actually pathogenic to this crop. No isolates belonging to AG-1 and AG-3 were found; the absence of the latter is probably related to the lack of signs of the foliar syndrome (target spot) in Italy. Unlike the above-mentioned groups, at present there are few reports of AG-5 on any crop in Europe. Two isolates that anastomosed with AG-5 testers were recovered from plants with sore shin symptoms; this finding has been recorded as the first report of such AG on tobacco (Nicoletti and Lahoz, 1995a).
Pathogenicity tests (Tab. 1) did not allow to state which group is the most virulent to tobacco. In fact the degree of virulence was not related to the AG belonging. Anyway it is very important to consider that diseased plant can undergo either an acute (damping-

off) or a chronic course (sore shin); therefore light symptoms just after inoculation of the plants do not necessarily imply a low virulence of the isolates tested.

In a previous study (Lahoz *et al.*, 1996) we compared virulence of AG-4 isolates from tobacco on several host species, demonstrating that those isolates were polyphagous since they were able to infect a broad range of plant species. Further investigations are in progress on this matter.

The observation of anastomosis revealed that isolates of AG-2 recovered from tobacco (including two received from France) showed an high fusion frequency when compared each other (FF>50%), and low (FF<30%) or medium (30%<FF<50%) frequency when compared with respectively AG-2-2 or AG-2-1 testers.

Electrophoretic patterns of isopolygalacturonases showed distinct clusters for each anastomosis group recovered from tobacco.

Isozyme analysis evidenced that AG-4 isolates from tobacco can be separated into at least two distinct subgroups differentiated by the presence on the gel of bands with high R_f values.

According to the high fusion frequency and the electrophoretic patterns (Fig. 1), isolates classified as AG-2-1 may possibly represent an homogeneous group slightly different from the testers.

FIGURE 1. Electrophoretic patterns of isopolygalacturonases of AG-2 isolates

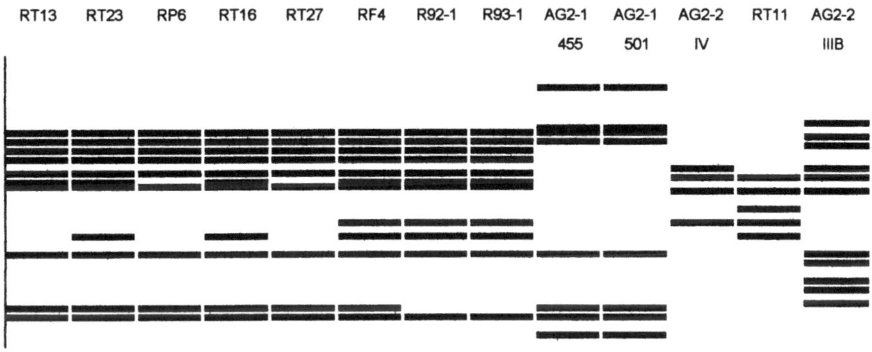

4. References

Cruickshank, R.H. and Wade, G.C. (1980) Detection of pectic enzymes in pectin-acrylamide gels, *Analytical Biochemistry* **107,** 177-181.

Lahoz, E., Nicoletti, R., Contillo R., Luongo L., and Corazza, L. (1996) Caratterizzazione biologica e patogenicità di isolati di *Rhizoctonia solani* AG-4 da tabacco, *Proceedings of "Giornate Fitopatologiche" Workshop* **2,** 443-448.

Meyer, J.C., Van Wyk, R.J., and Phillips, A.J.L. (1990) Rhizoctonia leaf spot of tobacco in South Africa, *Plant Pathology* **39,** 206-207.

Nicoletti, R. and Lahoz, E. (1995a) Recovery of *Rhizoctonia solani* AG-5 from tobacco in Italy, *Plant Disease* **79,** 540.

Nicoletti, R. and Lahoz, E. (1995b) Gruppi di anastomosi di *Rhizoctonia solani* patogeni su tabacco in Campania, *Informatore Fitopatologico* **45 (12),** 57-60.

Parmeter, J.R., Sherwood, R.T., and Platt, W.D. (1969) Anastomosis grouping among isolates of *Thanatephorus cucumeris, Phytopathology* **59,** 1270-1278.

Shew, H.D. and Lucas, G.B. (1991) *Compendium of tobacco diseases*, American Phytopathological Society, St. Paul.

Stevens Johnk, J., Jones, R.K., Shew, H.D., and Carling, D.E. (1993) Characterization of populations of *Rhizoctonia solani* AG-3 from potato and tobacco, *Phytopathology* **83,** 854-858.

USE OF FUNGICIDE-RESISTANCE FOR THE CHARACTERISATION AND DIFFERENTIATION OF *VENTURIA INAEQUALIS* STRAINS - SUITABILITY FOR COMPETITIVITY STUDIES

F. ORTEGA, U. STEINER AND H.-W. DEHNE
Institut für Pflanzenkrankheiten, Universität Bonn,
Nußallee 9, 53115 Bonn

1. Introduction

Increasing insensitivity to sterol biosynthesis inhibiting fungicides (SBI) is becoming a widespread threat for the effective control of apple scab (*Venturia inaequalis* (Cke.) Wint.) with this group of fungicides. SBI-resistant genotypes of the pathogen are present in almost all apple growing areas over the world. The evaluation of the relative parasitic and the reproductive capabilities of different resistant isolates is important for the managment of resistance to SBI-fungicides. Furthermore, competition studies on plants treated with non-SBI-compounds, i.e. other fungicides or resistance inducers, provide important information on the reversibility of resistance to SBI-fungicides. In this contribution a method that can be used for the evaluation of competition between SBI-resistant and SBI-sensitive isolates of *V. inaequalis* is presented.

2. Material and methods

Germ tube length was assessed 46 h after incubation on PD Agar plugs amended with different concentrations of the fungicide 'flusilazole' (Benocap 20 WG, Dupont de Nemurs). *In vivo* SBI-sensitivity was determined on apple seedlings 'Golden Delicious' in greenhouse experiments. 'Flusilazole' was sprayed in different concentrations on seven days old apple seedlings. After nine days infected leaf surface was assessed. *In vitro* and *in vivo* efficacy values (Abbott) were fitted to sigmoid curves with the formula $Y=(1-EXP(-(lg(X)/X_o)^{ß}))*100$ by non linear regression ('X' being the fungicide concentration; 'X_c' and 'ß' regression-parameters) [3]. The germ tube length distribution patterns for spores from every strain and their combinations was determined at different 'flusilazole' concentrations and normality was proofed using the Shapiro-Wilk test. Germ tube length discrimination and further statistical evaluation were carried out with the SAS statistical computer package (SAS Institute, 1995).

H.-W. Dehne et al. (eds.), Diagnosis and Identification of Plant Pathogens, 329–332.

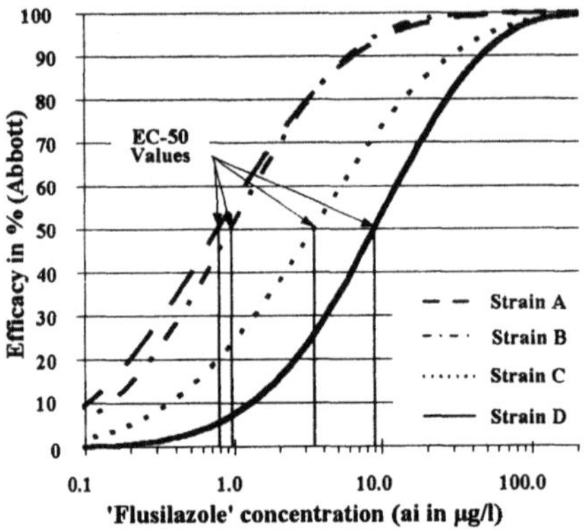

FIGURE. 1: Dose response curves from the tested strains of *V. inaequalis* against 'flusilazole'

TABLE 1:EC50-values of the tested strains with different 'flusilazole' sensitivity. Values in each colum with same letter are not significantly different at α=0.05; inverse prediction; non linear regression

Strain	EC 50 *in vivo* µg/l	EC 50 *in vitro* µg/l
A	1000^{a}	0.77^{a}
B	1600^{a}	0.96^{a}
C	7000^{b}	3.4^{b}
D	35700^{c}	8.7^{c}

3. Results

3.1. DIFFERENTIATION OF *V. INAEQUALIS* STRAINS BY THEIR SBI-SENSITIVITY

The shape of the dose response curves and the EC-50 values derived from them by inverse prediction were used to characterise and differentiate different strains of *Venturia ineaqualis* [1]. Four strains with different SBI-sensitivity were used in this work (Figure 1 and Table 1)

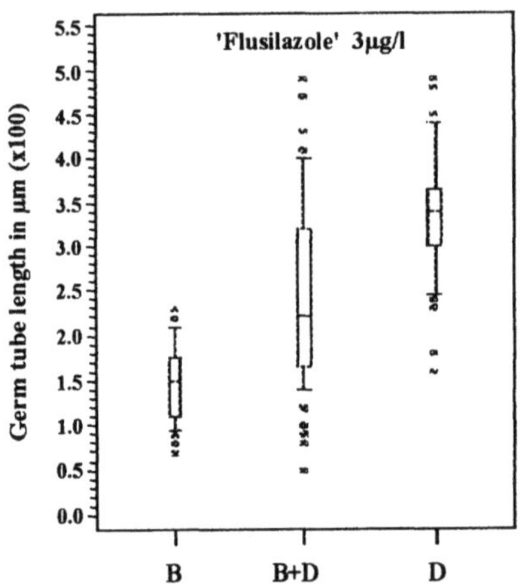

FIGURE 2: Germ tube length distribution for the sensitive strain B and the resistant strain D. Edges of the box depict the 25th and the 75th percentiles, the central horizontal line the median and the whiskers the 10th and the 90th percentiles. The red dots represent the extreme values.

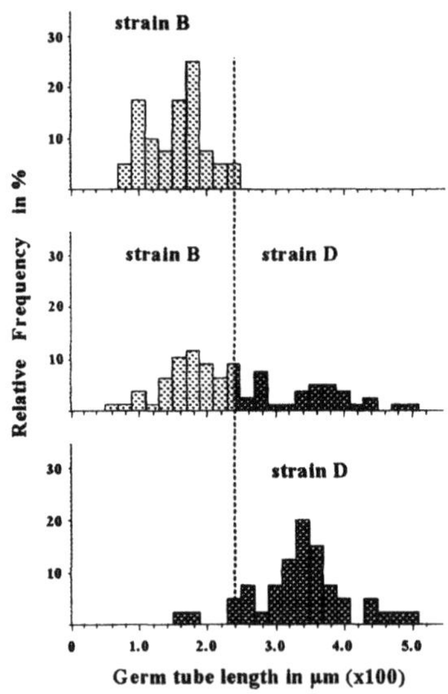

FIGURE 3: Histogram for the germ tube length of spores from two strains of *V. inaequalis* on agar with 'flusilazole' (3µg/l).

TABLE 2: Application of the germ tube length discrimination test to the recognition of spores from strains combined in known rates. *)Prior rate values were corrected with the spore germination rate.

Strains	Mixing ratio (%)		
	original*	germ tube discrimination	
		3μg/l	10μg/l
A : C	40 : 60	43 : 56	38 : 62
A : D	40 : 60	46 : 54	26 : 74
B : C	50 : 50	50 : 50	55 : 45
B : C	67 : 33	60 : 40	65 : 35
B : D	50 : 50	52 : 48	50 : 50
B : D	67 : 33	70 : 30	70 : 30

3.2. DIFFERENTIATION OF THE *V. INAEQUALIS* STRAINS BY THEIR GERM TUBE LENGTH DISTRIBUTION PATTERN

The distribution pattern of the germ tube length of spores from strains with different 'flusilazole' sensitivity was clearly distinct on certain concentrations of the fungicide (Figure 2) This feature can be used to recognise spores from strains with different 'flusilazole' sensitivity in spore samples taken from a mixed population of these strains (Figure 3).

3.3. DIFFERENTIATION OF *V. INAEQUALIS* STRAINS IN SPORE MIXTURES

The results obtained when the germ tube length discrimination test is applied to spore suspensions of a SBI-sensitive and a SBI-resistant strain, combined in known mixing ratios, are shown in Table 2. In almost all cases the results obtained with the germ tube length discrimination test corresponded well with the original combination rates.

4. Discussion

The discrimination of the germ tube length is a useful tool for the differentiation of spores from strains with different sensitivity to SBI-fungicides. Spores can be distinguished, when the overlapping of germ tube length distribution is low enough to keep discrimination error acceptable. Best discrimination is achieved when there is no overlapping (<5-10%). The method is applicable to the quantitative attribution of spores to their strains in spore samples taken from plants previously inoculated with a mixture of two or more strains with different sensitivity to the SBI-fungicide. Therefore, it can be used to determine the relative fitness of the pathogen strains. This test is suitable to be applied to other fungus/fungicide combinations and allows the evaluation of the effect of non-SBI-compounds on the persistence of SBI-resistant strains in pathogen populations. The non disruptive nature of SBI-sensitivity and the slow growth of the fungus *in vitro* makes the method suitable to determine the presence of defined strains quantitatively when compared to other methods [1,2,4].

5. References

1. Cohen, Y., Kadish, D. (1992) Determination of sensitivity of *Phytophthora infestans* to phenylamides: a quantitative tuber disc method, EPPO Bulletin **22**, 311-314.
2. Nuninger, C., Bolton, N., De Wijs, J.J., Staub, T. (1992) Determination of the sensitivity of *Phytophthora infestans* to phenylamides: a semi-quantitative test method, EPPO Bulletin **22**, 309-311.
3. Ortega, F., Steiner, U., Dehne, H.W. (1996) Differentiation of SBI-resistance and parasitic fitness components in German isolates of *Venturia inaequalis* (Cke.) Wint. and interest of induced resistance for antiresistance strategies, Med. Fac. Landbouww. Univ. Gent **61**, (in press).
4. Smith, F.D., Parker, D.M., Köller, W. (1989) Sensitivity distribution of *V. inaequalis* to the sterol demetilation inhibitor flusilazole: Baseline sensitivity and implications for resistance monitoring, Phytopathol. **82**, 425-429.

Acknowledgements
This work was supported by a grant of the 'Fundación La Caixa' and the DAAD.

ISOLATION OF MICROMYCETES FROM PEA SEEDS

E. PROKINOVÁ
Department of Plant Protection, Czech Agriculture University
Prague 6 - Suchdol, PC 165 21,
Czech Republic

Introduction

Seed's quality is one of the basic assumption for healthy growth. Next to other factors it is determined by health status of the seed. On the seed come up multitude of pathogens as well as saprophytic micromycetes (Neergaard 1979, Richardson 1979). For investigation of their influence upon seeds, mutual relations and their impact upon health condition of seeds and final recommendation for seed practise exact diagnostics of seed-borne fungi is necessary. Very accurate methods were elaborated recently for some important phytopathogenic fungi (Hansen and Wick 1993, Fox 1993, Okrouhlá 1994) and their research and development go forward very quickly. But these methods are relatively demanding upon laboratory equipment and their use in current practice is not substantial (except some exceptions - e.g. rapid disease kids for *Phytophthora* spp., *Pythium* spp., *Rhizoctonia* spp.). Here it is good to perceive the difference between detection and diagnosis - these denominations are interchanged even in professional community and many of detection methods are presented as diagnostic methods. It is actually well-known that mere presence of pathogen is not enough for objective evaluation of origin and development of the disease. Therefore most widespread remains classical detection methods, in fungi case isolation on agar plates and following determination of isolates according morphological or biochemical characteristics. The isolation method of pathogens on agar plates gives also partial information about relations between present microscopic fungi and plant.

In this work classical methods of isolation were used and followed by determination of acquired isolates.

Goal of the work:

1. Compare occurrence of micromycetes on pea seeds with different origin in relation to seed's germination.
2. Check influence of media, temperature and previous processing of the seed upon isolation results (and consider use of pea seeds from this point of view as model for teaching diagnostics - demonstration of conditions impact upon observed results).

H.-W. Dehne et al. (eds.), Diagnosis and Identification of Plant Pathogens, 333–336.

Materials and methods

In the first stage (1994) was used for isolation of micromycetes from seeds Sabouraud agar (pH 5,8), half of the seeds was superficially sterilized (70% "ethanol" exposure 5min.), half of them was put into Petri dishes without external sterilization. Cultivation in 24°C, for more easy isolation of fungi of genus Fusarium were dishes put for 48 hours into temperature 8°C. 9 cultivars (Alan, Bohatýr, Janus, Komet, Olivín, Profi, Romeo, Smaragd, Tyrkys) originating from three places (Jaromìøice, Libìjovice, Staòkov) were evaluated and there are stated average results. Information about precipitation was given by meteorology station. Acquired isolates were determined in co-operation with Czech collection of micro-organisms and Department of botany of Charles University in Prague. In the second stage (1995) of experiment was used cultivar Lantra from two places (Staòkov, Šumperk). Seeds both with and without surface disinfection were placed into Petri dishes, half of seeds in dry conditions, half after 24 hours of swelling in sterile, condensed water. There were used 4 agar media and filter paper, cultivation at 22 - 24°C. Seeds were taken from closed pods in full maturity.

Results

1. Comparison of representation of micromycetes on seeds with different origin in relation to their germination

A quantitative difference was found out in presence of micromycetes on seeds with different origin. After recalculation on average for 1 seed, on the seed from Jaromìøice grew up 0,58 colony, on seed from Staòkov 1,36 and from Libìjovice 1,38 colony of micromycetes. The lowest germination (30%) was determined by seeds from Jaromerice, seeds from Staòkov had germination of 73,6 % and highest germination (88.3%) showed seeds from Libìjovice in 1994. In 1995 on the seed from Staòkov grew up 3,74 colonies (germination 78,6%), on the seed from Šumperk 0,72 colony of micromycets (germination 65,8%). Representation in percentage of each isolated genera (species) is shown in Fig.1. Temperature and precipitation during head up period (when comes to main attack on the seeds) was different among each location.

The germination is influenced by presence of given micromycetes. As the diagram shows, lowest germination demonstrated seeds from which was isolated bigger volume of representatives of Aspergillus and Penicillium genera. Comparing with other two locations in this sample was recorded higher multitude of Paecilomyces sp., too. From this sample of seeds the lowest amount of micromycetes colonies were isolated. The highest temperature and the biggest precipitation was at this place throughout the head up period.

Fig. 1. Micromycetes isolated from pea seeds (%)

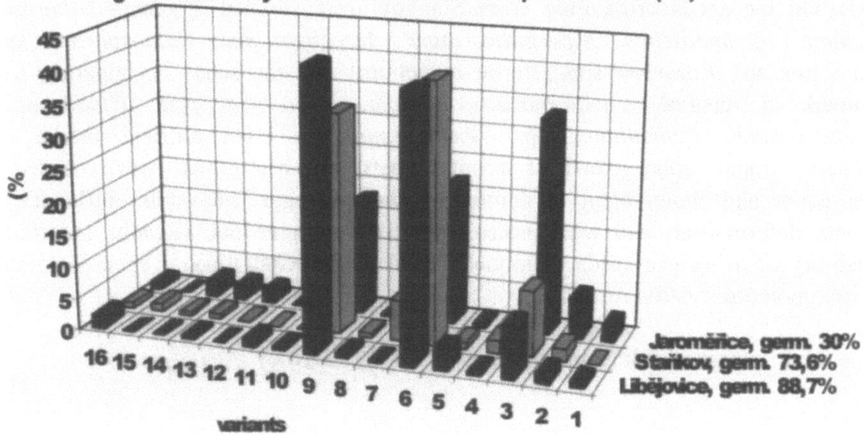

Variants:

1	*Aspergillus flavus*	9	*Cladosporium spp.*
2	*Aspergillus niger*	10	*Trichoderma harzianum*
3	*Penicillium spp.*	11	*Gliocladium roseum*
4	*Fusarium sp.*	12	*Acremonium sp.*
5	*Botrytis cinerea*	13	*Aspergillus sp.*
6	*Alternaria alternata*	14	*Paecilomyces sp.*
7	*Alternaria tenuissima*	15	*Chaetomium sp.*
8	*Epicoccum sp.*	16	*Sterile mycelium*

2. Influence of cultivation media, temperature and seed treatment upon results of microcmyetes isolation

Frequency of fungi colonies was induced by medium as well as by seed status: maximum colonies of fungi grew on Sabouraud agar, Potato-glucose and Pea agar was corresponding. On filtration paper and water agar grew - according presumption-minimum of fungi colonies. Temperature influence did not manifested itself at all Seeds, which was swelled before putting out into Petri dishes, had inhibit influence upon growth of micromycets in the variation without surface disinfection, whereas by seeds with surface disinfection the difference did not come out.

From the seeds being put out as swelled, were isolated bacteria in higher quantity, some with phytopathogenic influence. Most obviously occurred this phenomenon on water agar . There was distinctive difference between frequency of fungi on seeds from

Staòkov (aver.3,75 colonies on 1 seed) and from Šumperk (aver. 0,72 colony on 1 seed). On the seeds originating from Staòkov were isolated 7 genera (*Alternaria alternata, A. tenuissima, Aspergillus niger, Ascochyta pisi, Cladosporium* spp., *Epicoccum* sp., *Fusarium* spp., sterile mycelium) on the seeds originating from Šumperk 5 genera (*Alternaria alternata, Ascochyta pisi, Cladosporium cladosporioides, Penicillium* spp., *Trichoderma* sp.) Used nutrient media had distinctive impact upon growth of sprout rootlets - the best growth was recorded on filter paper and Water agar, in contrary Sabouraud agar had inhibit influence on growth, deformations and weak necrosis of rootlets came out. (Results manifested suitability of using pea seeds as model for visual demonstration of specific method impact upon observation results.)

(This work is part of task from GA CR No. 521/96/0616)

References

CzyŸewska S.(1976) Studies on Fungi Infesting Seeds of Pea (*Pisum sativum* L.).*Bull. warzywniczy XIX, IV. Ochrona roœlin:* 271 - 288

Fox R. T. V. (1993) *Principles og Diagnostic Techniques in Plant Pathology.* International Mycological Institute, UK Press, Cambridge

Hansen M. A. and R. L. Wick, (1993) Plant Disease Diagnosis: Present Status and Future Prospects. In:Andrew J. H. and I. C. Tommerup (eds.) (1993) *Advances in Plant Pathology, vol. 10:* 65 - 126

Neergaard P.(1979) *Seed Pathology*, Academic Press

Okrouhlá M., (1994) *Rychlá diagnostika chorob rostlin.* Studijní informace, Rostlinná výroba, No.3. UZPI Praha

Richardson M. J. (ed.) (1979) An Annotated List of Seed-Borne Pathogens. *Phytopathological Papers, no 23, ISTA Seed health Testing Handbook, Section* 1.1:186 - 192

ISOLATION AND IDENTIFICATION OF *FUSARIUM* SPP. AND *MICRODOCHIUM NIVALE* FROM WINTER WHEAT

A. SCHÜTZE, E.-C. OERKE, and H.-W. DEHNE
Institut für Pflanzenkrankheiten
Rheinische Friedrich-Wilhelms-Universität Bonn
Nußallee 9, 53115 Bonn

1. Introduction

Fusarium spp. are important pathogens of winter wheat, causing yield reduction and decreasing grain quality due to mycotoxin production. Therefore, the occurrence of *Fusarium* spp. at different locations with two systems of cultivation was investigated. Different isolation conditions were tested in order to validate the sometimes low isolation frequency.

2. Materials and methods

During the growing season 1994/95 *Fusarium* spp. and *Microdochium nivale* were isolated from winter wheat. For the isolation the selective "Czid" agar (Czapek dox with "Iprodione" and "Dicloran") was used (Abildgren et al. 1987). For microscopical identification each isolate had to be transferred onto potato-dextrose agar (PDA) and synthetic nutrient agar (SNA).

3. Results

Investigating 7 locations with a total of 81 plots, 5 to 37 % of the stem-bases were infected with *M. nivale* (Table 1). *F. avenaceum, F. graminearum and F. culmorum* were less prevalent. *F. sporotrichioides* and *F. poae* were isolated only occasionally.

After seed inoculation with *F. culmorum* this species could not be detected at the stem-bases at growth stage 59/60, however, at harvest time 44% of kernels were contaminated with *F. culmorum*. Similar results were obtained for an inoculation with *F. avenaceum* (Figure 1).

The highest isolation rate of *Fusarium* spp. was obtained by using the "Czid" agar and UV-light at 23°C under a light/dark regime (14/10h) (Figure 2). By using malt agar and daylight conditions the isolation rate of *Fusarium* spp. was only 12%

H.-W. Dehne et al. (eds.), Diagnosis and Identification of Plant Pathogens, 337–340.

compared to the maximum of 33% with "Czid".

The comparison between the microbiological detection method using isolation and microscopic identification and the nucleic acid based technique showed very promising results (Table 2). Identical results were obtained for close to 80 % of the samples investigated.

TABLE 1. Isolation of *Fusarium* spp. and *M. nivale* from stem-bases of winter wheat in organic and conventional farming systems, in 1995.

Pathogen	Isolation [%]						
	Organic farming			Conventional farming			
	VE	BL	HE	KB	ZÜ	ME	PO
M. nivale	28	34	18	24	5	7	37
F. avenaceum	5	7	3	7	9	5	12
F. graminearum	0.3	2	3	0	11	0	5
F. culmorum	2	1	4	1	5	7	0
F. sporotrichioides	1	2	1	0	0	0	3
F. poae	0	0	0	0	0	0	2

Sample size: VE, Velbert = 17, BL; Blankenheim = 16; HE, Hennef = 27, KB, Kerpen-Buir = 11; ZÜ, Zülpich = 4; ME, Meckenheim = 3, PO, Poppelsdorf = 3;

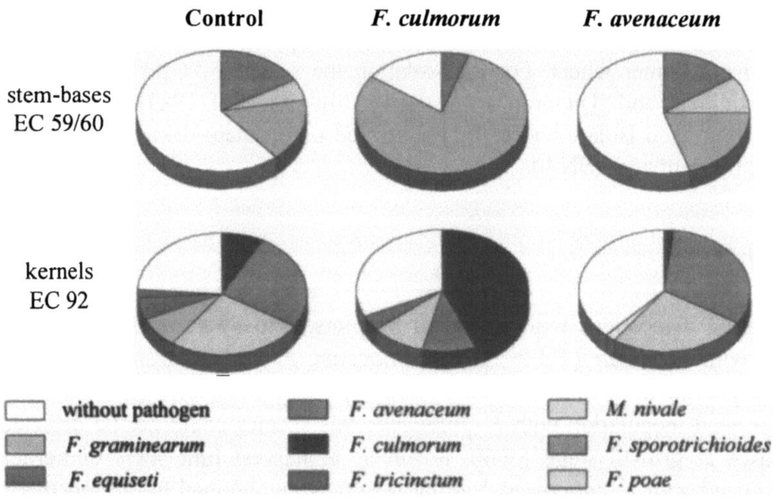

FIGURE 1. Isolation of *Fusarium* spp. and *Microdochium nivale* from stem-bases (EC59/60; n= 20) and kernels (EC 92, n= 50) without / after seed inoculation with *F. culmorum* / *F. avenaceum*.

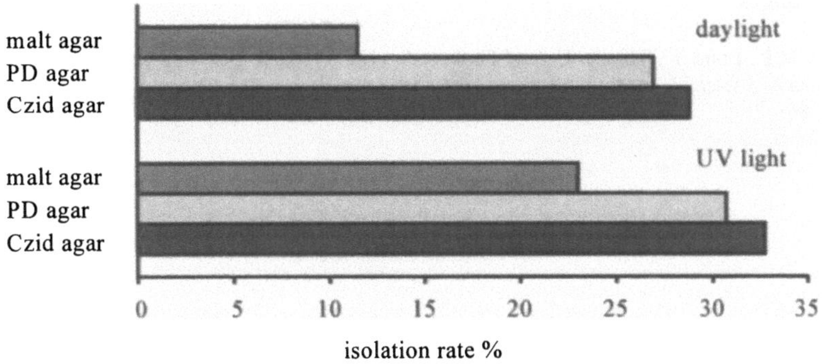

FIGURE 2. Influence of isolation medium and light conditions during incubation on the frequency of isolation of *Fusarium* spp. and *Microdochium nivale* from wheat kernels (n = 50).

TABLE 2. Comparison of *Fusarium* spp. detection by microbiological and PCR based techniques.

Sample	Method	Detection of *Fusarium* spp. [1]		
		F. avenaceum	*F. graminearum*	*F. culmorum*
I	Czid	0	5	0
	PCR	1	1	0
II	Czid	10	5	0
	PCR	2	1	0
III	Czid	15	10	0
	PCR	2	0	0
IV	Czid	0	5	0
	PCR	0	1	0
V	Czid	0	0	35
	PCR	0	1	0

[1] Frequency / PCR signal: not detected / 0; infection < 10% / 1; infection > 10% / 2

4. Conclusions

The ear infection of winter wheat by *Fusarium* spp. and *M. nivale* during the period from flowering until harvest is not correlated to the inoculum potential at the stem-bases. The success of isolating *Fusarium* spp. and *M. nivale* from plant material was influenced by the isolation conditions (media, etc.). This was found especially at low infection levels. The reliability and specifity of the PCR method provide a promising tool for the identification of *Fusarium* species and *M. nivale*.

5. References

Abildgren, M.P., Lund, F., Thrane, U., and Elmholt, S. (1987) Czapek-Dox agar containing iprodione and dicloran as a selective medium for the isolation of *Fusarium* species, Letters in Applied Microbiology 5, 83-86.

QUANTIFICATION OF *PHYTOPHTHORA INFESTANS* DURING EARLY STAGES OF PATHOGENESIS

R. Stierl, Ulrike Steiner and H.-W. Dehne
Insitut für Pflanzenkrankheiten,
Rheinische Friedrich-Wilhelms Universität Bonn,
Nußallee 9, D-53115 Bonn

1. Introduction

Phytophthora infestans is the pathogen causing the highly destructive disease late blight affecting both tomato and potato. The fungus attacks all aboveground parts of the plants and additionally in potato the tubers. Leaf lesions first appear as water-soaked spots, which may enlarge rapidly into brown necrotic lesions cover large areas of the leaves. Later, the fungus may be observed occasionally as mouldy growth around the undersides of large brown lesions. Therefore, the assessment of disease severity is usually done indirectly by the estimation of the amount of damaged area of the plant leaves. In research, i. e. breeding of resistant cultivars or development of new fungicides, techniques are essential which allow the evaluation of the development of *P. infestans* before any damage is visible. Microscopical and serological detection methods for a direct quantification of *P. infestans* in the leaf tissue, also during the early stages of pathogenesis, are presented.

2. Material and Methods

Tomato plants were cultivated under greenhouse conditions. Young, but fully developed leaflets were inoculated with a zoospore suspension of *P. infestans*, containing $6x10^4$ zoospores / ml.

2.1 MICROSCOPY

Zoospore germination and formation of appressoria were determined after staining with Blankophor (0,1 %) in ethanolic solution (10 %) by fluorescence microscopy. For fungus quantification in the tissue leaves were fixed and cleared with chloralhydrate (2,5 g / ml) at least for 4 weeks. After staining in acid fuchsin (0,05 %) in lactophenol for 24 to 48 hours the samples were differentiated in chloralhydrate for some hours, mounted in clear lactophenol and evaluated with Normaski interference contrast.

H.-W. Dehne et al. (eds.), Diagnosis and Identification of Plant Pathogens, 341–343.

2.2 ELISA TEST

Polyclonal antibodies against *P. infestans* were produced in rabbits by Eurogentec, Belgium. ELISA was performed according to Beckman et al. (1994).

3. Results

Zoospores, germ tubes and appressoria of *P. infestans* on the surface of tomato leaves showed bright fluorescence after staining with the optical brightner Blankophor without any further preparation allowing a quick determination of the initial stages of fungal development.

The clearing of the leaves with chloralhydrate resulted in an extreme high transparency of the leaf tissue. The growth of the hyphae and haustoria could be followed through the whole leaf from the upper epidermis cells through the mesophyll and down to the lower epidermis (Figure 1). Therefore, during the first 30 hours after inoculation it was possible to quantify the development of the fungus at each infection site and calculate the average quantity per infection site.

Additionally, defence reactions of the epidermis cells to invasion could be observed. Starting with a slight granulation the content of the infected cells collapsed totally (Figure 2). Parenchymatic cells, damaged by *P. infestans*, stained intensively red and their number increased rapidly preventing an exact evaluation of fungal biomass later than 30 hours after inoculation.

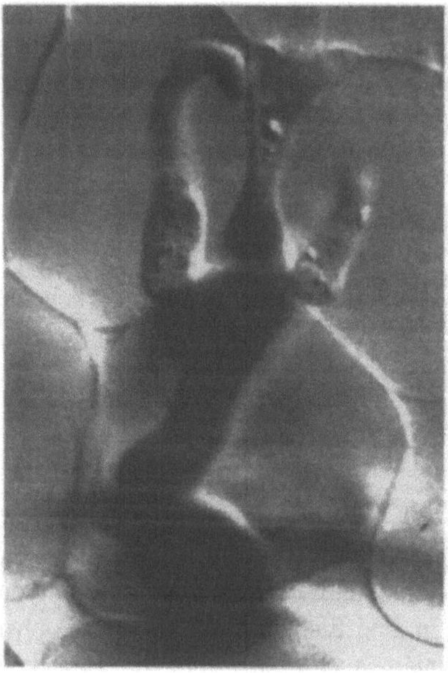

FIGURE 1. Fungal structures of *P. infestans* in the mesophyll of a tomato leaf 24 hpi

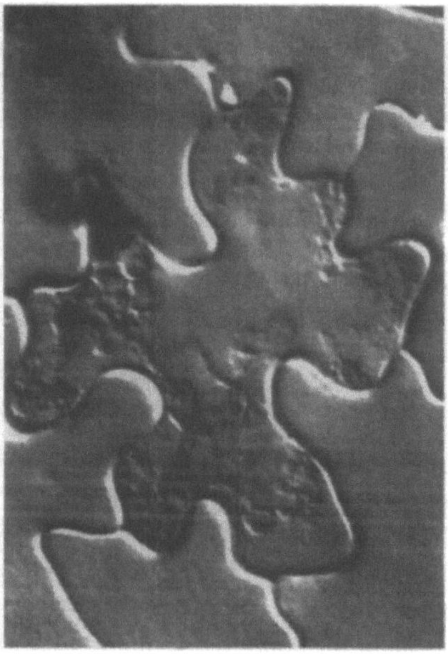

FIGURE 2. Cell collapse of an infected epidermal cell of a tomato leaf 8 hpi with *P. infestans*.

The quantity of fungal structures in the plant tissue 48 hours after inoculation were determined using an ELISA test (Table 1). Differences in fungal biomass determined with this technique were mostly related to the differences in the amount of damaged leaf area 5 days after inoculation in the respective samples.

TABLE 1. Quantification of disease severity with ELISA - Assay (48 hpi) and visual assessment (5 dpi)

ELISA-ASSAY mg mycelium / g leaf 48 hpi	Visual assessment % damaged leaf area 5 dpi
25	25
18	7
6	4
8	4

4. Conclusions

The microscopical estimations gave data about the quantity and distribution of the pathogen in the plant tissue during the early stages of pathogenesis. Additionally, they allowed the observation of defence reactions of plant cells during penetration and growth of the fungal structures. However, preparation of the samples and evaluation at the microscope are time consuming, whereas the ELISA assay is quick and can be run almost automatically. An disadvantage of the ELISA assay is that the distribution of mycelium or defence reactions of the plants can not be determined.

Both methods of direct quantification exclude a false assessment of the infection level, i.e. if plants react tolerantly or hypersensitively.

5. References

Beckman, K.B., Harrison, J.G. and Ingram, D.S. (1994) Optimization of a polyclonal enzyme-linked immunosorbent assay (ELISA) of fungal biomass for use in studies of plant defence responses. *Physiol. Mol. Pl. Path.* **44**, 19-32.

DESCRIPTION OF MICROBIAL COMMUNITIES OF RHIZOSPHERE SOIL BASED ON THE PHOSPHOLIPID FATTY - ACID SPECTRUM ANALYSIS OF MICROORGANISMS

JOSEF TERHARDT, VIVIAN VILICH, RICHARD A. SIKORA
Institute for Plant Diseases, Soil Ecosystem Pathology, University of Bonn, Nussallee 9, D-53115 Bonn, Germany

Abstract

We describe two approaches to measure microbial populations in bulk soil with or without prior culturing on media. The influence of different cropping systems on the microbial community was examined by: A) analysing the composition of fatty acids of the phospholipid fraction previously extracted from bulk soil samples and B) combining fatty acid analysis with the dilution plate count method. Fatty acid spectra obtained by both methods were analysed by gas chromatography using the Microbial Identification System (MIDI, Inc.). Data were calculated by principal component analysis and cluster analysis. Both extraction, purification and analysing systems were able to distinguish between microbial communities of rhizosphere soil as affected by cropping system. This technique could be a valuable instrument in classifying bulk soils or detecting treatment based shifts in rhizosphere microbial populations for plant health purposes. Therefore, data on the correlation of FAME spectrum patterns with the level of soil antagonistic potential toward soil-borne pests and diseases must be gathered.

1. Introduction

The composition of microbial populations has often been investigated on an individual organism level. Culture techniques used to enumerate and identify bacteria have several disadvantages. Estimates of specific populations are often variable and underestimated (Kluepfel, 1993). It has been suggested that "segments" of microbial communities are associated with enhanced soil ecosystem stability and that they aid in suppressing soilborne plant pathogens (Vilich and Sikora, 1997). Methods used to measure microbial populations, their activity or biomass by physical and chemical techniques, did not always offer satisfying results in characterizing communities (Kennedy and Smith, 1995). Lipids are components in relatively constant levels in biological membranes. Thus, the lipids of a given organism, as reflected in the composition of the associated fatty acids (FA), are regarded as a 'fingerprint' of the investigated organism. Fatty acid spectrums of unknown bacterial isolates can be used for identification by comparing them with the FA spectrum of known isolates. Because fatty acids can be readily volatilized following methylation, they can be readily analysed by gas chromatography (Moss *et al.*, 1980; Moss 1981; Vestal and White 1981). The value of fatty acid methyl ester (FAME) analysis arises from the fact that

H.-W. Dehne et al. (eds.), Diagnosis and Identification of Plant Pathogens, 345–348.

there are a great number of different kinds of fatty acids in the lipids of microorganisms and that different organisms have different combinations of these fatty acids. The commercial Microbial Identification System (MIDI, Inc; Newark, USA) is based on preparation of FAME`s out of pure bacterial isolates, followed by gas chromatographic (GC) analysis. The system is equipped with a complex software and fatty acid libraries to compare bacterial fatty acid extracts of unknown microorganisms and to determine their identity on the bases of similarity indices. White *et al.*, (1979) suggests that the lipid composition can be used to define the microbial community structure. Phospholipid fatty acids (PLFA) are essential components of bacterial cells and are not found in storage lipids (Kluepfl, 1993). Phospholipids show rapid loss from nonviable cells, because phosphate is rapidly hydrolyzed by exoenzymes upon cell death (Korner and Laczko, 1992). Some fatty acids, especially those in the phospholipid fraction have been found to be specific ("biomarkers") for some microorganisms. Zelles *et al.*, (1992) and Janzen *et al.*, (1994) concluded that FA extracted from soil were closely linked to soil microorganisms and that this method can be used in the study of soil microbial ecology. The objectives of our studies were: 1) to separate and analyze only the phospholipid fraction out of the soil lipids in order to concentrate on living cells versus entire fatty acid contents in the bulk soil samples and 2) to utilize the standard dilution plate method of describing microbial communities in combination with fatty acid analysis and MIDI system.

2. Materials and methods

2.1. PHOSPHOLIPID EXTRACTION FROM SOIL

Lipids were extracted and phospholipids separated according to the method of Zelles and Bai (1993) as modified below. 5 g field moist soil (n = 10) was transferred to an erlenmeyer flask and 25 ml methanol (CH_3OH), 12 ml chloroform ($CHCl_3$) and 10 ml 0.05 M phosphate buffer (K_2HPO_4 plus KH_2PO_4; pH 7,4) added to each flask and agitated for 2 hours. Afterwards 12 ml distilled water and 12 ml chloroform were added and agitated again for one minute. The liquid was transferred to a clean flask and the soil residue washed again with 25 ml methanol and 12 ml chloroform. The supernatant of both washing was combined and the soil discarded. The upper phase (water / methanol) was removed and discarded. The lower organic phase (chloroform) was separated, dried over anhydrous sodium sulfate (Na_2SO_4) and was passed through a filtration column (10 ml) containing 2 cm Celite 545 (Fluka Chemie AG) by rinsing the column with 20 ml chloroform. The lipids were then subdivided into neutral, glyco- and phospholipids using a silicia-bonded phase column (Mega Bound Elut; Internationale Chemie-Technik GmbH, Bad Homburg, size: 2g / 12 ml; Part. Nr. 1225-6018). The column was conditioned with 24 ml (2 x 12 ml successive) of pure chloroform. The lipid sample in chloroform was then added and elusion was done by rinsing the column with 12 ml of chloroform (to elute the neutral lipids) and 12 ml of acetone (to elute the glycolipids). These fractions were discarded. The column was then rinsed with 4 x 12 ml methanol to recover the phospholipid fraction. Liquid of this fraction was reduced to zero in a rotary evaporator and suspended in 1 ml methanol. Afterwards, the samples were transferred to 10 ml glass tubes and dried under a gentle stream of nitrogen. The extraction of the fatty acids out of the phospholipids was done according to the MIDI system standard protocol as described by Sasser (1990).

2.2. CELLOPHANE COLONY HARVEST (CCH) METHOD

One sterile, semipermeable square sector (2 x 4 cm) of cellophane (deti, 74909 Meckesheim, Einmach-Folie Art. Nr. 5050) was placed on Tryptic Soy Agar (TSA) media one day before the 10 soil subsamples of a treatment were diluted (10^{-2} w:v) and plated onto the media. After 24 hours of incubation at 25° C, colonies were harvested by transferring each square from each plate into separate 10 ml glass tubes. The bacterial community in each tube was then used for FA extraction according to the MIDI standard protocol.

2.3. STATISTICAL ANALYSIS

We used principal components analysis (PCA) to compare the resulting fatty acid retention time profiles among field soil samples that were detected by PL extraction of either samples of the entire microbial community direct from bulk soil or after cellophane colony separation on selective media. Additional cluster analysis was conducted from FA spectra detected with both analytical approaches using an unweighted pair - group method. Statistical analysis was conducted with the MIDI software provided.

3. Results

The principal component analysis (PCA) and the cluster analysis of the fatty acid spectra of the microbial community in each sample clearly distinguished between the two field plots. The 2-D plot (PCA) revealed two distinct groups of fatty acids. Each cluster was clearly related to one of two investigated field plots. This could be seen in both analytical extraction procedures. However, PL extraction from bulk soil led to a more intimate sample accumulation within each group compared to the CCH procedure. The dendrogram results of the cluster analysis also revealed the closeness of pairs of samples from one field over the other. The cluster analysis demonstrated that the micro flora found in the replications from bean field soil were more closely related to one another than those of the potato field soil samples. In the dendrograms from PL extraction from bulk soil and the cellophane colony harvest method, 18 out of 20 samples exhibited a intimate relationship to their field of origin.

4. Discussion

The results obtained from bulk soil extracts and from the CCH method demonstrated clearly that the extraction, purification and analysing methods used produced distinct fatty acid fingerprints for the two fields. Cavigelli *et al.* (1995) compared FAME profiles between lipids from soil extracts and lipids from agar plated microbial communities. They concluded that plated community profiles are not only the result of selection by the growth medium but reflect the distribution *in situ*, of the dominant, cultivable soil microbial population. They reported that some communities on culture media were distinct due to some of the same profile characteristics that identified soil community FAME profiles as distinct. This may be another reason for the clear grouping of the 2 soils in our study. The FAME spectra obtained need to be analysed with other statistical approaches (Schuster, 1996) because the MIDI software package is only designed to compare FAME spectra from pure isolates. If further data analysis of the fatty acids of microbial communities support the hypothesis of Cavigelli *et al.* (1995), the CCH method could be an improved, inexpensive and expedient alternative technique to the more time consuming and expensive PL extraction of whole *in situ* microbial communities in bulk soil. The occurrence of "marker fatty acids" as enumerated and discussed by Zelles *et al.* (1995) and Cavigelli *et al.* (1995) suggests that this technique could be a valuable instrument in classifying bulk soils or detecting treatment based shifts in rhizosphere microbial populations for plant health purposes (Terhardt *et al.*, 1995). The antagonistic potential regulating pathogen and nematode population densities, which is strongly influenced by microbial ecology in soils, could be used in risk assessment if a good fingerprinting technique can be developed.

5. References

Cavigelli M. A., Robertson G.P. and Klug M.J. (1995). Fatty acid methyl ester (FAME) profiles as measures of soil microbial community structure, *Plant and Soil* **170**; 99.

Janzen R. A., Raverkar K. P., Rutherford P.M. and Mc Gill W.B. (1994). Decreasing amounts of extractable phospholipid-linked fatty acids in a soil during decline in numbers of pseudomonads, *Canadian Journal of Soil Science*, 277 - 284.

Kennedy A.C. and Smith K.L. (1995). Soil microbial diversity and the sustainability of agricultural soils. The significance and regulation of soil biodiversity, 1995; 75-86; 1995 Kluwer Academic Publishers.

Kluepfel Daniel A. (1993). The behavior and tracking of bacteria in the rhizosphere, *Annu. Rev. Phytopathol.* 1993 **31**; 441 - 472.

Korner J. and Laczko E. (1992). A new method for assessing soil microorganism diversity and evidence of vitamin deficiency in low diversity communities, *Biol. Fertil. Soils* **13**, 58-60.

Moss C. (1981). Gas - liquid chromatography as an analytical tool in microbiology. *J. Chrom.* **203**, 337 - 347.

Moss C., Dees S. and Guerrant G. (1980). Gas - liquid chromatography of bacterial fatty acids with a fused - silicia capillary column, *J. Clin. Microbiol.* **12**, 127 - 130.

Sasser M. (1990). Identification of bacteria through fatty acid analysis. In: Klement Z., Rudolph K., Sands D.C. (eds.). *Methods in Phytobacteriology*, Akademiai Kiado, Budapest, pp 199-204.

Schuster R. P. (1996). Effect of crop on the diversity and the potential fungal egg pathogens of cyst nematodes, European Foundation for Plant Pathology (EFPP). 4 th International Symposium September 9-12 at Bonn, Germany; Proceedings.

Terhardt J., Vilich V., Sikora R.A. (1995). Untersuchungen zum Einfluß pflanzengesundheitsfördernder Rhizosphärebakterien an Tomaten durch Blattapplikation mit verschiedenen Substanzen., *Mitteilungen der Deutschen Phytomedizinischen Gesellschaft* **3**, 43.

Vestal J. and White D. (1989). Lipid analysis in microbial ecology *Bioscience* **39**, 535.

Vilich V. and Sikora R. A. (1997). Diversity in soil-borne microbial communities - a tool for biological system management of root health, In: *Plant - Microbe Interactions and Biological Control.*; G.J. Boland ,L.D. Kuykendall (eds.).Pub.: Marcel Decker; in press.

White D. C., Davis W.M., Nickels J. S., King J. D. and Bobbie R. J. (1979). Determination of sedimentary microbial biomass by extractable lipid phosphate, *Oecologia* (Berl.) **40**, 51 - 62.

Zelles L., Bai Q.Y., Beck T. and Beese F. (1992). Signature fatty acids in phospholipids and lipopolysaccharides as indicators of microbial biomass and community structure in agricultural soils, *Soil Biol. Biochem.* **Vol. 24**, No. 4, 317 - 323.

Zelles L. and Bai Q. Y. (1993). Fraction of fatty acids derived from soil lipids by solid phase extraction and their quantitative analysis by GC - MS, *Soil. Biol. Biochem.* **25**; 495 - 507.

Zelles L., Rackwitz R., Bai Q.Y., Beck T. and Beese F. (1995). Discrimination of microbial diversity by fatty acid profiles of phospholipids and lipopolysaccharides in differently cultivated soils, *Plant and Soil* **170**, 115 - 122.

SCANNING ELECTRON MICROSCOPY OF *VERTICILLIUM DAHLIAE* ON THE COTTON ROOT SURFACE

T.A. VLASSOVA

Moscow State University, Faculty of Biology, Department of Plant Physiology, Vorobjovy gory, 119899, Moscow, Russia

1. Abstract

The initial events of *Verticillium dahliae Kleb.* - *Gossipium hirsutum L.* interaction were examined by scanning electron microscopy. The sterile root cultures of susceptible and resistant cotton cultuvars inoculated with *V. dahliae* were used. It was shown that mycelium of *V. dahliae* grew intensively on the roots of both susceptible and resistant cotton cultivars showing no directional pattern. Direct penetration of rhizodermis by hyphae was observed. No visual difference in the ultrastructure and growth character of *V. dahliae* on the roots differed in susceptibility was determined.

2. Introduction

The initial stage of Verticillium wilt like many another plant diseases is considered as deciding one for the establishment of infection. The pathogen detection on this stage is also very important.

Present study was initiated to reveal growth character and to compare ultrastructure of virulent strain of *Verticillium dahliae Kleb.* on roots of susceptible and resistant cotton plants.

The sterile root cultures were used to avoid the foreign infection.

3. Materials and methods

Two cotton cultivars, C-4727 (highly susceptible to Verticillium wilt) and Tashkent-1 (relatively resistant) were used. The roots of aseptically grown seedlings were detached, dipped into conidial suspension of virulent strain of *V. dahliae* (biotype Chl-288, race 2) grown on Czapek Dox medium and placed onto the solidified Smirnov's medium. For scanning electron microscopy material was fixed in 2,5% glutaraldehyde, postfixed in 1% osmium tetroxide, dehydrated, critical-point dried, coated with gold-palladium and examined in S-405 a microscope.

H.-W. Dehne et al. (eds.), Diagnosis and Identification of Plant Pathogens, 349-351.

4. Results

The conidia of *V. dahliae* germinated on the root surface within 24-48 h, producing the germ tubes. The hyphae developed from the germ tubes grew very extensively in different directions producing dense masses on the root surface (Fig. 1). They branched monopodially, moved frequently and had stuck together or were connected together with the short strands in some sites if they grew in parallels. The hyphae varied in diameter significantly. Any hyphae growing immediately on the rhizodermal cell surfaces followed cell form and curved in conformity with it. However, the common growth of fungal mycelium showed no especial directional pattern. In the cases of close hyphal contact with the cell wall the adhesion was obtained; sometimes the hyphae were connected with to the cell surface via the strands like ones connecting the hyphae

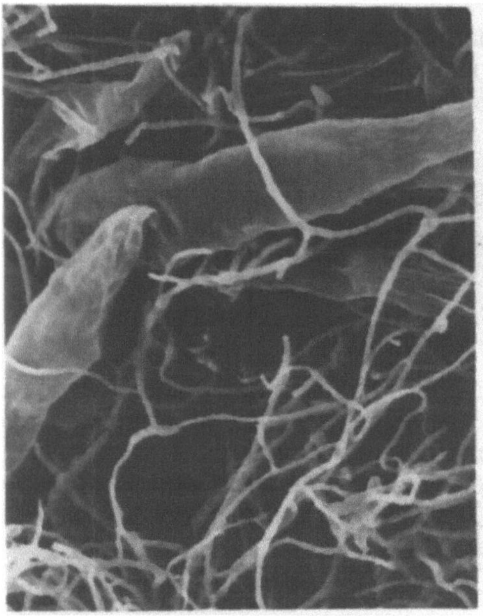

Figure 1. Irregular growth of *V. dahliae* mycelium on the cotton root surface and root hairs. 600 x

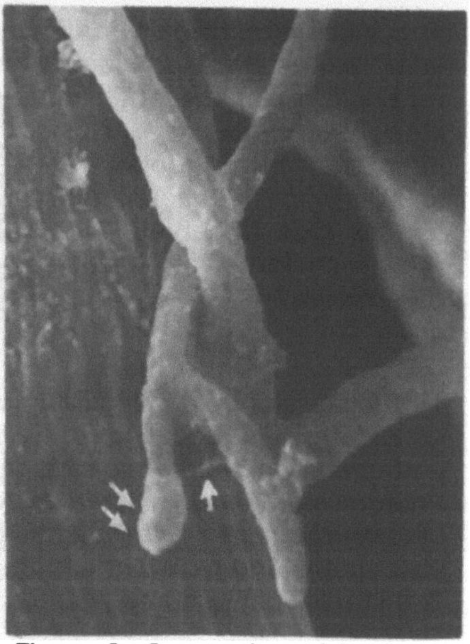

Figure 2. Connection of hyphae together and with the root surface via strands (🡇). Note the thickening of hyphal tip (🡇🡇). 5000 x

The root penetration by fungus directly through rhizodermal cell wall as well as through intercellular spaces was observed. Before penetration, some hyphal tips became appressoria-like swollen (Fig.2); however, many hyphae penetrated the cells without thickening. The ultrastructure and growth character of *V. dahliae* mycelium on the surfaces of roots differed in susceptibility demonstrated no qualitative difference.

5 Conclusion

The growth character of *V. dahliae* on cotton roots and colonization of root surface by this pathogen are typical for facultative parasite. The absence of visually detected difference in growth character and ultrastructure of *V. dahliae* on the roots of susceptible and resistant cotton cultivars has been confirmed by the results of light and transmission electron microscopy.

ISOLATION AND IDENTIFICATION OF ARBUSCULAR MYCORRHIZAL FUNGI (AMF)

B. WEICKEL[1], J.C.DODD[2] and H.-W. DEHNE[1]
[1] *Institut für Pflanzenkrankheiten, Nußallee 9, D-53115 Bonn, Germany*
[2] *The International Institute of Biotechnology, P.O. Box 228, Canterbury, Kent CT2 7YW, UK*

Introduction

Arbuscular mycorrhizal fungi (AMF) in the order *Glomales* colonize the root systems of the large majority of land plant species. They often have beneficial influence on growth and survival of their host due to better mineral nutrition and increased tolerance in abiotic and biotic stress situations. However, the functional and taxonomic characterization of individual AMF isolates is limited because of their obligate biotrophy. Morphological characteristics of the relatively large, asexual spores have been used for the identification and taxonomy of glomalean fungi. Difficulties in the identification of spores arise from the frequently poor state of spores isolated from the field and the few morphological characters available to discriminate species. Recent advances in biochemical and molecular techniques provide a promising tool to overcome these problems. In this study, AMF from different sites were characterised according to their morphological features as well as using isozymes as molecular markers.

Materials and Methods

Soil samples from two different sites (Germany J55 / Israel I85) were collected and mixed in autoclaved sand. Maize seedlings were used as "trap plants" to induce sporulation of indigenous AMF. After 2-4 months, viable spores were isolated to produce pure cultures with single and/or multispore inoculum for further investigations. The identification of the species was based on a) Spore morphology: Individual spores were examined microscopically on diagnostic slides. The main taxonomic parameters like spore color, spore size and shape, number of spore walls, hyphal attachment were assessed using the Bank of European Glomales (BEG)-Expert System (DODD et al., 1994, DODD and ROSENDAHL, 1996a). In addition, biochemical properties of the isolates were analysed: b) Spore isozyme patterns:

H.-W. Dehne et al. (eds.), Diagnosis and Identification of Plant Pathogens, 353-356.

Proteins extracted from spores (20-100) were separated electrophoretically and malate dehydrogenase (MDH) patterns were compared with those of well-described cultures (ROSENDAHL and SEN 1992; DODD et al., 1996b). Differences between individuals were identified in the variation of their banding patterns.

Results

SPORE MORPHOLOGY

J 55: (Fig 1) Spores were yellow to yellow-brown and the spore wall consisted of a white outer layer and a pigmented inner layer. Spore size was between 210-290 µm and the hyphal attachment was occluded by two septa characteristic for *Glomus caledonium.*

I 85: (Fig 2) Spores were maroon to dark-brown with a thick pigmented wall. Spore size was between 150-200 µm and the hyphal attachment was sometimes straight as in *Glomus geosporum*, but in most spores funnel-shaped characteristic for *Glomus mosseae.*

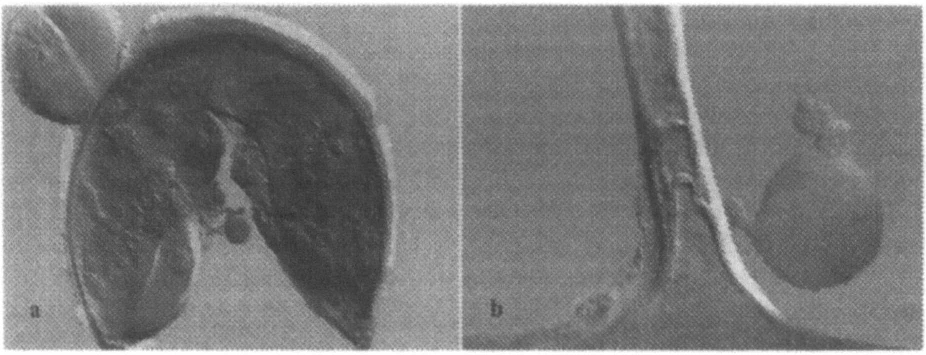

Figure 1: Crushed spore (a) and hyphal attachment (b) of J 55

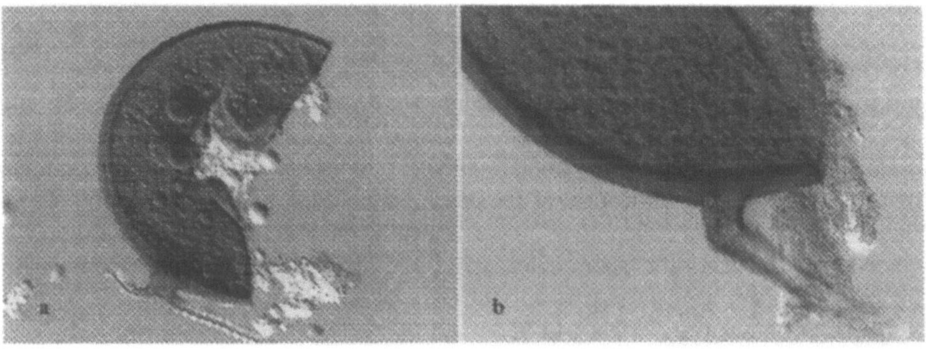

Figure 2: Crushed spore (a) and hyphal attachment (b) of I 85

SPORE ISOZYME PATTERNS

J 55: (Fig 3) The MDH pattern of J 55 (lane 9) clearly differed from *Glomus coronatum*-like (1, 3) and *Glomus mosseae* isolates (5-7). J 55 was distinct in a spore mixture with *Glomus mosseae* (8). Correspondence was found in *Glomus caledonium*/ UK (not shown). J 55 was finally classified as *Glomus caledonium*.

I 85: (Fig 3) The MDH pattern of I 85 (lane 1) indicated a clear distinction from *Glomus geosporum* (2) and *G. mosseae* isolates (5-7), but corresponded with the *G. coronatum*-like isolate (lane 3). I 85 was finally classified as *Glomus coronatum*, but this AM fungal culture has to be investigated in more detail.

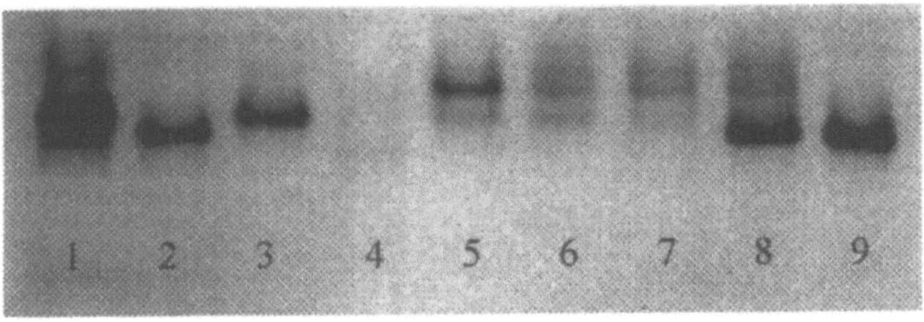

Figure 3: Malate dehydrogenase (MDH) patterns of different spore isolates

Discussion

J 55 was clearly classified as *Glomus caledonium* according to the morphological data. Stability of spore morphological characters as well as of spore isozyme patterns for this species have been reported previously (ROSENDAHL and SEN, 1992; MORTON, 1996). In the case of I 85, however, exclusive morphological characters to discriminate species were not available (DODD et al., 1996b). The use of MDH isozyme profiles revealed a close relationship to the *Glomus coronatum*-like isolate and clear distinction from morphologically-similar AMF like Glomus *mosseae*. LLOYD-MACGILP et al. (1996) who sequenced the internal transcribed spacer (ITS) region of *Glomus mosseae* isolates as well as of *Glomus coronatum*, also reported clear diversity between these species according to the molecular data. These results indicate, that a combined approach to get more information about related AMF species is needed.

References

Dodd, J.C., Gianinazzi-Pearson,V., Rosendahl S. and Walker, C. (1994) European Bank of Glomales - an essential tool for efficient international and interdisciplinary collaboration, in: S.Gianinazzi and H. Schuepp (eds.), *Impact of Arbuscular Mycorrhizas on Sustainable Agriculture and Natural Ecosystems*, Birkhauser Press, Basel, Switzerland, pp. 41-46.

Dodd, J.C. and Rosendahl, S. (1996a) The BEG Expert-System - a multimedia identification system for arbuscular mycorrhizal fungi, Mycorrhiza **6**, 275-278.

Dodd, J.C., Rosendahl, S., Giovannetti, M., Broome, A., Lanfranco, L., and Walker, C. (1996b) Inter- and intraspecific variation within the morphologically-similar arbuscular mycorrhizal fungi *Glomus mosseae* and *Glomus coronatum*. New Phytol. **133**, 113-121.

Lloyd-Maccilp, S. A., Chambers, S. M., Dodd, J. C., Fitter, A.H., Walker, C. and Young, J. P. W. (1996) Diversity of the ribosomal internal transcribed spacers within and among isolates of *Glomus mosseae* and related mycorrhizal fungi, New Phytol. **133**, 103-111.

Morton, J. B. (1996), Redescription of *Glomus caledonium* based on correspondence of spore morphological characters in type specimens and a living reference culture, Mycorrhiza **6**, 161-166.

Rosendahl, S. and Sen, R. (1992) Isozyme Analysis of Mycorrhizal Fungi and their Mycorrhiza, *Methods in Microbiology*, **24**, 629-654.

ANALYSIS OF AMF-*GLOMUS CALEDONIUM* J 55-HOST INTERACTION — ACTIVITY, ISOZYME PATTERNS AND PLANT RESPONSE

B. WEICKEL[1], J.C. DODD[2] and H.-W. DEHNE[1]
[1] *Institut für Pflanzenkrankheiten, Nußallee 9, D-53115 Bonn, Germany*
[2] *The International Institute of Biotechnology, P.O. Box 228, Canterbury, Kent CT2 7YW, UK*

Introduction

Arbuscular Mycorrhizal Fungi (AMF) associated with higher plants are common in most ecosystems but some plant species are consistently more mycorrhizal than others and benefit from the symbiosis. In addition to various environmental factors influencing colonisation, host plant species plays an important role in controlling the rate and extent of mycorrhizal formation. The interaction of host plant/fungal species determines the potential of AMF for practical use in plant breeding programs, screening and selection for field application as well as inoculum production. In this study, the symbiotic activity of *Glomus caledonium* J 55 was analysed in four different plant species.

Materials and Methods

Barley (*Hordeum vulgare* cv. Voldagsen), maize (*Zea mays* cv. Marshall), onion (*Allium cepa* cv. Ailsa Craig) and parsley (*Petroselinum crispum* cv. Einfacher Schnit 2) were inoculated with *Glomus caledonium* J 55 and grown in a mixture of sterile sand and expanded clay (5:1) under controlled conditions. Plants were harvested four and eight weeks after emergence to analyse: (1) Total mycorrhization (using trypan blue staining) and active mycorrhization (succinate dehydrogenase staining) (GIANINAZZI and GIANINAZZI-PEARSON, 1992) by calculating the frequency of colonisation (C%), the colonisation intensity (I%) and arbuscule development (A%). (2) The presence of the inoculated fungal species in the plant roots using diagnostic malate dehydrogenase (MDH) isozyme patterns after polyacrylamide gel electrophoresis (ROSENDAHL and SEN, 1992). Protein extracts of spores of *G. caledonium* J55 were used as the reference for the fungus-specific enzyme band in the gel. (3) The influence of the symbiosis on plant biomass (shoot fresh weight, root fresh weight, root length).

H.-W. Dehne et al. (eds.), Diagnosis and Identification of Plant Pathogens, 357–360.

Results

TOTAL AND ACTIVE MYCORRHIZATION

Mycorrhization (total and active) was generally very low in both grass species, but high in onion four weeks after plant emergence with a subsequent decrease after eight weeks (Tab 1). In parsley, however, mycorrhization was constantly high over time.

TABLE 1. Colonisation (C %), Colonisation intensity (I %) and arbuscule development (A %) of different plants inoculated with G. caledonium J 55

		weeks after plant emergence					
		4 weeks			8 weeks		
		C %	I %	A %	C %	I %	A %
barley							
	total	24.0	2.2	1.8	26.0	8.9	6.4
	active	*2.0*	*0.1*	*0.1*	*6.0*	*0.7*	*0.3*
maize							
	total	10.0	3.4	2.6	28.0	7.9	6.4
	active	*4.0*	*0.6*	*0.5*	*2.0*	*0.1*	*0.1*
onion							
	total	16.0	11.4	9.4	28.0	7.5	3.7
	active	*12.0*	*7.2*	*6.0*	*12.0*	*0.9*	*0.3*
parsley							
	total	16.0	8.0	6.7	50.0	17.3	12.7
	active	*14.0*	*3.1*	*2.3*	*30.0*	*6.4*	*5.8*

MALATE DEHYDROGENASE (MDH) ISOZYME PATTERNS

The analysis of isozyme patterns revealed that no fungus-specific isozyme band was visible in the roots of both grass species (Fig 1). In onion, a strong fungal MDH signal was detected four weeks (a) after plant emergence and a weak signal after eight weeks (b). Fungal signals were consistently strong in parsley (Fig 1).

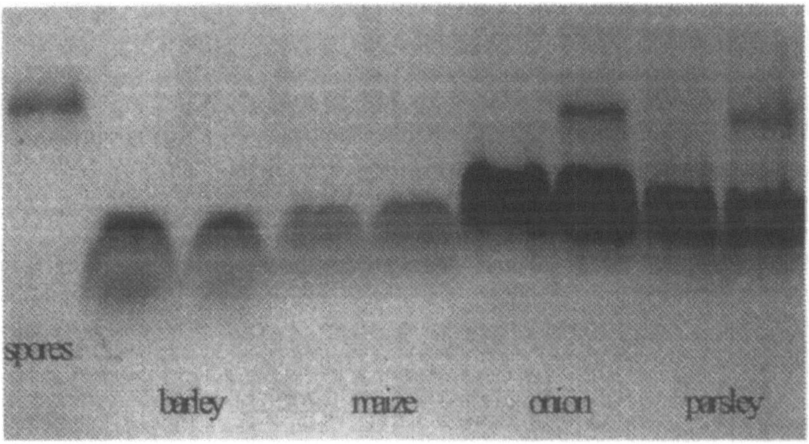

a) four weeks after plant emergence

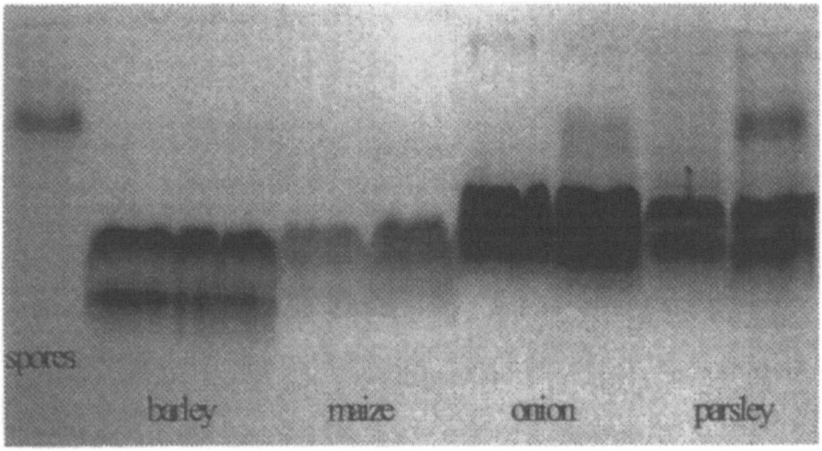

b) eight weeks after plant emergence

FIGURE 1. Malate dehydrogenase (MDH) isozyme patterns of root extracts from different plant species four weeks (a) and eight weeks (b) after plant emergence

PLANT BIOMASS

Root fresh weight and root length were almost the same in non-mycorrhizal and mycorrhizal plants for all hosts (Tab 2), but marked differences in growth response (% increase of shoot fresh weight) were detected (Fig 2). In all cases, inoculation resulted in increased shoot fresh weights, but differed between hosts (Fig 2).

TABLE 2. Root fresh weight (rfw) and root length (rl) of different plant species inoculated with *Glomus caledonium* J 55

		weeks after plant emergence			
		4 weeks		8 weeks	
(n=4)		rfw (g)	rl (m)	rfw (g)	rl (m)
barley					
	non-mycorrhizal	0.71	7.5	1.61	15.7
	J 55	0.68	7.3	1.54	16.7
maize					
	non-mycorrhizal	2.88	11.3	7.55	32.9
	J 55	2.74	12.4	7.80	36.9
onion					
	non-mycorrhizal	0.11	0.58	0.37	0.83
	J 55	0.13	0.60	0.43	0.80
parsley					
	non-mycorrhizal	0.10	1.13	0.40	1.95
	J 55	0.12	1.38	0.92	4.50

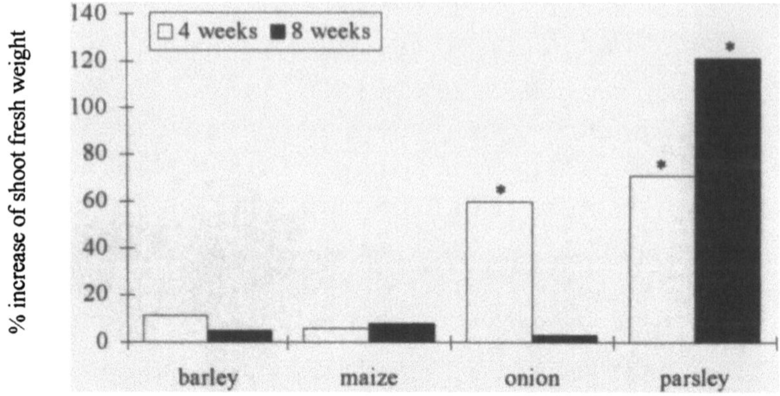

Figure 2. Growth response of different plant species inoculated with *G. caledonium* J 55 (% increase of shoot fresh weight compared to non-mycorrhizal plants, * significant difference, a =0,05, students t-test)

Conclusions

* Plant growth response showed good correlation with mycorrhizal formation and activity of *Glomus caledonium* J 55 in different host plant species.
* Diagnostic MDH isozyme patterns confirmed the presence of the inoculated fungal species in the root systems depending on colonisation intensity and metabolic state of the fungus.
* The results indicate that the investigated parameters are suitable to analyse the activity of the fungal symbiont and the responsiveness of the plant over the time course of the symbiosis.

References

Gianinazzi, S. and Gianinazzi-Pearson, V. (1992) Cytology, Histochemistry and Immunochemistry as Tools for Studying Structure and Function in Endomycorrhiza, *Methods in Microbiology*, **24**, 109-139.

Rosendahl, S. and Sen, R. (1992) Isozyme Analysis of Mycorrhizal Fungi and their Mycorrhiza, *Methods in Microbiology*, **24**, 629-654.

ANTAGONISM BETWEEN MYCORRHIZAL FUNGI AND *HETEROBASIDION ANNOSUM* (Fr.) Bref. - SCANNING ELECTRON MICROSCOPICAL STUDIES

A. WERNER and A. NAPIERAŁA
Polish Academy of Sciences, Institute of Dendrology
Parkowa 5, 62-035 Kórnik, Poland

1. Abstract

Scots pine seedlings individually inoculated in vitro with several strains of mycorrhizal fungi were additionally inoculated with *Heterobasidion annosum* (Fr.) Bref. Scanning electron microscopical studies were carried out in order to observe the behaviour of the pathogen's hyphae and their ability to overcome "mechanical barrier" created by the mantles. The differences in morphology of mantles and some details of the hyphae appeared to be proper tools helping in diagnosis of infection. The results suggest the lack of ability to complete the infection by the pathogen in areas of roots occupied by the hyphae of mycorrhizal fungi.

Key words: *Heterobasidion annosum*, Scots pine, scanning electron microscopy, mycorrhiza, protection

2. Introduction

Morphology and structure of the mantles have been studied for a proper classification of the mycorrhizas and identification of unknown mycosymbionts (Masicotte et al. 1987; Agerer 1991). In studies on the antagonism between symbiotic or saprophytic and pathogenic fungi inhabiting rhizosphere of host-plants SEM techniques are helpful in an understanding of the mutual influence of the organisms (Elad et al. 1983). The appearance of the outer mantles and some peculiar structurae of *H. annosum*, can permit to distinguish its hyphae on the surface of mycorrhizal roots.

The aim of the study was to describe the structural characteristics of the mycorrhizae formed by several symbiotic fungi with Scots pine in pure culture conditions and to evaluate the effectiveness of the mantles as structural barriers against infection caused by *H. annosum*.

3. Materials and methods

Scots pine (*Pinus sylvestris* L.) seedlings growing in pure culture conditions according to a method of Pachlewska (1968) were inoculated with several strains of ectomycorrhizal

H.-W. Dehne et al. (eds.), Diagnosis and Identification of Plant Pathogens, 361–363.

fungi representing species, such as *Suillus bovinus* (Fr.) Knutze, *S. luteus* (Fr.) Knutze, *Paxillus involutus* (Batsch) Fr., *Laccaria laccata* (Scop. Fr.) Berk. et Br., *Hebeloma crustuliniforme* (Bull.) Quél. and a strain of an ectendomycorrhizal fungus (Mrg X) having ascomycetous affinities. After establishing the mycorrhiza the seedlings were additionally infected by *H. annosum* representing P intersterility group.

Samples of roots were fixed in 2.5% glutaraldehyde in cacodylate buffer (0.05M, pH 7) for several hours and postfixed with 2% OsO_4 in the same buffer. After dehydration in a graded series of ethanol and acetone, critical point dried specimens were coated with gold and observed with a Philips 515 scanning electron microscope.

4. Results

At early stage of mantle formation, the arrangement of hyphae was similar when roots were colonized by both *H. annosum* and the mycorrhizal fungi. Parallelly orientated, poorly ramified hyphae were usually observed at the root surface. Some of them penetrated between cortical cells. Later on, branching and anastomosing of hyphae led to formation more compact mycelium of prosenchymatous structure.

Three weeks after inoculation of pine seedlings with *H. annosum,* numerous lateral roots were nearly completely covered with mycelium. The surface mantle hyphae were loosely organized. They were often devoid of cytoplasm and clamp connections. The hyphae of the inner mantle formed a net prosenchyma with frequent anastomoses and clamps. Infrequent conidiophores bearing spores were structural characteristics which permit to distinquish the pathogen on the mantles formed by the mycorrhizal fungi.

Initially, the hyphae of *Laccaria laccata* formed at the root surface a loose weft. Later on, the hyphae became densely organized. The outer prosenchymatous mantle consisted of tightly interwoven hyphae with evident clamp connetions and frequent anastomoses. A mature ectomycorrhizae showed hyphal strands. Inner mantle hyphae formed an irregular synenchyma. Two months after inoculation the mycorrhizal roots with the pathogen, very frequent crystals of calcium oxalate along outer mantle hyphae were observed.

The hyphae of *Paxillus involutus* slowly colonized the surface of the roots forming at first a subtle weft. Ectomycorrhizae formed by *P. involutus* had a well-developed mantles consisted of loosely interwoven hyphae. The outer mantle was prosenchymatous, while the inner layers had a synenchymatous structure. The outer mantle was characterized by frequent emanating hyphae and rhizomorphs. Mature stages of the association were characterized by abundant extracellular matrix. The smooth surface of old mantles consisted of hyphae embedded in gelatinous matrix which make hard to distinquish the individual hyphae. Similarly to mycorrhizae formed by *Laccaria laccata,* the hyphae of the outer mantles bore frequent calcium oxalate crystals after inoculation with the pathogen.

Suillus bovinus formed a dense, wooly multilayered mantles. The outer, prosenchymatous, layer consisted of tightly interwoven hyphae characterized by frequent ornamentations. The inner mantle was synenchymatous. The free spaces between the hyphae were filled by gelatinous matrix. Frequent drops of liquid substance of unknown origin and chemical composition were observed on the surface of the mantles. The ectomycorrhizae were characterized by infrequent emanating hyphae.

Mycorrhizae formed by *Suillus luteus* had abundant wooly mantles with the outer hyphae forming a felt prosenchyma. Some of the hyphae showed a stellate pattern. The inner part of mantles consisted of tightly woven hyphae forming a net synenchyma.

Pinus sylvestris - Hebeloma crustulinoforme mycorrhizae were abundant. The surface of the mantle was a loose to dense prosenchymatous web of branched hyphae bearing frequent clamps. The emanating hyphae often formed dense interwoven veils and rhizomorphs. The inner layer was composed of an irregular synenchymatous tissue with compactly arranged hyphae.

The most striking feature of ectendomycorrhiza formed by the strain Mrg X was a lack of clamp connections on the hyphae and the presence of bacteria. Initially, the fungus formed a subtle web-like prosenchyma on the root surface. As the mycorrhizae matured the mycelium became more compact due to ramification of the hyphae and formation frequent anastomoses.The matured mycorrhiza had thin outer layer of loosely interwoven hyphae and infrequent emanating hyphae, while the deeper layer displayed higher proportion of synenchymatous elements.

Only sporadically the hyphae of the pathogen interacted with the hyphae of the mantles. It was in the case of ectomycorrhizae formed by *Hebeloma crustuliniforme.* In this instance the conidiophores were seldom observed on the surface of the mantles.

5. Discussion

The comparative SEM study of Scots pine roots colonized by several mycorrhizal fungi and *H. annosum* showed the existance of differences in structure of the outer mantles and some unique features such as, ornamentations, incrustations, presence of gelatinous matrix, emanating hyphae and rhizomorphs or lack of clamps, which appeared to be useful in recognition the pathogen's mycelium on the surface of the mycorrhizae, thus can serve as proper tools of diagnosis. The results suggest lack of ability to overcome the mechanical barrier created by the mantles.

The accumulation of calcium oxalate crystals along hyphae of outer mantles formed by *P. involutus* and *L. laccata,* observed only after inoculation with the pathogen, may be recognized as a kind of reaction of the mycorrhizal fungi to immediate vicinity of the pathogen mycelium.

6. References

Agerer, R. (1991) Characterization of ectomycorrhiza, in J. R. Norrin, D. J. Read, and A. K. Varma (eds.), *Techniques for the study of mycorrhiza. Methods in Microbiology vol. 23*, Academic Press, London, pp 25-73.

Elad, X., Chet, I., Boyle, P., and Henis, X. (1983) Parasitism of *Trichoderma* spp. on *Rhizoctomia solani* and *Sclerotium rolfsii* - Scanning electron microscopy and Fluorescence microscopy, *Phytopathology* **73**, 85-88.

Massicotte, H. B., Melville, L. H., and Peterson, R. L. (1987) Scanning electron microscopy of ectomycorrhiza Potential and Limitations, *Scanning microscopy* **1**, 1439-1454.

Pachlewska, J. (1968) Badania nad syntezą mikoryzową (*Pinus sylvestris L.*) w czystych kulturach na agarze. *Prace IBL* **345**, 1-76.

PATHOGENICITY OF THE ISOLATES OF *RHIZOCTONIA CEREALIS* AND *R. SOLANI* FROM THE POZNAN REGION TOWARDS THE WHEAT SEEDLINGS

Ewa Zoltanska
University of Agriculture, Department of Phytopathology,
ul.D'browskiego 159, 60-594 Poznań, Poland

There were many studies to characterize *Rhizoctonia* spp. associated with diseased wheat seedlings and root rot of older plants and determine the relative virulence of isolates [3,4]. This taxonomic genus is divided into several species or anastomosis groups [5,6] and isolates were charakterized and their relative virulence was determined [2,7]. The aim of this research was the selection from the isolates of *R. cerealis* and *R. solani* the least and the most pathogenic ones towards wheat seedlings for the use in further investigations.

Methods

Isolates of the fungi belonging to the *Rhizoctonia* genus obtained from wheat growing in the Poznań region were divided into two groups - binucleate (*R. cerealis*) and multinucleate (*R. solani*) by Bandoni's method [1].
Pathogenicity of the isolates towards wheat seedlings was investigated in a laboratory experiment on Petri dishes and in a greenhouse experiment. The investigations were done on 2 varieties of wheat: Almari - winter wheat and Eta - spring wheat.
In the laboratory experiment filter paper soaked in distilled water was placed on Petri dish, next pieces of PDA overgrown with the isolates of the fungus under investigation were placed on top. For the control, pieces of sterile PDA were used. 10 seeds were placed on every dish. After one week the effect of the investigated isolates on wheat seedlings was assayed. The following characteristics were assayed: the percentage of infected stems and roots, the degree of stem and root infection. 22 isolates of *R. solani* and 17 isolates of *R. cerealis* were tested.
In the greenhouse experiment six seeds of wheat were placed in pots 10 cm in diameter with sterile soil. A piece of PDA overgrown with the isolate of the fungus under investigation was placed under each seed. After 1 month the effect of the isolates on the growth and health of plants was tested. The following characteristics were assayed: the height of the plants (after 2 weeks), the weight of the roots, the percentage of infected stems and the degree of stem infection. During the experiment 17 isolates of *R. solani* and 10 isolates of *R. cerealis* were investigated.

H.-W. Dehne et al. (eds.), Diagnosis and Identification of Plant Pathogens, 365–367.

Results

Of the 57 isolates of *Rhizoctonia* spp 54.4 % belonged to the *R. solani* species while 45.6 % belonged to the *R. cerealis* species (Table 1).

TABLE 1. Pathogenicity of *R. solani* and *R. cerealis* towards two varieties of wheat: Almari (A) - winter wheat, Eta (E) - spring wheat

Patho-genicity	Culti-vars	Laboratory experiment				Greenhouse experiment			
		stem infection		root infection		stem infection		growth of plants (cm)	weight of roots (g)
		%	degree	%	degree	%	degree		
R. solani weak	A	0.0-3.2	0.0-2.5	7.0-9.1	1.1-1.6	1.0-9.4	1.0-2.9	18.0-18.9	5.1-6.6
		isolates:							
		6, 34, 35, 40, 41, 45, 47, 50, 55		6, 30, 35, 38, 53, 55		12, 26, 31, 32, 34, 35, 41, 45, 52, 53, 57, 59, 60		26, 35, 52, 57, 60	26, 31, 52, 57
	E	0.0-10.4	0.0-2.3	9.2	1.1	11.2-19.1	2.3	23.9-25.6	4.2-4.8
		isolates:							
		35, 40, 41, 45, 47, 55		55		51, 52		32, 35, 59	34, 53
strong	A	31.6-97.4	1.6-3.0	35.5-93.5	1.0-1.6	-	-	14.9-16.9	3.3-4.0
		isolates:							
		5, 12, 26, 31, 59		5, 12, 26, 31, 39, 59		-		12, 31,34, 41,46, 51, 59	12, 34, 38, 51
	E	30.8-89.6	2.0-3.0	33.6-91.2	1.0-1.6	31.6-44.4	2.2-3.0	18.2-20.8	2.3-3.6
		isolates:							
		5, 12, 26, 30, 39, 52, 53, 59, 60		5, 6, 12, 23, 26, 31, 39, 40, 41, 45, 47, 50, 52, 53, 59		5, 12, 32, 34, 39, 45, 59, 60		34,45,51, 52,57	5,26,32, 35,38,39, 41,46,52, 57,59
R. cerealis weak	A	0.0-4.3	0.0-3.0	7.1-10.0	1.0-1.4	0.0-8.6	0.0-2.4	17.2-18.9	7.1
		isolates:							
		3, 10, 24, 29, 42		10, 24, 61		8, 11, 14, 24, 25, 36, 61		7, 11, 37, 61	11
	E	0.0-7.8	0.0-2.3	-	-	-	-	24.7	4.1
		isolates:							
		3, 8, 29, 42		-		-		14	36
strong	A	31.2-94.4	1.9-3.0	30.0-68.5	1.0-2.4	36.2	1.9	15.3-15.9	2.9-3.5
		isolates:							
		1, 2, 4, 7, 9, 14, 36, 61		2, 4, 7, 14, 36, 37		37		4, 25, 36	7, 36, 37
	E	39.1-89.0	2.4-3.0	30.3-77.9	1.0-1.7	33.7-62.7	1.9-3.0	20.7	2.4-3.6
		isolates:							
		1, 2, 4, 7, 9, 36, 61		1, 2, 3, 4, 9, 10, 11, 14, 24, 29, 36, 37, 61		7, 14, 37, 61		8	8, 11, 14, 37, 61

The least pathogenic isolates towards wheat seedling were RS 35, 55 and RC 24, 61. The most pathogenic isolates towards wheat seedling were RS 12, 59 and RC 14, 37. Eta - variety was a little more susceptibility to infection with the isolates under investigation

References

1. Bandoni R.J.: *Safranin as a rapid nuclear stain for fungi*, Mycologia 71 (1979), 873-874
2. Findels C.E., Nabben D.J.: *Charakterization and pathogenicity of anastomosis groups of Rhizoctonia solani isolated from Beta vulgaris*, Phytopathology 79 (1989), 83-88.
3. Lipps P.E. and Herr L.J.: *Etiology of Rhizoctonia cerealis in sharp eyespot of wheat*, Phytopathology 72 (1982), 1574-1577.
4. Lucas P., Smiley R.W. and Collins H.P.: *Decline of Rhizoctonia root rot on wheat in soils infested with Rhizoctonia solani AG-8*, Phytopathology 83 (1993), 260-265.
5. Ogoshi A., Oniki M., Araki T. and Ui T.: *Anastomosis groups of binulceate Rhizoctonia in Japan and North America and their perfect states*, Trans. Mycol. Soc. Jpn. 24 (1983), 79-87.
6. Ogoshi A., Oniki M., Sakai R. and Ui T, *Anastomosis grouping among isolates of binucleate Rhizoctonia*, Trans. Mycol. Soc. Jpn. 20 (1979), 33-79.
7. Ploetz R.C., Mitchell D.J. and Gallaher R.N.: *Charcterization of pathogenicity of Rhizoctonia species from a reduced - tillage experiment multicropped to rye and soybean in Florida*, Phytopathology 75 (1985), 833-839.

Serological Assays

THE PRESENCE OF *SPONGOSPORA SUBTERRANEA* F.SP. *SUBTERRANEA* IN THE NORTHERN AREAS OF PAKISTAN CONFIRMED BY MICROSCOPY, SEROLOGY AND BIOASSAY

I. Ahmad, S. Iftikhar[1] and U. Merz[2]
[1]*Crop Disease Research Institute, PARC, Islamabad, Pakistan,* [2]*Institute of Plant Sciences, Pathology Group, Swiss Federal Institute of Technology, CH-8092 Zurich, Switzerland*

During a survey in 1994 in the Skardu region of Northern Areas of Pakistan, potato plants were found in Astak valley with root galls and tuber lesions similar to those caused by *Spongospora subterranea* f.sp. *subterranea* and these were analysed further. Four months after replanting one plant again showed root galls on new roots and 'cauliflower' stage lesions on young tubers, typical of powdery scab. Microscopic preparations of root galls showed the presence of the characteristic spongy-like spore balls. The soil gave a high root infection score for the presence of zoosporangia in a bioassay test. It also reacted very strongly in ELISA. Preparations from root galls and the `cauliflower' stage lesions also gave high reactions in ELISA.

1. Introduction

Powdery scab was reported for the first time in Pakistan in 1986 by Turkensteen (1987). He found tubers with what he called 'deep-pitted' lesions and confirmed the presence of spore balls of *S. subterranea* by microscopic observations. Surveys of potato crops in the subsequent years in different regions showed that this type of lesion was much more common than previously expected but there were increasing doubts as to the true identity of the causal agent of these lesions. During a survey of potato diseases in the Skardu region of Northern Areas of Pakistan in 1994, potato plants were found in a field in Astak valley showing root galls and tuber symptoms typical for powdery scab. The plants together with soil probes from root zones were taken to CDRI, replanted and investigated further using visual methods, bioassay and serology.

2. Materials and Methods

2.1. CULTIVATION OF PLANTS

Three plants were replanted in the original soil in pots and cultivated in a growth room

H.-W. Dehne et al. (eds.), Diagnosis and Identification of Plant Pathogens, 371–373.

at 20°C with a 16 hour light (10,000 lux) / 8 hour dark regime. The plants were watered when the soil started to dry out. After four month the plants were harvested and the roots and tubers washed for further examination.
A portion of original roots with their galls was dried and stored.

2.2. MICROSCOPY

Transverse sections of fresh galls (from new roots) were made, stained with cotton blue and examined under the light microscope at 40fold magnification.

2.3. BIOASSAY DETERMINATION OF THE LEVEL OF *S. SUBTERRANEA* INFESTATION OF SOIL

A modified tomato bait plant test system as described by Merz (1989) was used. The original soil together with two other Pakistani soils (from fields where tubers with 'deep-pitted' lesions were found) and two Swiss soils (highly infested with *S. subterranea* / uncontaminated) were incubated for 10 days and baited for one day.

2.4. ELISA

A polyclonal antiserum to the spore ball stage of *S. subterranea* (γ-globulin fraction) was used in plate-trapped antigen (PTA)-ELISA in all experiments. Soil samples and spore ball preparations were prepared as described by Walsh and Merz (these proceedings). The same soil samples tested in the bioassay were also tested by ELISA. Samples from fresh and dried root galls and tuber lesion scrapings (of the original plant) were prepared and compared with scrapings of 'deep-pitted' lesions and Swiss spore ball material (scrapings of fresh lesions and dried spore ball preparations).

3. Results

After a cultivation period of 4 months one plant had new roots with galls and lesions in the early 'cauliflower' stage on young tubers showing white, wartlike outgrowths of infected host cells. Parts of the galls were still white whereas others had already turned to a dark-brown colour. When the galls were examined under the light microscope at 40fold magnification, spore balls typical for *S. subterranea* were seen. Their size was in the normal range (40-80μm). Each consisted of numerous single spores forming the characteristic honeycomb-like structure with hollows.
Both dried and fresh galls gave the same A_{405} values as Swiss lesion scrapings and spore ball preparations in PTA-ELISA (Fig. 1). The scrapings of the 'cauliflower' stage lesions produced a weaker but still much higher reaction compared to scrapings of the 'deep-pitted' lesions. When the original Astak soil was baited with tomato plants a root infection score similar to that of a highly infested Swiss soil was obtained (Fig. 2). No infection was found in the bait plant roots of the other soils. In PTA-ELISA the A_{405} of the Astak soil was about half of the Swiss infested soil whereas the other soils gave only

a weak reaction (Fig. 2).

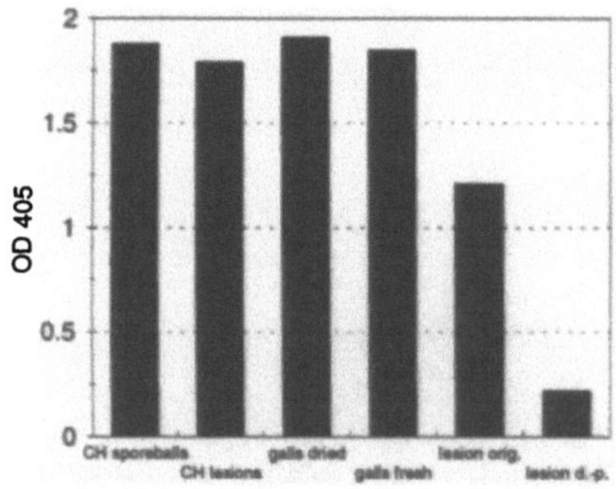

Fig. 1 A_{405} values of *S. subterranea* spore ball preparations and lesion scrapings from Switzerland, samples of galls and lesion scrapings from a Pakistani potato plant with powdery scab symptoms and scrapings of 'deep-pitted' lesions from Pakistani tubers in PTA-ELISA.

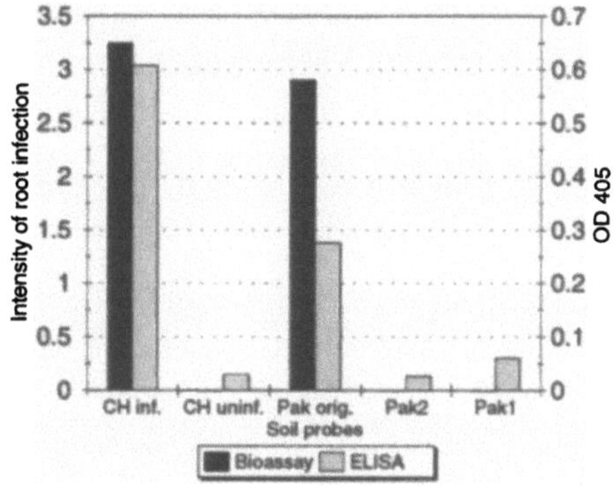

Fig. 2 Comparision between bioassay scores and A_{405} in PTA-ELISA of two Swiss soils (infested with *S. subterranea*/healthy), one Pakistani soil where potato plants with powdery scab symptoms have been found and two other Pakistani soils where tubers with 'deep pitted' lesions were found.

4. Discussion

Based on our observations and tests, we confirmed that the disease found in Astak was powdery scab caused by *Spongospora subterranea* f.sp. *subterranea*. The combination of tests (bioassay/ELISA) and microscopical observation is uniquely characteristic for the identification of this pathogen. This is the first substantiated report of *S. subterranea* in Pakistan.

Both bioassay scores and A_{405} values of 'deep-pitted' lesions and soils where tubers with this type of lesion have been found did not gave any indication for the presence of *S. subterranea*. Further investigation is therefore needed to identify the organism reported to cause powdery scab-like or 'deep-pitted' lesions on tubers.

5. References

Merz, U. (1989) Infectivity, inoculum density and germination of *Spongospora subterranea* resting spores: a solution-culture test system, *EPPO Bulletin* **19**, 585-592.

Turkensteen, L.J. (1987) Survey on bacterial and fungal diseases of potato in Punjab Province and Pappi area of NWFP, Dec. 1986, PSPDP, PARC, Islamabad, Pakistan, pp 24.

DIFFERENTIATION AND IMMUNOCHEMICAL CHARACTERISATION OF VARIOUS POTATO VIRUSES AND THEIR STRAINS BY MEANS OF MONOCLONAL ANTIBODIES

ČEŘOVSKÁ NOEMI, MORAVEC TOMÁŠ, DĚDIČ PETR AND [2]ŠUBR ZDENO
Institute of Experimental Botany, Czech Academy of Sciences,
Na Karlovce 1a, 160 00 Prague, Czech Republic,
[1]Potato Research and Breeding Institute,
Havlíčkův Brod, Czech Republic,
[2] Institute of Virology, Slovak Academy of Sciences,
Dubravská cesta 9, 842 46 Bratislava, Slovak Republic

Monoclonal antibodies (MAbs) produced in our laboratory against various potato viruses were used for detailed characterisation of immunochemical variability and, antigenic properties of potato virus A (PVA), potato virus S (PVS) and potato virus Y (PVY).

PVS (genus Carlavirus) has been divided into two main strains: PVS^0 (ordinary strain) and PVS^A (Andean strain). Both strains are serologically indistinguishable by polyclonal antibodies. The possibilities of detection of PVS^A by monoclonal antibodies were investigated.

Four MAbs were selected for limiting dilution cloning and further study from 14 hybridomas secreting antibodies to PVS^A obtained in the fusion experiment. MAbs were selected by indirect double antibody sandwich ELISA (IDAS-ELISA) with non-purified PVS^A. All four MAbs were of the IgG_{2a} isotype. Each of these MAbs reacted strongly with homologous isolate of PVS^A and detected three of the PVS^A isolates obtained from Germany. MAbs I and IV did not react detectably with five different isolates of PVS^0 (2 German and 3 Czech isolates) and no significant reactions were observed with six other carlaviruses. On the other side MAbs II and III were not strain specific. Reactivity of individual MAbs was compared in various types of ELISA. Two MAbs readily detected purified PVS^A particles in double antibody sandwich ELISA formats when used as coating antibodies or when conjugated to enzyme. Both

H.-W. Dehne et al. (eds.), Diagnosis and Identification of Plant Pathogens, 375–376.

these MAbs detected PVS^A in plate-trapped antigen (PTA) ELISA. Other MAbs reacted weakly in DAS ELISA formats and they did not react in PTA ELISA. The panel of four MAbs reacted with at least two different epitopes on PVS^A particles or polypeptide subunits. Epitope discrimination and topological relationship was checked in a series of competitive binding tests. Immunoblots of proteolytically cleaved PVS^A capsid protein fitted close proximity of epitopes recognised by MAbs I and IV.

PVY and PVA are both the members of potyvirus group. The antigenic properties of six MAbs to PVA and their reactivity were studied. The results of competitive binding tests combined with the reaction patterns of individual MAbs to different potyviruses indicate that the MAbs are specific for six distinct epitopes. The type of ELISA as well as the methods of MAbs purification played an essential role in selecting MAbs for routine diagnosis. PVA capsid protein migrated as two bands in SDS-PAGE. In immunoblot two of our MAbs reacted with the slower band, one MAb with the faster one and the other MAbs did not react.

The MAbs prepared to PVY^{NN} have been tested against different strains and isolates of this virus. The further investigation of immunochemical properties of these MAbs are continued.

Some of selected hybridoma lines produced MAbs with narrow (sharp) specificity, i.e. they were strain or isolate specific (e.g. PVS). On the contrary other hybridomas produced MAbs with sufficient broad activity, suitable for large scale utilisation.

IN*SIG*HT® - A NEW DIAGNOSTIC TOOL FOR *MYCOSPHAERELLA SPP.* IN BANANA LEAVES

L. ETIENNE, C. STEDEN & J. SUTER
CIBA-GEIGY Ltd., Division Crop Protection, CH 4002 Basel, Switzerland

1. Abstract

A specific enzyme-linked immunosorbent assay (ELISA) based on polyclonal antibodies has been developed to detect *Mycosphaerella spp.*, the causal agent of both yellow and black Sigatoka leaf diseases infecting banana. The quantitative immuno detection correlated well with the visual disease rating according to the traditional French system. Yet the immuno diagnostics method allows the quantitative detection of the very early stages of infection on the first leaf, whereas the visual assessment is only suitable to assess clear disease symptoms in later stages. Furthermore, the disease has been quantified several weeks before symptom occurrence, which shows the high potential of the IN*SIG*HT® assay for improved Sigatoka management.

2. Introduction

Originally the banana leaf-spot disease yellow Sigatoka, caused by *Mycosphaerella musicola* (anamorph *Pseudocercospora musae*) was first recognized in the Sigatoka valley in Fiji 1912. A second form of the disease named black Sigatoka, causal agent *Mycosphaerella fijiensis var. difformis* (anamorph *Paracercospora fijiensis*) was first observed in Fiji in 1964 (Meredith & Lawrence, 1969). Both forms of the disease are of economic importance and can cause important defoliation resulting in uneven and premature ripening of the fruit. However, black Sigatoka is generally considered to be far more damaging to the banana crops than yellow Sigatoka due to a more rapid disease development. The inoculum of both diseases is often present at high levels in banana plantations during the whole year.

For the control of black Sigatoka disease, traditionally the symptom severity is visually rated on several leaf levels e.g. according to the French system (Fouré, 1982). This visual method, which requires a high level of experience, is well suited for the assessment of advanced disease symptoms on leaf 2 and older leaves. We therefore decided to develop an easy, rapid and quantitative assay to measure the pathogen in early stages for routine disease monitoring and to optimize timing of fungicide

H.-W. Dehne et al. (eds.), Diagnosis and Identification of Plant Pathogens, 377–383.

applications. Since 1980, several different types of immunoassays have been developed to detect and quantify many plant pathogens. This includes diagnostics of plant pathogenic fungi (Miller *et al.*, 1988, Harrison *et al.*, 1990, Mittermeier et *al.*, 1993, Spire, 1995).

In this paper, we report on an enzyme-linked immunosorbent assay (ELISA) based on polyclonal antibodies (Pabs), which is suitable for quantitative and pre-symptomatic diagnosis of Sigatoka diseases.

3. Materials and Methods

3.1. POLYCLONAL ANTISERUM PREPARATION

A sheep was immunized using 1 ml of a *Mycosphaerella fijiensis var. difformis* mycelium extract grown in liquid culture (Phytone-Dextrose Broth) for 7 days, which was homogenized prior to injection using a Dyno-mill glass bead grinder. For each injection 1 ml of fungal extract was mixed in an emulsion with 1 ml of Freund's complete adjuvant. In total the sheep received 6 boost injections in total, at 1 month intervals.

3.2. ELISA

A double antibody sandwich ELISA assay (DAS-ELISA, Clark & Adams, 1977) has been developed based on immunoglobulins G (IgG) purified by affinity (Pharmacia C) and checked for cross reactivity against several fungi which are frequently present in banana plantations (Table 2). 50 μg IgG's are currently used for 96 multiwell plate coating. IgG's were conjugated with horseradish peroxidase (Boehringer/Mannheim).

3.3. PLANT MATERIAL AND VISUAL RATING

Leaf samples were collected from several banana plantations in Central-South America. Healthy and Sigatoka diseased samples from several leaf levels were collected in fungicide treated and untreated plots respectively. Disease symptoms were rated according to the French standard method, which classifies symptom severity according to the scale displayed in table 1.

In practice the leaf tip part is removed from the plant, and the disease severity is rated by observing the left-hand side tip leaf part viewed from the pseudostem. In addition each symptom stage is subdivided in 2 classes, -) when less than 50 symptoms of the dominating stage are present and +) representing more than 50 symptoms of the dominating stage.

TABLE 1. French classification for Sigatoka symptoms rating (Fouré, 1982)

Stage/ class	Symptom description
0	no symptom
1	first chlorotic points (only visible on lower leaf side), turning to a brown color
2	streaks with diffuse margins with reddish brown color
3	elongated streaks visible on both leaf sides, with dark brown to black color
4	streaks coalescing turning to elliptical spot lesion, with light brown border
5	stage 4 symptoms with depressed lesion center
6	numerous spots with dried out center, leaf collapsing

3.4. SAMPLE PREPARATION PROCEDURES

In the plantation a piece of of the leaf (5 X 15 cm) was removed from the left side (tip part) and placed in a plastic bag together with moist cheese cloth. In the laboratory, samples were weighed and extraction buffer (40 mM tris, 30 mM HCL, 150 mM NaCl, 0.1 % BSA, 0.02 % Nan3, at pH 7.7) added in a ratio (1 g / 2 ml, wt/v), then macerated with an Homex 6 (Bioreba) homogenizer. Extracts were loaded into a 96 vial transfer-plate.

3.5. TEST PROCEDURE

Each plant extract was tested in two replicates (100 µl/ well) on a Pabs pre-coated multiwell plate and incubated at 24 °C on an orbital shaker (MTS 2 at 900 rpm) for 10 minutes. Unspecific proteins were removed by washing the plate 5 times with a buffer containing 20 mM tris, 150 mM NaCl, 0.5 % Tween 80, 0.01% thimerosal, pH 7.9. Then the conjugate (Pab/ horseradish peroxidase, 100 µl/ well) was loaded and incubated on the shaker for 10 minutes. After washing, 3, 3', 5, 5', tetramethyl-benzidine substrate (TMB) was added (100 µl/ well) and incubated on the shaker for 10 minutes. Finally a stop solution of 1.5 % sodium fluoride was added (50 µl/ well) and shaken for 15 seconds. Substrate hydrolysis was followed by optical density (OD) measurements at 650 nm using a Biotek 312e plate reader linked to a computer interfaced with Kineticalc software which allows automatic data calculation. The plates were blanked against substrate buffer (OD: 650 nm $\leq$ 0.090).

4. Results & Discussion

Both *Mycoshaerella fijiensis* and *musicola* were well recognized by the IgG's even though some other fungi also reacted positively (Table 2). As high cross-reactivity occured with *Cladosporium herbarum,* we decided to cross-absorb IgG's prior to the enzyme-conjugation step. This resulted in lower detection of *Cladosporium herbarum* without affecting the specificity for the detection of *Mycosphaerella* antigen.

TABLE 2. ELISA results for polyclonal antibodies against various fungi that may be present in banana plantations (IgG concentration: 5.0 µg/ml and 0.23 µg/ml for coating and conjugate, respectively, ATCC = American Type Culture Collection).

Fungal species	Source	optical density (OD) at 650 nm
Acremonium stomaticum	ATCC 32187	0.069
Cercospora hayi	ATCC 28246	0.514
Cladosporium herbarum [a]	ATCC 28987	1.345
Cladosporium musae	ATCC 36952	0.209
Colletotrichum musae	ATCC 44422	0.144
Curvularia lunata	ATCC 14595	0.105
Deightoniella torulosa	ATCC 14090	0.130
Diplodia gossypina	ATCC 16391	0.513
Fusarium oxysporum cubense 1	ATCC 34661	0.042
Fusarium oxysporum cubense 4	ATCC 38741	0.041
Fusarium pallidorosum	ATCC 32605	0.047
Fusarium roseum	ATCC 26100	0.164
Guignardia musae	ATCC 28563	0.783
Nigrospora sphaerica	ATCC 32606	0.329
Pyricularia grisea	ATCC 64557	0.085
Verticillium thebromae	ATCC 14197	0.159
Mycosphaerella fijiensis	ATCC 36055	3.000
Mycosphaerella fijiensis var. difformis	ATCC 36054	2.918
Mycosphaerella musicola	ATCC 22115	1.640
Mycosphaerella musicola	ATCC 36143	1.782
Pseudocercospora musae [b]	ATCC 12565	1.674

[a] Isolate used for futher cross-absorbtion; [b] anamorph of *Mycosphaerella musicola*

4.1 COMPARISON OF DIAGNOSTICS KIT READINGS VERSUS VISUAL ASSESSMENT

The comparison in Fig. 1 shows clearly that the diagnostics method detects *Mycosphaerella spp.* in visually non infected leaf tips. Samples with slight and unclear visual symptoms at stage -1 were tested and showed high readings in the immunoassay. These results suggest that the ELISA assay correlates well with visual assessment and that it can detect the disease quantitatively. For baseline determination in healthy leaf tissue by the diagnostics kit, samples of the leaf tips (earliest unrolled part) were taken. The background of the ELISA test was defined (Cut-off: OD = 0.35).

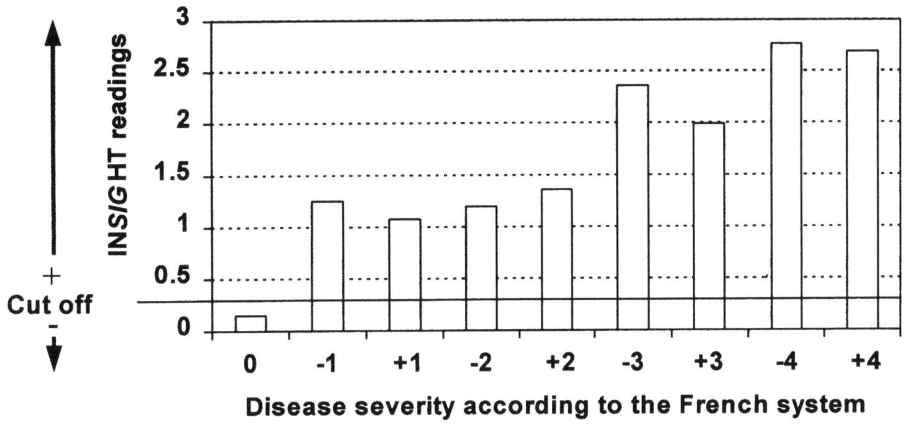

FIGURE 1: Correlation between ELISA readings at A 650 nm and visual assessment in various leaf levels of banana leaves. Each bar represents the average of 4 leaf samples (for description of visual ratings see Tab. 1).

4.2 ASSAY USE FOR PRE-SYMPTOMATIC DETECTION

The experiment shown in Fig. 2 was carried out by collecting samples from an area with low Sigatoka disease pressure following a dry period oftwo months. The presence of Sigatoka was monitored visually as well as by using the ELISA on all leaf levels which were available on the banana plant. The low level of Sigatoka infection was confirmed by both methods. However the disease could be detected positively from leaf 4 onwards by the kit, whereas it was possible to assess the first symptoms visually only on leaf 10 and 11. From these results it can be concluded that the disease was quantified by the assay at a presymptomatic stage of the disease infection, when mycelium was developing within the leaf tissue.

4.3 DISEASE CONTROL EFFICIENCY

The objective of the experiment presented in Figure 3 was to check the disease development of Sigatoka in a plantation treated with fungicides in comparison with an untreated plantation located in the same area. Leaf levels 1 to 8 were sampled in both plantations and then tested. As there was no influence of fungicide application on the disease in the untreated check, the amount of pathogen dramatically increased from the top to the bottom part of the plant (L 1 - L 8). This was confirmed both visually (data not shown) and by ELISA measurements. The lower curve in the figure shows the disease development in presence of fungicide sprays. In addition the upper curve represents the natural disease development under specific local conditions (e.g. climate). From the results presented we conclude that the differences in immunoassay measurements between the treated and untreated plantation can be used to estimate the efficacy of the fungicide program which was used to control the progression of Sigatoka.

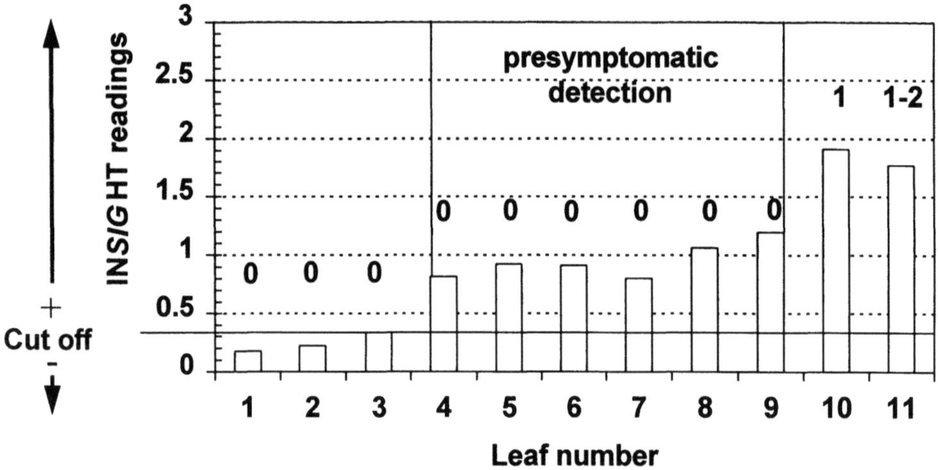

Figure 2: Comparison of ELISA measurements and visual disease ratings on all leaf levels of a system (see Table 1). Single banana plant (numbers above columns represent visual ratings according to the French system.

5. Conclusions

The results reported here show that a rapid and sensitive immunoassay, specific for *Mycosphaerella spp.* has been developed which is suitable for quantitative detection of Sigatoka disease in banana leaves. Polyclonal antibodies in the ELISA kit were able to detect low levels of infection in plant tissue with high accuracy, even when no visual symptoms could be observed. This indicates a potential for pre-symptomatic detection. The IN*SIG*HT® assay gives an opportunity for monitoring of pre-symptomatic infection levels of *Mycosphaerella spp.* in banana plantations, as well as providing an estimate of the disease pressure. Therefore it can be a useful tool for taking judicious decisions for the choice and timing of fungicides on the basis of Integrated Crop Protection of Sigatoka diseases in bananas.

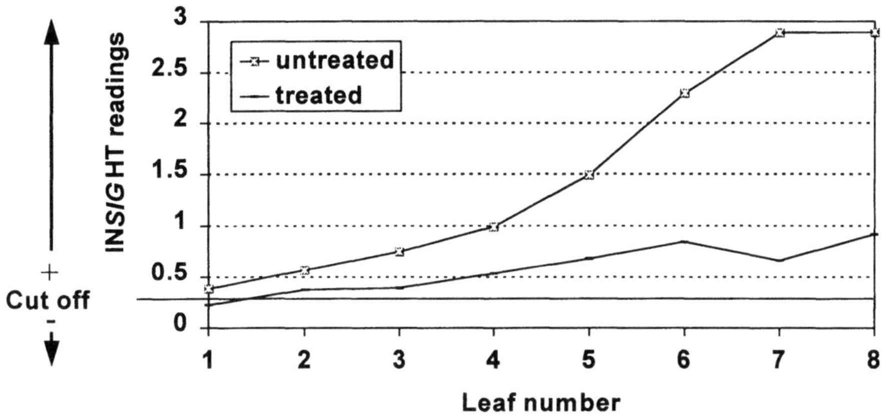

Figure 3: Sigatoka disease monitoring under low disease pressure using INS*I*GHT®. Treatments were carried out according to commercial plantation management in C. America

6. References

Clark, M F & Adams, A N (1977) Characteristics of the microplate method of enzyme-linked immunosorbent assay for detection of plant viruses. *Journal of General Virology* **34**, 475-483.

Fouré, E (1982) Les cercosporioses du bananier et leur traitements. *Fruit.* vol. **37**, 739-771.

Harrison, J G; Barker, H; Lowe, R & Ress, E A (1990) Estimation of amounts of *Phytophthora infestans* mycelium in leaf tissue by enzyme-linked immunosorbent assay. *Plant Pathology* **39**, 274-277.

Meredith, D S & Lawrence J S (1969) Black leaf streak disease of bananas (*Myco-sphaerella fijiensis*): symptoms of disease in Hawaii, and notes on the conidial state of the causal fungus. *Transactions British Mycological Society.* **52 (3)**, 459- 476.

Miller, S A; Rittenburg, J H; Petersen, F P & Grothaus, G D (1988) Application of rapid field, field-usable immunoassays for the diagnosis and monitoring of fungal pathogens in plants. *Brighton Crop Protection Conference - Pests and Diseases* **2**, 795-803.

Mittermeier, L; Dercks, W; West, S J E; Miller, S.A. (1990) Field results with a diagnostic system for the identification of *Septoria nodorum* and *Septoria tritici*. *Brighton Crop Protection Conference - Pests and Diseases* **2**, 757-762.

Spire, D (1995) Le diagnostic sérologique des champignons: quelques exemples pour une stratégie d'utilisation. Bulletin OEPP/EPPO Bulletin **25**, 19-29.

FEASIBILITY OF A DIAGNOSTIC KIT FOR MULTIPLEX DIG-CHEMILUMINESCENT DETECTION OF TOMATO VIRUSES

FINETTI SIALER M., BARBAROSSA L. AND GALLITELLI D.
Dipartimento di Protezione delle Piante dalle Malattie, Università degli Studi and Centro di Studio del C.N.R. sui Virus e le Virosi delle Colture Mediterranee, Via Amendola 165/A, 70126 Bari, Italy.

Abstract

Six specific DIG-labeled riboprobes have been synthesized to the following viruses: alfalfa mosaic alfamovirus (AMV), cucumber mosaic cucumovirus (CMV), potato Y potyvirus (PVY), tomato mosaic tobamovirus (ToMV), tomato spotted wilt tospovirus (TSWV), and tomato yellow leaf curl geminivirus (TYLCV). The probes were stored for several months as dried pellet at 4°C or in ethanol at -70°C and used separately or in mixture in nucleic acid hybridization tests. Samples were applied to Nylon membranes either as spots of crude plant extract prepared in 50 mM NaOH-2.5 mM EDTA or as tissue imprints made in the field on a dried, alkali-treated membrane. The different storage conditions did not affect the limit of sensitivity. The DIG-cRNA probes stored at 4°C were used also to detect virus infection in the imprints of different virus-host combinations, including woody and herbaceous plants. The dried pellet stored at 4 °C seems a suitable format for supplying ready-to-use DIG-labeled riboprobes in a kit which could also contain a number of ready-to-use alkali-treated Nylon membranes already spotted with control samples.

Introduction

Control of plant virus diseases capitalizes on availability of sensitive, reliable and rapid methods of diagnosis. Nucleic acid hybridization (NAH) technology has been successfully applied in this field as it permits to overcome some limitations of immunological techniques. Among the mixed-phase hybridization formats, dot blot represents the most suitable for routine testing because it allows highly specific and sensitive detection of sequences of unfractionated nucleic acid in crude plant sap applied directly onto solid supports. Despite this, however, NAH is still not as widespread in the routine testings as it could be. This is mainly due to: i) lack of standardization; ii) use of reporter systems (radioactive) that are incompatible with most of the diagnosis laboratories devoted to routine testings; iii) unavailability of reagents (probes) that, like antibodies, are ready-to-use.

A great effort is currently being made in adapting diagnostic techniques to field use by minimizing the manipulation of the collected samples. Ideally, the test itself should be done in the field but methods that overcome the transfer of samples to the laboratory, would already represent a great practical improvement. Squash blot, tissue print or whole leaf press are examples of such applications. Printed membranes prepared in the field, can then be mailed to a central laboratory equipped for the hybridization.

In this paper we give a summerized account of the work carried out in our laboratory for the preparation of a diagnostic kit for multiplex detection of the following viruses infecting tomato: alfalfa mosaic alfamovirus (AMV), cucumber mosaic cucumovirus (CMV), potato Y potyvirus (PVY), tomato mosaic tobamovirus (ToMV), tomato spotted wilt tospovirus (TSWV), and tomato yellow leaf curl geminivirus (TYLCV).

Materials and Methods

GENERATION OF DIGOXIGENIN-LABELED MINUS-SENSE RIBOPROBES

Six recombinant plasmids containing cloned cDNA sequences of AMV, CMV, PVY, ToMV, TSWV and TYLCV genomic RNA were used for digoxigenin-labeled RNA (DIG-cRNA) probe synthesis. To give minus-sense riboprobes, one microgram of the each plasmid was linearized

H.-W. Dehne et al. (eds.), Diagnosis and Identification of Plant Pathogens, 385–389.

by digestion with the appropriate restriction enzymes and *in vitro*-transcribed with the DIG RNA Labeling kit (Boehringer Mannheim, GmbH), following the manufacturer's instructions.

STORAGE CONDITIONS

The average yield of each transcription reaction was 10 μg of DIG-cRNA from 1 μg of linearized plasmid. The six DIG-cRNA probes were stored separately or in mixture under the following conditions: i) ethanol-precipitated dried pellet at 4°C; ii) in ethanol at -20°C; iii) in ethanol at -70°C. The probes were used in TE buffer (10 mM Tris, 1 mMEDTA, pH 7.4) containing 0.5% SDS and kept at -70°C.

NON-RADIOACTIVE DOT-BLOT HYBRIDIZATION ASSAY

Positively charged Nylon membranes (HYBOND N^+, Amersham, U.K.) with spots or tissue prints were pre-hybridized for 1h at 55°C in a hybridization mix (150μl/cm^2) containing 5X SSC (20X SSC = 3M sodium chloride, 300mM sodium citrate), 50% deionized formamide, 0.02% sodium dodecyl sulphate (SDS), 0.1% sodium lauryl sarcosine, 2% blocking reagent (Boehringer Mannheim, GmbH) and hybridized overnight with a DIG-labeled mixture of the six riboprobes, containing approximatively 100 ng/ml of each probe. Filters were washed three times for 30 min at 65°C in 0.1X SSC, 0.01% SDS, then incubated for 30 min at room temperature in 2X SSC containing 1μg/ml RNase A to remove aspecifically bound riboprobe, washed again in 2X SSC for 5 min and used directly for chemi-luminescent detection with the DIG luminescent detection kit (Boehringer Mannheim, GmbH) with the following modifications: (i) the blocking step was prolonged up to 60 min; (ii) anti-DIG IgG alkaline phosphatase conjugate was diluted to 1:5000 in the blocking solution and membranes were incubated for 60 min; (iii) blocking and antibody reaction steps were performed in plastic bags, using 250 μl solution/cm^2 of membrane. Damp membranes, sealed in transparent polyester sheets, were exposed to X-ray film for 1-2 h.

SAMPLE PREPARATION

Tomato samples were applied onto the Nylon membrane either as plant extract or as tissue imprint. In the first case, tissues of mechanically infected 'Rutgers' tomato plants were ground in a plastic bag with a roll press in the presence of 3 - 6 vol of 50 mM NaOH, 2.5 mM EDTA, pH 8.0. The extract was incubated at room temperature for 5 min and then 5 μl were spotted onto the membrane. For tissue imprinting, membranes were soaked in 50 mM NaOH, 2.5 mM EDTA, pH 8.0 then air-dried and stored at room temperature. Transversal sections of tomato stems and fruits were made as soon as they were collected from fields and imprinted directly onto the membrane.

Results

Using freshly prepared homologous probes in separate reactions, all six viruses were easily detected up to 17 μg of infected tissue/spot. In simultaneous hybridization. The limit of detection was represented by TSWV with a signal clearly visible up to 21 μg of infected tissue/spot while for the other five viruses the test retained the same sensitivity showed in separate reactions (i.e. 17 μg of infected tissue/spot). Heterologous hybridizations were not observed (not shown).
The different storage conditions (i.e. dried pellet at 4°C or in ethanol at -70°C) did not affect the limit of sensitivity. Fig. 1 (a, b) shows the results obtained with DIG-cRNA probes after four months storage. Comparable results were obtained with probes stored in ethanol at -20°C (not shown).
The DIG-cRNA probes stored at 4°C detected virus infection also in the imprints of different virus-host combinations, including herbaceous and woody plants (fig. 2, 3).

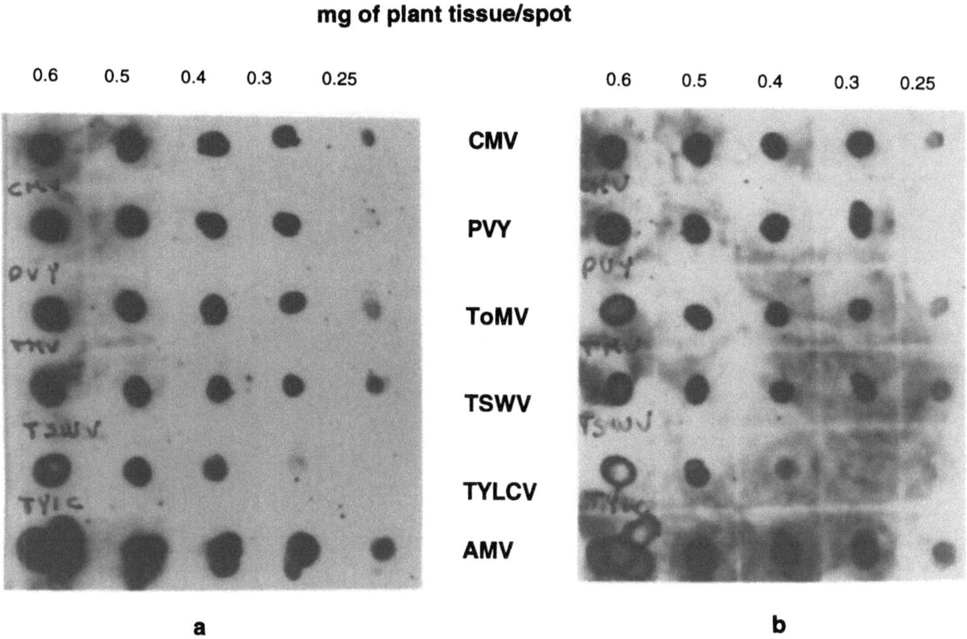

Figure 1. Chemiluminescent detection of six dot blot hybridization assays of AMV, CMV, PVY, ToMV, TSWV and TYLCV infected tomato plants with the homologous ribropobe stored for four months either at 4°C as dried pellet (panel a) or at -70°C in ethanol (panel b).

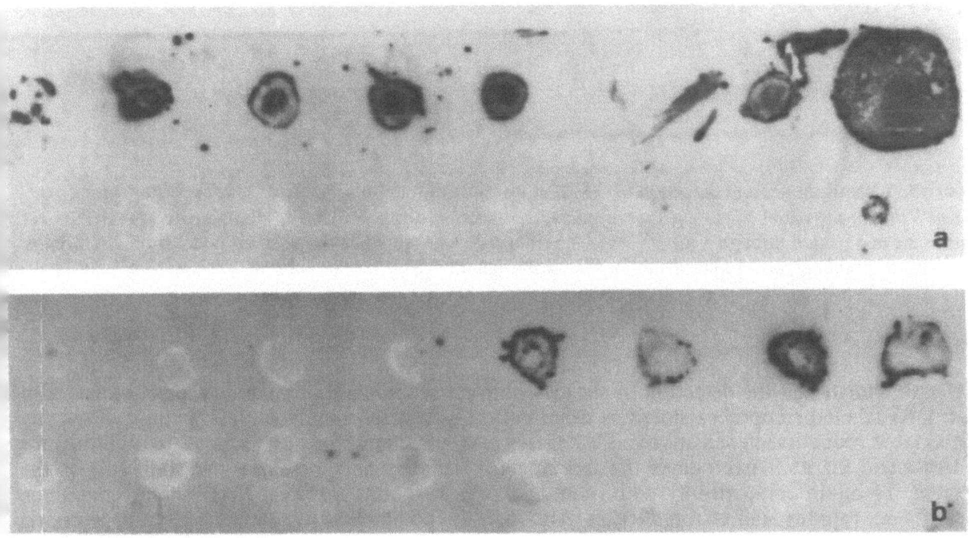

Figure 2. Chemiluminescent detection of CMV in stem and fruit sections of naturally infected tomato plants (panel a) and of PVY in stems of naturally infected potato plants (panel b). The spots were prepared in the field by imprinting freshly made cuts onto a dry, alkali-treated, Hybond N+ membrane. The riboprobe was stored at 4°C as dried pellet for eight months.

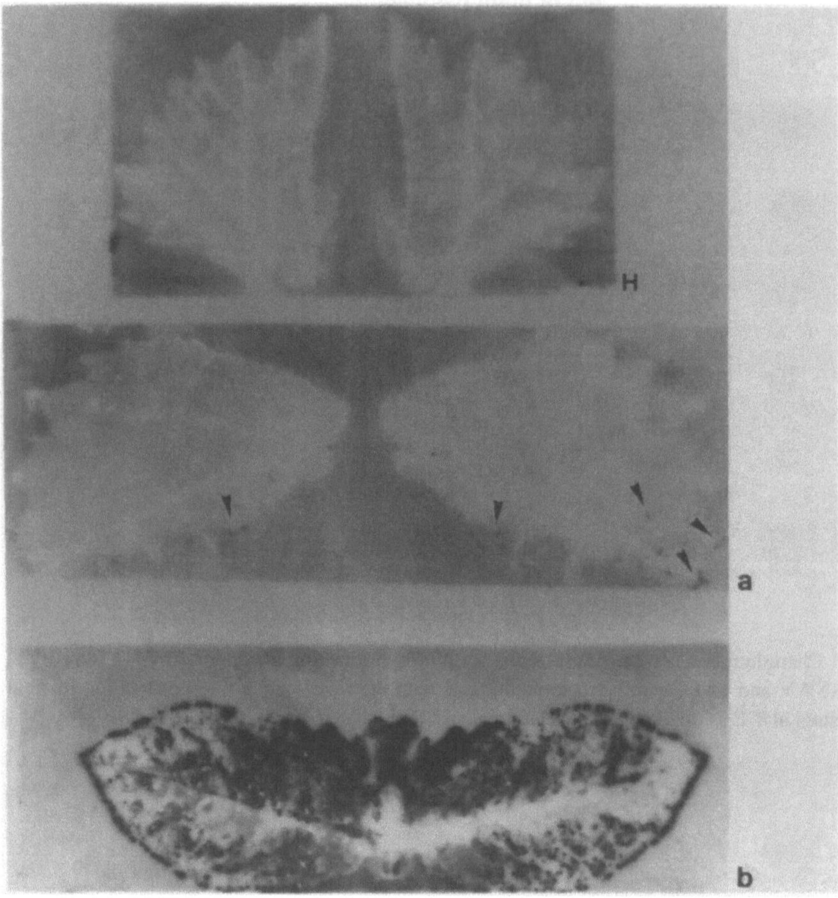

Figure 3. Chemiluminescent detection of AMV in imprints of naturally infected leaves of Grapevine (panel a) and *Viburnum* (panel b). Leaves were placed between two pieces of dry, alkali-treated HYBOND N^+ membrane and passed through a roller press. The riboprobe was stored at 4°C as dried pellet for eight months. Arrowheads point signals of hybridization. H = healty plant.

Discussion

We have shown that the detection of six different viruses infecting tomato can be achieved also with DIG-labeled riboprobes stored as dried pellet for several months at 4°C without observing appreciable reduction in sensitivity. This seems a very suitable form of storage, particularly for a diagnostic kit as it overcomes the freezing for delivery and increases the shelf-life of the reagent. To attain competitivity with immunoassays in routine testings, NAH should guarantee ease-of-use, rapidity and cost reduction. Although the commercially available kits for synthesis and labeling of the probe warrant sure results, their use is still reserved to specialized personnel quite skilled in molecular techniques. This represents a major constraint to application of NAH methodology to routine testings. Providing ready-to-use probes, the kit we propose here represents a good improvement.

Depending on the finality of the test, the DIG-labeled riboprobes can be used separately or in mixture. During a phytosanitary certification campaign for tomato viruses carried out in Apulia (southern Italy) (Saldarelli *et al,* 1996) the multiplex detection was applied. In this particular instance, there was no need to identify which of the six viruses was present since the standards fixed from Apulian Regional Extension service for Plant Protection for rejection of lots of tomato seedlings were 10% of the sample units tested regardless of the virus detected. During the phytosanitary campaign, four operators were able to collect and analyse 400 to 500 sample units/day and the possibility of detecting up to six viruses in a one-step assay allowed to carry out the complete protocol in three days, from sample collection to turning out of the results. The possibility to use a dried, alkali-treated membrane for direct tissue imprinting in the field seems to be a second improvement.

In addition to the dried DIG-labeled riboprobes, the kit could contain a number ready-to-use alkali-treated Nylon membranes already spotted with control samples (i.e. spots of sap of healthy plants with or without the plasmid DNA containing cDNA clones of the viruses). These membranes could be used either to apply aliquots of plant extract prepared in the laboratory or to imprint whole plant tissues. The imprint hybridization, probably represents the most suitable format for routine testings since it can be performed directly in the field thus overcoming the transfer of several samples to the laboratory. The small amount of plant sap coming out from the freshly cut surface is sufficient to solubilize the alkaline solution already present on the membrane that fixes the nucleic acid. Out of its rapidity, tissue print presents the following advantages: i) the viral nucleic acid sequences are more concentrated in the prints than in crude plant extract prepared in any extraction medium; ii) the interference of highly concentrated plant constituents that can be obtained, for example, with slot blot technique is greatly reduced. This method can be particularly useful for detection of quarantine viruses avoiding the introduction of harmful pathogens with the samples to be tested.

The results we have obtained show that a kit for nucleic acid-based detection of viruses infecting tomato is feasible and could represent a user-friendly, specific and sensitive assay, hopefully soon available in the agricultural market.

References

Saldarelli, P. Barbarossa L., Grieco, F. and Gallitelli D. (1996). Digoxigenin-labeled riboprobes applied to phytosanitary certification of tomato in Italy. *Plant Disease* **80** (in press).

Acknowledgements

This work was supported by a grant of Italian Ministry of Agriculture and Forestry, D.M. 346/7240/91 del 11/11/91 prog. 1 "Uso di tecniche sierologiche, molecolari e della PCR per la messa a punto di metodi diagnostici rapidi e precoci". Grateful thanks are also due to Agritest S.r.L., Valenzano (BA) Italy for the generous financial support.

DEVELOPMENT OF MONOCLONAL ANTIBODIES FOR THE DETECTION AND IDENTIFICATION OF *MONILINIA* SPP. CAUSING BROWN ROT OF STONE AND POME FRUIT.

BANKS, J.N, RIZVI, R.H, LANE, C.R, HUGHES, K.J.D, AND COOK, R.T.A.
Central Science Laboratory, MAFF, Sand Hutton, York, YO4 1LZ, UK.

1. Introduction

Monilinia brown rot is the most economically important disease of stone (*Prunus*) and pome (*Malus* and *Pyrus* spp.) fruit. It affects both ripe fruit (brown rot) before and after harvest and also blossom and twigs (blight) early in the season. However, it can result in latent infections, symptoms only developing after fruit ripening.

Three species of *Monilinia* are involved. *M. laxa* and *M. fructigena* occur widely within Europe, whilst the third, *M. fructicola* does not, being found mainly in North and South America, Australasia and South Africa. If *M. fructicola* was to become established in Europe, economic losses would be considerable. Outbreaks could not be easily eradicated because fungicide resistance is a problem in countries where *M. fructicola* occurs. *M. fructicola* is also considered a greater threat to EU growers than either *M. laxa* or *M. fructigena* as it sporulates more abundantly. It also seems more variable and adaptable which may be linked to the greater occurrence of the sexual stage and the increased potential for genetic diversity. Because of this serious phytosanitary risk to EU growers, *M. fructicola* has been listed as a quarantine organism.

Identification of *Monilinia* species often presents a problem. The necessary observations about cultural and morphological differences are slow and may be inconclusive, often relying on variable characters. Attempts have previously been made to improve these protocols, but they still require considerable time and expertise and none compare all three species directly. Inevitably, this has led to considerable difficulty in monitoring these diseases.

Immunological methods are being increasingly used for the detection and identification of fungi. However, most of these methods are laboratory based and are not suitable for rapid diagnosis at the point of inspection. There is an urgent need for the development of a user-friendly, rapid, reliable diagnostic kit for plant health inspectors, traders and growers.

H.-W. Dehne et al. (eds.), Diagnosis and Identification of Plant Pathogens, 391–393.

2. Production of monoclonal antibodies (MAb) for identification of *Monilinia* spp.

2.1 ANTIGEN PRODUCTION

Soluble fungal antigen preparations were prepared from shake cultures as described in Banks *et. al.* (1992) except that the liquid medium was Czapek Dox with 10% sucrose. These preparations were used for immunisation and cross-reactivity studies.

2.2 IMMUNISATION, MAb PRODUCTION AND INDIRECT ELISA

Soluble antigen preparations, derived from a typical culture of *M. fructicola* (CSL CC 778) were used for immunisation. Further details about protocols have been previously described (Banks *et. al.* 1994, Banks *et. al.* 1996). The resulting MAbs were then used in the indirect ELISA at a dilution of 1 : 2.

2.3 CROSS REACTIVITY TESTING ON SIX WEEK SERUM AND MAb

Blood samples were collected at 6 wk to monitor immune response and to select the spleens most likely to yield specific MAbs to *M. fructicola.* Antisera and MAbs were screened by indirect ELISA with soluble fungal extracts obtained from fungi likely to be present on fruit. These were categorised into three groups - *Monilinia* spp., related fungi (*Botrytis* and *Sclerotinia*) and unrelated fungi (*Alternaria, Cladosporium, Penicillium* and *Rhizopus*).

3. Results

Initial screening of the MAbs from three fusions has been completed. The cross-reactivity of 'typical' MAbs is given in Table 1.

Table 1. ELISA absorbance values of a selection of MAbs raised against *M. fructicola* to establish their cross-reactivity

	MAb number						
Fungus	**1**	**2**	**3**	**4**	**5**	**6**	**7**
M. fructicola	1.16	2.19	1.72	1.98	2.01	1.50	2.03
M. fructigena	0.43	1.85	0.17	0.28	0.35	0.29	0.56
M. laxa	0.79	0.49	0.17	0.27	0.68	0.46	0.19
S. sclerotiorum	0.30	0.21	1.34	0.14	0.23	0.26	0.29
B. cinerea	0.33	0.20	0.23	0.55	0.40	1.49	0.43
A. alternata	0.23	0.71	0.20	0.22	0.29	0.12	0.21
C. herbarum	0.25	0.36	0.11	0.16	0.21	0.18	0.29
P. expansum	0.24	0.10	0.23	0.25	0.31	0.27	0.16
R. stolonifer	0.24	0.10	0.17	0.16	0.14	0.12	0.39

All MAbs showed higher absorbance values (reactivity) with *M. fructicola* than with other fungi listed in Table 1. Three MAbs (4, 5 and 7) had very high values for *M. fructicola* (about 2.00) with comparatively low values for the other fungi (0.14-0.68). However, each showed a moderate cross-reactivity with one other fungus e.g. MAb 4 gave an absorbance of 0.55 for *B. cinerea*, MAb 5 against *M. laxa* (0.68), and MAb 7 against *M. fructigena* (0.56). With the other MAbs (1, 2, 3 and 6), apart from high absorbance values for *M. fructicola,* strong reactions also occurred with other heterologus fungi e.g. MAb 2 against *M. fructigena* (1.85), MAb 3 against *S. sclerotiorum* (1.34) and MAb 6 against *B. cinerea* (1.49) and moderate reactions with MAb 1 against *M. laxa* (0.79) and MAb 2 against *A. alternata* (0.71). During the intensive screening and selection stages in MAb production, some cell lines, that were discarded, also reacted with at least one of the species of *Alternaria*, *Cladosporium*, *Penicillium* or *Rhizopus* tested.

4. Discussion

Three MAbs, 4, 5 and 7, showed a high degree of specificity to *M. fructicola,* but each showed some cross-reactivity with another fungus. As the absorbances were relatively low for these heterologous fungi (0.55 to 0.68) compared to the homologous fungus (about 2.0) it may be possible to dilute out these unwanted cross-reactions. These initial studies also highlighted the importance of screening for cross-reactivity to related genera e.g. MAb 3 and 6 reacting to *Sclerotinia* and *Botrytis* and also taxonomically distinct fungi e.g. MAb 2 reacting to *A. alternata.* It was not surprising that with some of the MAbs (e.g. 1 and 2) there were appreciable cross-reactions with other *Monilinia* spp. It may be possible to use these MAbs in combination to detect *Monilinia* at the genus and/or species level.

In conclusion, initial results have shown that it is possible to develop MAbs to *Monilinia* spp. that cause brown rot primarily on stone and pome fruit. However, these require further screening to establish if they react with a wide range of *Monilinia* isolates. The next stage is to screen promising MAbs against known infected material. Previous experience in developing serological assays for diagnosis of fungi has shown the importance of using infected host material at an early stage in screening. This will also show whether there is likely to be a problem with inhibitors in host material. Further work may also be necessary to produce additional MAbs with greater specificity to *M. fructicola* and other *Monilinia* spp. in order to develop rapid diagnostic kits.

5. Acknowledgements

The authors wish to acknowledge the financial support of Plant Health Division, Ministry of Agriculture, Fisheries and Food and the EU (FAIR Project 1-0725). We also wish to thank the EU project co-ordinator Dr Luciana Corazza and partners, and Beverley Northway and Steve Freeman at CSL for their technical assistance.

6. References

Banks, J. N., Cox, S. J., Clarke, J. H., Shamsi, R. H. & Northway, B. J. (1992) Progress towards the immunological detection of field and storage fungi, in Modern Methods in Food Mycology (Samson, R. A., Hocking, A. D., Pitt, J. I. & King, A. D., eds.), Elsevier, Amsterdam, pp. 247-252.

Banks, J. N., Cox, S. J., Northway, B. J. & Rizvi, R. H. (1994) Monoclonal antibodies to fungi of significance to the quality of foods and feeds. Food and Agricultural Immunology. 6 (3) 321-327.

Banks, J. N., Barker, I., Turner, J. A., Northway, B. J., Rizvi, R. H. & Rahman, S. (1996) Immunological techniques for the identification of fungi with particular reference to *Fusarium* species. In proceedings of workshop on 'Fungal Identification Techniques', Barcelona 5-8th April 1995 (Rossen, L., Rubio, V., Dawson, M. T. & Frisvard, J. eds.) European Commission EUR 16510, Luxembourg. pp138-152.

DETERMINATION OF *CLAVIBACTER MICHIGANENSIS SUBSP. INSIDIOSUS* INFECTION IN LUCERNE SEEDS BY IMMUNOFLUORESCENCE STAINING AND DIRECT ISOLATION USING SELECTIVE MEDIUM

NÉMETH J., KOVÁCS A.N.
Plant Health and Soil Conservation Station
Bacteriological Laboratory, Pécs
7615 Pécs P.O.Box 13. Hungary

Introduction

The bacterial wilt pathogen *Clavibacter michiganensis subsp. insidiosus* is an EPPO A2 organism occurring in some countries of the EPPO region. There are several ways of transmission of the pathogen, but the infected seed is practically the only way to introduce the bacteria into areas free from the disease.

Hungary being free from the pathogen is interested in to keep its territory free from the pathogen. On the other hand seed importers require bacterial wilt pathogen free seeds, which have to be proven. For this purpose it is possible to carry out visual field inspection during the growing season, but recently the seed testing became accepted and widespread.

For the seed testing practice a reliable and sensitive method is required.

In the past decade a method including IF screening, subsequent isolation on nutrient media identification of he suspected isolates and finally artificial inoculation test of the host plant was used in the bacteriological laboratory of the Hungarian Plant Health Organization (Németh et al., 1991).

Based on experiences some modification of the applied assay was carried out in order to improve the recovery of bacteria by introducing selective media for dilution plating.

Materials and methods

The bacterial strains tested in the experiments were obtained from Research Institute of Crop Production, Praha-Ruzine and Research Institute for Plant Protection, Budapest.

Seed extracts were prepared from commercial lucerne seed lots. The seed extract were spiked with bacterial suspension of known CFU/ml concentration. King's B, GYCA and CNS base were used as nutrition media (Vidaver and Davis, 1988). Nalidixic acid (10, 25 mg/l), polymyxin B sulphate (100.000 U/l), cycloheximide (40, 200 mg/l) and nystatin (100, 200 mg/l) were used as components for selectivity.

H.-W. Dehne et al. (eds.), Diagnosis and Identification of Plant Pathogens, 395-397.

Results

From nutrition media used or proposed for isolation of *C.m insidiosus* King's B was found to give the highest recovery of bacterial wilt pathogen (100 %). Recovery of the studied strains on GYCA varied between 7 and 13 %, while on CNS base (without antibiotics) between 35-49 %. Because of the very poor and low recovery of the pathogens on GYCA and CNS subsequently, these media were not used in further experiments.
In order to improve the recovery of the pathogen on King's B, 3 g/l yeast extract was added. In contrary to the expected result recovery on yeast extract supplemented King's B decreased to 80-88 %.
Supplementing King's B medium with nalidixic acid + polymyxin B sulphate + cycloheximide (King's B PLUS), the recovery of *C.m insidiosus* strains decreased to 75 - 82 %, and resulted in 62-95 % selectivity. If King's B was supplemented with yeast extract + antibiotics, not only the recovery was lower, but selectivity decreased as well. The 40 ppm cycloheximide concentration was not always enough to suppress the growth of some fungi, increasing its concentration to 200 ppm satisfactory fungistatic effect was achieved but at the same time a certain decline in recovery of the pathogen was observed. To overcome this problem nystatin was tested, which had nearly the same negative effect on the recovery of *C.m insidiosus*, but a 100 ppm concentration served as a good compromise because of its satisfactory suppressing effect on fungi.

Discussion

The applied method for detection and identification of *C.m insidiosus* used in a past was based on immunofluorescnt staining of the suspension prepared from the seed samples, followed by isolation after dilution plating of the samples found positive in the IF screen. The King's B medium was found to be more suitable than the standard nutrient medium used before. Lucerne seed extracts however usually contain a high number of saprophytic fungal and especially bacterial microflora, which hinders the growth and recovery of the target pathogen on the King's B media. To overcome this problem antibiotics were added, to provide selectivity for the King's B, which assured higher recovery of the target pathogen than GYCA and CNS base. The low recovery of the pathogen on GYCA medium is not especially unexpected, some pathovars of *Xanthomonas campestris* and *Pseudomonas syringae* species react similarly on this medium (Németh, unpublished). CNS proposed for selective isolation of some *Clavibacter michiganensis* subspecies (Vidaver and Davis, 1988) gave considerable lower recovery of the bacterial wilt pathogen, than King's B.
King's B PLUS (King's B supplemented with 25 mg/l nalidixic acid, 100.000 U/l polymyxin B sulphate and 100 mg/l nystatin) under the experimental conditions studied, provided good recovery and selectivity. Adding yeast extract to King's B did not improve recovery of the target pathogen, but decreased the selectivity of the medium, probably due to the promotion of growth of some saprophytes.
Based on experiments and experiences, the following scheme is proposed for the testing of lucerne seed lots for *C.m insidiosus* infection:

Testing scheme for detection and identification of *Clavibacter michiganensis subsp. insidiosus* in lucerne seed

1. IF staining of seed extracts (from 5 x 2.000 seeds/sample) → negative → Sample considered free from *C. m. insidiosus*

↓ positive

2. Dilution plating, on King's B and King's B PLUS selective medium, selection of typical colonies

↓ positive

3. IF staining of bacteria, isolated → negative → Sample considered free from *C. m. insidiosus*

↓ positive

4. Plating on King's B, identification by morphological and biochemical tests → negative → Sample considered free from *C. m. insidiosus*

↓ positive

5. Pathogenicity test → negative → Sample considered free from *C. m. insidiosus*
   → **positive** → **Sample is infected by *C. m. insidiosus***

References

Németh, J., László, E. and Emődy, L. (1991) Clavibacter michiganensis ssp. insidiosus in lucerne seeds. *EPPO Bulletin* 21, 713-718.

Vidaver, A.K. and Davis, M.J. (1988) Coryneform plant pathogens, in Schaad, N.W. (ed.) *Laboratory Guide for Identification of plant pathogenic bacteria*. 2nd edition, APS Press, St. Paul, Minnesota

DETECTION OF RYEGRASS MOSAIC VIRUS BY MONOCLONAL ANTIBODIES AND IMMUNOCAPTURE PCR

F. RABENSTEIN & J. SCHUBERT
Federal Centre for Breeding Research on Cultivated Plants
Institute for Resistance Research and Pathogen Diagnostics
P. O. Box 1505, D-06435 Aschersleben, Germany

1. Introduction

Ryegrass mosaic virus (RGMV) is currently classified as the type member of the genus *Rymovirus* of the family *Potyviridae*. This genus contains not only RGMV but also four other members, namely Agropyron mosaic virus (AgMV), Hordeum mosaic virus (HoMV), oat necrotic mottle virus (ONMV), and wheat streak mosaic virus (WSMV), and the two tentative members, brome streak mosaic virus (BrSMV) and Spartina mottle virus (SpMV) (Murphy *et al.*, 1995).

Sensitive and specific detection methods are required especially for RGMV detection in its vector as well as for resistance evaluation of *Lolium* breeding material and transgenic plants. Because polyclonal antibodies (PABs) against purified RGMV from *Lolium* spp. may contain also antibodies to ryegrass cryptic virus, we produced mouse monoclonal antibodies (MABs) specific to RGMV. In this paper we compare the sensitivity of RGMV detection by different ELISA methods utilising these MABs, and that of immunocapture RT-PCR (IC-PCR).

2. Materials and Methods

2.1. VIRUS PURIFICATION AND ANTIBODIES

A Bulgarian isolate of RGMV (RGMV-B) was propagated in *L. multiflorum* and purified according to Schubert & Rabenstein (1995). PAPs and MABs against RGMV-B were produced and triple antibody sandwich ELISA (TAS-ELISA) was performed as already described (Rabenstein *et al.*, 1994). Several PABs were produced in rabbits against all members of the genus *Rymovirus* except for SpMV and used as coating antibodies in TAS-ELISA.

Two MABs (RGMV-4-E6 and RGMV-4-G12) were selected from a fusion experiment using RGMV-B stimulated mouse B lympocytes and the FOX-NP myeloma cell line. Both MABs were purified from tissue culture supernatants and labelled with alkaline phosphatase.

H.-W. Dehne et al. (eds.), Diagnosis and Identification of Plant Pathogens, 399-402.

The chromogenic substrate p-nitrophenyl phosphate (pNPP) was compared with the ELISA Amplification System (GIBCO) and the fluorogenic substrate AttoPhos (Merlin). Absorbance was measured at 405 nm for pNPP-TAS-ELISA and at 490 nm for the Amp-ELISA. Fluorescence was measured in a Fluoroscan II (Merlin) at excitation and emission wavelengths of 444 nm and 555 nm, respectively after substrate incubation of 15 min.

2.2 IMMUNOCAPTURE PCR (IC-PCR)

IC-PCR was performed in CovaLink plates (NUNC) using polyclonal IgGs for trapping of RGMV particles from infected plant material. The primer pair RG1 (5'-TTAACCAGGTGTACTCTTGGCTCTGA-3') (downstream) and RG2 (5'-AGACCAATCCAAAAGGATCAAGTG-3') (upstream) were derived from the viral NIa gene. For cDNA synthesis AMV-RT (USB) was used at 40 °C, for PCR ExTaq (Takara) with addition of RNase H. After 30 cycles 20 µl of the reaction mixture (100 µl) were used for gelelectrophoresis. The expected size of the insert is 800 bp.

3. Results and Discussion

In DAS-ELISA as well as in TAS-ELISA (using polyclonal rabbit IgGs as coating antibodies) MAB RGMV-4-G12 reacted strongly with all tested 10 RGMV isolates originating from different geographic regions. In TAS-ELISA this MAB also showed a weak reactivity with four isolates of WSMV. MAB RGMV-4-E6 only recognised 8 out of 10 RGMV isolates, but not the RGMV isolates from the Netherlands and the Czech Republic. This MAB did not react with five other members of the genus *Rymovirus* (AgMV, BrSMV, HoMV, ONMV, and WSMV) (data not shown).

ELISA results were confirmed by Western blotting experiments. Using prestained marker proteins and freshly prepared extracts from infected *L. multiflorum* the molecular weight of the RGMV coat protein (CP) was approximately 45 kD.

The estimated molecular weight of the RGMV CP is in accordance with our sequencing data (Schubert *et al.*, 1996a). The nucleotide sequence of the 3'-terminal genome part of a Danish and South African isolate of RGMV were determined. Two hitherto undescribed putative NIb/CP cleavage sites (Q/L or E/A) have recently been proposed for both RGMV isolates and a corresponding Q/M site in AgMV and HoMV (Schubert *et al.,* 1995, Salm *et al.*, 1996). If we assume E/A as cleavage site we can deduce a CP size of 44,5 kD for RGMV.

In contrast to PABs both MABs did not react with purified RGMV preparations stored for several month at 4 °C and with leaf extracts stored for more than 5 d at room temperature (Fig. 1b), indicating that the epitopes detected are only present in the intact CP of RGMV.

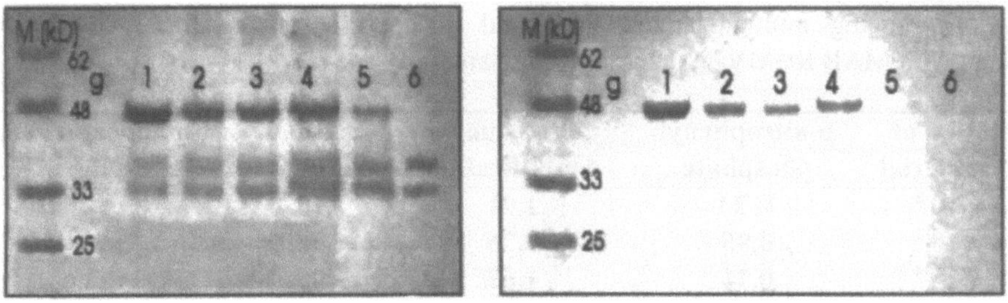

FIGURE 1: Western blot analysis of RGMV-Denmark after storage of *Lolium multiflorum* leaf sap at room temperature. Samples were taken after 1 to 6 days and CP was detected with polyclonal antiserum (left) or MAB RGMV-4E6 (right). M marker proteins, g healthy control.

The appearance of smaller proteins increased with the degradation of the intact capsid protein. These fragments were detectable with PABs only. CP degradation of rymoviruses seems to be a common phenomenon and has been described for WSMV and AgMV (Brakke *et al.*, 1990, Seifert, 1992). This phenomenon may explain the smaller CP size estimates for RGMV by Paliwal & Tremaine (1976) and Salm *et al.* (1994).

All RGMV isolates tested are serological unrelated to AgMV, BrSMV, and HoMV, but do react with polyclonal or monoclonal potyvirus group-specific antibodies (Richter *et al.*, 1995). On the other hand PABs to WSMV or ONMV cross-reacted in Western blots with RGMV from infected plants (E. Proll & F. Rabenstein unpublished). ONMV and WSMV showed a high degree of sequence identity to each other (Merits *et al.*, 1995) and were clearly distinct from AgMV and HoMV which also shared a high level of amino acid sequence identity at their C-terminal ends of the CP (Salm *et al.*, 1996).

The detection limit of TAS-ELISA using MAB RGMV-4-G12 as detecting antibody was increased further by the use of the enzyme amplification technique or the fluorogenic substrate AttoPhos. In infected *Lolium* plants RGMV was detected up to a dilution of 5×10^{-6} (Table 1).

4. References

Brakke, M. K., Scopp, R. N. & Lane, L. C. (1990) Degradation of wheat streak mosaic virus capsid protein during leaf senescence. *Phytopathology* **80**, 1401-1405.

Merits, A., Andrejeva, J., Kibe, K., Rabenstein, F., Proll, E. & Järvekülg, L. (1995) 3'terminal sequence analysis and homologies for two members of genus Rymovirus (Potyviridae). *Züchtungsforschung* **1**, Heft 2, 120-123.

Murphy, F. A., Fauquet, C. M., Bishop, D. H. L., Ghabrial, S. A., Jarvis, A. W., Martelli, G. P., Mayo, M. A. & Summers, M. D. (1995) "Virus Taxonomy", Sixth Report of the International Committee on Taxonomy of Viruses, Springer-Verlag Wien, New York, 1995, 586 pp.

Paliwal, Y. C. & Tremaine, J. H. (1976) Multiplication, purification and properties of ryegrass mosaic virus. *Phytopathology* **66**, 406-414.

TABLE 1. Comparison of detection end points for RGMV-Denmark in *Lolium multiflorum* using immunocapture PCR and different substrates for TAS-ELISA reaction with MAB RGMV-4G12 as detecting antibody.

Dilution of leaf material	p-nitrophenyl phosphate	Enzyme-Amplification	Atto-Phos	Immunocapture-PCR
5×10^{-3}	**1, 24**	**1,91**	**2,25**	+++
1×10^{-3}	**0,95**	**1,76**	**1,36**	+++
5×10^{-4}	**0,71**	**1,57**	**1,05**	+++
1×10^{-4}	**0,33**	**0,95**	**0,67**	+++
5×10^{-5}	**0,11**	**0,42**	**0,28**	+++
1×10^{-5}	0,08	**0,22**	**0,17**	++
5×10^{-6}	0,03	**0,14**	0,08	++
1×10^{-6}	0,03	0,07	0,05	+
5×10^{-7}	0,03	0,05	0,05	(+)
healthy	0,03	0,05	0,05	-

Rabenstein, F., Proll, E., Proeseler, G. & Lellbach, H. (1994) Entwicklung von Enzymimmunoassays zur Bewertung der Resistenz gegen das ryegrass mosaic virus. *Vorträge für Pflanzenzüchtung* **28**, 303-305.

Richter, J., Rabenstein, F., Proll, E. & Vetten, H.-J. (1995) Use of cross-reactive antibodies to detect members of the *Potyviridae*. *Journal of Phytopathology* **143**, 459-464.

Salm, S. N., Rey, M. E. C. & Wolfson, M. M. W. (1994) A South African isolate of ryegrass mosaic virus. *Plant Pathology* **43**, 708-712.

Salm, S. N., Rey, M. E. C., Robertson, N. L., French, R., Rabenstein, F. & Schubert, J. (1996) Molecular cloning and nucleotide sequencing of two members of the Rymovirus genus in the family Potyviridae. *Archives of Virology*. accepted.

Slykhuis, J. T. & Paliwal, Y. C. (1972) Ryegrass mosaic virus. C. M. I./A. A. B. Descriptions of Plant Viruses No 86, 4 pp.

DETECTION OF POPLAR MOSAIC CARLAVIRUS (POPMV) BY IMMUNO-CAPTURE RT-PCR

[1]R. Werner, [2]H.-P. Mühlbach and [1]C. Büttner
University of Hamburg
[1]Institute of Applied Botany, Dept. of Plant Protection, Marseiller Strasse 7, D-22305 Hamburg, [2]Institute of General Botany, Dept. of Genetics, Ohnhorststrasse 18, D-22609 Hamburg, Germany

1. Introduction

Poplar mosaic virus (PopMV) is a widespread disease of poplar trees (*Populus* sp.), where it can causing severe growth losses. Typical symptoms are a mosaic or diffuse spotting in mature leaves and, in severely responding clones, necrosis of veins and leaf stalks or splits on the stem (BIDDLE and TINSLEY, 1971). Viruses of forest trees are often difficult to detect due to their low virus concentration, their inhomogenous distribution in the plant and phenolic compounds in plant extracts. For reliable diagnosis several samples from each plant have to be tested either as single samples by ELISA or gathered as one sample by RT-PCR. Immuno-capture RT-PCR is a highly sensitive method for the detection of specific viruses. We applied this sensitive method in the detection of CLRV (BÜTTNER et al., 1996) and PopMV. Immuno-capture RT-PCR is sensitive enough to detect minute amounts of a virus in small samples e.g. CLRV in single birch seeds or bugs (*Kleidocerys resedae* PANZER), sampled from infected trees and in highly diluted plant extracts (WERNER et al., 1996). The advanced sensitivity of this diagnostic tool allows the combination of several samples in one charge for virus detection in woody plants. We show that the very inhomogenous distribution of PopMV in poplar leaves makes this procedure nessessary to obtain reliable diagnostic results.

2. Material and Methods

2.1 PLANT MATERIAL

Poplar mosaic virus infected leaf material was sampled in forested areas in Hamburg. Another PopMV isolate propagated in *Nicotiana glutinosa* as well as the antibodys against PopMV were kindly supplied by Dr. Hamacher, Bonn.

H.-W. Dehne et al. (eds.), Diagnosis and Identification of Plant Pathogens, 403–405.

2.2 IMMUNO-CAPTURE RT-PCR AND SEQUENCING

Immuno-capture was carried out as described for CLRV detection by Büttner et al. (1996). For PopMV detection P4 (5′-TGG CAT CTC AAT CAG CAA GTT A-3′) was choosen as the upper strand primer and P3 (5′-AGT CAA AAG TGT CAA AGG CGG C-3′) as the RT- and lower strand primer. Both primers match a fragment of the PopMV coat protein sequence. After washing the immunocaptured virions PopMV-RNA was reverse transcribed using 3U AMV reverse transcriptase and 100 pmol P3 in 20 µl of 50 mM Tris-HCl pH 8.5, containing 8 mM $MgCl_2$, 30 mM KCl, 1 mM DTT, 1 mM dNTP′s and 10 U RNase inhibitor for 1h at 55°C without further disruption of the virions. Reverse transcriptase was heat inactivated for 3 min at 95°C. Aliquots of 2 µl were amplified in a subsequent polymerase chain reaction in 10 mM Tris-HCl, pH 8.3, containing 1.5 mM $MgCl_2$, 50mM KCl, 20 pmol of each primer, 200 µM dNTP′s and 2.5 U Taq-DNA-Polymerase in a total volume of 100 µl. Viral cDNA was amplified in an Autogene II thermocycler (Grant) with following thermal cycling scheme for 35 cycles: primer annealing at 55.5°C for 60 s, chain elongation at 72°C for 60 s, denaturation at 95°C for 30 s. Amplification products were analysed on 2% (w/v) agarose gels and detected by ethidium bromide staining. For sequencing, PCR products were purified from remaining primer molecules with Jet Pure PCR product purification kit (Genomed). Purified amplicons were sequenced using one of the amplification primers and a cycle sequencing kit with dye-terminators. Sequences were analysed on an automated DNA-sequencer (Applied Biosystems).

3. Results and Discussion

PopMV often induces a very inhomogenous symptom expression. Figure 1 shows a poplar twig of a tree infected with PopMV. Some of the leaves showed very strong mosaic pattern, while others indicated no symptoms. We have gathered three symptomless leaves as one sample and three symptom developing leaves as another one. Samples were ground with sterile seasand in PBS-T-PVP (phosphate buffer, pH 7.4, 135 mM NaCl, 0.05% Tween20, 2% (w/v) Polyvinylpyrrolidone 40) in a ratio of 1:10 (w/v) and diluted in steps of 1:10 in PBS-T-PVP. PopMV was detected by IC-RT-

Figure 1: Heterogenous symptom expression on leaves of a poplar twig infected with PopMV.

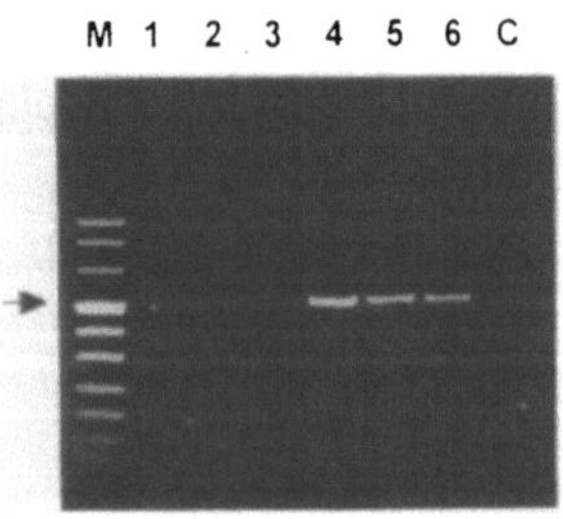

Figure 2: Detection of PopMV by IC-RT-PCR. Lanes **1-3**: symptomless leaves; **4-6**: symptom developing leaves. Plant sap dilutions: **1** and **4**: 10^{-1}; **2** and **5**: 10^{-2}; **3** and **6**: 10^{-3}; **M**: molecular weight marker; **C**: PCR-Control

PCR as described. As given in figure 2 only the leaves developing symptoms showed a clear viral fragment apparently of the expected size of 512 bp. To prevent unspecific priming and to avoid problems with secondary structure of the RNA, we used AMV-reverse transcriptase at an incubation temperature of 55°C. The PCR products of the amplified viral genome fragments were sequenced. Sequencing of two independent isolates revealed that the amplified fragments of the PopMV coat proteine gene consist of only 509 bp as compared to 512 bp of the sequence previously published by HENDERSON et al. (1992). Both fragments contained a double frame shift mutation, which results in an exchange of 33 amino acids within the coat protein (Fig. 3). Interestingly the coat protein turns back into the previous published reading frame two amino acids in front of the conserved signature pattern of the potex-and carlaviruses, which could be detected by the program PROSITE (accession no.:PS00418).

```
PopMV_HH1     MAFIEAEFQSLGVPKENLAKVMWTLTRYCVGASSSQYLDPKGEEEKLCGGVTRAALIACI
              ::::::::::::::::::::::::::::::::::::::::::::::::::::::::::::
PopMV_COAT    MAFIEAEFQSLGVPKENLAKVMWTLTRYCVGASSSQYLDPKGEEEKLCGGVTRAALIACI
                    150       160       170       180       190       200

PopMV_HH1     KKRSTL-RKVCRLYAPIVWNYMLVNNVPPEDWQSKGYTEETRFAAFDTFD
              ::::::                                 ::.::::::::
PopMV_COAT    KKRSTLSVKCADFMRPSCGITCWSTTFLQKIGSPRATLSETKFAAFDTFD
                210       220       230           240       250

                          consensus pattern: (R/K)FA(G/A)FDXFXXVXXXAA
```

Fig. 3: Comparison of the deduced amino acid sequence of a german PopMV isolate (PopMV_HH1) with the coat protein sequence published by Henderson et al. (1992)(PopMV_coat).

Since the nucleotide sequence of both German isolates of PopMV only show silent or conservative mutations, they may represent a separate strain of PopMV.

4. Literature

Biddle,P. G. and Tinsley, T.W. (1971). Poplar mosaic virus. C.I.M./A.A.B. Descriptions of Plant Viruses, No 75.

Büttner, C., Führling, M., Werner, R., Mühlbach, H.-P. and Lukàcs N. (1996) Plant viruses in deciduous trees of public gardens and nurseries as well as in soils and waters - a guide of diagnostic methods, *Gesunde Pflanzen* **48**, 95-103.

Henderson, J., Gibbs, M. J., Edwards, M. I., Clarke, V.A., Gardner, K. A. and Cooper, J.I. (1992) Partial nucleotide sequence of poplar mosaic virus RNA confirms its classification as a carlavirus, *J. Gen. Virol.* **73**,1887-1890.

Werner, R., Mühlbach, H.-P. and Büttner, C. (1996) Anwendung der PCR zum Nachweis von Cherry Leaf Roll Virus (CLRV) in *Betula pendula* ROTH, *Phytomedizin* **26**, 64.

The nucleotide sequence of the PopMV coat protein fragments are available in the EMBL database under the accession numbers X97683 and X97765.

The project was supported by BMBF grant no: 0317535A and NEW LAB Diagnostic Systems GMBH.

INDIRECT IMMUNOFLUORESCENCE - AN USEFUL METHOD IN STUDIES ON SOME FUNGAL PATHOGENS

M. WIWART*, Z. MIERZWA**
* *Department of Plant Breeding and Seed Production, University of Agriculture and Technology, 10-724 Olsztyn,*
***Potato Institute, Gdańsk, Poland*

Abstract
The research was aimed at evaluating the usefulness of indirect immunofluorescence test (IF) for an identification of fungi that are most often isolated from the seeds of legumes and cereals. Five polyclonal antisera (for 2 isolates of *Ascochyta fabae, Fusarium culmorum, F. poae* and *Microdochium nivale*) with antibody titres from 1:4000 to 1:12000 were obtained. The results of heterologous reactions indicate a high specificity of the obtained sera - no or only slight cross reactions with spores of different fungus species were observed.

1. Introduction

An immediate identification of fungal microorganisms in the tissues of host plants is of primary importance as far as both successful plant protection treatment and breeding experiments aimed at obtaining new genotypes resistant to pathogens are concerned. Immunological methods, including immunofluorescence (IF), create here certain possibilities. The technique in question, applied so far mainly to immunodiagnosis of bacterial pathogens, enables not only to make an analysis of serological relationships between examined organisms, but also to carry out their morphological observations, which makes it a useful research instrument (Stead 1992). The above mentioned method, used among others for an identification of *Corynebacterium sepedonicum* cells (Malec, Mierzwa 1986), may be also employed in the course of research on some fungal pathogens, as it was in the case of studies on *Phoma exiqua var. foveata* (Hingand et al. 1983, Malec et al. 1992) and *Fusarium avenaceum* (Mierzwa et al. 1996).

2. Material and Methods

Antigenic preparations applied to immunization were made of pycnospores of two *Ascochyta fabae* isolates (no. 1 and 2), which differed in the rate of growth and sporulation intensity, and of conidia: *Fusarium culmorum, F. poae* and *Microdochium nivale*. The fungal cultures were growing on Petri's dishes at the temperature of 25^0C, on PDA (bioMérieux) enriched with glucose in the amount of 10 g/l. After 10 days of incubation, the mycelia were drenched with sterile distilled water (15 ml/dish) and the spores were rinsed throughoutly. Suspensions whose concentration was equal to ca. 10^7 conidia/ml were obtained in this way. The preparations were then centrifuged

H.-W. Dehne et al. (eds.), Diagnosis and Identification of Plant Pathogens, 407–409.

(15min., 7000 n·min^{-1}, centrifuge Sorval RC 5B, rotor SS 34) and washed three times with buffer solution (PBS, pH=7.4). The washed spores were placed in an analogous solution, keeping the concentration of ca. 10^6 conidia/ml. Afterwards, when glutaric aldehyde was added to the suspension up to the concentration of 2%, it was left for 1 hour at the temp. of 4^{0}C and later on subjected to a dialysis against the solution in which the spores were placed. The dialysis was carried out for 24 hours and the buffer was changed three times. The prepared in this way suspension constituted an antigenic preparation that was used to immunize rabbits weighing ca. 3 kg each. Intramuscular injections were performed together with incomplete Freund's adiuvant. The first series included 6 injections (2 per week). 1 ml of antigen was administered each time. After 10 days, 3 more intramuscular and 1 intravenous injections were given. 4 days after the last immunization, blood was taken to analyses and the obtained serum was stored at the temp. of -25^{0}C.

The test of indirect immunofluorescence was conducted according to the method elaborated by Malec and Mierzwa (1986) and used for an identification of *Corynebacterium sepedonicum*. Goat antiserum conjugated with fluorescein was applied against rabbit IgG (Biomed, Poland). The observations were made using a microscope Nikon Labofot 2A equipped with a device to epifluorescence EFD-3, at the magnification of ca. 600*x*.

Titre denotation of serum antibodies was performed observing the results of homologous reactions at the following dilutions of sera: 1:200, 1:500, 1:1 000, 1:2 000, 1:2 500, 1:4 000, 1:8 000, 1:12 000, 1:18 000. Preparations in the case of which the serum of a non-immunized rabbit replaced the examined antiserum constituted a control combination. Heterologous reactions were observed in the combinations with spores of the following species: *Ascochyta fabae* (both isolates), *Fusarium culmorum, F. poae, Alternaria alternata, Epicoccum nigrum, Botrytis fabae*. The dilutions 1:200 and 1:500 were applied.

RESULTS

All the obtained antisera were characterized by a high antibody titre. It amounted to 1:4 000 for homologous reactions of *F. culmorum* and *F. poae*, and to 1:12000 for homologous reactions of both *A. fabae* isolates (Table 1, Figure 1).

TABLE 1. Reactions with the use of obtained antisera (dilutions 1:200). The bold numbers placed diagonally indicate the values of antiserum titres

Antierum / Spores	anti- *A. fabae* 1	anti- *A. fabae* 2	anti- *F.culmorum*	anti- *F.poae*	anti- *M.nivale*
Ascochyta fabae 1	**1:12 000**	+++	±	+	-
Ascochyta fabae 2	+++	**1:12 000**	-	-	-
Fusarium culmorum	-	-	**1:4 000**	-	-
Fusarium poae	+	+	++	**1:4 000**	±
Microdochium nivale	-	-	-	±	**1:8 000**
Alternaria alternata	-	-	-	-	±
Epicoccum nigrum	-	-		±	-
Botrytis fabae	-	-	-	-	-

fluorescence:+++ very strong; ++ moderate; + slight; ± very slight; - lack of fluorescence

Celadon blue fluorescence of conidia in homologous reactions was most intensive at the dilutions 1:200 and 1:500 (with no difference); it was becoming slighter and slighter with dilution growing. Heterologous reactions carried out for all of the obtained sera revealed their good specificity - no cross reactions were observed, except for slight fluorescence in the tests with *F. poae*, which disappeared at the dilution 1:2000. Fluorescence of *F. poae* spores may indicate that an inconsiderable amount of non-specific antigens could be present on their surface.

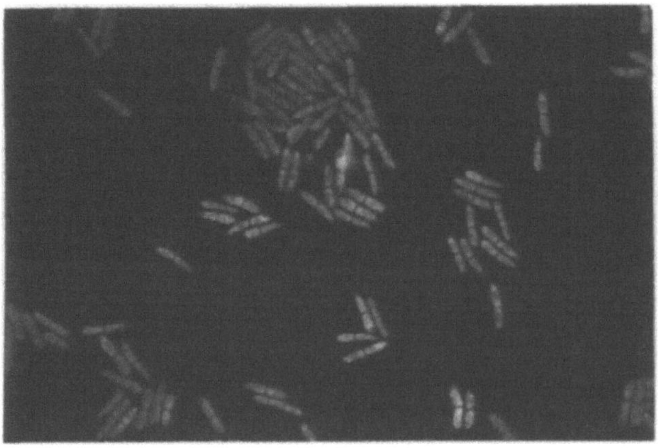

Figure 1. Homologous reaction of anti *Ascochyta fabae* 1 serum. Dilution 1:200, magnification ca. 600*x*

CONCLUSIONS

1. Obtained antisera manifested good specificity in relation to spores of studied fungi
2. Presented results indicate the usability of indirect immunofluorescence in the mycological investigations

REFERENCES

Hingand L., Le Coz S., Kerlan C. Jouan B. (1983). Application de l'immunofluorescence a'la detection de *Phoma exigua var. foveata,* agent de la gangrene de la pomme de terre. Agronomie **3/1**, 51-55

Malec K, Malinowska E, Mierzwa Z, Sas-Piotrowska B, Stefan K. (1992) Identyfikacja grzyba *Phoma exigua var. foveata* sprawcy gangreny ziemniaka, metodą immunofluorescencji pośredniej [Identification of fungus *Phoma exigua var. foveata,* the causing agent of potato gangrene, using indirect immunofluorescence] Biul. Instytutu Ziemniaka **41**, 101-106

Malec K, Mierzwa Z (1986) Metoda immunofluorescencji pośredniej i identyfikacja *Corynebacterium sepedonicum* (Spieck et Koth) Skaptason et Burkholder. Szczegóły postępowania [The method of indirect immunofluorescence and identification of *Corynebacterium sepedonicum* (Spieck et Koth) Skaptason et Burkholder. Details of procedure] Biul. Instytutu Ziemniaka **34**, 75-79

Mierzwa Z. Wiwart M. Fordoński G. (1996). Otrzymanie surowicy odpornościowej przeciwko zarodnikom konidialnym *Fusarum avenaceum.* [The production of antiserum against conidia of *Fusarium avenaceum*]. Biul Instytutu Hodowli i Aklimatyzacji Roślin **197**, 147-150

Stead D.E. (1986). Techniques for detecting and identifying plant pathogenic bacteria, in J.M Duncan and L Torrance (eds) *Techniques for the Rapid Detection of Plant Pathogens.* Blackwell Scientific Publications, pp 76-111

THE USE OF PHYSIOLOGICAL AND BIOCHEMICAL PARAMETERS OF PLANTS FOR THE ESTIMATION OF THE RESISTANCE TO STEM RUST

T.P. Yurina[1], E.V. Yurina[1], V.A. Karavaev[2] and M.K. Solntsev[2]
Faculty of Biology ([1]), Faculty of Physics ([2]), Moscow State University, Vorobjovy, Moscow, 119899, Russia

We have studied the effects of the antibiotic trichothecin *(Trichothecium* roseum*)* and the systemic fungicide triadimefon on the parasitic activity of *Puccinia graminis* and on metabolic processes of plants. Trichothecin is an antifungial antibiotic formed by *Trichothecium roseum*. This antibiotic is a ketone with empirical formula $C_{19}H_{24}O_5$. Triadimefon is a systemic fungicide (3,3-dimethil-1/1H-1,2,4,-trizolil/4-chlorophenoxi/buthanol-2). The experiments were performed on isogenic lines of wheat *Triticum aestivum* L. cv. Markiz with different resistance to stem rust. Two strains of wheat were studied: the resistant strain with gene Sr Tt and susceptible strain with gene Sr 9B. Physiological and biochemical tests were performed on the first leaves of wheat seedlings of uninfected and infected plants. The concentrations of trichothecin and triadimefon were selected on the basis of data of inhibitory effect on activity of *P.graminis* uredospores germinating in moist camera on the surface of 1% water. The solution concentration of trichothecin and triadimefon which strongly suppressed the uredospores germinating on water agar were chosen for the experiments with plants. The influence of the preparation on the degree of defeat was controlled by the number of pustules on leaves, which appeared om 8-10-th day after the invasion. The number of pustules was counted on 20-30 plants.

H.-W. Dehne et al. (eds.), Diagnosis and Identification of Plant Pathogens, 411-413.

It was found that the preparations have efficient fungitoxic influence at low concentrations. Trichothecin at concentration of 0,5 μg/ml decreased the uredospores germinating by 20-30 % and at concentration of 4μg/ml the germinating was suppressed completely. Triadimefon possessed the strong inhibitory effect on uredospores germinating at the concentration 0.02-0.05 % and completely stopped the germinating at 0.1 %. The spraying of wheat plants by trichothecin solution with concentration 2 μg/ml decreased the degree of defeat of susceptible line ten fold approximately as compared with test plants and at concentration of 4.0 μg/ml there was no defeat at all. Table 1 demonstrates the positive influence of trichothecin and triadimefon on protein content in wheat leaves. The degree of activation of protein in tissues of resistant strain is higher than in susceptible line. The analogous behaviour is observed for cataslitically active protein (peroxidase), giving the evidence of activation of oxidative processes, which play an important role in increasing the resistance to pathogens. It was found that triadimefon had lower activation ability than trichothecin (table 2).

Table 1. The influence of trichothecin and triadimefon on protein content in leaves of resistant and susceptible lines of wheat cv. Markiz.

Concentration	Protein		content	
	Resistant line Sr Tt μg/g of fresh mass, % of control		Susceptible line Sr 9B μg/g of fresh mass, % of controll	
		Trichothecin		
Control-water	16.3 ± 1.3	100	15.6 ± 1.2	100
0.01μg/ml	19.7 ± 1.0	120	18.0 ± 1.3	115
0.5	19.2 ± 1.2	118	17.1 ± 1.1	110
1.0	22.0 ± 0.9	134	17.8 ± 1.3	114
2.0	20.5 ± 1.5	125	17.3 ± 0.8	111
		Triadimefon		
Control-water	17.1 ± 1.0	100	16.3 ± 0.9	100
0.01%	18.8 ± 1.1	110	17.4 ± 1.0	107
0.03	19.5 ± 1.1	114	19.0 ± 1.1	116
0.05	21.0 ± 1.6	118	18.4 ± 1.2	113
0.10	18.5 ± 1.2	108	18.0 ± 1.0	110

Table 2. The changes of peroxidase activity of wheat plants under trichothecin and triademefon treatment

Concentration	Resistsant line Sr Tt		Susceptible line Sr 9B	
	arb. units	% of control	arb. units	%of control
		Trichothecin		
Control	11.6	100	9.4	100
0.2μg/ml	17.0	146	10.2	108
0.5	15.8	136	11.0	117
1.0	17.5	151	10.8	114
2.0	16.3	141	11.3	120
		Triadimefon		
Control	12.1	100	10.2	100
0.01%	13.2	108	10.7	105
0.03	14.0	116	11.7	115
0.05	14.4	119	11.6	114
0.10	13.3	111	11.1	109

The general influence of trichothecin and triadimefon on wheat leaves consists in essential increase of oxygen respiration intensity (table 3).

Table 3. The influence of trichothecin and triadimefon on the respiration intensity of wheat cv. Markiz.

Concentration	Resistant line Sr Tt		Susceptible line Sr 9B	
	μl O_2	% of control	μl O_2	% of control
		Trichothecin		
Control	325±16	100	273±15	100
0.2μg/ml	366±16	115	300±11	110
0.5	435±25	136	307±15	112
1.0	450±20	140	328±16	120
2,0	440±21	138	318±13	116
		Triadimefon		
Control	336±17	100	265±15	100
0.01%	364±18	108	281±11	106
0.03	393±16	117	303±13	114
0.05	406±20	121	292±14	110
0.10	380±19	113	295±12	111

The reported physiological and biochemical parameters with fast response to preparations with fungitoxic action give the efficient means for selection and screening of fungicides.

Nucleic Acid Based Hybridisation Techniques

DETECTION AND DIFFERENTIATION OF THREE POTYVIRUSES INFECTING SWEET POTATO BY PCR

D.COLINET[1], F. ZU XIA[2], P.LEPOIVRE[1]
[1] Faculté Universitaire des Sciences Agronomiques, Passage des Deportes, 5030 Gembloux, Belgium
[2] UCRI, Guangshu, PR China

1. Introduction

Sweetpotato viruses are generally difficult to mechanically transmit and to purify from sweetpotato due to phenolic compounds and other substances and are often restricted to *Convolvulaceae*. Moreover, mixed infections are frequent and a number of diseases are the result of synergistic complexes of viruses, among which sweetpotato feathery mottle (SPFMV) is frequently identified hindering the identification of unknown associated viruses (Moyer and Salazar, 1989).

Owing to the difficulties inherent to the isolation and purification of sweetpotato viruses frequently occuring in mixed infection, a reverse transcription-polymerase chain reaction (RT-PCR) strategy was proposed to analyse the variability of potyviruses infecting 18 sweetpotato clones from the Guangdong province of China.

2. Identification and characterisation of distinct potyviruses in mixedly-infected sweet potato

2.1. USE OF POTYVIRUS-SPECIFIC PCR

The amplifications were carried out using degenerate primers Pot1 and Pot2 (Colinet *et al.*, 1994) designed to amplify the variable 5' terminal part of the coat protein cistron, together with the 3' terminal part of the NIb protein cistron of potyviruses. Total RNA was isolated from symptomatic leaves of *Ipomoea setosa* grafted with the sweetpotato clones by means of Trizol (Gibco BRL). RT-PCR was performed on mRNA purified with Oligotex-dt (Qiagen).

Amplification of the variable 5' terminal region of the coat protein cistron yielded one to three fragments (1.30 kb, 1.35 kb and 1.45 kb) according to the clone. No PCR products were visible for 3 clones. The PCR products were cloned and the sequences of the fragments covering the 5' terminal region of the coat protein cistron were compared to those of SPFMV (-RC and -C) (Alad *et al.*, 1992) and of a Taiwan (-T)

H.-W. Dehne et al. (eds.), Diagnosis and Identification of Plant Pathogens, 417–419.

isolate of the sweetpotato latent virus (SPLV) (Colinet *et al.*, 1994). These sequences allowed to identify five strains from three distinct potyviruses. SPFMV (-CH and -CH2) (yielding a 1.35 kb fragment) were identified as closely related to the russet crack (-RC) and the common (-C) strains of SPFMV respectively, whereas sequence data demonstrated that SPLV-CH (yielding a 1.30 kb fragment) was very close to SPLV-T. A previously non-reported virus giving rise to a 1.45 kb fragment was tentatively named sweetpotato virus G (Colinet *et al.*, 1994) (SPVG-CH and -CH2). Analysis of the repartition of these viruses in the sweetpotato clones from China allowed to detect SPVG in nearly all the clones investigated, whereas SPFMV(CH and CH-2) was detected in 8 clones and SPLV in only one clone.

2.2. TAXONOMIC RELATIONSHIPS BETWEEN SPFMV, SPLV AND SPVG

RACE (Rapid Amplification of cDNA Ends) method (Frohman, 1990) was used to amplify a cDNA fragment corresponding to the 3' terminal region of SPLV and SPVG genome. The nucleotide sequences of the 3' terminal part of the coat protein cistron and of the 3' untranslated region of SPLV (-CH and -T) and SPVG (-CH and -CH2) were combined with those previously determined for the 5' terminal part of the SPLV and SPVG coat protein cistrons. The deduced SPLV and SPVG coat protein amino acid sequences were aligned with the corresponding regions of SPFM-RC and -C and selected potyviruses. The dendrogram of sequence relationship generated from the alignment of the conserved coat protein core of strains of SPVG, SPLV,SPFMV and those of selected potyviruses allowed to rapidly determine the taxonomic position and the distinct/closely related/ strain status of SPFMV, SPLV and SPVG and strains of these viruses. These sequence analyses indicated that SPVG is closely related to SPFMV, though it should be considered as a distinct virus. In contrast, SPLV was only distantly related to SPFMV and SPVG.

2.3. DIFFERENTIATION AMONG SWEET POTATO INFECTING POTYVIRUSES IN MIXED INFECTIONS

The 3' genomic nucleotide sequences of SPFMV-RC and -C, SPVG-CH and -CH2 and SPLV-CH were compared and six primers were designed. The primer MPCOM (GGCTCGATCACGAACCAAAAAGGC) is specific of either strain of SPFMV and SPVG. The primer MPVG (TATTGGATCATGCTGTCCCCACA) is specific of both -CH and -CH2 strains of SPVG whereas MPFMVCH1 (TGTTGACGA-CACAGGAATGGGG) and MPFMVCH2 (AACCTAGA-GTCAATGGA-ATCTCTG) were designed to specifically amplify SPFMV-RC and -C respectively when combined with MPCOM. Two primers, MPLV1(GACTAA-CACACCAACCCAGCCA) and MPLV2 (CCATTCA-TCAAGATTGACAT-CTGCTG) were designed to amplify either strain of SPLV. The length of the amplified fragments was predicted as follows : 581 pb for SPVG (MPCOM + MPVG), 700 bp for SPFMV-RC type (MPCOM+

MPFMVCH1), 589 bp for SPFMV-C type (MPCOM + MPFMVCH2) and 337 bp for SPLV (MPLV1 + MPLV2).

3. Conclusion

RT-PCR revealed as a powerful tool allowing the detection of different members of the Potyvirus genus in sweetpotato by one relatively rapid and simple, yet sensitive, assay. This polyvalent approach does not require any preliminary work of separation and purification of the components of the viral complexes, and, depending of the choice of primers, allows the detection of a single virus or many members of a group or family of related viruses. Broad-spectrum PCR is particularily useful when the primary objective of tests performed for quarantine or certification purposes, to identify potyvirus-infected samples, is regardless of the identity of the involved viruses. These potyvirus-specific primers were further successfully applied to the identification of bract mosaic virus (Anceau *et al.*, 1995) and to characterize six isolates of yam mosaic potyvirus (Duterme *et al.*, 1996).

In further work, the amplified fragments gave rise to molecular characterisation of members of the Potyvirus infecting sweetpotato which is of primary importance for their subsequent classification and taxonomical studies.

4. References

Abad J.A., Conkling M.A. and Moyer J.W. (1992) Comparison of the capsid protein cistron from serologically distinct strains of sweetpotato feathery mottle virus (SPFMV), *Arch. Virol.* **126**,147-157.

Anceau C., Iskra-Caruana M.L., Colinet D., Kummert J., Lepoivre P. (1995) The use of PCR for the characterization of banana bract mosaic agent, *Eur. J. Plant Pathol.* XIII Int. Pant Protection Cong., P623.

Colinet D., Kummert J., Lepoivre P. and Semal J. (1994) Identification of distinct Potyviruses in mixedly-infected sweetpotato by the Polymerase Chain Reaction, *Phytopathology* **84,** 65-69.

Duterme O., Colinet D., Kummert J., Lepoivre P. (1996) Determination of the taxonomic position and characterization of yam mosaic virus isolates based on sequence data of the 5' terminal part of the coat protein cistron, *Arch. Virol.* **141**, 1067-1075.

Frohman M.A.(1990) RACE: Rapid amplification of cDNA ends, in Innis M.A., Gelfand D.H., Sninsky J.J., White T.J. (eds.),*PCR protocols: A guide to methods and applications*, Academic Press, San Diego, pp. 28-38.

Moyer J.W. and Salazar L.F. (1989) Virus and virus-like diseases of sweetpotato, *Plant Dis.* **73**, 451-455.

DIFFERENTIATION OF *FUSARIUM* SPECIES BY RESTRICTION FRAGMENT LENGTH POLYMORPHISM ANALYSIS OF PCR-AMPLIFIED RIBOSOMAL DNA

V. EDEL, C. STEINBERG, N. GAUTHERON, C. ALABOUVETTE
INRA-CMSE, Laboratoire de Recherches sur la Flore Pathogène dans le Sol, 17 rue Sully, 21034 Dijon Cedex, France.

1. Introduction

Fusarium species are ubiquitous soil-borne fungi, responsible for numerous plant diseases. They form an heterogeneous genus and their identification based on morphological characteristics remains difficult (Nelson, 1991). Our objective was to develop an additional and rapid procedure for the differentiation of *Fusarium* species, based on polymerase chain reaction (PCR) and restriction fragment length polymorphism (RFLP) analysis of ribosomal DNA (rDNA). In this study, we estimated the variation in internal transcribed spacers (ITSs) and variable domains of the 28S rDNA among 18 *Fusarium* species. Moreover, for *F. oxysporum* and two closely related species, the results were compared with those obtained from PCR/RFLP analysis of ribosomal intergenic spacer (IGS) sequences.

2. Materials and methods

This study included 92 strains representing 18 *Fusarium* species (Table 1). A fragment of rDNA including ITSs and variable domains of the 28S rDNA was amplified by PCR with oligonucleotide primers ITS1 and P3 and digested with eight different restriction enzymes (Table 1) (Edel *et al.*, 1996). *F. oxysporum* and two related species were further characterized by using a method based on PCR/RFLP analysis of IGS, described for the discrimination of strains at the intraspecific level (Edel *et al.*, 1995).

3. Results and discussion

Restriction analysis of rDNAs amplified with primers ITS1 and P3 revealed 7 to 14 distinct restriction patterns for each enzyme used (Table 1). With the eight restriction enzymes, a total of 148 fragments was analyzed within the sample strains. Twenty-three different combinations of patterns representing 23 rDNA types were identified among the 92 strains. Most *Fusarium* species could be clearly identified from their specific rDNA type. Only two rDNA types were common to two or three species. On the other hand, intraspecific polymorphism was observed in some *Fusarium* species.

H.-W. Dehne et al. (eds.), Diagnosis and Identification of Plant Pathogens, 421–424.

TABLE 1. *Fusarium* species and rDNA types revealed by RFLP analysis of rDNA sequences amplified by PCR 01with oligonucleotide primers ITS1 and P3.

Fusarium species	IGS	Restriction patterns of rDNA fragments digested with								Number
	type	*Alu*I	*Ava*I	*Bso*FI	*Hha*I	*Mbo*I	*Mse*I	*Msp*I	*Nci*I	of strains
F. oxysporum	1	A	A	A	A	A	A	A	A	19
F. oxysporum var. *redolens*	2	L	C	A	B	B	C	L	A	6
F. subglutinans	3	A	A	B	A	A	A	A	A	2
F. anthophilum	3	A	A	B	A	A	A	A	A	4
F. sacchari var. *sacchari*	4	A	A	B	A	A	B	A	A	1
F. moniliforme	5	A	B	B	A	A	B	A	A	6
F. proliferatum	6	B	C	C	B	B	C	L	A	1
F. proliferatum	7	B	C	C	B	B	C	B	B	6
F. culmorum	8	H	F	I	G	A	I	I	H	4
F. culmorum	9	H	F	I	G	A	B	I	H	2
F. graminearum	9	H	F	I	G	A	B	I	H	8
F. crookwellense	9	H	F	I	G	A	B	I	H	2
F. sambucinum	10	I	F	J	A	F	B	I	H	4
F. sporotrichioides	11	A	F	K	G	G	B	I	H	2
F. poae	12	K	F	J	H	F	K	I	H	3
F. poae	13	J	F	M	H	F	B	I	H	1
F. chlamydosporum	14	C	D	L	F	B	J	J	I	2
F. equiseti	15	F	E	B	G	A	F	F	F	3
F. equiseti	16	F	E	B	G	A	G	G	F	2
F. lateritium	17	E	C	G	F	B	E	E	E	1
F. avenaceum	18	G	D	H	F	B	H	H	G	3
F. decemcellulare	19	C	G	N	I	H	C	K	J	2
F. solani	20	C	C	D	C	C	D	C	C	2
F. solani	21	D	D	E	D	D	D	D	D	2
F. solani	22	D	D	F	D	E	D	D	D	2
F. solani	23	D	D	D	E	E	D	D	D	2

A combination of four restriction enzymes (*Ava*I, *Bso*FI, *Mse*I, and *Msp*I or *Nci*I) had the same discriminating power as the combination of eight enzymes and was sufficient to resolve the strains into species. Thus, the PCR-RFLP method described provides a rapid tool for the differentiation of *Fusarium* strains at the species level.

Cluster analysis revealed variable levels of intraspecific and interspecific divergence (Figure 1). Grouping obtained among *Fusarium* strains was, on the whole, in agreement with other morphological and molecular classification criteria (Guadet *et al.*, 1989; Donaldson *et al.*, 1995) and reflects the known heterogeneity of the genus *Fusarium*.

However, *F. oxysporum* var. *redolens* was found more distantly related to *F. oxysporum* than species from section *Liseola* (Figure 1). The grouping obtained among these strains was well correlated with the grouping obtained with the second method, restriction analysis of the IGS region (Figure 2). The high level of divergence found between *F. oxysporum* and *F. oxysporum* var. *redolens* was in agreement with results of Waalwijk *et al.* (1996) and supports the assignment of the latter to another species.

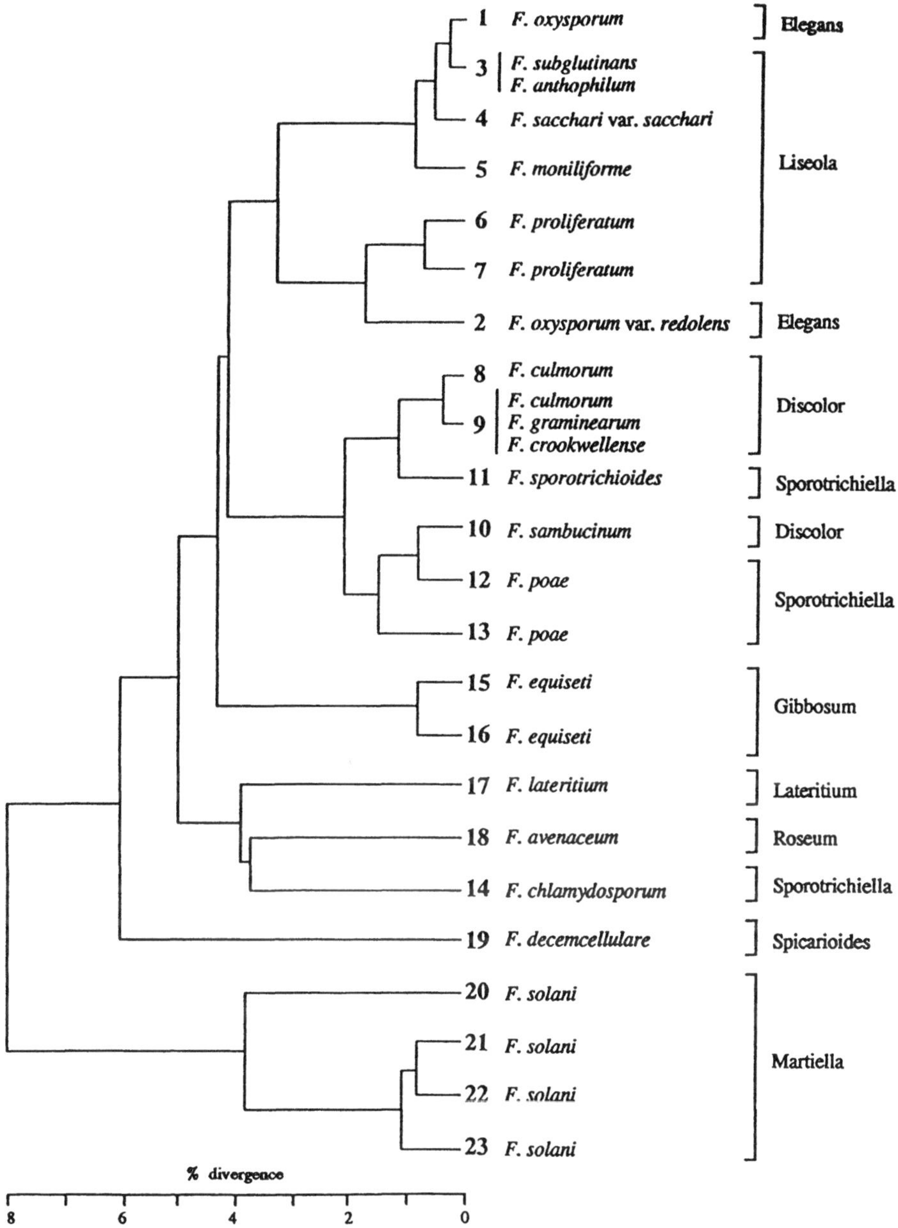

Figure 1. UPGMA dendrogram showing the relationships among rDNA types defined by RFLP analysis of rDNA sequences amplified with oligonucleotide primers ITS1 and P3.

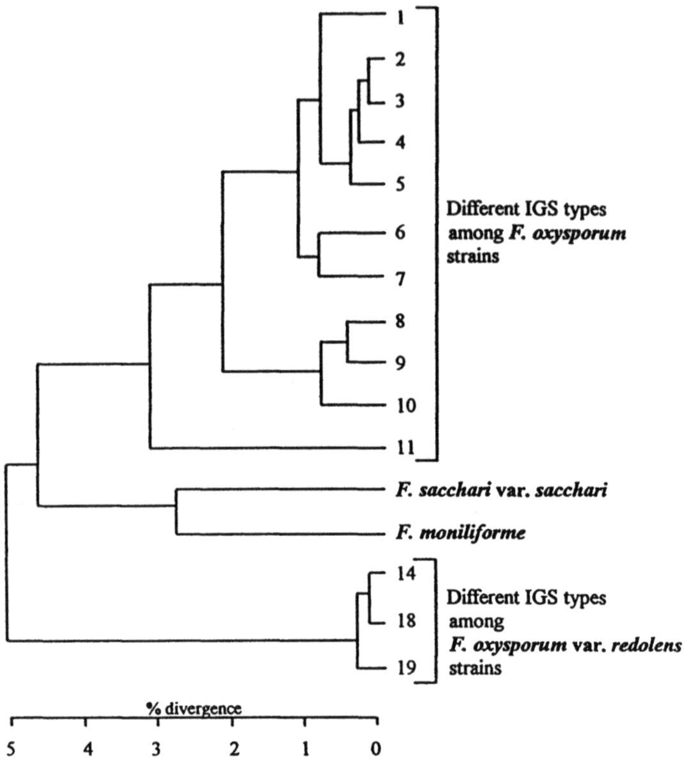

Figure 2. UPGMA dendrogram showing the relationships among IGS types defined by PCR-RFLP analysis

References

Donaldson, G.C., Ball, L.A., Axelrood, P.E., and Glass, N.L. (1995) Primer sets developed to amplify conserved genes from filamentous ascomycetes are useful in differentiating *Fusarium* species associated with conifers. *Applied and Environmental Microbiology* **61**, 1331-1340.

Edel, V., Steinberg, C., Gautheron, N., and Alabouvette, C. (1996) Evaluation of restriction analysis of polymerase chain reaction (PCR)-amplified ribosomal DNA for the identification of *Fusarium* species. *Mycological Research,* in press.

Edel, V., Steinberg, C., Avelange, I., Laguerre, G., and Alabouvette, C. (1995) Comparison of three molecular methods for the characterization of *Fusarium oxysporum* strains. *Phytopathology* **85**, 579-585.

Guadet, J., Julien, J., Lafay, J.F., and Brygoo, Y. (1989) Phylogeny of some *Fusarium* species as determined by large-subunit rRNA sequence comparison. *Molecular Biology and Evolution* **6**, 227-242.

Nelson, P.E. (1991) History of *Fusarium* systematics. *Phytopathology* **81**, 1045-1048.

Waalwijk, C., de Koning, J.R.A., Baayen, R.P., and Gams, W. (1996) Discordant groupings of *Fusarium* spp. from sections *Elegans, Liseola* and *Dlaminia* based on ribosomal ITS1 and ITS2 sequences. *Mycologia* **88**, 361-368.

DIAGNOSIS AND IDENTIFICATION OF PATHOGENS CAUSING APPLE STORAGE ROTS, BY USING PCR-RFLP AND ALLELE SPECIFIC AMPLIFICATION

URS W. HILBER, MAJA HILBER-BODMER,
HANNES SCHÜEPP AND JÜRG E. FREY
Swiss Federal Research Station, 8820 Wädenswil

1. Abstract

The incidence of apple fruit rots such as *Venturia inaequalis, Botryotinia fuckeliana, Monilinia-, Penicillium-, and Gloeosporium*-species varies depending on region, weather condition and spray programme. So far in Switzerland two additional fungicide sprays have to be applied to protect apples against storage rot. Early detection of latent infections is an essential tool to monitor and forecast the development of storage rot fungi. Fungicide applications could thus be recommended according to the occurrence of fungal infection. For genotype identification the internal transcribed spacer (ITS) sequence was amplified by PCR. These sequences were then digested with restriction enzymes to produce species-specific fragment patterns (RFLP). Fingerprinting considerably contributes to better understand the epidemiology of these pathogens.

2. Introduction

Storage rot infections are mostly due to infections in the field. To avoid outbreak of infections during storage, in Swiss apple orchards two preventive fungicide treatments are routinely applied shortly before harvest (Anonymous 1996). This is necessary because so far it has not been possible to predict the occurrence and the species of storage rot fungi at this stage. We developed a PCR-RFLP assay based on internal transcribed spacer (ITS) regions of the ribosomal genes that allows reliable differentiation between the most common apple fruit rots. Early detection of latent infections will be used as an essential tool to monitor and forecast the development of storage rot fungi. This is an important step towards sustainable use of fungicides as applications can be recommended according to the occurrence of latent fungal infections and selection pressure is reduced minimizing the risk of resistance build up.

H.-W. Dehne et al. (eds.), Diagnosis and Identification of Plant Pathogens, 425–428.

3. Materials and Methods

Fruit rots on Golden delicious were used as sources to isolate *Venturia inaequalis, Botryotinia fuckeliana, Monilia-, Penicillium-,* and *Gloeosporium*-species. Five different isolates of each fungus were used in this study.

For DNA preparation we used a protocol modified after Kawasaki (1990): Mycelium was ground in liquid nitrogen, homogenised in 10-50 µl of lysis buffer (10 mM Tris-HCl, pH= 8.0, 1 mM EDTA, 0.5 % Tween 20, 100 µg/ml Proteinase K [Appligene]) and supplemented by the same amount of sterile water. The lysate was then heat denatured at 85°C for 10 min, then centrifuged for 1 min at 6000 rpm. Three µl of the supernatant were taken for PCR in a total volume of 50 µl (10 mM Tris-HCl, pH= 8.3, 50 mM KCl, 1.5 mM $MgCl_2$, 0.01% gelatine, 200 µM dNTP's, 1.7 units of Taq polymerase [Supertaq, Stehelin, Switzerland], 1-5 ng template DNA and 0.5 µM of the primers ITS1 [5'-TCCGTAGGTGAACCTGCGG-3'] and ITS4 [5'-TCCTCCGCTT ATTGATATGC-3']) using the following protocol: 3 min denaturation at 95°C, then 35 cycles of 1 min at 93°C, 1 min annealing at 56°C, and 2 min extension at 72°C. The amplification was completed by a final 7 min extension step at 72°C. Ten µl of the amplification product was electrophoresed on a 1.5% agarose gel run at 9 V/cm for 60 min.

Restriction analysis was performed according to the manufacturers instructions (Boehringer Mannheim) using 10 units of restriction enzyme (CfoI, MspI) in 50 µl total volume, incubated for 2 hours at 37°C.

Sequencing of the amplification product was performed on a automated laser fluorescent sequencer (ALF Pharmacia) and the selected internal ITS primer sequence was synthesized by Microsynth, Switzerland.

4. Results and Discussion

PCR amplification using the primers ITS1 and ITS4 resulted in a single fragment of ca. 540 bp in *Botryotinia, Monilinia, Venturia* and *Gloeosporium*. The corresponding fragment size for *Penicillium*, ca. 560 bp, was clearly larger than that of the other 4 species, allowing direct identification upon PCR amplification (Fig. 1).

Restriction fragment patterns of the enzymes CfoI and MspI allowed for easy differentiation between *Gloeosporium, Venturia,* and the *Botrytinia/Monilinia*-group (Fig. 2). These latter two species could not be differentiated with these nor any other enzymes tested. Therefore, we partially sequenced the ITS fragment of these two species and designed primers matching diagnostic sequences. Using these species-specific internal ITS primers resulted in a shorter, ca. 420 bp fragment in the corresponding species only, thus allowing for differentiation between *Botryotinia* and *Monilinia* (Fig. 3).

We have established a method for species identification of storage rot fungi that requires only minute amounts of DNA (< 1-5 ng). We will proceed with testing the suitability of the method for early detection of latent infections of storage rot fungi under practice conditions by applying this method to stored apples. We expect that the PCR

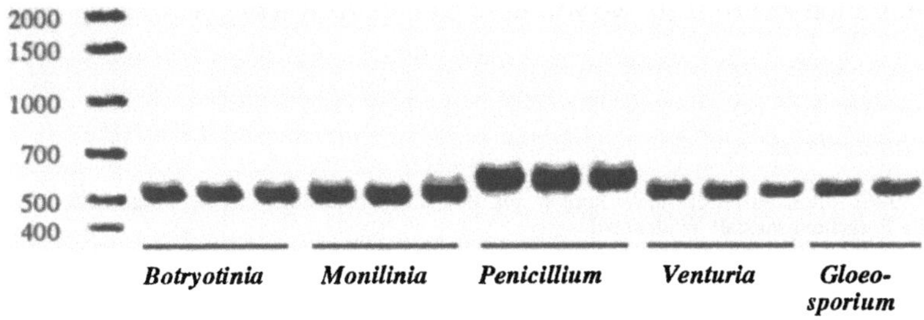

Fig. 1 : Amplification products of different strains of five storag rot fungi using primers ITS1 and ITS4. Penicillium can be identified based on its larger amplification fragment size.

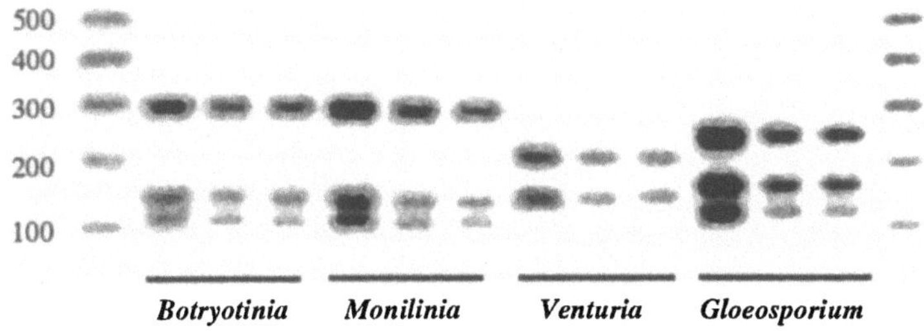

Fig. 2 : Restriction fragment pattern of ITS amplification products of three different strains of four storage rot fungi after CfoI digestion. Differentiation between Gloeosporium, Venturia, and the Botryotinia/Monilinia -group is possible.

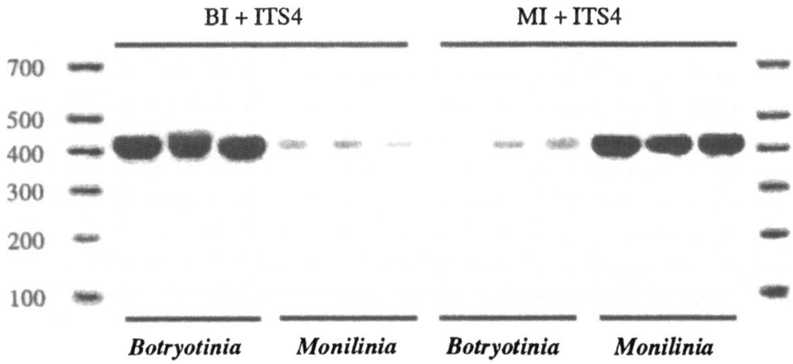

Fig. 3 : Amplification of three different strains of Botrotinia and Monilinia with internal primers BI (Botryotinia, 5'-ATGCTCG-CCAGAGAATACCA-3') and MI (Monilinia, 5'-ATGCTCGC-CAGAGAATAATC-3'). Each primer allows species identification by species-specific amplification.

amplification is capable to provide sufficient DNA for our analysis even in cases of very low infestation levels. If so, we will have an important tool for reducing pesticide use.

References

Anonymous (1996) Pflanzenschutzempfehlungen für den Erwerbsobstbau 1996, Flugschrift Nr. 121, Eidgenössische Forschungsanstalt Wädenswil.

Kawasaki, E.S. (1990) Sample preparation from blood, cells, and other fluids, in: M.A. Innis, D.H. Gelfand, J.J. Sninsky and T.J. White (eds.), PCR Protocols, Academic Press, New York; pp. 146-152.

Development of DNA-assays for *in vitro* differentiation and *in planta* detection of *Fusarium culmorum* and *Fusarium graminearum*

Birger Koopmann, Petr Karlovsky* and Gerhard A. Wolf
*Institut für Pflanzenpathologie und Pflanzenschutz, Universität Göttingen, Grisebachstr. 6, 37077 Göttingen, * Institut für Phytomedizin, Universität Hohenheim, Otto-Sander-Str.5, 70593 Stuttgart*

1. Introduction

Fusarium culmorum (W. G. Sm.) Sacc. (teleomorph unknown) and *F. graminearum* Schwabe (teleomorph *Gibberella zeae* (Schwein.) Petch) are causal agents of seedling rots, foot rots, leaf spots and head blights of small grain cereals (COOKE 1981). Their potential to produce mycotoxins harmful to the health of men and animals has drawn the attention to *Fusarium* seed contaminations. Simple and reliable diagnostic methods for both detection and quantification of *Fusarium* spp. are prerequisites for health risk evaluations. The diagnosis of *Fusarium* infections is based on visual assessment, microscopic or immunological techniques (ELISA). Simultaneous *in planta* quantification of distinct species is still not possible. Differentiation of *Fusarium* species relies on highly variable traits such as size and shape of conidia and conidiophores. Immuno- and DNA techniques are suitable to differentiate the forementioned species by species-specific antibodies or DNA probes and have the potential of simultaneous quantification. The objective of our work was the development of a reliable DNA-method for both *in vitro* differentiation and *in planta* detection of *F. culmorum* and *F. graminearum*.

2. Material and Methods

Fusarium isolates were obtained from T. Miedaner, University of Hohenheim (H), `Deutsche Sammlung von Mikroorganismen´ (DSM) and from our collection of fungal isolates (SAGW). Isolates were grown in Czapek-Dox liquid media supplemented with 0,5 % peptone. Total DNA was isolated according to MOELLER et al. (1992). DNA probes were obtained by shot-gun cloning. Total DNA of *F. culmorum* was digested with *Sau*3AI. Fragments were cloned into pUC19 by standard procedures (SAMBROCK et al. 1987). Inserts were labeled with DIG-dUTP using a vector-PCR protocol as described previously (KOOPMANN et al. 1994). Probes were tested with size separated, immobilized *Eco*RI digested total DNA of plant pathogenic fungi. Southern hybrization and detection were performed according to the manufacturer´s recommendations (Boehringer Mannheim, Germany). Nucleotide sequences were determined by the chain termination method (SANGER 1977) using the `sequenase kit´ of Pharmacia (Freiburg,

H.-W. Dehne et al. (eds.), Diagnosis and Identification of Plant Pathogens, 429–432.

Germany). Oligonucleotides were synthesized by Pharmacia customer service (Freiburg, Germany).

3. Results

Specifity of twentyeight *F. culmorum* DNA probes was tested by Southern hybridization with *Eco*RI digested total DNA of 31 isolates of *F. culmorum*, 13 isolates of *F. graminearum* and 3 isolates each of *F. avenaceum*, *Microdochium nivale*, *F. oxysporum*, *F. sulphureum*, *Rhizoctonia cerealis*, *Septoria nodorum* and *Pseudocercosporella herpotrichoides*. 50 % of the probes did not differentiate *F. culmorum* and *F. graminearum*. 12 probes were shown to differentiate the species based on RFLP patterns. One of the probes was specific for *F. culmorum* (p220-24, KOOPMANN et al. 1994). Another probe was characterized which was specific for *F. culmorum* and *F. graminearum* (p220-21, KOOPMANN et al. 1994). These probes were sequenced and two sets of primer pairs were designed (CUL3/CUL4 and CG1/CG2). Based on these oligonucleotides we developed a PCR assay for *in vitro* differentiation and *in planta* detection of *F. culmorum* and *F. graminearum*. The PCR assay can be used for differentiation of isolates with extracts of total DNA prepared according to the method of MOELLER et al. (1992) (figure 1) or both mycelium and spore lysates. It is also suitable for *in planta* detection of artificially inoculated juvenile wheat plants and artificially inoculated wheat kernels (figure 2).

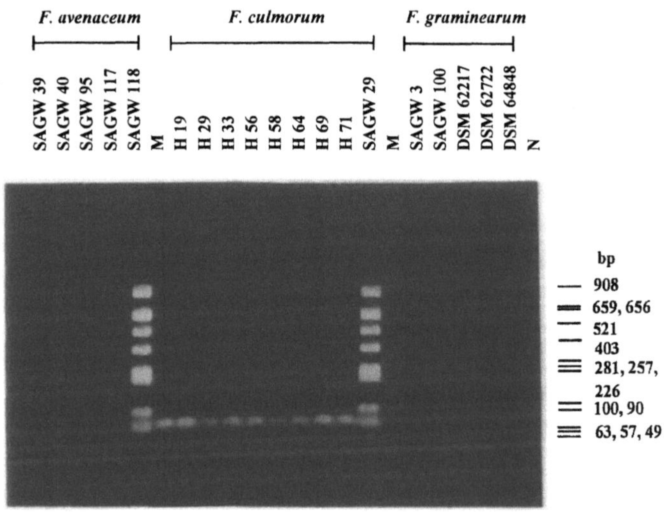

Fig. 1: PCR products obtained with the primer pair CUL3/CUL4 and total DNA of several *Fusarium* isolates of different species

We determined an *in vitro* detection limit of 500 templates with both primer sets and within a multiplex PCR assay.

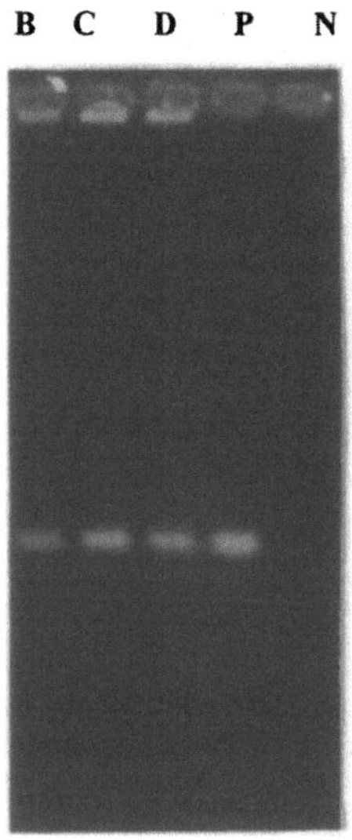

Fig. 2.: *In planta* detection of *F. culmorum* with the primer set CUL3/CUL4 (B,C,D: different samples of infected wheat kernels, P- positive control , N- negative control)

4. Discussion

F. culmorum and *F. graminearum* are considered to be closely related. Our results are consistent with these findings. Most of the tested probes were common to both species (93%). Nevcrtheles the differentiation of *F. culmorum* and *F. graminearum* was shown to be possible by DNA techniques. QUELLET & SEIFERT (1993) and SCHILLING et al. 1994 used the RAPD technique for differentiation.
We selected a species-specific probe and a probe specific for *F. culmorum* and *F. graminearum* by the shot-gun cloning approach. These probes were useful for *in vitro* differentiation by Southern hybridization but not suitable for *in planta* detection using non-radioactive detection protocols. Two characters of the specific probes may be responsible for low sensivity: (i) the probes represent single copy sequences of the genome and (ii) low molecular weight of the probes (55 and 85 bp). Sensitivity was improved 10000 fold by the development of a PCR assay. Primer pairs obtained from probes were suitable for both *in vitro* differentiation and *in planta* detection.

Development of internal standards will be the basis for simultaneous quantification of the two *Fusarium* species in plant tissue by competitive PCR.

5. References

COOK, R. J. (1981): *Fusarium* diseases of wheat and other small grains in North America. In: NELSON P.E., TOUSSOUN T. A., COOK R. J. (Hrsg.): *Fusarium* diseases, biology and taxonomy. Pennsylvania State University Press.

KOOPMANN B., KARLOVSKY P., WOLF G. A.(1994): Differentiation between *Fusarium culmorum* and *Fusarium graminearum* by RFLP and with species-specific DNA probes. In : DEWEY M., SCHOTS A., OLIVIER, R.: Modern detection assays for plant pathogenic fungi.CAB International, Oxford.

MÖLLER, E. M., BAHNWEG, G., SANDERMANN, H., GEIGER H. H. (1992): A simple and efficient protocol for the isolation of high molecular weight DNA from filamentous fungi, fruit bodies, and infected plant tissues. Nucleic Acid Res. **20**, 6115-6116.

QUELLET T., SEIFERT K. A. (1993): Genetic characterization of *Fusarium graminearum* strains using RAPD and PCR amplification. Phytopath. **83**, 1003-1007.

SAMBROOK, J., FRITSCH E. F., MANIATIS, T. (1989): Molecular cloning: a laboratory manual. Cold Spring Habor Laboratory Press, New York.

SANGER F., NICKLEN S., COULSON R. (1977): DNA sequencing with chain termination inhibitors. Proc. Natl. Acad. Sci. USA **74**, 5463-5467.

SCHILLING A. G., MÖLLER E. M., GEIGER H. H. (1994): RAPDs of *Fusarium* species: Application for genotyping and species identification. in : DEWEY M., SCHOTS A., OLIVIER, R. (Hrsg.):`Modern detection assays for plant pathogenic fungi´. CAB International, Oxfordt.

SOUTHERN E. M. (1975): Detection of specific sequences among DNA fragments separated by gel electrophoresis. J. Mol. Biol. **98**, 503-517.

ISOLATION AND CHARACTERIZATION OF A NEPO-TYPE VIRUS FROM REVERTED BLACKCURRANT (*Ribes nigrum* L.)

S. LATVALA[1,2], A. LEMMETTY[1] and K. LEHTO[2]

[1]*Agricultural Research Centre, Institute of Plant Protection, 31600 Jokioinen, Finland* [2]*Department of Biology, Laboratory of Plant Physiology and Molecular Biology, University of Turku, 20500 Turku, Finland*

Abstract

An unknown virus was isolated from a reverted blackcurrant which showed malformed flowers. A mechanically transmissible virus was purified and partially characterized. The particle size and morphology, the genome structure and the size of the coat protein found suggest that although the virus is very similar to nepovirus group, it is distinct from any other reported virus.

1. Introduction

Blackcurrant reversion is the most serious disease of blackcurrants, occurring world-wide with the exception of the Americas. Despite studies over many years the causal agent has remained obscure (Adams and Tresh, 1987). Three forms of reversion are reported by Jones (1995). At least two forms of the reversion disease occur in Finland. The severe form of the disease is easy to distinquish on the basis of severely malformed flowers with the style elongated, stamens absent and the petals sepal-like in appearance. In this study we have isolated an unknown virus from a reverted blackcurrant plant and have partially characterized the virus.

H.-W. Dehne et al. (eds.), Diagnosis and Identification of Plant Pathogens, 433–435.

2. Materials and methods

Plant material originated from a breeding line of reverted bushes of blackcurrant growing in an experimental field at the Agricultural Research centre, Institute of Horticulture, Piikkiö, Finland. Mechanical inoculation to *Chenopodium quinoa* Willd. was done by grinding blackcurrant leaves in 2% nicotine solution, pH 9.5. The virus was maintained on *C. quinoa* and purified using a slight modification of the method described by Frison and Stace-Smith (1992) for arabis mosaic nepovirus. The purified virus particles were negatively stained with 1.5% phosphotungstic acid (PTA) (pH 7.0) and examined in a JEOL JEM-100SX electron microscope.
Coat protein was isolated from purified virus particles according to Laemmli (1970). Total viral RNA was extracted from purified virus particles by treatment with phenol and SDS, and the sizes of viral RNAs were determined by size separation on an agarose gel stained with ethidium bromide according to Sambrook *et al.* (1989). cDNA from the viral RNA was synthesized with oligo(dT) primers using a cDNA Synthesis kit (Pharmacia Biotech). cDNA clones were sequenced by automated sequencing (ALF Manager system, version 2.5) and sequence data were analyzed using several of the Genetics Computer Group (GCG) programs.

3. Results and discussion

During rooting of herbaceous cuttings of a reverted bush of blackcurrant breeding line, clear line-pattern and ringspot symptoms appeared. Repeated attemps to mechanically transmit an agent from these leaves to *C. quinoa* were successful on only one occasion. One week after inoculation, a few chlorotic/necrotic lesions developed in the inoculated leaves of *C. quinoa*; systemic chlorotic or necrotic flecking and malformation of leaves appeared one week later. Yields of the virus were relatively small.
The purified particles were isometric, c.a. 27 nm in diameter and sedimented as two nucleoprotein components comprised of a protein species of M_r 55K and two RNA molecules (RNA1 and RNA2) c.a. 6700 and 7700 nucleotides in length; both RNAs posessed poly-A tails. Sequence analysis of 3' terminal cDNA clones of one of the genomic RNAs revealed that this RNA posessed a long 3' terminal non-coding sequence. The particle size and morphology,

structure of the genome and the size of the coat protein suggest that this new virus is very similar to nepoviruses (Mayo and Robinson, 1996).

Short homologous sequences between the new virus studied and several nepoviruses and two comoviruses were found after sequence analysis of the 3' terminal cDNA clones. Apart from these short homologous regions, the viral sequence was distinct from those reported for other viruses.
Using an Immuno-capture reverse transcriptase PCR assay we have detected the virus from blackcurrants affected both with the severe and the common strain of the reversion disease. Results demonstrating the close association between the virus and the reversion disease are reported in this volume by Lehto *et al.*

5. References

Adams, A.N. and Tresh, J.M. (1987) Reversion of Blackcurrant. Pages 133-136 in: *Virus Diseases of Small Fruits.* R.H. Converse, ed.USDA Agriculture Handbook No. 631.

Frison, E.A. and Stace-Smith, R. (1992) Cross-reacting and heterospecific monoclonal antibodies produced against arabis mosaic virus. *J. Gen. Virol.* **73**, 2525-2530

Jones, A.T. (1995) Recent developmets affecting the control of two important diseases of small fruit crops in the UK. *Acta Hortic.* **385**, 64-69

Laemmli, U.K. (1970) Cleavage of structural proteins during the assembly of the head of bacteriophage T4, *Nature* (London) **227**, 680-685

Mayo, M.A. and Robinson, D.J. (1996) Nepoviruses:Molecular biology and replication. Pages 139-186 in: *The Plant Viruses,* volume 5. Polyhedral virions and bibartite RNA genomes. B.D. Harrison and A.F. Murant, eds. Plenum Press

Sambrook, J., Fritsch, E.-F. and Maniatis, T. (1989) *Molecular Cloning. A Laboratory Manual*, 2nd edn. New York: Cold Spring Harbor Laboratory

DIFFERENTIAL DETECTION OF *ERWINIA AMYLOVORA* ISOLATES FROM FRUIT TREES AND *RUBUS* SPP. BASED ON RIBOSOMAL DNA AND PCR.

M. MAES and C. CREPEL
Research Station for Plant Pathology, Centre for Agricultural Research, Belgium

1. Abstract

A 23S ribosomal DNA primer sequence, specific for *Erwinia amylovora* bacteria, was combined in PCR with a conserved 16S rDNA primer, and only produced a PCR banding pattern on agarose gel with *E. amylovora* cells. No PCR signal was observed with other plant-associated organisms or with taxonomically related bacteria. These specific PCR bands represented the different types of 16S-23S spacers present in the bacterial ribosomal operons. Variation in detection patterns was observed between *E. amylovora* isolated from *Rubus* and fruit tree host plants, and enabled direct ribotype discrimination of the detected fire blight pathotype. In *E. amylovora* isolates from *Rubus*, the PCR detection signal was composed of two ribosomal spacer bands, whereas in strains from apple, pear and Japanese plum (*Prunus salicina*), three major bands were produced. The latter ribotype detection pattern was also generated with *E. amylovora* strains derived from ornamental host plants, as *Cotoneaster*, *Crataegus* and *Cydonia*.

2. Introduction

E. amylovora is the known fire blight pathogen on apple, pear and different ornamentals, which all belong to the *Malaceae* plant family. Other *E. amylovora* strains are isolated from naturally diseased raspberry and blackberry, which arc *Rosaceae* plants. Moreover, *E. amylovora* bacteria have recently been found on diseased Japanese plum (Mohan & Thomson, 1996). Plum belongs to a third family of the *Rosales*, the *Amygdalaceae*. Some phytopathological differentiations have been observed between these *E. amylovora* strains. *E. amylovora* strains isolated from naturally infected *Rubus* species (EaR) have been reported not to be infective on *Malaceae* seedlings and shoots (Evans, 1996; Garbeva et al., 1996; Norelli, 1984), whereas *E. amylovora* isolates from *Malaceae* plants (EaM) show an attenuated or suppressed virulence towards the tested raspberry (Garbeva et al., 1996; Evans, 1996) and blackberry plants (Starr et al., 1951). Preliminary pathological tests indicate that the isolates from the *Prunus* host (EaP) can infect apple and pear upon artificial inoculation (Mohan & Thomson, 1996). The *Prunus* trees with fire blight were located in the neighbourhood of severely diseased apple trees, and most probably resulted from a cross

H.-W. Dehne et al. (eds.), Diagnosis and Identification of Plant Pathogens, 437–439.

infection with EaM (Mohan & Thomson, 1996).
The application of molecular typing and detection of bacteria can be useful for the tracing of the existing pathways responsible for the dissemination of the different *E. amylovora* types. Mc Manus and Jones (1995) have used polymerase chain reaction (PCR)-ribotyping to discriminate the pathotypes EaM and EaR. In our laboratory, a ribosomal DNA sequence has been designed for the specific PCR detection of all strains of the *E. amylovora* species, directly in plant extracts (Maes et al., 1996). Based on this sequence information, we propose a one-step PCR detection of the different *E. amylovora* ribotypes.

3. Materials and methods

Bacterial strains:
The *E. amylovora* isolates from *Prunus* (ATCC 51852 and ATCC 51853) were obtained from S. K. Mohan (University of Idaho, Parma, ID). The isolates from *Rubus* and *Malaceae* were from the Belgian Coordinated Collections of Microorganisms, Laboratory of Microbiology, University Gent, Belgium (LMG), and contained LMG 2069 and LMG 2084 (*Rubus*), and LMG 1933 (*Cotoneaster*), LMG 1885, LMG 1879 (*Crataegus*), LMG 2069, LMG 2022, LMG 2078, LMG 2074 (*Pyrus*), LMG 1983, LMG 1969 (*Malus*) and LMG 1877 (*Cydonia*).

Bacterial DNA preparation and PCR amplification:
E. amylovora cells, grown in rich medium or extracted from plant samples, were lysed by heating a 5 µl aliquot for 7 min at 95 °C, quickly cooled on ice and immediately used in PCR. The PCR reaction ingredients were as described before (Maes et al., 1996), and a 95°C/60°C/70°C PCR cycling program was run on a Thermal Cycler PHC-3 (Techne, Cambridge, UK). The PCR products were analysed by electrophoresis on a 1.8 % agarose gel.

4. Results and discussion

E. amylovora strains that were isolated from *Malaceae*, *Rubus* and *Prunus* plants were introduced in PCR to amplify the 16S-23S rDNA spacer region. The two PCR primers were flanking these spacer regions and were located in the 16 rDNA-3' end (*E. coli* position numbering 1493-1513) and the 23S rDNA-5' end (position 23-43). Two *E. amylovora* ribotypes were observed. The tested EaM and EaP strains produced the same fragment pattern on gel. In the EaR isolates the PCR pattern was composed of two main bands and was clearly discriminated from the EaM/EaP profile. In these PCR experiments the primers were situated in conserved ribosomal gene sequences. They produced banding patterns with most phytobacteria and are therefor not suitable for direct detection of *E. amylovora* bacteria directly in plant extracts, without previous bacterial isolation.
When the reverse primer was replaced by an *E. amylovora*-specific 23S rDNA primer, only the *E. amylovora* bacteria reacted in PCR. This reverse primer was designed as a segment of the EaF sequence, which was described in Maes et al. (1996), and located in the 5' region of

the 23S rDNA of *E. amylovora* bacteria. In this new PCR, the EaM and EaP strains produced the same PCR profile with three major bands, and EaR a profile with two major bands. The PCR products ranged in size from 700 to 1150 bp. With all other phytobacteria no PCR product was observed on gel. All the tested isolates from fruit trees, as well as from *Cotoneaster*, *Crataegus* and *Cydonia* (EaP and EaM) had the same ribotype profile, although the strains from *Prunus* are from US-origin (Oregon), and the strains from *Malaceae* from European, Canadian and New-Zealand Origin. The produced EaM/EaP profile could be compared to the ribotype 3, as described by Mc Manus and Jones (1995).
The derived ribotype classification was in agreement with the grouping of the *E. amylovora* isolates based on the observed phytopathological specialization, suggesting a phylogenetic distance between the EaR and EaM/EaP pathotypes.
By introducing an *E. amylovora*-specific ribosequence in the PCR-ribotyping, a pathotype-specific detection is possible with a single PCR procedure. This system and possible variants could have their application in the epidemiology of the fire blight pathogen.

5. References

Evans, I.R. (1996) Fire blight of raspberries in Alberta. Acta Horticulturae **411**, 69-72.

Maes, M., Garbeva, P. and Crepel, C. (1996) Identification and sensitive endophytic detection of the fire blight pathogen *Erwinia amylovora* with 23S ribosomal DNA sequences and the polymerase chain reaction. Plant Pathology, in press.

Mc Manus, P.S. and Jones, A.L. (1995) Genetic fingerprint of *Erwinia amylovora* strains isolated from fruit-tree crops and *Rubus* spp. Phytopathology **85**, 1543-1553.

Mohan, S.K. and Thomson, S.V. (1996) An outbreak of fire blight in plums. Acta Horticulturae **411**, 73-76.

Norelli, J.L., Aldwinckie, H.S. and Beer, S.V. (1984) Differential host and pathogen interactions among cultivars of apple and strains of *Erwinia amylovora*. Phytopathology **74**, 136-139.

Starr, M.P., Cardona, C. and Folsom, D (1951) Bacterial fire blight of raspberry. Phytopathology **41**, 915-919.

van der Zwet, T. and Beer, S.V. eds. (1995) Fire Blight - Its Nature, Prevention and Control: A Practical Guide to Integrated Disease Management. U.S. Department of Agriculture, Agriculture Information Bulletin N° 631.

APPLICATION OF POLYMERASE CHAIN REACTION (PCR) FOR THE POSITIVE IDENTIFICATION OF TWO PHYTOPLASMA DISEASES IN POLAND.

T. MALINOWSKI, B. KOMOROWSKA, M. KAMIŃSKA
AND B. ZAWADZKA

Research Institute of Pomology and Floriculture,
96-100 Skierniewice, Poland

1. Introduction

Detection and identification of phytoplasmas (mycoplasma like organisms, MLOs) by conventional techniques such as DAPI staining followed by fluorescence microscopy, ELISA or biological indexing is limited by relatively low sensitivity and unspecificity of observed symptoms (Seemüller, 1989). Recently, rapid and sensitive methods of phytoplasma detection, based on the enzymatic amplification of pathogen DNA, were developed (Deng and Hiruki, 1990; Ahrens and Seemüller, 1992; Lee et al., 1993; Kirkpatrick et al., 1994). These methods were used successfully in our study for the diagnosis of two diseases observed in Poland.

2. Materials and methods

2.1. PLANT MATERIAL

Premature foliage reddening and slow decline of pear trees (*Pyrus communis* L.; Kiparyjska, William's Bon Chretien and Patten cvs) were observed in some orchards during the last few years. Quick decline of pear occurred in 1995 in two other orchards (Fig.1). Samples of leaves and roots were collected from affected trees and healthy ones.

Leaves and fragments of stems were also collected from *Helichrysum bracteatum* Willd. showing abnormal production of secondary shoots, flower malformation and dieback as well as from healthy plants.

2.2. ISOLATION AND AMPLIFICATION OF PATHOGEN DNA

DNA was isolated using MLO enrichment procedure as described by Ahrens and Seemüller (1992) and served as a template for PCR reaction (Saiki et al., 1988). Primers fU5/rU3 were used for all samples. These primers were reported to be

H.-W. Dehne et al. (eds.), Diagnosis and Identification of Plant Pathogens, 441–443.

universal for phytoplasma detection and initiated amplification of DNA fragment (about 874 bp) from 16S rRNA gene region of phytoplasma DNA. Additionally, for pear samples PCR was conducted with primers fPD-rO1 and fAT-rAS (Lorenz et al., 1995). Expected product lengths were 923 bp and 530 bp, respectively.

2.3. ANALYSIS OF PCR PRODUCTS

Aliquots of post reaction mixture were resolved by agarose gel electrophoresis or precipitated and used for *Alu*I digestion. The products of enzymatic hydrolysis were then resolved in 5.6% polyacrylamide gel and the length of obtained DNA fragments was estimated. The sequences of 16S rRNA gene of several MLOs representing six of seven RFLP groups were retrieved from Gene Bank and predicted length of PCR/*Alu*I fragments were calculated for these sequences. The calculated lengths of DNA were compared with the lengths estimated from the experiments.

3. Results and discussion

PCR products of the expected sizes were obtained in the experiments with fU5-rU3, rAT-fO1 and rAS-fAT primer pair, for samples containing DNA isolated from pear trees showing symptoms of slow or quick decline. These bands were not observed after PCR conducted with DNA isolated from healthy plants. Gel electrophoresis pattern obtained for PCR products digested with *Alu*I corresponded to the one expected for pear decline phytoplasma. Both the presence of PCR products of expected sizes and the observed restriction fragment length patterns confirmed the presence of pear decline phytoplasma in the trees showing symptoms of the disease.

DNA of the expected size was also observed in PCR product amplified from diseased *H. bracteatum.* Three main bands were present when PCR product was digested with *Alu*I and resolved by electrophoresis. Positions of these bands corresponded well to the lengths of the main DNA fragments of American aster yellows phytoplasma group (**352**bp, **275**bp, **191**bp, 56bp, 5bp).

From these results we draw a conclusion that two phytoplasma diseases were identified in Poland using polymerase chain reaction:

- pear decline
- American aster yellows

Application of RFLP analysis to PCR products allowed us to increase the reliability and accuracy of identification

4. Literature

Ahrens U., Seemüller E., 1992. Detection of DNA of Plant Pathogenic Mycoplasmalike Organisms by a Polymerase Chain Reaction that Amplifies a Sequence of the 16S rRNA Gene. Phytopathology 82:828-832.

Deng S. and Hiruki C., 1990. Enhanced detection of a plant pathogenic mycoplasma-like organism by using polymerase chain reaction. Proc. Japan Acad. 66:140-144.

Kirkpatrick B., Smart C., Blomquist C., Guerra L., Harrison N., Ahrens U., Lorenz K.H., Scheider B., Seemüller E., 1994. Identification of MLO strain-specific primers obtained from 16/23S spacer sequences. IOM Letters 3:261-262.

Lorenz K.H., Schneider B., Ahrens U., Seemüller E., 1995. Detection of the Apple proliferation and Pear decline phytoplasmas by PCR Amplification of Ribosomal and Nonribosomal DNA. Phytopathology 85:771-776.

Lee I.M., Hammond R.W., Davis R.E., Gundersen D.E., 1993. Universal Amplification and Analysis of Pathogen 16S rDNA for Classification and Identification of Mycoplasmalike Organisms. Phytopathology 83:834-842.

Saiki R.K., Gelfand D.H., Stoffel S., Scharf S.J., Higuchi R., Horn G.T., Mullis K.T., 1988. Primer-Directed Enzymatic Amplification of DNA with a Thermostable DNA Polymerase. Science 239:487-491.

Schneider B., Ahrens U., Kirkpatrick B.C., Seemüller E., 1993. Classification of plant-pathogenic mycoplasma-like organisms using restriction-site analysis of PCR-amplified 16S rDNA. J.Gen.Microbiol. 139:519-527.

Seemüller E., 1989. Pear Decline. In: Virus and viruslike diseases of pome fruits and simulating noninfectious disorders. Ed. P.F. Fridlund, pp.188-201. Washington State University, Cooperative Extension, Special Publication No. SP0003.

Fig.1 Quick decline observed on the pear tree of Patten cv.

SILICACAPTURE-REVERSE TRANSCRIPTION-POLYMERASE CHAIN REACTION (SC-RT-PCR): APPLICATION FOR THE DETECTION OF SEVERAL PLANT VIRUSES.

TADEUSZ MALINOWSKI
Research Institute of Pomology and Floriculture,
96-100 Skierniewice, Poland
e-mail: ISKVIR@PLEARN.EDU.PL

1. Introduction

Application of polymerase chain reaction (PCR) for the detection of plant RNA viruses is considerably limited by problems with the first step (RT = reverse transcription) - synthesis of cDNA complementary to RNA of pathogen present in the infected tissue. Not only the concentration of plant viruses may be very low but also plant cells may contain many components which can attribute to degradation of RNA or/and inhibition of enzymatic reactions. Several approaches have been tried to overcome these problems, among them immunocapture (IC) of virus particles seemed to be the most successful one. However, an antiserum is necessary to perform IC-RT-PCR, which limits the application. The virus particle should also remain intact during the immunocapture step which is not always true.

The alternative solution might be SC-RT-PCR (SC-RT-PCR: silica capture or silica coarse - RT-PCR) based on the method of total nucleic acid isolation developed by Boom et al. (1990). This method has been tried in our lab for several plant viruses.

2. Materials and methods

2.1 HEALTHY AND INFECTED PLANT MATERIAL

The following viruses were included in this study: plum pox virus (PPV), apple chlorotic leaf spot virus (ACLSV), prune dwarf virus (PDV), apple stem pitting virus (ASPV) = pear vein yellows virus (PVYV). Both woody hosts and inoculated herbaceous plants were used as a source of the virus with the exception of PVYV=ASPV for which no herbaceous host was tested. Plant samples of leaves or bark were collected from the trees kept at the collection of our laboratory.

H.-W. Dehne et al. (eds.), Diagnosis and Identification of Plant Pathogens, 445–448.

2.2 ISOLATION OF NUCLEIC ACIDS

The procedure used for total nucleic acids (NA) isolation was performed at room temperature. Tested tissue (leaves, bark, flowers) was homogenised in a suitable buffer (usually PBS-TPO) treated by UV for 15 min. Homogenate was centrifuged in Eppendorf tubes at 12,000 rpm for 2-4 min. Fifty microliters of the supernatant was then added to preassembled tubes containing 40 μl of the suspension of silica particles in 900 μl of L6 buffer (Figure 1). Tubes were then vortexed briefly and left for ten min. at room temperature. Tubes were then vortexed briefly again, centrifuged 15 min. at 12,000 rpm, supernatant was removed by suction. Pellet was washed two times with L2 buffer (1 ml of L2 was added to the tube, tube was vortexed and centrifuged; supernatant was removed), twice with 70% ethanol and once with acetone. Pellet was allowed to dry for 10 min. at 56 °C. Then 50 μl of elution buffer (or Milli-Q water, UV treated) was added, tubes were closed, vortexed, and incubated at 56 °C for 10 min. After vortexing and centrifuging for 2 min. at 12,000 rpm. few microliters of the supernatant (NA) were carefully removed and used for reverse transcription.

2.3 RT-PCR

One to ten microliters of extracted NA were aliquoted to the tubes, RT-mix prepared from GeneAmp RNA PCR kit (Perkin-Elmer) was added and RT-PCR reaction was conducted in Perkin-Elmer 480 Thermocycler. Primers and thermal profiles of PCR reactions were used as described for PPV by Wetzel et al. (1992), for ACLSV by Nemchinov et al. (1995), for PDV by Parakh et al. (1995), for ASPV were designed in our lab, based on the sequences obtained by Jelkman (1994).

3. Results and discussion

The procedure of NA isolation took about one hour. PCR products of the expected sizes were obtained after RT-PCR reactions with NA isolated by SC method from plants infected with PPV, ACLSV, PDV, ASPV and PVYV (Fig. 1). It was possible to perform reverse transcription without additional denaturation of isolated nucleic acid. Detection of PPV by SC-RT-PCR directly in the plum trees was possible even during winter months. Successful reactions were also performed with total nucleic acid extract stored for several weeks at -70°C under 70% ethanol.

The sensitivity of SC-RT-PCR and IC-RT-PCR was usually comparable in both the experiments with PPV and ACLSV. However, more unspecific bands were sometimes observed after SC-RT-PCR than after IC-RT-PCR.

4. Conclusion

Total nucleic acid extracts prepared in about one hour by reversible binding on silica particles in the presence of guanidinum thiocyanate (GTC) proved to be suitable for RT-PCR detection of plant viruses: PPV, ACLSV, PDV, ASPV and PVYV belonging to the different virus groups. Therefore, silica capture (SC) seems to be useful and potentially universal method for the first step of RT-PCR based detection of plant viruses.

5. Literature

Boom R., Sol C.J.A., Salimans M.M.M., Jansen C.L., Wertheim-Van Dillen P.M.E. and Van Der Nordaa J. (1990) Rapid and simple method for purification of nucleic acids. J. Clin. Microbiol. 28:495-503.

Jelkmann W., (1994) Nucleotide sequences of apple stem pitting virus and of the coat protein gene of a similar virus from pear associated with vein yellows disease and their relationship with potex- and carlaviruses. J. Gen. Virol. 75:1535-1542.

Nemchinov L., Hadidi A., Candresse T., Foster J.A., Verderevskaya T., (1995) Sensitive detection of Apple chlorotic leafspot virus from infected apple and peach tissue using RT-PCR, IC-RT-PCR or multiplex IC-RT-PCR. Acta Horticulturae 386:51-62.

Parakh D.R., Shamloul A.M., Hadidi A., Waterworth H.E., Scott A.S., Howell W.E., Mink G.I., (1995) Detection of prune dwarf ilarvirus from infected stone fruits using reverse transcription-polymerase chain reaction. Acta Horticulturae 386:421-430.

Silica particles suspension:
Sixty grams of silica powder (SIGMA S5631) were suspended in water (total volume 500 ml) and left in 500 ml cylinder. After 24 hours 430 ml of supernatant was removed by suction, water was added to 500 ml, solution was shaked and left again for 5 hours. Finally 440 ml of supernatant was removed and 600 µl of 32% HCl was added to remaining 60 ml suspension to adjust pH to 2. Suspension was aliquoted, autoclaved 20 min. at 121°C and stored in darkness at room temperature up to 6 months.

Buffers used in SC protocol

L6: 120 g of guanidinum thiocyanate (GTC) was dissolved in
100 ml of 0.1 M Tris hydrochloride pH 6.4,
then 22 ml of 0.2 M EDTA pH 8.0
and 2.6 g of Triton X-100 were added

L2: 120 g of guanidinum thiocyanate (GTC) was dissolved in
100 ml of 0.1 M Tris hydrochloride pH 6.4

Buffers should be prepared in a fume hood. They may be stored in darkness for up to three weeks. For detailed recommendations, see Boom et al. (1990).

Figure 1. Preparation of the silica particles suspension and buffers used in SC protocol.

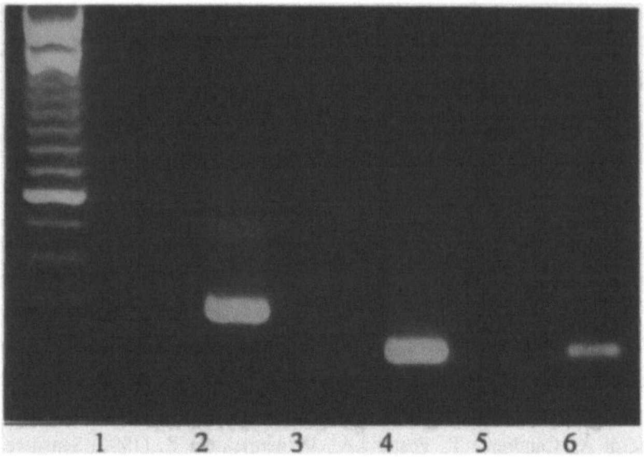

Figure 2. Detection of PVYV by SC-RT-PCR
Samples: 1,3,5) healthy pear leaves
2,4,6) leaves of pear infected with PVYV
Primers: 1,2) ASPf5/ASPr1
3,4) ASPf1/ASPr1
5,6) nested PCR: ASPf5/ASPr6,
then ASPf1/ASPr2

NON-RADIOACTIVE PROBE FOR THE DETECTION OF STRAWBERRY VEIN BANDING VIRUS

I. MRÁZ, J. HONETŠLEGROVÁ, M. ŠÍP and K. PETRZIK
Institute of Plant Molecular Biology, Department of Plant Virology, Academy of Sciences of the Czech Republic, Branišovská 31, 370 05 České Budějovice

Abstract

The first non-radioactive digoxigenin labeled cDNA probes were prepared from part of coat protein gene of the American (No.45058, American Type Culture Collection), Czech (A52/E6), German (Chandler/E6) and Norwegian (Mimek/Alpine) isolates of strawberry vein banding virus (SVBV). All five American different air-dried SVBV leaf samples originating from National Clonal Germplasm Repository (N.C.G.R., Corvallis, U.S.A.) and six from twelve Czech strawberry isolates exhibiting SVBV-like infection symptoms reacted positively in dot blot hybridization with American probe. Twenty six Czech strawberry SVBV samples were used in dot blot with probes prepared against the Czech, American, German and Norwegian isolates. Our results showed that the samples reacted identically, independently on the origin of the probe. They confirm the spread of SVBV in Central Europe and introduce the first reliable screening method for this virus.

1. Introduction

Strawberry vein banding virus (SVBV) has been observed on cultivated strawberries in North America, Australia, Brazil, Japan (Frazier and Morris, 1987) and recently in Europe by Blystad and Schia (Converse, 1992) and Honetšlegrová (Honetšlegrová *et al.*, 1995). The virus may reduce both yield and fruit quality. SVBV usually occurs in host plants at low concentrations and in complexes with other viruses which tend to obscure SVBV symptoms (Frazier and Morris, 1987). Therefore the diagnosis of this virus is very difficult. The classical detection methods which have been used, such as grafting onto indicator clones or serological tests (ELISA) are either time consuming or unreliable. So far the antiserum against SVBV has not been prepared.

H.-W. Dehne et al. (eds.), Diagnosis and Identification of Plant Pathogens, 449–451.

2. Materials and methods

A nonradioactive cDNA probe labeled with digoxigenin was prepared from part of coat protein gene of genomic DNA of the American SVBV isolate. Freeze - dried *Escherichia coli* JM83 containing clone pSVBV-E3 No.45058 was obtained from American Type Culture Collection. The pUC8 plasmid with SVBV DNA insert was isolated, genomic SVBV DNA was excised by *EcoR*I, cleaved by *Hind*III and recloned to pBS-. The nonradioactive probe was prepared from fragments isolated from pBS- and purified on low melting agarose. The *Hind*III fragments were labeled with Klenow enzyme in the presence of DIG-dUTP (Boehringer). The Czech (A52/ E6), German (Chandler/E6) and Norwegian (Mimek/Alpine) non-radioactive probes labeled with digoxigenin were obtained from strawberry plants infected with SVBV using isolation of genomic SVBV DNA based on alkaline lysis method.

Twelve Czech virus isolates from strawberry plants which revealed SVBV-like symptoms after grafting onto indicator clones *Fragaria vesca* and five American air-dried strawberry isolates of SVBV originated from N.C.G.R. from the collection of strawberry viruses originally assembled by R.H. Converse were tested in dot blot by American probe (45058). *Fragaria vesca* UC4 cultivated from tissue culture acquired from N.C.G.R. in Corvallis was used as healthy control.

Twenty six Czech SVBV samples from strawberry plants were tested in dot blot with probes prepared against American (45058), Czech (A52/E6), German (Chandler/E6) and Norwegian (Mimek/ Alpine) SVBV isolates. DNA isolation from leaf tissue was based on alkaline lysis method. Five μl of isolated DNA SVBV were used per dot in dot blot hybridization.

3. Results and discussion

Six of twelve Czech strawberry isolates exhibiting SVBV-like infection symptoms reacted positively in dot blot hybridization with the American probe. The other six Czech isolates did not react although they revealed marked symptoms of vein banding. All air-dried American isolates of SVBV reacted positively in dot blot with American nonradioactive probe.

The results of these studies seem to indicate that some SVBV-like symptoms could be induced by complex of viruses or by another virus, which did not react with the probe prepared against the American SVBV isolate. Indeed, in these Czech samples isometric virus particles of about 35 nm in diameter were observed in addition to 41 nm caulimovirus particles (Honetšlegrová *et al.*, 1995). Comparison of dot blot hybridization of the American sample 9044.001, three Czech samples I8, A52, A89K and healthy control (*Fragaria vesca* UC 4) shows fig.1.

Sensitivity of the dot blot assay using different fresh sample dilutions from 0,25 mg of sample/dot to 0,01 mg of sample/dot was tested.

Best performance in dot blot was obtained for 0,15 mg of fresh sample per dot although the minimum quantity required for reliable assay was only 0,10 mg of fresh sample per dot.

Results of tests performed on American and Czech strawberry samples with American probe prove the applicability of dot blot hybridization using a nonradioactive probe for reliable detection of SVBV in leaf extracts. Positive reactions in dot blot hybridization on samples originating from Czech Republic confirm the presence of SVBV in the Czech Republic.

Twenty six Czech strawberry SVBV samples were tested in dot blot with four probes prepared against American, Czech, German and Norwegian SVBV isolates. Our results from dot blots with these probes were independent on the origin of the probes, which proves a high degree of homology of four SVBV isolates.

Acknowledgements

This research work was supported by the grants of the Grant Agency of the Czech Republic No.513/93/2490 and No.522/96/0854.

References

Converse, R.H. (1992) Modern approaches to strawberry virus research, *Acta Horticulturae* **308,** 19-30.
Frazier, N.W. and Morris, T.J. (1987) Strawberry vein banding, in R.H. Converse (ed.), *Virus diseases of small fruits,* p. 16-20.
Honetšlegrová, J., Mráz, I. and Špak, J.(1995) Detection and isolation of strawberry vein banding virus in the Czech Republic, *Acta Horticulturae* **385,** 29-32.
Stenger, D.C., Mullin, R.H. and Morris,T.J. (1988) Isolation, molecular cloning and detection of strawberry vein banding virus DNA, *Phytopathology* **78,** 154-159.

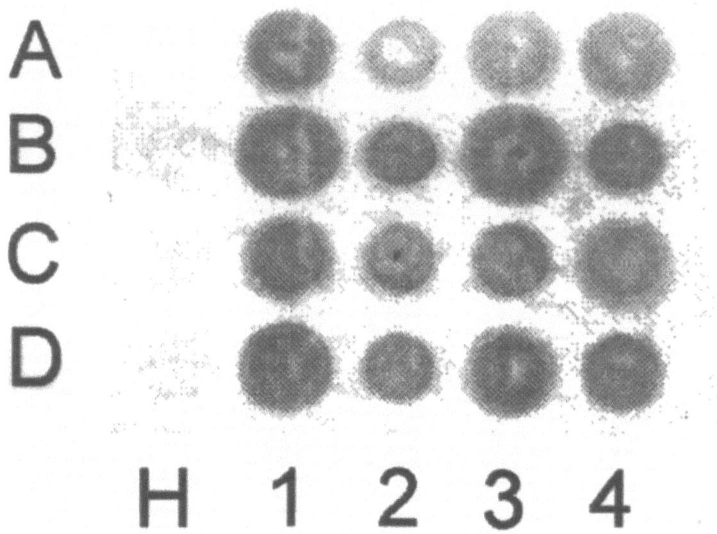

Figure 1. Dot blot reaction of Czech SVBV isolates I8 (column 1 A-D), A89 (2 A-D), A52 (3 A-D) and American SVBV isolate 9044.001 (4 A-D). Healthy control *F. vesca* UC4 clone (H A-D).

DETECTION OF BACTERIAL DISEASES OF TOMATOES BY THE POLYMERASE CHAIN REACTION (PCR) AND SEROLOGY

NABIZADEH-ARDEKANI, FERIDON; KOOPMANN, BIRGER and RUDOLPH, KLAUS
Institut für Pflanzenpathologie und Pflanzenschutz der Universität, Grisebachstr. 6, 37077 Göttingen, Germany

1. INTRODUCTION

Tomatoes contribute a large portion of vegetables grown in Turkey. Production is still increasing and bacterial diseases are causing high yield losses [1]. Because effective bactericides are not available and tomatoes are mainly grown on small farms with limited budgets, disease control is limited [2,3]. The present strategy to reduce yield losses by bacterial diseases is to use healthy seeds and resistant cultivars in combination with chemical, biological or cultural measures [4,5]. Therefore, highly sensitive detection methods are required for improved indexing of seed by the plant protection service in Turkey.

2. MATERIALS AND METHODS

Bacteria were grown overnight in Rhodes′ medium [6]. Total DNA was extracted according to Ausubel *et al.* [7].The oligonucleotides used for PCR based diagnosis were derived from a region of the *cfl* gene of *Pseudomonas syringae* pathovars as described by Bereswill *et al.* [8].

For PCR amplifications 10 ng of total DNA were used. 50 µl PCR mixture containing 0.8 U Taq-polymerase, 25 pmol of each primer, 0,2 mM dNTPs, 10 mM Tris-HCl pH 8.8, 50 mM KCl, 0.08% Nonidet P40, 5% DMSO, 10 mM 2-mercaptoethanol, and 100 µg bovine serum albumine. After the first denaturation the polymerase was added (hot start). The PCR profile consisted of 35 cycles each 2′ 92 ° C; 2′ 59 ° C and 30′′ 72 ° C. Serological methods were used as described by Baharuddin *et al.* [9].

3. RESULTS

PCR technique: the primers [8] used for the amplification of a region from a gene coding for an enzyme involved in coronatine biosynthesis (*cfl*) led to a successful amplification of a 650 bp fragment from *Pseudomonas syringae* pv. *tomato* (*Pst*) strain

H.-W. Dehne et al. (eds.), Diagnosis and Identification of Plant Pathogens, 453–455.

GSPB-no. 487. After optimizing the reaction conditions the identity of the amplification product with the sequence published by Bereswill *et al.* [8] was checked by both restriction analysis and sequencing. The sequencing of the amplification product confirmed the identity of the product (98% identity).

To determine the minimum number of cells to be detected by PCR, a culture of *Pst* (strain GSPB 487) was serially diluted in distilled water. A clear signal of *Pst* could be detected down to 5×10^2 cells/PCR reaction mixture (50 µl). Even at 5×10 cells/PCR reaction mixture a very weak signal was visible.

The specificity of primers was checked with a collection of *Pst* strains originating from various geographic locations of Turkey. All strains tested so far resulted in a positive PCR-signal (Fig. 1).

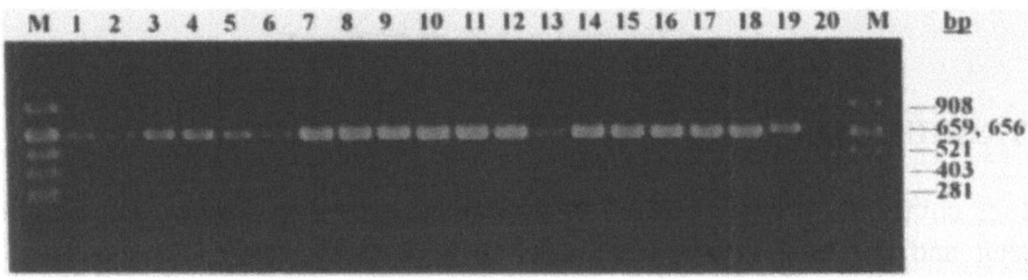

Fig. 1 Specifity test of the *cfl*-primers for the detection of various *P. tomato* isolates from different regions of Turkey. (M, marker; lanes 1-17, *Pst* strains from Turkey; lane 18, + control with *Pst* strain GSPB 493; lane 19, + control with total DNA from *Pst* GSPB 493; lane 20, - control).

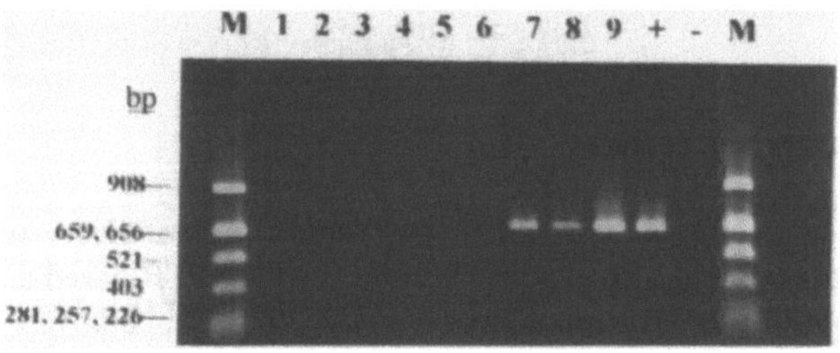

Fig. 2. PCR products amplified from various pathogenic and epiphytic bacteria (M, marker; lane 1, *P. solanacearum*; lane 2, *P. s.* pv. *syringae*; lane 3, *P. fluorescens*; lane 4, *X. c.* pv. *vesicatoria*; lane 5, *C. m.* ssp. *michiganensis*; lane 6, *P. corrugata*; lane 7, *P. s.* pv. *glycinea*; lane 8, *P. s.* pv.*morsprunorum*; lane 9, *P. s.* pv. *tomato*; +, positive control with total DNA from *Pst*; -, negative control).

Furthermore, *in vitro* tests showed successful amplification with other coronatine producing plant pathogenic bacteria. Non-producing bacteria showed no signal (Fig. 2).

Serology: the antiserum against *Xanthomonas campestris* pv. *vesicatoria* was tested with three strains (GSPB-nos. 2050, 2045, 224). Cross reactions were tested with *Pseudomonas corrugata, Pseudomonas s.* pv. *tomato, Burkholderia (Pseudomonas)* *solanacearum, Clavibacter m.* ssp. *michiganensis, Pseudomonas fluorescens.* The three *X. c.* pv. *vesicatoria* strains showed an identical protein band of 28 kDa which did not occur in any other bacterial species or pathovar tested.

4. DISCUSSION

The objective of our work was to develop diagnostic methods for sensitive detection of bacterial pathogens of tomato by serological as well as DNA techniques. The PCR technique was successfully applied to amplify part of a DNA sequence (*cfl*) coding for coronafacate ligase [8]. This enzyme catalyses the coupling of coronafacic acid and coronamic acid. Our results revealed that non-coronatine-producers (*P. solanacearum, P. s.* pv. *syringae, P. fluorescens, X. c.* pv. *vesicatoria, C. m.* ssp. *michiganensis* and *P. corrugata*) did not give a positive PCR signal. Therefore, this technique might be useful for sensitive detection of *P. s.* pv. *tomato* in tomato plants, especially since all virulent *P.s.* pv. *tomato* strains seem to produce coronatine [10]. The *in planta* detection of the pathogen in artificially inoculated plant material (seed, leaves) was also successful.
The incitant of bacterial spot, *Xanthomonas campestris* pv. *vesicatoria,* could be characterized by a specific protein band of 28 kD. Further experiments have to reveal whether this protein band can be used to prepare a highly specific antiserum.

5. REFERENCES

1. Karaca, I., Saygili, H.: Investigations on disease rate, causal agents and symptoms of bacterial diseases of tomatoes and sensitivity of the host varieties in some parts of western Turkey. J. Turkish Phytopath. **11**, (1982), 120-121.
2. Özaktan, H.: Investigations on control means of tomato bacterial canker caused by *Clavibacter michiganensis* subsp. *michiganensis.* Dissertation, (1991), University of Ege (Izmir).
3. Pilavci, O., Ulukus, I.: Studies on possibilities of using Troleandomycin as a seedling treatment chemical against tomato bacterial canker. II. In vivo effectiveness of the antibiotic and its comparison with streptomycin. J. Turkish Phytopath. **16,** (1987), 67-70.
4. Öktem, Y. E.: Studies on determination of susceptibility of tomato varieties against *Corynebacterium michiganensis* pv. *michiganensis.* IV. Türkiye fitopatoloji kongresi, 8-11 Ekim, (1985), Izmir, Turkey.
5. Jones, J. B., Jones, J. P., Stall, R. E., Zitter T. A.(edts): Compendium of Tomato Diseases. The American Phytopathological Society St. Paul, Minnesota 55121, USA, (1991), 73 pp.
6. Rhodes, M. E.: The characterization *of Pseudomonas fluorescens.* J. Gen. Microbiol. **21**, (1959), 221-263
7. Ausubel, I., Frederik, M.: Preparation of genomic DNA from bacteria. Current protocols in molecular biology: (1987), Unit 2.4.
8. Bereswill, S., Bugert, P., Völksch, B., Ulrich, M., Bender, C., Geider, K.: Identification and relatedness of coronatine-producing *Pseudomonas syringae* pathovars by PCR analysis and sequence determination of the amplification products. Appl. Environ. Microbiol. **60**, (1994), 2924-2930.
9. Baharuddin, B., Rudolph, K., Niepold, F.: Production of monospecific antiserum against the blood disease bacterium affecting banana and plantain. Phytopathology **84**, (1994), 570-575.
10. Bender, C. L., Stone, H. E., Sims, J. J., Cooksey, D. A.: Reduced pathogen fitness of *Pseudomonas syringae* pv. *tomato* Tn5 mutants defective in coronatine production. Physiol. Mol. Plant Pathol. **30**, (1987), 273-283.

DETECTION OF *Agrobacterium tumefaciens* IN SOIL USING PCR

J.PULAWSKA
Research Institute of Pomology and Floriculture, Pomologiczna 18
96-100 Skierniewice, Poland

1. Introduction

A. tumefaciens [8] is a soil-borne bacterium which causes crown gall of various dicotyledonous plants. Tumor formation occurs by transfer of T-DNA located on Ti plasmid of bacterium to the plant cell. Heavily infected plants, especially those with disease symptoms on main roots and root collar are disqualificated and should not be planted. The pathogenesis of crown gall is specific and differs from other plant diseases [1].The classical detection methods of *A. tumefaciens* are time consuming, not sufficiently sensitive and difficult to perform because putative bacteria must be isolated, purified and tested individually for pathogenicity. The PCR (polymerase chain reaction) allows detection of variety of plant pathogenic bacteria in short time and with high a sensitivity and specificity. Some PCR-based methods for detection of *A. tumefaciens* use primers corresponding to different plasmid sequences e.g. *tmr* [4], *vir*D, *vir*G [9], *vir*D [7], *vir*C [13], *6a, virA* [5] or to chromosomal DNA *pehA*, [5].

The purpose of the presented study was to determine the specificity of primers designed to amplify a fragment of *tms2* gene located in T-DNA and to develop the procedure of extraction of *A.tumefaciens* DNA from soil. The sequences of primers were worked out at Technical University in Gdańsk (Poland) [11].

2. Materials and methods

2.1. BACTERIAL STRAINS

All bacterial strains used in this study are listed in Table 1. Their pathogenicity was tested on sunflowers seedling according to Sobiczewski [14].

2.2. SOIL CHARACTERIZATION

Three types of soil with similar content (15 - 20%) of silt and clay particles (<0.02mm) and different carbon contents (0.5, 1.2, 2.1 %) were used in the experiment. pH (KCl) of soils was 6.3. All three types of soil were colonized by saprophytic bacteria at concentration of about 10^6 cfu/g.

2.3. SOIL PREPARATION

Soil samples (1g) placed in Eppendorf tubes were autoclaved at 121°C for 30 min during 3 consecutive days. After that sterility was checked using plating method on the King's B medium.

Sterile and nonsterile soil samples (1g) were inoculated with 10^4, 10^3, 10^2 and 50 cfu of

H.-W. Dehne et al. (eds.), Diagnosis and Identification of Plant Pathogens, 457-460.

Table 1. Bacterial strains tested, their pathogenicity to sunflower and results of PCR

Species and strains	Source and location	Pathogenicity on sunflower	PCR
Agrobacterium tumefaciens biovar1			
ACH5	Unknown; M van Montagu (Belgium)	+	+
AT4	Tart cherry; S.A. Piotrowski (Poland)	+	+
B6	Tomato; P.Manigault (France)	+	+
C58	Sweet cherry; R.Dickey (USA)	+	+
T37	Unknown; T.Tempe (France)	+	+
13B	Unknown; J.H.Haas (Israel)	+	+
Agrobacterium tumefaciens biovar2			
K27	Poplar; A.Kerr (Australia)	+	+
31a/b, 32a,32b,39, 40,	Cherry; P.Sobiczewski (Poland)	+	+
88, 101, 107, 125, 126,		+	+
129, 133, 137, 301,		+	+
387, 389, 392, 400		+	+
60, 67, 69, 71, 74, 77	Sweet cherry; P.Sobiczewski (Poland)	+	+
81, 83, 84, 89, 221b, 365		+	+
46, 182, 192, 262	Pear; P.Sobiczewski (Poland)	+	+
P-Ag-1,2,3,4,6	Pear; H.Sawada (Japan)	+	+
131, 307, 580, 15b, 21	Apple; P.Sobiczewski (Poland)	+	+
32/1	Apple; S.Süle (Hungary)	-	-
103, 104, 104	Peach; P.Sobiczewski (Poland)	+	+
30/2, 9	Peach; S.Süle (Hungary)	+	+
Pch-Ag-2,3,4,5,6	Peach, H.Sawada (Japan)	+	+
122, 568,	Plum; P.Sobiczewski (Poland)	+	+
8/1, 51, 9/2	Raspberry; S.Süle (Hungary)	+	+
65, 7/1	Raspberry; S.Süle (Hungary)	-	-
Ch-3,6,11,12	Chrysanthemum; P.Sobiczewski (Poland)	+	+
Agrobacterium radiobacter			
G114-3-2 biovar1	Unknown; J.H.Haas (Israel)	-	-
K84 biovar2	Peach; A.Kerr (Australia)	-	-

Species and strains	Source and location	Pathogenicity on sunflower	PCR
Agrobacterium vitis biovar 3			
AB4, 2/3S5R, Tm4	Grapevine; E.Szegedi (Hungary)	+	+
CG49	Grapevine; T.J.Burr (USA)	+	+
Ag162	Grapevine; C.Panagopoulos (Greece)	+	+
G114-5-IV-1	Grapevine; J.H.Haas (Israel)	+	+
G115-1-2	Grapevine; J.H.Haas (Israel)	-	-
0	Grapevine; S.Süle (Hungary)	+	+
K-Ag -1	Kiwifruit; H.Sawada (Japan)	+	+
Agrobacterium tumefaciens unclassified			
K-Ag -3	Kiwifruit; H.Sawada (Japan)	+	+
Ch-Ag -4	Cherry; H.Sawada (Japan)	+	+
Agrobacterium rhizogenes			
LMG150	Unknown; M.Maes (Belgium)	+[1)]	-
P-Ag-5	Unknown; H.Sawada (Japan)	+[1)]	-
Arthrobacter globiformis	Unknown; A.Skorupska (Poland)	-	-
Arthrobacer pascens	Unknown; A.Skorupska (Poland)	-	-
Arthrobacer ramosus	Unknown; A.Skorupska (Poland)	-	-
Bradyrhizobium japonicum	Soybean; A.Skorupska (Poland)	-	-
Rhizobium leguminosarum	Pea; A.Skorupska (Poland)	-	-
Rhizobium loti	Lotus; A.Skorupska (Poland)	-	-
Rhizobium meliloti	Alfalfa; A.Skorupska (Poland)	-	-
Pseudomonas fluorescens	Unknown; A.Skorupska (Poland)	-	-
Erwinia amylovora	Apple; P. Sobiczewski (Poland)	-	-
Erwinia chrysantemum	Chrysantemum; J.Bogatko (Poland)	-	-
Xanthomonas campestris	Unknown; J.Bogatko (Poland)	-	-
unidentified soil bacteria	Soil; J.Pulawska (Poland)		
G1, G8, G9, G10		-	-
G11, G15		-	-

[1)] Bacteria cause hairy roots

A.tumefaciens, strain C58, suspended in 100µl of sterile water. Concentration of inoculum was determined by dilution plating method on King's B medium.

2.4. PREPARATION OF TEMPLATE DNA FROM PURE CULTURES

Total bacterial DNA was extracted by using a modified method described by Sambrook et all. [12]. Strains of *Rhizobium* and *Arthobacter* were cultured in 79CA medium, whereas other strains in Nutrient Broth (Difco). For DNA isolation 1.5 ml of overnight culture was centrifuged for 10 min at 5000 g. The pellet was resuspended in 100µl of TE buffer (10mM TRIS-HCl, 1mM EDTA pH 8.0). To lyse the cells 200µl of 0.5% SDS, 0.1M NaOH buffer was added and then the mixture was incubated at 65°C for 15min. Next 200µl of 5M NaCl solution was added, the mixture was carefully mixed and incubated in ice for 15 min. After centrifugation (10min at 15000 g) DNA was precipitated with 2 volumes of EtOH at -20°C. The pellet was washed with 70% EtOH, dried and dissolved in 100µl of 10 mM TRIS-HCl buffer, pH 8.2.

2.5. EXTRACTION OF BACTERIAL DNA FROM SOIL

After inoculation, soil samples were incubated at room temperature for 1 hour, suspended in 10 ml of sterile water and shaken in orbital shaker at room temperature for 30 min. 3 ml of resulting suspension was transferred to 20 ml of a liquid a medium. Nutrient Broth medium supplemented with 1.25 % of glucose was used in case of sterile soil whereas in case of nonsterile soil modified Brisbane & Kerr medium (20mM arabitol, 25mM erythritol, 2mM NH_4NO_3, 4mM KH_2PO_4, 6mM K_2HPO_4, 1mM $MgSO_4$ x $7H_2O$, 0.5 mM sodium taurocholate, 0.5 mM Na_2SeO_3.) was used [3]. Bacteria suspended in the above media were incubated in an orbital incubator at 26°C. After 22h 1.5 ml of the culture was centrifuged and a pellet was resuspended in a mixture of 100 μl TE and 200 μl LT buffers. Bacterial cells were lysed in four cycles of freezing at -55°C and boiling followed by proteinase K (30µl 20mg/ml) treatment at 37°C for 1 hour. DNA was purified on column (Genomic DNA Prep Plus, A&A Biotechnology, Gdańsk, Poland) and than eluted with 400µl of hot 10mM TRIS-HCl, pH 8.2 buffer. DNA was precipitated with two volumes of ethanol and 1/10 volume of 3M sodium acetate and the pellet was redissolved in 20 µl of 10mM TRIS-HCl, pH 8.2 buffer.

2.6. PCR

PCR amplification was performed in a total volume of 25µl under the layer of mineral oil in Perkin-Elmer DNA Thermal Cycler 480. PCR experiment was done in 1 x PCR amplification buffer (10 mM TRIS-HCl, pH 8.8; 1.5mM $MgCl_2$; 50mM KCl; 0.1% Triton X-100) with 200µM of each dNTP-s; 1µM of each primer and 1.5 U of thermostable DNA polymerase (Biometra, Göttingen, Germany). An initial 2 min denaturation step (93°C) was followed by 42 cycles of a three-step PCR program (93°C, 15s; 67°C, 15s; 72°C, 10s) and a final 2 min extension step at 72°C for 2 min.

Extraction of bacterial DNA from soil and it's amplification using this method was performed at least in three replications.

3. Results and discussion

3.1. SPECIFICITY OF PRIMERS

Of 106 strains of phytopathogenic and saprophytic bacteria originated from various plants, geographical regions and classified to different taxons only tumor-inducing strains produced in PCR specific 220bp product (Table 1)

3.2. DETECTION OF *A.TUMEFACIENS* IN INOCULATED SOIL

Our method of nucleic acids extraction from soil facilitates detection of at least 50 cfu of *A. tumefaciens* in 1g either of sterile or nonsterile soil. Results obtained from each series of experiment were similar. The sensitivity of this method was irrespective of soil humus content. This finding is very important because humic acids are known as the inhibitors of PCR [16]. Another advantage of the method is omitting organic chemicals, e.g. SDS and phenol, which also inhibit polymerase activity [17]. Techniques described by other authors allow to detect 10^3-10^4 cfu/g of *A.tumefaciens* and other soil bacteria [6,10,15,16]. From practical point of view detection of possibly a lowest number of crown gall bacteria is of a great importance. For plant cell transformation only one bacterial cell is sufficient, whereas in case of other diseases higher population of pathogen is necessary for infection and typical symptom expression [2]. Incubation of bacteria in artificial medium allowed to increase their population, reduce humic acids concentration and to avoid extracellular and dead cell DNA detection. After extraction, DNA was so pure that it could be used directly for PCR without any further dilution.

Acknowledgements
I wish to thank P.Sobiczewski for helpful discussion and revising the text, P. Sachadyn for design the primers, J.H.Haas, M.Maes, A. Skorupska, S. Süle, H.Sawada, for bacteria strains.

1. Agrios G.N. (1988) *Plant Pathology*,3rd ed. Academic Press, New York
2. Billing E. (1987) *Bacteria as plant pathogens,* Van Nostrand Reinhold Co. Ltd, England
3. Brisbane P.G., Kerr A. (1983) Selective media for three biovars of *Agrobacterium, J.Appl. Bacteriol* **54**, 425
4. Dong L. -C., Sun C.-W., Thies K.L., Luthe D.S., Graves C.H. Jr. (1992) Use of polymerase chain reaction to detect pathogenic strains of *Agrobacterium, Phytopathology* **82**, 434-439
5. Eastwell K.C., Willis L.G., Cavileer T.D (1995) A Rapid and Sensitive Method to Detect *Agrobacterium vitis* in Grapevine Cuttings Using the Polymerase Chain Reaction, *Plant Dis.* **79**,822-827
6. Flemming C.A., Leung K.T., Trevors J.T., Greer C.W. (1994) Survival of *lux-lac*-marked biosurfactant-producing *Pseudomonas aeruginosa* UG2L in soil monitored by nonselective plating and PCR, *Appl. Environ. Microbiol* **60**, 1606-1613
7. Haas J.H., Moore L.W., Ream W.,(1995) Universal PCR primers for detection of phytopathogenic *Agrobacterium* strains, *Appl. Environ. Microbiol* **61**, 2879-2884
8. Lippincott J.A., Lippincott B.B., Starr M.P. (1983) The genus *Agrobacterium, Phytopathogenic bacteria,* Springer Verlag, New York, 842-855
9. Nesme X., Leclerc M.C., Bardin R. (1990) PCR detection of an original endosymbiont: the Ti plasmid of *Agrobacterium tumefaciens.* p. 47-50 W P.Nardon, V. Gianinazzi-Pearson, A.M. Greinier, L. Margulis, D.Smith *Endocytobiology* Institut National de la Recherche Agronomique, Paris
10. Picard C., Ponsonnet C., Paget E., Nesme X., Simonet X. (1992) Detection and enumeration of bacteria in soil by direct DNA extraction and polymerase chain reaction, *Appl. Environ. Microbiol* **58**, 2717-2722
11. Sachadyn P., Kur J. PCR detection of *Agrobacterium tumefaciens,* Abstracts of 3rd International Students Scientific Conference, 10-14.05.1995, Gdańsk, Poland,4
12. Sambrook J., Fritsch E.F., ManiatisT. (1989) *Mol. cloning, a lab. manual 2nd ed.* Cold Spring Harbor Press.
13. Sawada H., Ieki H., Matsuda I. (1995) PCR detection of Ti and Ri Plasmids from phytopathogenic *Agrobacterium* strains, *Appl. Environ. Microbiol,* **61**,828-831
14. Sobiczewski P. (1995) Studies on crown gall of fruit trees (*A. tumefaciens*), Res. Inst. Pomol. Floric. Skierniewice, 96 pp, Poland, (in Polish)
15. Tas E., Saano A., Lienonen P., Lindström K. (1995) Identification of *Rhizobium* spp. in peat-based inooculants by DNA hybidization and PCR and its application in inoculant quality control, *Appl. Environ. Microbiol* **61**, 1822-1827
16. Tsai Yu-Li, Olson B.H.,(1992) Detection of low numbers of bacterial cells in soils and sediments by polymerase chain reaction, *Appl. Environ. Microbiol* **58**, 754-757
17. Weyant P.S., Edmands P., Swaminathan B. (1990) Effect of ionic and nonionic detergents on the Taq polymerase, *Biofeedback* **9**, 308-309

SEQUENCE ANALYSIS OF THE RIBOSOMAL DNA SPACER REGION OF *PYRENOPHORA* SPP.; EVALUATION OF ITS POTENTIAL FOR PCR DETECTION IN A SEED HEALTH TEST

STEVENS, ELISABETH A., BLAKEMORE, EMILY J. A. AND REEVES, JAMES C.
National Institute of Agricultural Botany, MBD Section, Huntingdon Road, Cambridge. CB3 OLE. United Kingdom.

1. Introduction

The features of the three species of the *Pyrenophora* fungus which cause diseases of barley are summarised in Table 1. Seed transmission is an important feature of the disease complex and therefore seed testing is essential as part of an effective control strategy. Conventional seed health tests are time-consuming and labour-intensive. PCR-based techniques have the potential to overcome these problems. Here we evaluate the potential of the fungal intergenic region for development of a PCR-based detection strategy for these three pathogens. Sequencing of this region has been successfully used in other systems to design specific primers for PCR (Xue *et al.*, 1992).

TABLE 1. Summary of disease transmission and appropriate control

Fungal Species	Disease	Transmission	Appropriate action
P. graminea	leaf stripe	strictly seed-borne	seed treatment
P. teres	net blotch	mainly crop debris	foliar spray
P. teres f. sp. maculata	spot form of net blotch	mainly crop debris	foliar spray

2. Method

Sequencing of the ribosomal DNA spacer region was done using the universal primers designed for this purpose (White *et al.*, 1990). Several isolates of each species were sequenced, as well as some isolates of the related oat pathogen *P. avenae* (Table 2). An initial PCR was performed using primers ITS4 and ITS5. The fragment generated was then sequenced using primers ITS1-4 in reactions containing Amplitaq® polymerase FS enzyme (Perkin Elmer) as directed by the manufacturer. Sequence analysis was subsequently done using DNAstar software.

H.-W. Dehne et al. (eds.), Diagnosis and Identification of Plant Pathogens, 461-463.

TABLE 2. Sources and Isolates of *Pyrenophora* spp. used in this study

Species	No. of isolates	Countries of Origin	Original host
P. graminea	8	Denmark, UK, Norway	Barley
P. teres f. sp. maculata	3	Denmark	Barley
P. teres	6	UK, Norway	Barley
P. avenae	4	Norway	Oats

3. Results and Discussion

- a DNA fragment approximately 600 bp in length was generated by ITS4 & 5 amplification from all barley *Pyrenophora* spp. (Figure 1)

- size difference (approx. 650 bp) of DNA fragment amplified from *P. avenae* (Figure 1)

- some intra-specific sequence variation, particularly when isolates have diverse geographical origins (Table 3)

- there is some inter-specific sequence variation between the three barley pathogens (Table 3)

- however, this inter-specific sequence variation is too inconsistent to design primers able to distinguish between the three barley pathogens

- *P. avenae* has a much higher degree of intra-specific sequence variation

- the high degree of sequence variation betweeen *P. avenae* and the barley pathogens analysed may be due to their different ecological niches

- future analysis of the IGS region of these fungi may reveal information about their genetic relatedness and provide further evidence for a race structure of *Pyrenophora* spp. rather than species

TABLE 3: Summary of percentage intra- and inter- specific sequence similarity

	P. graminea	*P. teres*	*P. maculata*	*P. avenae*
P. graminea	96.1-100 %	-	-	-
P. teres	91.9-99.4 %	96.2-100 %	-	-
P. maculata	96.1- 100 %	93.1-97.2 %	100 %	-
P. avenae	23.0- 61.0 %	23.0-60.1 %	21.0-57.2 %	23.0-97%

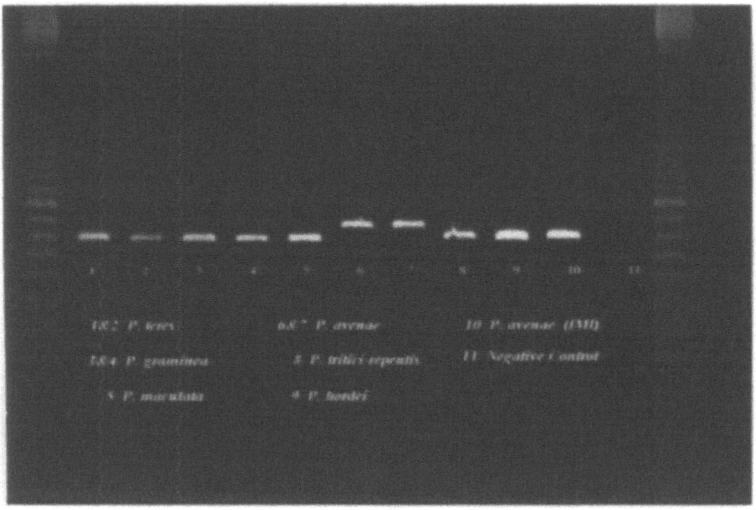

FIGURE 1: PCR amplification using ITS4 & 5 primers

4. References

White, T J; Bruns, T; Lee, S & Taylor J (1990) Amplification and direct sequencing of fungal ribosomal RNA genes for phylogenetics. In: *PCR Protocols: A guide to methods and applications*, M A Innis, D H Gelfand, J J Sninsky and T J White (eds.), USA: Academic press inc., pp. 315-322

Xue, B; Goodwin, P H & Annis, S L (1992) Pathotype identification of *Leptosphaeria maculans* with PCR and oligonucleotide primers from ribosomal internal transcribed spacer sequences. *Physiological and Molecular Plant Pathology* **41**, 179-188

5. Acknowledgements

This research was supported by a MAFF grant. We would like to thank Robert Feakes at the University of Cambridge for his assistance with sequencing, Dr Karen Husted and the International Mycologocal Institute for providing some of the *Pyrenophora* spp. isolates, and the members of the Molecular Biology and Diagnostic Section, NIAB for their technical support.

IDENTIFICATION OF POTYVIRUSES INFECTING *ALSTROEMERIA*

VAN DER VLUGT, RENÉ A.A. AND BOUWEN, INGE
DLO Research Institute for Plant Protection (IPO-DLO), Wageningen, The Netherlands.

Introduction

Alstroemeria is worldwide an important ornamental crop. It has been propagated vegetatively for many years, and contains several hitherto poorly defined viruses. Virus-infected plants usually show chlorotic leafstreaking and sometimes flower-colour breaking. Virus infections can cause serious yield losses and reduced vaselife of flowers. Two viruses of *Alstroemeria* have recently been investigated in more detail: Alstroemeria mosaic potyvirus (AlMV) and Alstroemeria streak potyvirus (AlSV) (Van Zaayen *et al*, 1994). Immuno-electron microscopy (IEM) studies have also revealed the presence of other potyviruses in *Alstroemeria* that show no serological relationship with AlMV (De Blank *et al.*, 1994).
At this moment both AlMV and AlSV can be detected and distinguished in ELISA using a polyclonal antiserum reacting with both AlMV and AlSV, and potyvirus groupspecific monoclonal antibodies that react only with AlMV. However, this situation is clearly not satisfactory and does not provide detection methods for the other potyviruses present in *Alstroemeria*. In order to clarify the virus status of *Alstroemeria* in the Member States of the European Union (EU) the European Committee (EC) has funded a project that should provide standardized detection methods for all major viruses present in this crop and tissue culture methods to clean up *Alstroemeria* germplasm and stocks. Within this project, IPO-DLO will identify, characterize, and develop detection methods for the potyviruses present in *Alstroemeria*.

Materials and Methods

The potyviruses often seem to occur in mixtures in virus infected *Alstroemeria*. For this reason a methodology was adopted that aimed at separation of these viruses and their subsequent individual characterization.
Single virus isolates were obtained from *Alstroemeria* showing typical virus symptoms by four serial passages through a local lesion host (e.g *Chenopodium quinoa*) and subsequently inoculated on a systemic host (e.g *Nicotiana benthamiana*). When a serological reaction was observed in IEM, virus was purified from infected plant material

H.-W. Dehne et al. (eds.), Diagnosis and Identification of Plant Pathogens, 465–467.

by an immunocapture procedure. Otherwise virus was partially purified by centrifugation through a sucrose cushion.
cDNA was synthesized from immunocaptured or purified virus using oligo-dT as first strand primer and subsequently amplified in a polymerase chain reaction (PCR) using oligo-dT as the downstream primer and a degenerate upstream 'potyvirus primer' matching the highly conserved M(V/I)WCI(E/D) motif in the potyviral coat protein (Fig. 1).

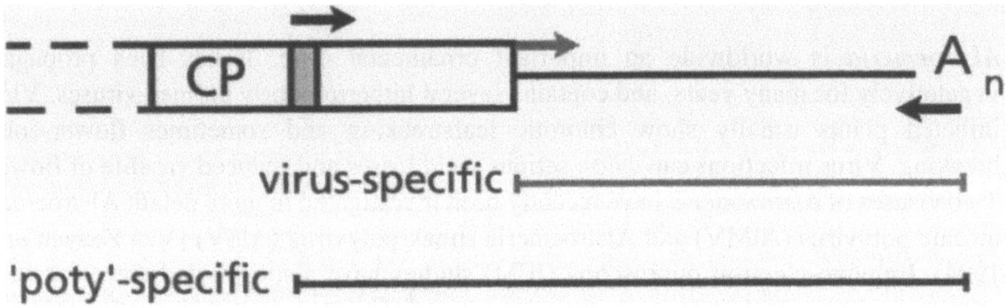

Figure 1. Position of primers on the potyvirus genome

Results

Results from test plants suggested the presence of at least four different viruses. PCR amplification generated several 'poty'-specific DNA fragments which were each cloned and sequenced. A homology search in the EMBL sequence database and an unpublished sequence of AlMV (G. Adam, personal communication) with these sequences indeed identified four distinct potyviruses i.e. AlMV, lily mottle virus (LMoV), Ornithogalum mosaic virus (OrMV) and a sofar unidentified potyvirus (AV-W024/1) that had no sequence homology with any of the registered potyviruses.
Based on the virus-specific 3'-nontranslated region (3'NTR), for each virus a specific primer was designed which, in combination with an oligo-dT primer, generated a virus-specific PCR fragment (Fig. 2).
RT-PCR on RNA isolated from 75 *Alstroemeria* plants in the IPO-DLO collection with typical virus symptoms, showed that both the degenerate 'poty'-primer (Fig. 2A) as well as the virus-specific primers (Fig. 2B) allowed specific detection of the individual viruses, even in mixed infections.

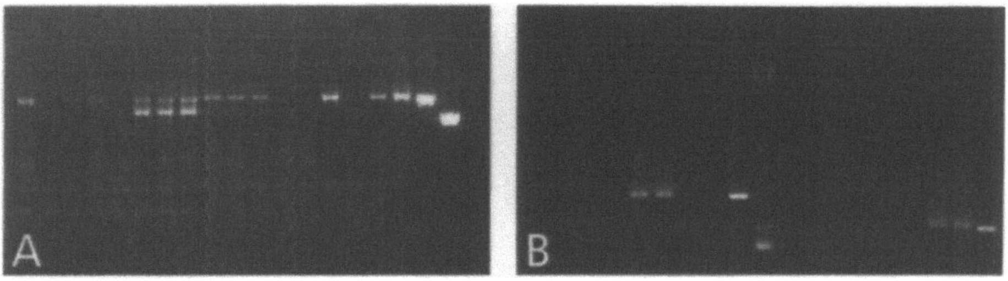

Figure 2. RT-PCR detection of *Alstroemeria* potyviruses using a degenerate 'poty'-primer (A) or a virus-specific primer (B).

Conclusion

Using an RT-PCR approach with a degenerate 'poty'-primer on *Alstroemeria* plants showing typical virus symptoms four potyviruses were identified. Apart from AlMV, this is the first report of LMoV, OrMV and the new potyvirus AV-W024/1, in *Alstroemeria*. Using both the degenerate 'poty'-primer as well as the virus-specific primers each of the individual viruses, can be detected in *Alstroemeria*, even in mixed infection.
The approach described above seems to offer good opportunities for the identification of additional potyviruses in *Alstroemeria* and the development of virus-specific detection methods.

References

De Blank, C.M., Van Zaayen, A. and Bouwen, I. (1994) Towards a reliable detection of Alstroemeria mosaic virus, *Acta Horticulturae* **377**, 199-208.

Van Zaayen, A., De Blank, C.M. and Bouwen, I. (1994) Differentiation between two potyviruses in *Alstroemeria*. *European Journal of Plant Pathology* **100**, 85-90.

DIFFERENTIATION OF *ALLIUM* POTYVIRUSES BY POLYMERASE CHAIN REACTION

H.J. VETTEN, A. BUTGEREITT and S. WINTER*
*Institute for Biochemistry and Plant Virology, Federal Research Centre for Agriculture and Forestry, BBA, Messeweg 11/12, 38104 Braunschweig, Germany * AG Pflanzenviren, German Collection of Microorganisms and Cell Cultures, DSMZ, c/o BBA Messeweg 11/12, 38104 Braunschweig, Germany*

1. Introduction

Cultivated *Allium*, such as onion, shallot and garlic are infected by various viruses, most predominantly by potyviruses, carlaviruses and mite-borne filamentous viruses. Only potyviruses generally cause conspicuous symptoms in *Allium* spp. particularly when present in mixed infection with viruses of other taxa, e.g. carlaviruses. Virus diagnosis is especially complicated when taxonomically related viruses occur in mixed infections or, when virus concentration in singly infected *Allium* spp. remains extremely low.

The problems of virus diagnosis in *Allium* spp., the partly unclear taxonomic position of viruses infecting *Allium* spp. and the putative occurrence of host adapted strains of viruses in garlic and onion, have prompted this study.

This paper focuses on onion yellow dwarf virus (OYDV) and addresses some of the problems pertaining to OYDV taxonomy, differentiation of OYDV strains and their diagnosis. To elucidate the relationship of the OYDV strains infecting onion and garlic, the 3' region of the viral genomes were determined and compared with sequences available for other *Allium* potyviruses.

2. Methods

RNA was extracted either from purified preparations of OYDV from onion (OYDV-O) or from garlic (OYDV-G) using a standard proteinase K, Phenol/Chloroform extraction protocol or, from infected plants using the protocol described by Chomczynski and Sacchi (1987). Virus RNA served as template for

H.-W. Dehne et al. (eds.), Diagnosis and Identification of Plant Pathogens, 469–471.

oligo $dT_{(12\text{-}18)}$ primed cDNA synthesis, and subsequent cloning (Sambrock *et al.*, 1989). Alternatively, a PCR-based cloning strategy was followed using total RNA extracts from infected plants as template for oligo $dT_{(12\text{-}18)}$ primed reverse transcription and subsequent PCR with the conserved potyvirus primer PO3 (Fig. 1) as upstream primer. PCR products were cloned using a TA cloning kit (Novagen). Nucleic acid sequencing was done using the ΔTaq™cycle sequencing kit (USB) and analysed using the DNASIS 2.1 for Windows (Hitachi) sequence analysis program.

3. Results

The sequence of the coat protein gene (CP) and 3' non translated region (3'-NTR) were determined for OYDV-O and for OYDV-G. A comparison of the deduced amino acid sequences of the CPs revealed a sequence similarity (identical amino acids) of 89%. The nucleotide sequence of their respective 3'-NTR had an identity of 79%. This striking similarity in CP amino acid and 3'-NTR sequence support the conclusion that OYDV-O and OYDV-G are strains of one virus. Comparing CP sequences of those of OYDV with sequences of those of other *Allium* infecting potyviruses, leek yellow stripe virus and shallot yellow stripe virus, sequence identities of only 50% and 67% were obtained indicating a distant relationship of OYDV to other *Allium* potyviruses.

Alignment of the nucleic acid sequences of the two OYDV strains revealed highly conserved regions (e.g. CP core region) and less conserved regions (3'-NTR). From the aligned sequences, oligonuclotides were chosen as primers for differential amplification of OYDV-O and OYDV-G. The primer sequences and their position in the 3' part of the genome are given in Fig. 1.

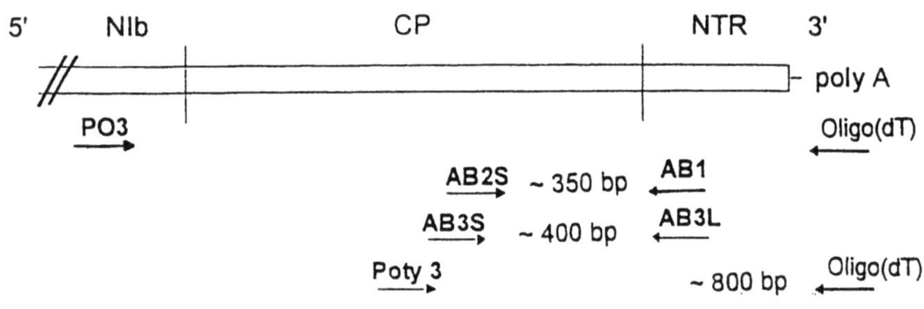

Figure 1. Map of 3' terminal part of the OYDV genome indicating the positions and product sizes of oligonucleotide primers for cDNA synthesis and for RT-PCR. The sequence of the primers is:
PO3: 5'-AAAGCGCC(AG)TTTATCTCTGAGGCA-3'
Poty 3: 5'-TGAGGATCCTGGTG(CT)AT(ACT)GA(AG)AA(CT)GG-3'
AB1: 5'-GACTAGAGACACCAGCGTTA-3' AB2S: 5'-TGAAGCATACATTGAATATA-3'
AB3L: 5'-AAAACTAAAAACATACAACT-3' AB3S :5'-AAAACCAACATTCAGACAAA-3'

In RT-PCR using total plant nucleic acid as template and the primer combinations shown in Fig. 1, specific amplification products corresponding in size to the primer combination used were obtained (Fig. 2).

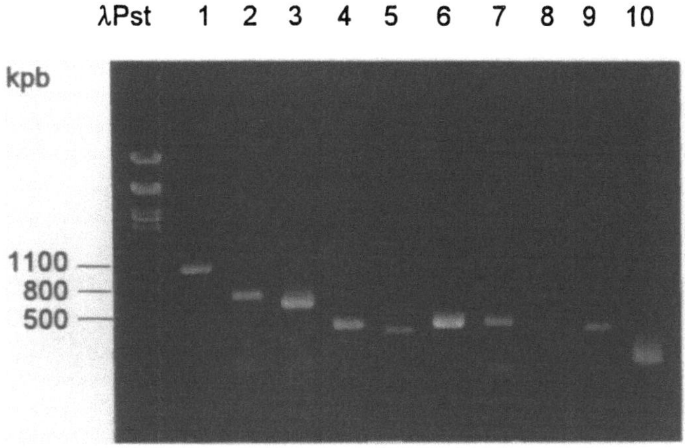

Figure 2. RT-PCR to detect and differentiate OYDV sequences in *Allium* spp. Lane 1, leek yellow stripe potyvirus amplification with the general potyvirus primer combination oligo $dT_{(12-18)}$ /Poty3); lane 2, OYDV-O amplification with Poty3/oligo $dT_{(12-18)}$ primers and lane 3, OYDV-O amplification with the OYDV specific primer combination oligo $dT_{(12-18)}$ /AB3S; lane 4, OYDV-O amplification with a primer combination AB3L/AB3S for differential amplification of OYDV-O; lane 5-9, *Allium* samples suspected for OYDV. Lane 5, 8,9, OYDV-G specific products amplified with the primer combination AB1/AB2S and lane 6 and 7, OYDV-O amplification with the specific primer combination AB3L/AB3S ; lane 10, virus-free *Allium cepa* sample as RT-PCR control.

4. Conclusion

Analysis of 3' terminal sequences of OYDV isolates infecting onion (OYDV-O) and garlic (OYDV-G) revealed that the viruses are strains of the same potyvirus despite their striking host specificities. Although very similar in their CP amino acid sequences, the two OYDV isolates differ considerably in their nucleotide sequences. Regions of similarity and of divergence were chosen to design primers for detection and differentiation of OYDV strains in *Allium*. The RT-PCR method was shown to be a powerful and robust tool for indexing for OYDV in *Allium* spp.

5. References

Chomczynski, P., and Sacchi, N. (1987) Single-step method of RNA isolation by acid guanidium thiocyanate-phenol-chloroform extraction. *Anal. Biochem.* **162,** 156-159.

Sambrock, J., Fritsch, E.F., Maniatis, T. (1989) Molecular cloning: a manual, 2nd ed. Cold Spring Harbor Laboratory Press, New York.

DETECTION AND DIFFERENTIATION OF SWEET POTATO CLOSTEROVIRUSES BY RT-PCR AND SINGLE-STRAND CONFORMATION POLYMORPHISM

S. WINTER, M. KÖRBLER, and U. KÖLLNER
AG Pflanzenviren, German Collection of Microorganisms and Cell Cultures, DSMZ, c/o BBA Messeweg 11/12, 38104 Braunschweig, Germany

1. Introduction

Sweet potato virus disease (SPVD) is a serious deterioration of sweet potato caused by the interaction of sweet potato feathery mottle potyvirus (SPFMV) and a whitefly transmitted closterovirus (SPVD-aC) [Winter *et al.*, 1992;]. SPVD was first reported from Nigeria (Schaefers and Terry, 1976) but diseases similar to SPVD were also found in other African countries, in Israel (Cohen *et al.*, 1992), Kenya, Brazil, Argentina, Taiwan and USA.

Adequate diagnosis for the viruses involved in the disease is rather difficult because sweet potato plants infected with only one of the viruses rarely express symptoms and virus concentrations generally remain very low. Especially in singly infected plants, the presence of the closterovirus can not reliably be proven since sensitive and fast testing procedures are unavailable.

This paper describes the development of a PCR based method for detection of the closterovirus (SPVD-aC) associated with sweet potato virus disease and the application of the method for detection and, in combination with single-strand conformation polymorphism (SSCP) analysis, for discrimination of closterovirus isolates from different geographic origins.

2. Methods

A partial sequence of RNA 2 of SPVD-aC was analysed for specific oligonucleotide sequences suitable for use as amplification primers. A primer pair positioned in the heat shock protein (HSP 70) gene analogue that is conserved in the genus closterovirus was chosen.

H.-W. Dehne et al. (eds.), Diagnosis and Identification of Plant Pathogens, 473–475.

The upstream primer 5'-ATCGGCGTATGTTGGTGGTA-3' and the downstream primer 5'-CAGCAGAAGGCTCGTTTAT-3' are flanking a 486 base portion of the HSP 70 gene.

For RT-PCR, total RNA from sweet potato was prepared following the method of Chomczynski and Sacchi (1987) and used as template for reverse transcription PCR (RT-PCR). This was done as a one-tube reaction with about 1 μg of total plant RNA as template for reverse transcription and standard thermocycling conditions with a 49 °C primer annealing temperature.

PCR products were further cloned using a TA cloning kit (Novagen) and sequenced by ΔTaq™cycle sequencing (USB) according to the manufacturers specifications. PCR fragment sequences were analysed using the DNASIS for Windows (Hitachi) version 2.1 and sequences were aligned using the clustal algorithm of Higgins *et al.* (1991).

For virus differentiation, single-strand conformation polymorphism analysis was performed essentially as described by Koenig *et al.* (1995) subjecting aliquots of the amplified PCR products to strand separation conditions and electrophoresis.

3. Results

In RT-PCR an approximately 480 bp dsDNA fragment amplified from total RNA extracts of sweet potato indicated for presence of the closterovirus SPVD-aC in infected plants. Compared with a nucleic acid hybridisation test using a labelled RNA transcript (Piero-Riebeiro et al., 1996) or the bio-assay with *Ipomoea setosa* indicator plants, this tests revealed superior results to dot-blot hybridisation and similar but faster results compared with the bio-assay. In addition, the RT-PCR diagnostic test could be performed with *in vitro* cultured material, a tissue that otherwise could not be examined directly for presence of sweet potato closteroviruses due to extremely low virus concentrations and adverse effects of tissue culture constituents preventing unequivocal virus detection.

The RT-PCR protocol was subsequently used for screening of sweet potato clones obtained from different geographical origins for presence of SPVD-aC. The results of the RT-PCR test suggest that the virus is distributed throughout the sweet potato growing areas. Furthermore, the subsequently conducted sequence analysis of the cloned PCR products revealed a close relationship among all closterovirus isolates with an overall nucleotide identity of the HSP 70 gene fragments sequenced of more than 90%.

Multiple sequence alignment clearly showed that virus isolates obtained from different African countries, USA and Israel form a cluster that is separated well from virus isolates from Taiwan. A clearly separate branch is formed by SPVD-aC isolates from Brazil and Argentina. These viruses share a sequence homology of 96%, however are less than 85% identical to virus isolates of the other branches.

In SSCP analysis, geographic isolates were also readily distinguished by the different electrophoretic mobilities of the + and - strands of their separated dsDNAs. The polymorphic banding pattern obtained compared well with the results obtained from multiple sequence alignment and clustal analysis thus confirming the assignment of isolates into geographic groups.

4. Conclusion

We have developed a RT-PCR method for diagnosis of the closterovirus SPVD-aC in infected sweet potato materials. The method proved to be powerful (sensitive) and robust for detection of this previously very difficult to diagnose plant virus, even when viral sequences were present only in minute amounts e.g. in single infections and in tissue cultured materials. Consequently, this test will significantly shorten sweet potato testing routines once incorporated into a sweet potato virus indexing scheme.

In combination with SSCP analysis, SPVD-aC sources could be assigned into geographic groups and new variants can now be recognized in a simple routine hence providing significant information for plant quarantine and resistance breeding.

5. References

Chomczynski, P., and Sacchi, N. (1987) Single-step method of RNA isolation by acid guanidium thiocyanate-phenol-chloroform extraction. *Anal. Biochem.* **162**, 156-159.

Cohen, J., Frank, A., Vetten, H.J., Lesemann, D.E., and Loebenstein, G. (1992) Purification and properties of closterovirus-like particles associated with a whitefly transmitted disease of sweet potato. *Ann. appl. Biol.* **121**, 257-268.

Higgins, D.G., Bleasly, A., and Fuchs, R. (1991) Clustal V: improved software for multiple sequence alignement. *CABIOS* **8**, 189-191.

Koenig, R., Lüddecke, P., and Haeberlé, A.M. (1995) Detection of beet necrotic yellow vein virus strains, variants and mixed infections by examining single-strand conformation polymorphisms of immunocapture RT-PCR products. *J. Gen Virol.* **76**, 2051-2055.

Pio-Ribeiro, G., Winter, S., Jarret, R., Demski, J.W., and Hamilton, R.I. H. (1996) Detection of sweet potato virus disease associated closterovirus in a sweet potato accession in the United States. *Plant Disease.* **80**, 551-554.

Schaefers, G.A., and Terry, E.R. (1976) Insect transmission of sweet potato disease agents in Nigeria. *Phytopathology* **66**, 642-645.

Winter, S., Purac, A., Leggett, F., Frison, E.A., and Hamilton, R.I.H. (1992) Partial characterization and molecular cloning of a closterovirus from sweet potato infected with the sweet potato virus disease complex from Nigeria. *Phytopathology* **82**, 869-875.

Disease Assessment and Quantification

THE USE OF VIDEO TIME-LAPSE RECORDINGS TO STUDY THE EFFECTS OF ANTIFUNGAL COMPOUNDS ON PHYTOPATHOGENIC FUNGI

G. Albert and H. Heinen
Cyanamid Forschung GmbH
P.O. Box 100, Zur Propstei, D-55270 Schwabenheim, Germany

Abstract

In addition to classical microscopic techniques, video-microscope systems and time-lapse recordings have proved to be important tools for observing the effects of antifungal compounds on fungal structures. Long-term recordings were made using a CCD colour video camera and a JVC VHS time-lapse video cassette recorder type BR-9060E or BR-S925E. Time-lapse recordings have been successfully applied in elucidating the fungicidal mode of action of dimethomorph, a fungicide for the control of downy mildews and *Phytophthora* diseases. The results of the video time-lapse studies revealed that the fungicidal activity of dimethomorph is related to cell walls or cell wall forming processes. Although the biochemical target of dimethomorph still remains unknown, the metabolic pathways at which the biochemist should look in his search for the target molecule are now more defined. An accompanying video presentation shows important parts of the time-lapse microscope study on the fungicidal effect of dimethomorph on *Phytophthora infestans.*

1. Introduction

When studying the effects of antifungal compounds on target pathogenic fungi, morphological investigations can often help both in identifying those biochemical pathways where the search for the target molecule is most likely to be successful and in the interpretation/confirmation of biochemical findings. In addition to classical microscopic techniques, video-microscope systems and time-lapse recordings have proved to be important tools for observing the effects of antifungal compounds on fungal structures. Time-lapse video recordings have been successfully applied in elucidating the fungicidal mode of action of dimethomorph (Thomas *et al.*, 1992), an apoplastically mobile fungicide for the control of downy mildews and *Phytophthora* diseases (Albert *et al.*, 1988). Important parts of a time-lapse microscope study showing the fungicidal effect of dimethomorph on different developmental stages of *Phytophthora infestans* in liquid culture are compiled in an accompanying video presentation.

H.-W. Dehne et al. (eds.), Diagnosis and Identification of Plant Pathogens, 479–481.

2. Materials and Methods

2.1. CULTURE MEDIA FOR *PHYTOPHTHORA INFESTANS*

Rye agar: Rye broth was prepared by boiling 200 g whole rye grains for 1 h in 1 l distilled water. Five g D - glucose and 20 g agar were added to the filtered supernatant which was made up to 1000 ml with distilled water. V8 liquid culture medium: 300 ml V8 juice (Campbell's Soups Ltd) + 4.5 g $CaCO_3$ were mixed and centrifuged. The supernatant was diluted with distilled water in the ratio 1:10.

2.2. FUNGICIDES

Dimethomorph was diluted in acetone at a concentration of 0.5 % AI *m/V*. Dilutions were prepared from this stock solution by adding distilled water or liquid culture medium. For most of the time-lapse studies, a dimethomorph concentration of 2.5 μg/ml was used.

2.3. CULTURAL CONDITIONS FOR MICROSCOPE STUDIES

Observations of cells or colonies by light microscopy were made using tissue culture plates (24 well Greiner No. 862 160). The wells were filled with 0.5 ml of an aqueous sporangia suspension (5×10^3 spores/ml) plus 0.5 ml of a fungicide suspension. Object slides with a hole bounded on both sides by a cover-slide served also as micro culture tool. Zoospore release, cyst formation and cyst germination was observed in distilled water; hyphal development in V8 liquid culture medium.

2.4. PRODUCTION OF PROTOPLASTS FROM SPORANGIA

Protoplast generation was carried out in tissue culture plates by digestion of sporangial walls with 20 mg/ml NovoZym 234 at 25°C. Dimethomorph (0.5 % in acetone) was diluted with 0.8 M KCl (osmotic stabilizer) to the desired final concentration in the culture plate. After the formation of protoplasts, 0.5 ml of a 1:3 diluted V8 liquid culture medium containing 0.6 M sucrose and the fungicide was added to the protoplasts to allow further growth. Protoplast viability was determined by staining with fluorescein diacetate.

2.5. LIGHT MICROSCOPY AND VIDEO EQUIPMENT

Observations were made using a Zeiss Axiovert 405 M Inverted Microscope. The objects were viewed by bright field and differential interference microscopy. Long - term recordings were made using a JVC colour video camera TK - 1085 E or a Sony 3CCD colour video camera with a MC-3215/PI (AVT-Horn) control unit and a JVC VHS time - lapse video casette recorder Type BR - 9060E and BR - S925E. Time - lapse recordings were made with time factors of 160 and 320. Light-sensitivity in the fungi can be compensated for by connecting a Simplified Time Base Corrector Sony MPU-F100P between the video camera and the recorder. This allows recordings at very low light intensities.

3. Results

Time-lapse recordings of germinating cysts, sporangia, growing germ tubes and hyphae treated with dimethomorph revealed that cell death was due to lysis at the point of germ tube formation or in hyphal tip zones. Rupture of the cell wall caused the death of the whole or part of the fungus. Spores and hyphae in which germination or hyphal elongation, and therefore cell wall growth, did not take place, were not affected. Germinating cysts were killed by a single ruption and the entire protoplasmic contents were destroyed. In larger, branched, hyphae a stepwise destruction of the protoplasm was observed. When dimethomorph was added at 2.5 µg/ml, growing hyphae burst at their tips but the protoplast was destroyed only in the hyphal tip zones. The remaining living protoplasm continued growth by forming new branches which also burst at their apices and so on until the entire mycelial protoplasm was destroyed. After an apex had burst, the remaining protoplasm reorganized and hyphal tip formation started again. Dimethomorph then caused lysis of this new tip. This pattern was clearly detectable because time-lapse recordings allowed clear recognition of living protoplasm due to its movement. Thus, dead and living protoplasmic compartments could be clearly distinguished in the coenocytic mycelium. Fungal cells without any cell wall, such as zoospores and isolated protoplasts in isotonic media, were not affected even at high concentrations (50 µg/ml) of dimethomorph.

4. Discussion

The results of the video time-lapse studies on the fungicidal activity of dimethomorph on *Phytophthora infestans* revealed that its fungicidal activity is related to cell walls or cell wall forming processes. Although the biochemical target of dimethomorph still remains unknown, the metabolic pathways at which the biochemist should look in his search for the target molecule are now more defined. The advantages of the time-lapse video technique in elucidating the fungicidal mode of action of antifungal compounds on phytopathogenic fungi can be summarized as follows:

- investigations can be carried out on living cells
- the fungicidal activity can be monitored from the application of the compound to the death of the cells
- it can help in distinguishing between primary and secondary morphological effects of the fungicide
- a dynamic view of the morphological changes allows a better interpretation of the observed symptoms.
- the typical plasma movement observed by time-lapse recordings is an indicator for the viability of cells or parts of the protoplasm.

5. References

Albert, G., Curtze, J. and Drandarevski, Ch.A. (1988) Dimethomorph (CME 151), a novel curative fungicide, *Proceedings of the British Crop Protection Conference - Pests and Diseases*, 1, 17-24.

Thomas, A., Albert, G. and Schlösser E. (1992) Use of video-microscope systems to study the mode of action of dimethomorph on *Phytophthora infestans*, *Mededelingen van de Faculteit Landbouwwetenschappen Rijksuniversiteit Gent*, 57, 189 - 197.

IDENTIFICATION AND QUANTITATIVE DETERMINATION OF PHYTOPHTHORA SPECIES INFECTING ORNAMENTAL PLANTS IN NURSERIES

CACCIOLA S.O.[(1)], PANE A.[(1)] and MAGNANO DI SAN GLIO G.[(2)]
[(1)]Istituto di Patologia vegetale, University of Catania, Italy
[(2)]Istituto di Difesa delle piante, University of Reggio Calabria, Italy

1. Introduction

Collar and root rot incited by species of *Phytophthora* are a serious problem in nurseries of ornamentals in Italy. New species of *Phytophthora* are introduced inadvertently with exotic ornamentals imported as potted plantlets. Some species of *Phytophthora* infecting ornamentals show a great intraspecific variability, possibly as a consequence of natural selection due to host plant specialization and geographical isolation (many ornamental plant species are native to tropical forests). Identification keys based on morphological characters appear inadequate to identify these species. Among molecular diagnostic techniques, the electrophoresis of mycelial proteins has proved to be a valuable tool for identification of *Phytophthora* species [4].

We report briefly the results of a survey supported by a grant of Ministero delle Risorse Agricole Alimentari e Forestali (Rome, Italy), research project “Biotecnologie vegetali”, aimed at identifying the species of *Phytophthora* causing collar and root rot in nurseries of ornamental plants, in Southern Italy.

2. Materials and Method

The BNPRA selective medium [4] was used for the isolation of *Phytophthora* species from infected tissues and soil. Species, such as *P. cactorum*, that are inhibited by hymexazol were isolated using baiting techniques. The population of *Phytophthora* was estimated quantitatively by the soil-dilution plating technique. Mycelial extracts of axenic cultures were prepared and analyzed by polyacrylamide slab gel electrophoresis according to the procedures described previously [1, 2, 3].

H.-W. Dehne et al. (eds.), Diagnosis and Identification of Plant Pathogens, 483–485.

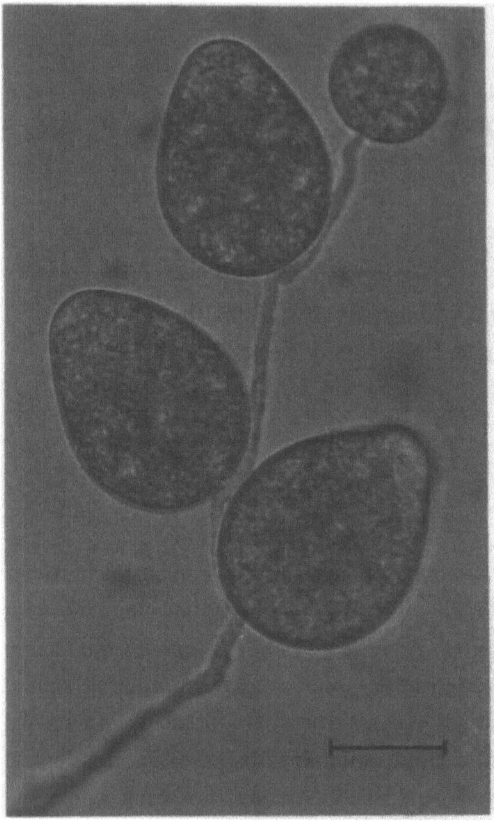

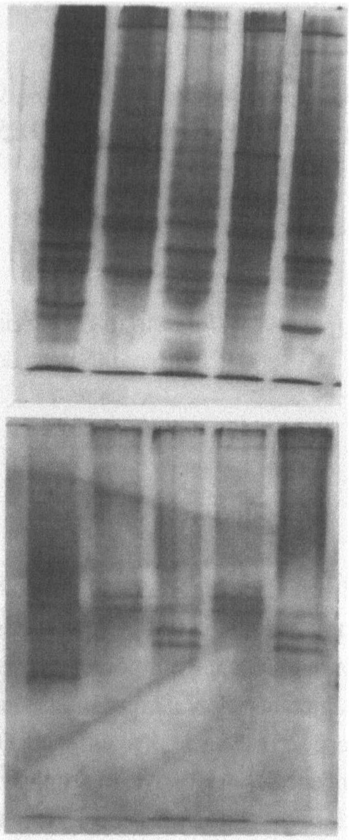

Fig. 1 Sporangia produced on a simple sympodium by an isolate of *Phytophthora cryptogea* recovered from silver-bush (*Convolvulus cneorum*). Scale bar = 20 µm.

Fig. 2 Electrophoretic banding patterns of total soluble mycelial proteins in a native system (upper panel) and arylesterase zymograms (lower panel) in 7.5% polyacrylamide slab gel. From left to right: lane 1, *P. citricola* from myrtle roots; lanes 3 and 5, isolates of *P. citrophthora* from roots of buckthorn and silver-bush, respectively; lanes 2 and 4, isolates from rhizosphere-soil of buckthorn, identified tentatively as *P. citrophthora*.

3. Results and Discussion

P. capsici (both A_1 and A_2 mating types), *P. cactorum, P. cinnamomi, P. citricola, P. citrophthora, P. cryptogea* (A_1 mating type), *P. drechsleri* (A_1 mating type), *P. nicotianae* (both A_1 and A_2 mating types), *P. palmivora* (A_1 mating type), as well as heterothallic non-papillate *Phytophthora* spp. and homothallic papillate *Phytophthora* spp. were isolated from rhizosphere soil, basal stem and roots of plants with symptoms of collar and root rot. The species were identified on the basis of both morphological characters and profiles of total mycelial proteins and isozymes. Electrophoretic protein and isozyme patterns of isolates of *P. nicotianae* were extremely uniform, although isolates of both mating types were recovered from taxonomically unrelated hosts. In contrast, isolates identified as *P. citrophthora* showed different electrophorethic phenotypes. This is not consistent with the results of a previous study where only isolates from citrus had been examined. *P. cryptogea* and *P. drechsleri* were separated into several discrete groups on the basis of the electrophoretic phenotype.

Isolates of *P. palmivora* MF4, now considered a variant of *P. capsici* [4], were recovered from silver-bush (*Convolvolus cneorum*). These isolates showed an electrophoretic phenotype very different from those of reference strains of *P. palmivora* obtained from pittosporum (*Pittosporum* sp.), *P. citrophthora* from buckthorn (*Rhamnus* sp.) and *P. capsici* from paprika pepper (*Capsicum annuum*), suggesting they belong to a distinct taxon. This is the first report of *P. palmivora* MF4, in Europe.

P. nicotianae was the species recovered most frequently from the following host-species: *Arbutus unedo, Callistemon citrinus, Leptospermium* spp., *Myrtus communis, Cistus* spp., *Escallonia laevis, Grevillea rosmarinifolia, Hebe x andersoni, Hedera helix, Hibiscus rosa-sinensis, Lavandula officinalis, Passiflora caerulea, Pistacia lentiscus, Pittosporum* spp., *Rhamnus* spp. and *Rosmarinus officinalis.*

The inoculum density of *Phytophthora* species in rhizosphere soil of symptomatic plants ranged between 0 and 91 CFU/g of soil and was directly proportional to the percentage of infected roots, but not to the incidence of collar rot.

The time of year was a critical factor in the isolation of *Phytophthora*: populations of *P. nicotianae* and *P. palmivora* peaked in summer.

4. References

1. Cacciola, S.O., and Magnano di San Lio, G.: Foot rot of prickly pear cactus caused by *Phytophthora nicotianae, Plant Disease* **72** (1988), 793-796.
2. Cacciola, S.O., Belisario, A., and Magnano di San Lio, G.: *Phytophthora italica* sp. nov. on myrtle, *Phytopathologia Mediterranea,* **35** (1996) (in press).
3. Cacciola, S.O., Belisario, A., Pane, A., and Magnano di San Lio, G.: Forsythia: a new host of *Phytophthora nicotianae* in Italy, *Plant Disease,* **78** (1994), 525-528.
4. Erwin, D.C., and Ribeiro, O.K.: *Phytophthora diseases worldwide,* APS Press, St. Paul, 1996.

EVALUATION OF THREE METHODS FOR THE DETECTION OF *SEPTORIA* SPP. IN WHEAT LEAVES

D. GINDRAT, P. FREI
Swiss Federal Research Station for Plant Production
CH- 1260 Nyon, Switzerland

Early detection of *Septoria (Stagonospora) nodorum* and *Septoria tritici* in wheat leaves is needed to assess the risk of glume blotch and speckled leaf blotch, respectively. Glume blotch fungicides are sprayed at growth stages (GS) 51-55 (beginning to mid-earing). Sprays to control speckled leaf blotch are not yet utilized in Switzerland. Experiments were carried out in four sites of Western Switzerland to compare three methods of detecting *Septoria* spp. in wheat leaves to assess the risk of glume blotch and of speckled leaf blotch. These methods were : the observation of symptoms on leaves, an ELISA assay, and the stimulation of growth of endophytic *Septoria* spp. on leaves treated with paraquat (Gindrat and Pezet, 1994).

1. Experiments in 1993

Severe glume blotch was observed on cv. Arina after an earlier *S. nodorum* ELISA assay (DuPont Advisor) detected the pathogen, or following the observation of symptoms on either one of the last three leaves at GS 41-49. No glume blotch was recorded after ELISA assay gave negative results or where symptoms were not visible. *S. tritici* was faintly observed.

2. Experiments in 1994

ELISA and visual assessement were compared at three field locations. At GS 41-45, 0, ≈0.1, and 68 - 100 % (at each field location) of the culms of the cv. Arina showed symptoms of *S. nodorum* on either one of the three last leaves. At GS 32-45, ELISA assays detected the pathogen at all three locations with antigen values lower than in 1993 and with a confusing saw-toothed pattern for the 2nd unfolded leaf between GS 37 and 45 in one site. The paraquat method was examined in two out of the three sites. It was more sensitive and provided more regular results than ELISA in detecting *S. nodorum.* No glume blotch was recorded at either location, perhaps because of unfavourable weather conditions during heading at the site (Peney-le-Jorat) where *S.*

H.-W. Dehne et al. (eds.), Diagnosis and Identification of Plant Pathogens, 487–489.

nodorum was consistently detected by the 3 methods. *S. tritici* was not observed. A summary of the results from the experiment in Peney-le-Jorat is presented in Table 1.

TABLE 1. Comparison of 3 methods for the detection of *S. nodorum* in wheat leaves (Peney-le-Jorat, 1994).

Growth stages	E	P	S
	Last unfolded leaf :		
14	< 5	13	6
21	< 5	0	0
23	5	30	0
30	< 5	57	0
31	< 5	93	0
	Leaf Nr. 5 :		
32	11	--	12
	Leaf Nr. 4 :		
32	5	--	0
37	130	--	54
	Leaf Nr. 3 :		
32	< 5	--	0
37	20	--	0
41	70	--	4
45 (spray decision)	150	--	100
	Leaf Nr. 2 :		
37	< 5	80	0
41	8	--	0
45	10	--	68
	Flag leaf :		
41	5	--	0
45 (spray decision)	5	53	2

E : ELISA; numbers = antigen units/ml of a sample. Detection threshold = 5 a.u./ml.
P : Paraquat assay; numbers = % of leaves yielding *S. nodorum*.
S : Observation of symptoms; numbers = % of leaves with symptoms of *S. nodorum*.
-- : not tested.
All values were obtained from samples of 30 leaves.

3. Experiments in 1995

Paraquat was used to detect *S. nodorum* and *S. tritici* on the last fully developped leaf (asymptomatic) of cv. Arina at two field locations. In one site, *S. nodorum* was detected on 40-50% of the culms at GS 32 and 45, but only sparsely detected at GS 37. At the other location, about 10-20% of the culms provided *S. nodorum* at GS 24, 32 and 41, but no isolation was obtained at GS 31. No glume blotch was observed at either location. Detection of *S. tritici* showed a pattern almost similar to that of *S. nodorum* up to GS 37, then isolations decreased markedly. Severity of speckled leaf blotch followed the rate of isolation of *S. tritici*.

4. Conclusions

The ELISA assay obviously gave promising results. However, we are of the opinion that the observation of symptoms of *S. nodorum* on either 3 top leaves shortly before heading is still the easiest mean to assess the risk of glume blotch. The paraquat method was the most sensitive and reliable method. However, it is time-consuming (10 days needed to obtain consistent results) and requires experience in identifying *Septoria* colonies on leaves and on agar.

In this study, we also confirm that other factors, which may include weather conditions influencing spore production on upper leaves and infection of ears, could reduce the reliability of assessing the risk of glume blotch by either method at the time of spraying. A system based on pycnidia formation and inoculum dynamics might be more reliable (Verreet and Hoffmann, 1990).

The usefulness of the three methods (visual assessment, paraquat, ELISA) for assessing the risk of speckled leaf blotch should be examined further, since this disease is only sporadically severe in Switzerland.

5. References

Gindrat, D. and Pezet, R (1994) Le paraquat, un outil pour la révélation rapide d'infections fongiques latentes et de champignons endophytes, *J. Phytopathology* **141**, 86-98.

Verreet, J.A. and Hoffmann, G.M. (1990) A biologically oriented threshold decision model for control of epidemics of *Septoria nodorum* in wheat, *Plant Disease* **74**, 731-738.

VARIATION FOR RESISTANCE TO FUSARIUM HEAD BLIGHT (*FUSARIUM CULMORUM*) IN SOMACLONES OF WINTER TRITICALE

T. GÓRAL, E. ARSENIUK
Plant Breeding and Acclimatization Institute
Radzików, 05-870 B³onie, Poland

1. Introduction

Presence of broad spectrum of genetic variation is necessary for resistance breeding. Wild types, sometimes unrelated and mutational techniques were applied to increase this variation. Cultures *in vitro* of plant cells or organs where somaclonal variation can be induced are another source of variation (Semal *et al.*, 1988). This may be done by two methods - *in vitro* selection with selecting factor (pathogen inoculum, mycotoxins or pathogen culture filtrate) or search for resistant types among regenerated plants obtained without *in vitro* selecting factor (Ahmed *et al.*, 1991, Van den Bulk, 1991, Arseniuk, 1994). However, the existence of stable and applicable somaclonal variation is still in question.

The purpose of the present study was preliminary assessment of somaclonal variation presence and range of resistance to Fusarium head blight. The disease was induced in regenerants (somaclones) and parental cultivars/lines of winter triticale by inoculation of heads with *Fusarium culmorum* isolates. Resistance of somaclones was compared to the resistance of their parental cultivars/lines to identify types of higher resistance or higher susceptibility. Differences in Fusarium head blight resistance were assessed among somaclones generated from various parental lines, among somaclones from the same parental line, and among somaclones generated from the same calli. Resistance of subsequent generations of somaclones was compared to verify stability of observed variation.

2. Materials and methods

Ten isolates of *Fusarium culmorum* used for the study were derived from *Fusarium* infected triticale seed and their pathogenicity towards triticale seedlings was determined in preliminary experiment. Isolates were grown on autoclaved triticale grain. Inoculum consisted of mixture of *F. culmorum* isolates. The conidial concentration in the suspension was adjusted to 5×10^6 spores/ml.

Resistance of 458 somaclones of R_1, R_2 and R_3 generations and 19 parental cultivars (Almo, Bolero, Dagro, Grado, Lasko, Malno, Moniko, Presto, Ugo) and lines (CHD

H.-W. Dehne et al. (eds.), Diagnosis and Identification of Plant Pathogens, 491–494.

470/85 = CHD, GRH 32-1, LAD 285, MAH 6353-108 = MAH 1, MAH 15668-1 = MAH 2, MAH 15833-1/4 = MAH 3, T/8, T/9, T/28, TSW 2.742.85 = TSW) of winter triticale was tested. Somaclone families (SE families) consisted somaclones obtained using immature embryos technique from callus of one embryo and/or from calli of different embryos of parental cultivar/line (Arseniuk, 1994).

Experiment was set up in a randomized complete block design, in 2 replications. Each replicate consisted of two parallel rows of plots 1 × 0.5 m in size. Inoculated plots were in one row and the control in the another one. Two opposite plots were planted with the same cereal entry.

Heads of triticale plants were inoculated by spraying with spore suspension at mid-anthesis. It was done 4 times, depending on mid-anthesis time of different cultivars/lines. The disease was rated first at about 14 days after the last inoculation. Four ratings were done at 7 days intervals. Mean percentage of blighted spikelets per infected head and the percentage of infected heads in a plot were used as disease parameters.

On the basis of both parameters Fusarium head blight severity (FHB) was calculated as follows:

$$\text{FHB (\%)} = (\text{\% blighted spikelets} \times \text{\% infected heads})/100\%$$

3. Results and conclusions

In this experiment mean severity of Fusarium head blight was FHB=13.6% for all entries and FHB=14.6% for parental cultivars and lines. Significant differences in resistance to Fusarium head blight between parental cultivars and lines were found (Fig. 1). Their reaction ranged from FHB=3.5 for Lasko cv. to FHB=27.6% for GRH 32-1 line. Reaction of all somaclones was at range FHB=1.3 - 35.9%. Analyses of variance performed separately for SE families revealed significant differences in resistance between somaclones generated from the same parental cultivar/line, except TSW SE. In the previous study broad range of variation for resistance to Septoria nodorum blotch in these somaclones was found by Arseniuk (1994).

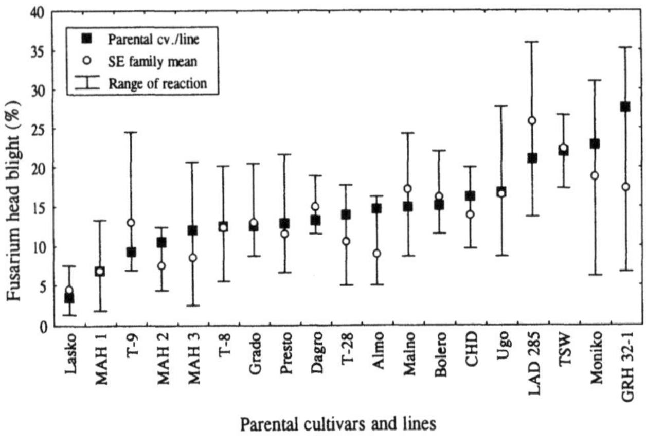

Figure 1. Disease reaction of winter triticale somaclones and their parental cultivars and lines to inoculation of heads with *Fusarium culmorum*.

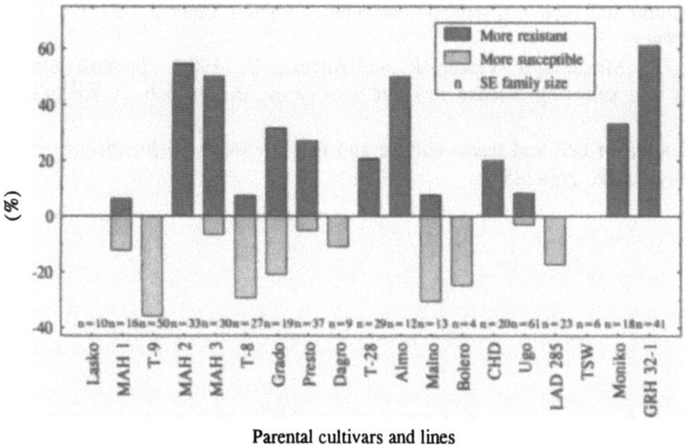

Figure 2. Proportion in SE families of somaclones with significantly improved or reduced resistance as compared to parental cultivars and lines.

In all SE families, except TSW SE and Lasko SE, somaclones significantly more and/or less resistant than parental cv./line were found (Fig. 2). It was for SE families generated from resistant parental cv./lines (MAH 2, T-9) as well as from susceptible parental cultivars/lines (GRH 32-1, Moniko, LAD 285).

Comparison of mean resistance to Fusarium head blight of successive generations proved differences between SE families. The largest differences in mean resistance as compared to parental cultivars/lines resistance were observed for R_1 generations. These differences decreased for the next generations (R_2, R_3), e.g. MAH 3 SE, Grado SE, Presto SE. However, mean resistances of R_1, R_2 and R_3 generations in three SE families (T-28 SE, LAD 285 SE, GRH 32-1 SE) were similar and different from parental cultivars/lines. In 10 SE families somaclones with resistance of all successive generations significantly different from parental cv./line resistance were found. Significantly more *Fusarium*-resistant somaclones of wheat in not selected *in vitro* population were found earlier by Ahmed *et al.* (1991).

These results indicated that:

a. Significant differences in Fusarium head blight resistance among somaclones can be found.
b. Parental cultivars and lines, despite their resistance, can be source of more resistant as well as more susceptible somaclones.
c. For the most somaclones a trend towards decreasing of differences in resistance between somaclones and parental cv./lines in generations R_1 - R_3 can be observed.
d. Somaclones with resistance significantly different from parental cv./line and stable in succeeding generations can be found.
e. Existence of somaclonal variation for resistance to Fusarium head blight require confirmation in the further research. Seed infection with *F. culmorum*, mycotoxin accumulation in seed and yield components reduction will be determined. Field and laboratory experiments will be carried out next year.

5. References

Ahmed, K.Z., Mesterhazy, A., and Sagi, F. (1991) *In vitro* techniques for selecting wheat (*Triticum aestivum* L.) for *Fusarium*-resistance. I. Double-layer culture technique. Euphytica **57**, 251-257.

Arseniuk, E. (1994) Septoria glume and leaf blotch of triticale and wheat. Annual Technical Report. Project PL-AES-241. IHAR Radzików

Semal, J., Kummert, P., Lepoivre, P., Meulemans, Viseur, J., and Anceau, C. (1988) *In vitro* cultures producing pathogen-free plants and selecting disease resistant genotypes. Bull. Rech. Agron. Gembloux **23**, 261-269.

Van den Bulk, R.W. (1991) Application of cell and tissue culture and in vitro selection for disease resistance breeding - a review. Euphytica **56**, 269-285.

QUANTIFICATION OF *PHYTOPHTHORA NICOTIANAE* ON TOMATOS GROWN WITH THE NUTRIENT FILM TECHNIQUE (NFT)

D. GROTE
Institut für Gemüse- und Zierpflanzenbau Großbeeren/Erfurt e. V.
Theodor Echtermeyer Weg 1, 14979 Großbeeren, Germany

1. Introduction

The aim of this investigation is to adapt the Enzyme-Linked-Immunosorbent Assay (ELISA) for quantification of *Phytophthora nicotianae* (*P. nicotianae*) in a hydroponic system. This pathogen causes foot and root rot on tomatoes especially when plants are grown with the Nutrient Film Technique (NFT). To evaluate the effectiveness of any control against *P. nicotianae*, an indirect ELISA technique was developed as a quantification method in this system. Values obtained with the ELISA technique were also compared to those measured with the Most-Probable Number (MPN) method. It is common to find mycelium, sporangia and zoospores on/in the roots, as well as zoospores in the recirculating nutrient solution during the growing period. Accordingly, three kinds of samples were taken: roots, nutrient solution and pea-baits. The antiserum was previously developed against mycelial extracts of another *P. nicotianae*-isolate (Richter and Gabler, 1991; Gabler and Urban, 1995; Gabler and Richter, to be published). The suitability of the assay for the different fungal stages of the used *P. nicotianae*-isolate was tested and standardized for the quantification in NFT system. For standardization, the influence of tomato cultivars and the plant stage, application of *Bacillus subtilis* as a bioagent on noninoculated plants, and the time of incubation of the enzyme substrate on the ELISA extinction value was established. Furthermore, the possibility of quantification of the pathogen with both methods was evaluated with the standard solutions, on tomato seedlings in a container culture and on tomato plants in a glasshouse NFT-system.

2. Material and Methods

The *Phytophthora nicotianae*-isolate was obtained from BBA Kleinmachnow and cultivated on oatmeal agar for the inoculation of the tomato seedlings and on carrot-shred agar for the inoculation of the tomato plants in the glasshouse trial.

2.1. STANDARDIZATION OF THE TEST-SYSTEM WITH ELISA

The antiserum was provided by BAZ Aschersleben and the ELISA-procedure was developed for *Phytophthora spp.* by Richter and Gabler (1991). For standardization the influence of tomato cultivars and plant stage with noninoculated plants, the treatments with the bioagent *Bacillus subtilis* and the incubation time of the enzyme substrate on the ELISA extinction values were investigated. Three cultivars showed no differences in the values (between 0.089 and 0.096), whereas for 'Suso', they were lower at 0.058, but also not significantly (Reader ht III, Fa. Anthos). Significant influence on the value showed the plant stage: the extinction value 110 days after inoculation was significantly higher than with 5 days. The threshold value for determining the sickness of plants was calculated after Gabler and Urban (1994): T= means of EXTINCTION of noninoculated plants + 3x standard deviation. The threshold ranged from 0.209 to 0.05, respectively, measured for the cultivar 'Harzfeuer'. For further investigations, the cultivar 'Counter' was used. With 18 measurements, no notable influence of the treatments with different concentrations of the bioagent on the ELISA extinction values was found. Values ranged between 0.06 and 0.1. The change of the extinction, depending on the duration of incubation of the substrate, was measured to calculate the corrected values based on an equal incubation time for all measurements. The equation for the mycelium was λ_{405}=-0.004+0.007*t, for the inoculum (mycelium+zoosporangia+zoospores), λ_{405}= 0.0004+0.002*t and for the zoospores, λ_{405}=0.01+0.003*t (λ_{405}-extinction value at 405 nm; t-time, min). For both assays, sample collection and preparation were standardized. The MPN Method from Couchran (1950) was modified to compare values. For this method, a selective liquid medium specifically for *Phytophthora spp.* was found. Previously frozen samples of roots and baits were ground

H.-W. Dehne et al. (eds.), Diagnosis and Identification of Plant Pathogens, 495–497.

in a 1:10 solution ratio in phosphate buffer (pH=6.8) for ELISA and 1.8% glucerol for MPN. The samples of the nutrient solution were concentrated 50-fold, and for ELISA mixed in the adequate buffer in a 1:2 ratio.

2.2. QUANTIFICATION IN CONTAINER- AND NFT- SYSTEMS

Quantification of *P. nicotianae* in standard solutions, on tomato seedlings and in a NFT-system was carried out with both methods. To prepare the standards, mycelia were harvested from an oat-bean liquid medium after 7 days, filtered and deep frozen. Zoospores were obtained by flooding Petri dishes containing a 7-day-old mycelial mat with sterile distilled water, exposing the dishes for 30 min at 4°C, then harvesting spores by filtration after another 30 min. The inoculum standard was prepared by flooding agar plates, containing 7-day-old *P. nicotianae* with 2 ml sterile distilled water per plate and exposing the plates for 30 min at 4°C for better zoospore production, then homogenizing the material for 1 min. Thus, the inoculum contained mycelia, zoosporangia and zoospores.

The trials in container culture lasted 5 or 7 days and were carried out in the following manner: Two tomato seedlings without any substrate were placed in a container with 500 ml nutrient solution. Aeration of the roots was maintained. Each trial consisted of a control without inoculaton of *P. nicotianae,* and 5 different concentrations of zoospores, with each treatment replicated 3 times.

The NFT glasshouse trial consisted of a control without inoculation and an infested variant, inoculated with an inoculum of $2{,}91 \times 10^6$ propagules (mycelium+zoosporangia+zoospores) /ml. The inoculum of *P. nicotianae* grown on carrot-shred agar was prepared in the same way as that from oatmeal agar. Treatments were replicated 3 times with separate NFT-systems for each trough. Ten tomato plants at the 5-6 leaf stage, pricked out in Grodan blocks, were planted in each trough. The nutrient solution in both trials was prepared after Voogt and Bloehmhard (1994) and the tomato cultivar 'Counter' was used. The inoculation was carried out 8 days after planting. Root samples were taken 5, 26, 48, 58 and 79 days after inoculation and samples of the nutrient solution and the pea-baits 19, 34, 54, 75 and 96 days after inoculation.

The statistical calculations were made with STATISTICA (5.0) at $p=0.05$.

3. Results

3.1. STANDARDIZATION OF THE TEST-SYSTEM WITH ELISA

For extinction values between approximately 2.0 and 0.05, quantification of *P. nicotianae* with ELISA gave the following three standard curves for mycelium (1), inoculum (2) and zoospores (3): (1) $\lambda_{405}=-0.06+0.79*\log M$, (2) $\lambda_{405}=-2.29+0.61*\log c$ and (3) $\lambda_{405}=-1.86+0.52*\log c$ (λ_{405}-extinction value at 405 nm; M-mycelium dry weight, µg; c-concentration, zoospores or propagule number /ml). A close correlation in all three standard forms was found between ELISA and MPN methods, with correlation coefficients (1) $r=0.851$, (2) $r=0.839$ and (3) $r=0.902$, respectively.

3.2. QUANTIFICATION IN CONTAINER- AND NFT- SYSTEMS

After the inoculation of tomato seedlings with zoospores, the correlation coefficient between the two methods for identification of the pathogen ($r=0.763$) was significant. The same methods were applied for the quantification of the pathogen in one NFT system of glasshouse tomato culture. For both methods the distribution of the pathogen in the system was nearly the same. The highest concentration found in the roots was above 10^5 cfu/g dry weight with MPN. In the nutrient solution, a concentration below 10^3 cfu/ml was found with MPN. On average, the baits showed a concentration 100-fold less than that on the roots. This relation did not change over the whole cultivation period in this experiment (Figure 1 and 2).

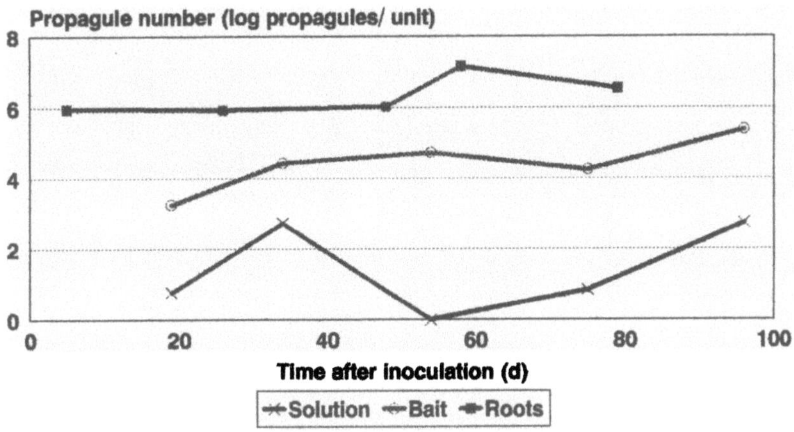

Figure 1: Propagules, found with the MPN method of *Phytophthora nicotianae* on inoculated tomato plants cultivated with NFT

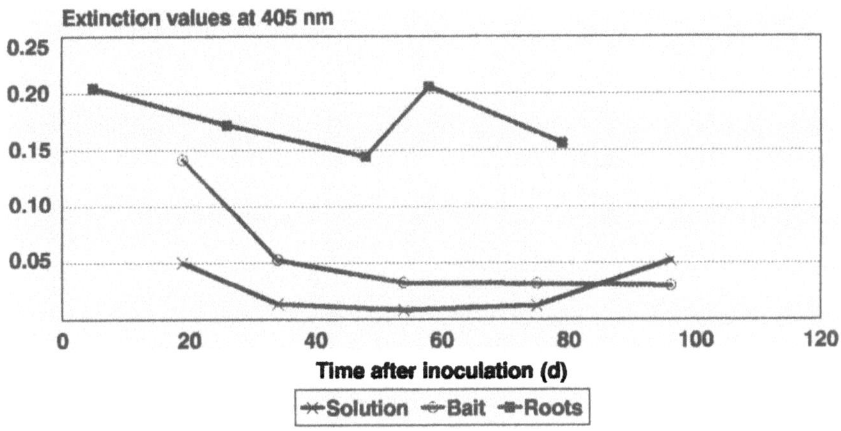

Figure 2: ELISA extinction values of *Phytophthora nicotianae* inoculated tomato plants in NFT-culture

4. Conclusions

The ELISA method is suitable for the quantification of the pathogen, for all tested fungal stages and for the isolat used in this study. The MPN method is also suitable for comparing the values. These are preliminary results and further research is required.

5. Acknowledgements

Thanks are due to the BAZ Aschersleben for providing the antiserum, to Dr. Gabler for supporting this work, and to Dr. J. L. Kovar for editing the English expression.

6. References

Cochran, W.G. (1950) Estimation of bacterial densities by means of the most probable number. *Biometrics* 6: 105-116.

Gabler, J. and Richter,J. (to be published) Kreuzreaktivität eines polyklonalen Antiserums gegen Phytophthora nicotianae. *Zeitschrift für Pflanzenkrankheiten und Pflanzenschutz*

Gabler, J. and Urban, M. (1995) Evaluation of resistance differences between tomato cultivars of *Phytophthora nicotianae* by indirect ELISA. *Zeitschrift für Pflanzenkrankheiten und Pflanzenschutz* 102, 3 275-283.

Richter, J. and Gabler, J. (1991) Nachweis von *Phytophthora nicotianae. Arch. Phytopathol. Pflanzenschutz* 27 3, 193-197.

Voogt, W. and Bloemhard, C. (1994) Voedingsoplossing voor de teelt van tomatenin gesloten systemen. *Proefstation voor Tuinbouw onder Glas, Naaldwijk* 17, 2. Aufl.

COMPARISON OF WILLOW LEAF RUST QUANTIFICATION METHODS

J. HAKULINEN
Department of Biology, University of Joensuu, 80101 Joensuu, Finland

1. Introduction

Leaf rust caused by *Melampsora* spp. is one of the most widely distributed diseases of willows (*Salix* sp., Salicaceae) [1]. Objective and rapid methods for quantification of rust severity are needed in selection of rust resistant willow species and clones for the breeding programs of short rotation forestry [3], and in ecophysiological studies, which investigate the phytochemical changes occurring in rust-infected willows [2].

Conventionally, foliar rust disease severity has been assessed visually [4]. Moreover, certain chemical methods, such as chitin analysis, may be used to estimate rust biomass in plant tissues [5]. Recently, also the application of digital image analysis to quantification of coffee leaf rust has been reported [4].

The aim of this study was to compare the accuracy, repeatability and rapidity of different willow rust quantification methods, and to correlate the results obtained using each method with levels of phytochemicals in willows. Moreover, results from a preliminary test using chitin analysis for the estimation of rust biomass in willow leaf tissues are briefly discussed.

2. Material and methods

2.1. LEAF SPECIMENS

Visual and image analysis methods were applied to a collection of 100 slightly pressed, rust-infected *Salix myrsinifolia* Salisb. leaves, which were previously analysed for the content of phenolic secondary metabolites (Hakulinen, unpublished). The assessments were made by one person on different occasions. To study the repeatability, 30 randomly chosen leaves were analysed with each method three times. Randomly collected fresh leaves were used for the chitin analysis.

2.2. VISUAL ASSESSMENTS, IMAGE ANALYSIS AND CHITIN ANALYSIS

Three visual scoring methods were applied: 1) Counting. The number of uredia per leaf was counted. 2) Ranking. The level of rust infection on each leaf was recorded according to an assessment key described by McCracken and Dawson [3]. 3) Rust frequency index (RFI). RFI was determined using a stereomicroscope and an ocular equipped with a grid of 225 squares. The percentage of uredia containing squares was counted from 675 squares per each leaf [2].

H.-W. Dehne et al. (eds.), Diagnosis and Identification of Plant Pathogens, 499–501.

The images of infected leaves were acquired using a colour CCTV camera (Panasonic Model WV-CL700/G) equipped with a TV zoom lens, and processed with Leica Quantimet 500+ Colour Option image analysis system connected to a 486 PC (QWIN VO2.OOA software). The number of uredia per leaf and the percentage of uredial area per leaf area were recorded.

Chitin analysis based on enzymatic hydrolysis of chitin and colorimetric determination of the released N-acetylglucosamine was applied to composite samples of leaves with different degree of rust-infection (heavy, light and no visible infection). The analysis followed the method reported by Wiese et al. [4].

3. Results and discussion

Visual estimation of disease severity is prone to errors caused by subjectivity, and it has been suggested that the application of IA to quantification of disease symptoms may increase objectivity and precision [4]. In this study, there were no significant differences among the tested methds. The results obtained with each method showed a significant and positive correlation with each others (coefficients not shown). Similar result was obtained when the number of samples was randomly reduced to 50. Nearly equal mean values were recorded when the number of uredial pustules per leaf was counted visually or with IA, and when the percentage of uredia area per leaf area was estimated visually or measured with IA (Table 1).

TABLE 1.The mean values of different willw rust assessments, the coefficient of variation (CV) of three separate repeated measurements (30 randomly chsen replicates), and the average number of samples assessed with each method in an hour.

	Image anaysis:		Visual assessments:		
	Counting (nr of uredia/leaf)	Uredia area per leaf area (%)	Counting (nr of uredia/leaf)	Ranking (%)	RFI
Mean (±SE, n=100)	44.1 (±3.77)	5.63 (±0.62)	43.8 (±4.78)	5.10 (±0.65)	22.2(±2.48)
CV (n=30)	0.26	0.15	0.15	0.18	0.39
Samples/one hour	30	25	20	100	30

The variation between repeated measurements was highest for the RFI method (see CV in Table 1). This may be due to the difficulty of positioning the grid of the ocular to exactly the same position over the leaf in repeated measurements. With IA, the distinction between leaf tissue and uredia remained vague when the leaf was covered with numerous small uredia. In these cases, uredia had to be added to the image in an interactive stage by the operator, which may have increased the variation in counting with IA (Table 1).

Ranking proved to be the fastest of the studied methods (see the average number of samples assessed per hour in Table 1). With the other methods, the time required to assess a sample depended considerably on the level of infection.Visual counting and RFI were faster than IA when the number of uredia was low (up to 20/leaf). The higher the infection level, the slower the visual counting and RFI were relative to IA. An additional time-saving property of IA systems is that the obtained data can simultaneously be saved in a form ready for further computerized processing.

Consistent correlations between rust and concentrations of potentially antifungal phenolic metabolites could be established using different rust quantification methods: the correlations between rust and (+)-catechin or chlorogenic acid content were significantly positive, whereas concentrations of salicylates were not in significant relation with rust in the analysed leaves (Table 2).

TABLE 2. Correlations between rust severity assessments and concentrations of phenolic metabolites in willow leaves. Shown are the Pearson correlation coefficients and significance level (* = P<0.05, ** = P<0.01, *** = P< 0.001, ns = not significant)

Phenolics	Image analysis: Counting	Uredia area per leaf area	Visual assessments: Counting	Ranking	RFI
(+)-catechin	0.354***	0.270**	0.326**	0.379***	0.386***
Chlorogenic acid	0.416***	0.290 **	0.359***	0.310***	0.339***
Salicortin	- 0.088 ns	- 0.102 ns	- 0.099 ns	- 0.117 ns	- 0.069 ns
Salicin	- 0.128 ns	0.099 ns	- 0.067 ns	- 0.064 ns	- 0.063 ns

Using the chitin analysis, 109 μg/g (fresh weight) of N-acetylglucosamine was detected in visibly non-infected , 216 μg/g in lightly infected and 312 μg/g in heavily infected willow leaves. Although further experiments are necessary before the value of chitin analysis as a willow rust quantification method can be thoroughly evaluated, these results suggest that it may be a promising approach.

Chitin analysis, although not specific for rusts, provides an estimate of the amount of fungal biomass inside the leaf tissue [5]. Thus, the use of chitin analysis, e.g. as a complementary methd to visual or IA assessments may increase the accuracy in detection of rust-infection related changes in phytochemistry of willows. Colour IA has a potential to accurately quantify tissue chlorosis and necrosis, which are observed in certain willow-rust combinations and are likely to concur with alterations in phytochemistry. Therefore, the use of colour IA may bring new aspects to phytochemical studies on willow-rust interaction.

References:

1. FAO 1979: *Poplars and willows*. Forestry series No 10. Rome, pp. 186-210.
2. Julkunen-Tiitto ,R., Hakulinen, J. and Meier, B. (1994) The response of growth and secondary metabolism to *Melampsora* rusts in field cultivated willow (*Salix*) clones. *Acta Horticulturae* 381,679-682.
3. McCracken, A.R. and Dawson, M. (1992). Clonal response in *Salix* to *Melampsora* rusts in short rotation coppice plantations. *Eur. J. For. Path.* 22, 19-28
4. Price ,T.V., Gross, R., Ho Wey, J. and Osborne, C.F. (1993). A comparison of visual and digital image-processing methods in quantifying the severity of coffee leaf rust (*Hemileia vastatrix*). *Aust. J. Exp. Agr.* 33, 97-101.
5. Wiese, G., Hugo-Wissemann, D., and Grambow, H.J. (1986). An improved procedure for the quantitative estimation of the rust fungus in infected plant tissue. *Z. Naturforsch.* 41c:1127-1130.

COMPARISON BETWEEN A DNA-BASED ASSAY AND A SOIL BIOASSAY IN QUANTIFYING THE AMOUNT OF *GAEUMANNOMYCES GRAMINIS* VAR. *TRITICI* IN SOIL

HERDINA, K. OPHEL-KELLER, D. ROGET AND P. HARVEY
Cooperative Research Centre for Soil and Land Management and CSIRO, Division of Soils, Glen Osmond, PMB 2, South Australia 5064, Australia

1. Abstract

A slot-blot hybridization assay for quantifying the amount of *Gaeumannomyces graminis* var. *tritici* (*Ggt*) in diseased roots and in soil has been developed, using a specific and sensitive DNA probe (pG158). The procedure involves separation of organic matter from soil, DNA extraction of soil organic matter, and a slot-blot hybridization assay. Soil of different types was collected to a depth of 10 cm and organic matter between 0.35-1.4 mm in size was separated from this soil. Using a slot-blot hybridization assay, it was found that *Ggt* was mostly present in the organic matter fraction greater than 0.5 mm and in soil at a depth of 5 cm.

The slot-blot hybridization assay correlated well with a soil bioassay, as measured by the percentage of seminal roots with *Ggt* lesions, in estimating the amount of *Ggt* inoculum in soil. Artificially inoculated soils and naturally infested soils were used for the comparison between the two assays. This rapid and reliable DNA-based assay is being used in conjunction with a model of take-all disease development to predict disease levels in the field due to *Ggt*.

2. Introduction

Gaeumannomyces graminis var. *tritici* (*Ggt*) is a soil-borne fungus and considered to cause one of the most important root diseases (take-all) of wheat and barley world-wide.

A soil bioassay is currently used for disease diagnosis of take-all fungus but it is subjective, space and time consuming. DNA based methods to quantify fungi in soil have recently been developed (Whisson, Herdina and Francis, 1995; Herdina, Harvey & Ophel-Keller, 1996). *Ggt* detection has been achieved using a specific probe, pG158 which is based on highly repeated sequences (Harvey, 1993; Herdina *et al.*, 1996). This DNA-based assay has advantages over other molecular methods because it is specific and quantifiable. Furthermore, correlation between inoculum amounts and disease severity may allow us to predict the disease severity caused by this fungus so that appropriate management and control practices can be improved to avoid crop losses.

3. Materials and Methods

The details of the methodology for quantifying *Ggt* in soil, both using the DNA-based assay (Fig. 1) and the soil bioassay, are as described by Herdina *et al.* (1996).

H.-W. Dehne et al. (eds.), Diagnosis and Identification of Plant Pathogens, 503–505.

Collect soil samples
(0-10 cm from the soil surface)

↓

Separate organic matter
(0.5-2.0 mm fraction)

↓

Extract and purify DNA

↓

Load DNA on membrane
(using slot-blot apparatus)

↓

Detect and quantify *Ggt* levels
(using pG158 probe)

Figure 1. The DNA-based assay for quantifying *Ggt* in soil.

Infected crown roots and infested organic matter were used as an inoculum in order to obtain soils with differing inoculum levels. Crown roots were collected from Mannum, South Australia (SA) soil by sieving and then ground to a smaller size. Organic matter inoculum was collected from Bute (SA) soil and separated by washing through two sets of sieves with 1.4 mm and 0.35 mm mesh. Sand was removed by decanting several times. The organic matter remaining on the 0.35 mm mesh was collected on filter paper by vacuum filtration and air dried.

4. Results and Discussion

In general, the slot-blot hybridization assay correlated well with a soil bioassay (expressed as a percentage of seminal roots with *Ggt* lesions), in estimating the amount of *Ggt* inoculum both in artificially inoculated soil (Fig 2 a) and in naturally infested soil (Fig. 2 b). Artificially inoculated soil was used so that the levels of inoculum could be varied within one soil type. There was a good correlation between the disease severity and the amount of inoculum added either using infected crown roots or infested organic matter. However, when Roseworthy soil was inoculated with 0.2 or 1.0% infected crown roots, the percentage of the diseased roots in 0.2% inoculated soil was higher than that in 1.0%. This may be due to the variability of the amount of *Ggt* in the crown roots, used as the inoculum source, although the crown roots were ground and mixed well.

The DNA-based assay is very sensitive, as it is able to detect a very low level of *Ggt* in soil which caused a corresponding low level of the disease. Generally, a *Ggt* DNA level of less than 0.1ng, between 0.1-0.4ng, and above 0.4ng in 100g soil corresponds to a low, medium and high levels of the disease, respectively.

The DNA-based assay is currently being used for rapid routine quantitative assessment of the take-all fungus in soil and for the improvement of a take-all prediction model. An improved take-all prediction model would consequently improve adoption of currently available disease control strategies. Furthermore, the DNA-based method provides a tool for studying the biology of this fungus and can be extended to other soil fungi associated with plant debris, once specific probes are available.

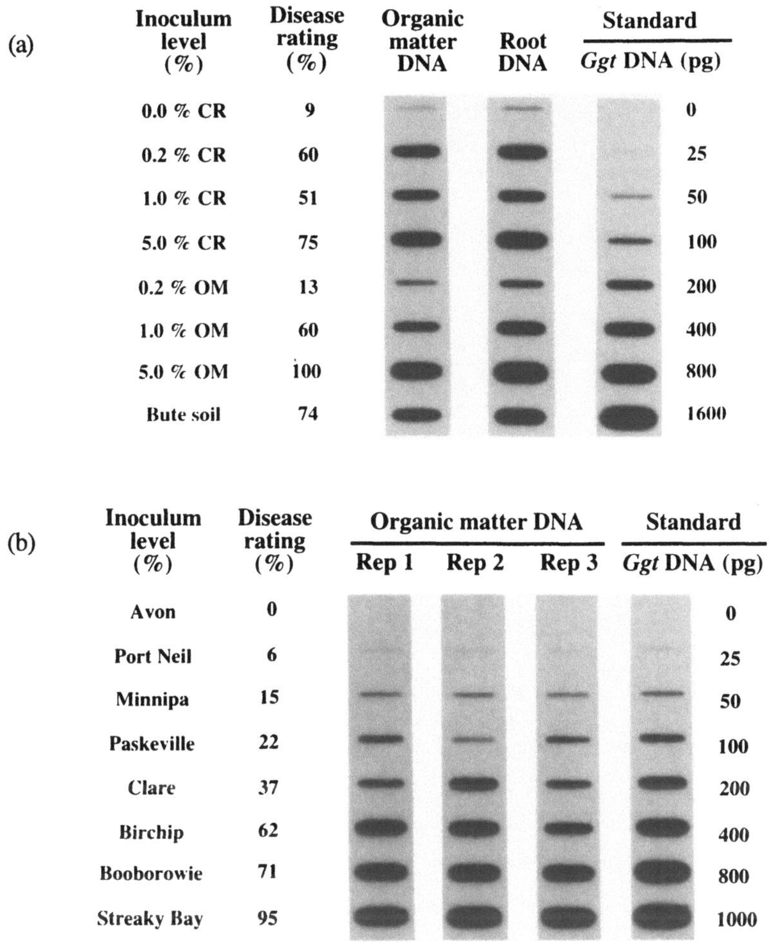

Figure 2. **Comparison between a DNA-based assay and a soil bioassay in quantifying the amount of** *Gaeumannomyces graminis* var. *tritici* in soil. (a) Roseworthy soil inoculated either with infected crown roots (CR) or infested organic matter (OM) of Bute soil at various levels and (b) field soils with various levels of natural inoculum. Three replicates from each field soil were assayed.

5. References

Harvey, P. (1993) Genetic diversity among populations of the take-all fungus, *Gaeumannomyces graminis*. Ph.D. thesis, University of Adelaide, Australia.

Herdina, Harvey, P. and Ophel-Keller, K. (1996) Quantification of *Gaeumannomyces graminis* var. *tritici* in infected roots and soil using slot-blot hybridisation, *Mycological Research* **100**, 962-970.

Whisson, D. L., Herdina and Francis, L. (1995) Detection of *Rhizoctonia solani* AG-8 in soil using a specific DNA probe, *Mycological Research* **99**, 1299-1302.

Roget, D. K. and Rovira, A. D. (1991) The relationship between incidence of infection by the take-all fungus (*Gaeumannomyces graminis* var. *tritici*), rainfall and yield of wheat in South Australia, *Australian Journal of Experimental Agriculture* **31**, 509-513.

IDENTIFICATION OF COMPLEX PATHOTYPES AND SEASONAL CHANGES OF THEIR FREQUENCY IN SUBREGIONAL POPULATIONS OF BARLEY POWDERY MILDEW (*ERYSIPHE GRAMINIS* F.SP. *HORDEI*)

T. HOFFSTADT *and* V. CAFFIER
INRA
Laboratoire de Pathologie Végétale
F-78850 Thiverval-Grignon, France

1. Introduction

A particular feature of barley powdery mildew is the occurence of diverse pathotypes, which are characterized by their combinations of virulence genes ("virulence formula"). These combinations are of great importance, as they enable the pathogen to attack varieties with corresponding resistances. Studies have already been done on this topic over European regions (Limpert, 1987; Andrivon and de Vallavieille-Pope, 1993), but they are seldom concerned with a defined local area (Hovmøller *et al.*, 1993) and investigations are conducted in most cases only on the basis of a single spore sampling per season. The objective of this study was to identify pathotypes of barley mildew in sub regional populations and to detect changes in their frequency over the growing season.

2. Materials and methods

2.1. SAMPLING OF ISOLATES AND VIRULENCE ANALYSIS

Airborne conidia of *Erysiphe graminis* f.sp. *hordei* were sampled in a single locality in northern France (Grignon, near Paris) over a growing season with a stationary spore trap (Schwarzbach, 1979). The samplings were carried out four times in spring (March, April, May and June). After sampling and multiplication, the isolates were tested on a differential set which comprised 12 near-isogenic lines of Pallas containing the most commonly used commercial resistance genes and one susceptible variety as control. After inoculation and incubation, the infection types resulting fom the host-parasite interactions were evaluated. The virulence frequencies of the four series of spore samples were compared by χ^2 tests of homogeneity. The various pathotypes in the four series of spore samples were identified on the basis of the virulence analysis.

2.2. DETERMINATION OF SELECTION PRESSURES

The distribution of the barley cultivars grown in this investigated region was known by a survey of farmers. After identification of the resistance genes used in the cultivars, the frequency of resistance genes used in winter and spring cultivars was calculated for each region. The barley area in this region consisted of only winter cultivars (carrying

H.-W. Dehne et al. (eds.), Diagnosis and Identification of Plant Pathogens, 507–509.

few specific resistance genes) in March-April and of both winter and spring cultivars (carrying diverse resistance genes) in May-June.

3. Results

The commonest pathotype in the first two series of spore samples (15-16%) had the simplest structure, possessing just one virulence gene (table 1, pathotype No. 1). As the growing period progressed, its frequency decreased considerably.

TABLE 1. Frequencies of the main pathotypes of *Erysiphe graminis* f.sp. *hordei* during the growing period

No.	Pathotypes (Virulence formula)	Frequency March	April	May	June
1	c	**14,8**	**15,9**	1,1	1,1
2	c+g	3,7	6,4	0,0	0,0
3	a12+c+g	3,7	6,4	0,0	1,1
4	**a10+a12+g**	0,0	6,4	2,2	0,0
5	a10+a12+k+g	3,7	3,2	4,4	1,1
6	a7+a10+a12+k+g	2,5	6,4	8,8	0,0
7	a7+a10+a12+c+k+g	4,9	3,2	0,0	0,0
8	a7+a9+a10+a13+k+g+La	0,0	0,0	8,8	9,0
9	a7+a10+a12+a13+k+g+La	1,2	0,0	6,6	**18,0**
10	**a7+a9+a10+a12+a13+c+k+g+La**	1,2	0,0	1,1	6,7

The pathotypes which predominate at the end of the season were complex. The most frequent pathotype towards the end of the season (18%) had seven virulences (pathotype No. 9). The complexity of the isolates increases between March (4,3 virulences per isolates) and June (7,1 virulences per isolates).

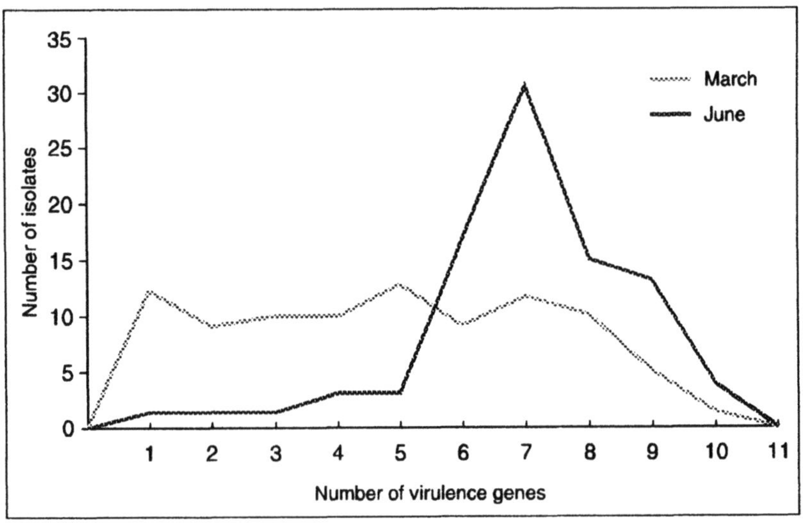

Figure 1. Complexity of isolates at the beginning (March) and the end (June) of the season (Hoffstadt *et al.*, 1995)

The number of pathotypes with a simple structure subsequently decreased in favour of pathotypes with more complex ones (*Figure 1.*). Finally, super races appeared, which possessed the appropriate virulence genes for all the resistance genes present in the region and as a consequence were able to attack every one of the varieties (pathotype No. 10).

The simple pathotypes, which at the start predominated, were later at a disadvantage because they did not have the virulence genes needed to be able to develop on spring barley varieties. In the previous year, the mildew populations at the end of the season were made up of complex pathotypes. At the beginning of the following season simple pathotypes predominated again. There are evidently two different population types: a simple population, present on winter barley varieties (which possesses only a few resistance genes), and a complex population, present on spring barley varieties (which possesses many different resistance genes). The existence of these two population types has never been pointed out before.

4. Discussion

This suggests that contraselection is exerting an effect on the complex pathotypes . There may be a cost of unnecessary virulences (Vanderplank, 1968), which, however, is controversal for E. g. hordei (Stähle *et al.*, 1984; Stähle, 1986); or there may be an effect of the genetic background of the complex pathotypes, which may be not so good adapted to winter cultivars or climatic conditions as the genetic background of the simple pathotypes.
This mechanism as well as a combination with a fungicide resistance could explain the presence of simple pathotypes; however, it is necessary here, to investigate the influence of the epidemics during a growing period on winter and spring barley. Spring barley is inoculated by spores from winter barley and since mature plants are more resistant to mildew than young ones, it is possible that the complex pathotypes selected by spring barley do not go on to re-infect winter barley and that simple pathotypes can remain predominant on winter barley, even if there is no contraselection at work.

References

Andrivon, D. and de Vallavielle-Pope, C. (1993) Racial diversity and complexity in regional populations of *Erysiphe graminis* f.sp. *hordei* in France over a 5-year period, *Plant Pathology* **42**, 443-64.

Hoffstadt, T., Caffier, V. and de Vallavieille-Pope, C. (1995) Entwicklung des Pathosystems Gerste-Gerstenmehltau (*Erysiphe graminis* f.sp. *hordei*) im Laufe einer Vegetationsperiode und phytosanitäre Konsequenzen - eine Studie in Nordfrankreich, *Pflanzenschutz-Nachrichten Bayer* **48**, 397-418.

Hovmøller, M.S., Munk, L. and Østergard, H. (1993) Observed and predicted changes in virulence gene frequencies at 11 loci in a local barley powdery mildew population, *Phytopathology*. **83**, 253-260.

Limpert, E. (1987) Frequencies of virulence and fungicide resistance in the European barley mildew population in 1985, *Journal of Phytopathology* **119**, 298-311.

Schwarzbach, E. (1979) A high throughput jet trap for collecting mildew spores on living leaves, *Phytopathologische Zeitschrift* **94**, 165-71.

Stähle, U., Kranz, J. und Hau, B. (1984) Untersuchungen zur Konkurrenzfähigkeit eines einfachen und komplexen Pathotypen von *Erysiph graminis*. f.sp. *hordei*, *Z. Pflanzenkr. Pflanzenschutz* **91**, 543-548.

Stähle, U. (1986) Untersuchungen zum Einfluß der Virulenzstruktur von Rassen des Gerstenmehltaus (*Erysiphe graminis*. f.sp *hordei*) auf ihre Fitness, *Z. Pflanzenkr. Pflanzenschutz* **93**, 177-182.

Vanderplank, J.E. (1968) *Disease Resistance in Plants*, Academic Press, New York.

CHARACTERIZATION AND ROLE IN THE PATHOGENESIS OF POTATOES OF A NOVEL PECTATE LYASE FROM *ERIWINIA CHRYSANTHEMI* 3937

[1]S.Jafra, [1]I. Figura, [2]N. Hugouvieux-Cotte-Pattat and [1]E. Lojkowska
[1]*Laboratory of Plant Pathology, Department of Biotechnology, University of Gdansk, Poland*
[2]*Laboratory of Molecular Genetics of Microorganisms, URACNRS 1486, INSA Villeurbanne, France*

1. Abstract

Erwinia chrysanthemi 3937 secretes five major isoenzymes of pectate lyase encoded by *the pelA, pelB, pelC, pelD, pelE* genes. A new isoenzyme of pectate lyase was identified in an *Erwinia chrysanthemi* mutant deleted of these *pel* genes. Regulation of *pelL* transcription in bacterial cells growing in the synthetic medium and in the potato tuber tissue was analysed using gene fusion *pel::GUS*. As shown for other pel genes, the transcription of *pelL* is dependent on various environmental conditions. The *pelL* protein is able to macerate plant tissue. *PelL* mutant displayed a reduced virulence on potato tubers, demonstrating the important role of this isoenzyme in soft rot disease. Study of *pelL* gene expression in potato tuber tissue, using fusion of *pelL::GUS* showed low level of expression of this isoenzyme 24, 48 and 72 hours after inoculation in comparison with other *pel* (100 times lower than *pelB* and *pelC* and 1000 times lower than *pelD* and *pelE*). In contrast to other *pel*, the expression of *pelL* in tuber tissue achieves its maximum 24 hours inoculation and then decreases.

2. Introduction

The enterobacterium *Erwinia chrysanthemi* caused soft rot disease in various plants. The maceration process involves the depolymerization of pectin in the plant cell walls. Strains of *E. chrysanthemi* produce and secrete a combination of enzymes: pectin methyl esterase (PME), pectin lyase (PNL), polygalacturonase (PEH) and pectate lyase (PL). The PL activity in *E. chrysanthemi* 3937 resulted from the cumulative action of the five major pectate lyases: PLa, PLb, PLc, PLd, PLe, coded by five separate genes (for review, Barras et al. 1994).

The deletion of *pelA, pelB, pelC, pelD* and *pelE* genes failed to totally eliminate tissue maceration. Analysis of the macerated tissue revealed the presence of a new set

H.-W. Dehne et al. (eds.), Diagnosis and Identification of Plant Pathogens, 511–514.

of Pls which were thought to be synthesised by the bacteria only in plant tissue. Because of their low activity in synthetic medium, they are described as a secondary PL, produced when the bacteria arc grown on plant material or a synthetic medium with the plant extract (Kelemu and Collmer 1993, Beaulieu et al. 1993). The structural gene for a secondary pectate lyase which we called *pelL*, was cloned and its molecular characterisation was carried out (Lojkowska et al. 1995). We also constructed mutants with transcriptional fusion of the *pelL* gene and *GUS* reporter gene. The purpose of the study was to study the pathogenicity of the mutant deprived of the *pelL* gene and analysis of the expression of *E. chrysanthemi pelL* gene in plants.

3. Materials and Methods

The derivatives of the Ech strain 3937 used in this study were a set of mutants with transcriptional fusion of *pelE* and *pelL* genes with reporter gene *GUS* (Hugouvieux-Cotte-Pattat et al. 1992). Assays of pectate lyase and β-glucuronidase were performed on toluenized extracts of infected tissue. Pectate lyase activity was determined by degradation of PGA to unsaturated products that absorb at 235nm (Moran et al. 1968). β-glucuronidase activity was measured following the degradation of p-nitrophenyl-6-D- glucuronide into p-nitrophenyl that absorbs at 405 nm (Bardonnet et al. 1992)

4. Results and Discussion

Isolation of *pelL* gene. A genomic library containing an insert of 3-6kb fragments was constructed in pTZ18 using DNA extracted from *E. chrysanthemi* 3937 derivative from which the five major *pel* genes have been deleted. Using pectate semisolid agar medium, we have identified pectinolylic transformants of *Escherichia coli* (Lojkowska et al. 1995).

Expression of *pelL* gene in *E. chrysanthemi* producing five major Pls.
A *pelL::GUS* transcriptional fusion was constructed by matching a *GUS::kan* cassette in the BssHII site. After recombination to *E. chrysanthemi* chromosome, the expression of the *pelL* fusion was tested in various conditions. The expression of the *pelL::GUS* fusion was followed during the growth under conditions including PL synthesis (Lojkowska et al. 1995). Its transcription was stimulated by a 2.5 factor in the presence of PGA and plant extract. Addition of a glucose to the medium inhibits the *pelL* expression. Induction of SOS system by addition of mitomycin C did not affect transcription of the *pelL* gene. Semiaerobic conditions had little effect on *pelL* expression. In conditions of nitrogen starvation or in a medium with high osmolarity, *pelL* expression was inhibited (Hugouvieux-Cotte-Pattat et al. 1994).

Expression of the *pelL* gene in *E. chrysanthemi* deleted of five major Pls. The expression of the *pelL::GUS* fusion in the delta pel strain was induced by PGA and by

plant extract, with 2-fold and 5-fold respectively, in comparison to the media with glycerol only. Therefore, plant extract appeared to induce better *pelL* expression in mutant deleted of five major Pls background than in a wild type background.
One can suppose that the plant inducer is rapidly degraded in wild type strain but not in the delta *pel* mutant. This degradation is either the result of the activity of the major PL(s) or the product of a gene located in the vicinity of the *pel* genes and deleted in *delta pel* strain.

Effect of various regulatory mutations on *pelL* transcription. We also checked the influence of different regulatory mutations on *pelL* transcription. The *pelL::GUS* fusion was transduced into strains containing regulatory mutations affecting the synthesis of the major Pls. In contrast to major Pls the *pelL* expression was not affected by kdgR or kdgK mutations. This suggests that kdg is not responsible for the induction of *pelL* or expression in the presence of PGA and that the kdgR gene is not involved in the regulation of *pelL* expression. A mutation in the pecS gene increased the expression of the fusion 6-fold, which became independent of PGA induction in this mutant. The pecS gene appears to be involved in *pelL* regulation. These regulatory systems are poorly known, except for that involved in induction by pectic derivatives, mediated by the kdgR repressor (Nasser et al. 1994).

Expression of the *pelL* gene in potato tuber tissue. The mean values of the rot after inoculation of potato tubers with the wild type, a mutant with fusion in *pelL* gene and a *delta pel* mutant were compared. Disease severity was estimated by the rot progression. Tubers were inoculated with pipette tips containing 50 μl of bacterial suspension and incubated in a dew chamber at 27°C. Every day width of tissue was measured. The rot progress was about 25% lower in tubers inoculated with the mutant with fusion in the *pelL* gene. The *delta pel* mutant exhibited less than 35% of the WT pathogenicity. The disease severity was determined 72 hours after inoculation with four different levels of inoculum. In contrast, we did not observe differences in disease severity after inoculation of potato tubers with a mutant deleted of single major PL (Lojkowska et al. 1993).

We assayed the PL activity in tuber tissue 8,12,16 and 24 hours after inoculation with WT, *pelE::GUS*, *pelL::GUS* and *delta pel* strains. Similar kinetics of PL specific activity were observed. The PL activity in WT and *pelE::GUS* strains was similar. However, PL activity in *pelL::GUS* mutant was not more than 10% of WT (Fig. 2). We analysed the expression of *pelL* in tuber tissue by measuring the β-glucuronidase activity. By comparison of the *pelL::GUS* fusion expression with *pelA* and *pelE* expression, we found that the expression of the PL1 isoenzyme is lower than PLa.

The rotting in tubers inoculated with *pelL:GUS* mutants is significantly lower than in tubers inoculated with WT. However, the differences in PL activity in tissue inoculated with WT and *pelL* strains is even more spectacular. This indicates that *pelL* expression may play an important role in regulation of the major *pel* genes transcription.

5. References

Bardonnet, N., Trautweter, A., Couchoux, G. and Blanco, C. (1988) Plasmids with the *uid*A reporter gene for the detection of promotors and transcription signal. Mol. Gen. Genet. **212,** 390-392.

Barras, F., Van Gijsegem, F. and Chatterjee, A.K. (1994) Extracellular enzymes and pathogenesis of soft-rot *Erwinia*. Ann. Rev. Phytopathol. **32,** 201-234.

Beaulieu, C., Boccara, M. and Van Gijsegem, F. (1993) Pathogenic behaviour of pectinase-defective *Erwinia chrysanthemi* mutants on different plants. Mol. Plant Microbe Interact. **6**:197-202.

Kelemu, S, and Collmer, A. (1993) *Erwinia chrysanthemi* EC16 produces a second set of plant-inducible pectate lyase isoenzymes. Appl. Envirom. Microbiol. **59,** 1756- 1761.

Hugouvieux-Cotte-Pattat, N., Dominguez, H. and Robert-Baudouy, J. (1992) Independent expression of pectinase genes of *Erwinia chrysanthemi* 3937. J. Bacteriol. **174,** 7807-7818.

Lojkowska, E., Dorel, C., Reignaul, P., Hugouvieux-Cotte-Pattat, N. and Robert- Baudouy, J. (1993) Use of GUS fusion to study the expression of *Erwinia chrysanthemi* pectinase genes during infection of potato tuber. Mol. Plant-Microbial Interat. **6,** 488-494.

Lojkowska, E., Masclaux, C., Boccara, M., Robert-Baudouy, J. and Bugouvieux-Cotte-Pattat, N. (1995) Characterisation of the gene encoding a secondary pectate lyases of *Erwinia chrysanthemi* 3937. Mol. Microbiol. **16,** 1183-1195.

Moran, F., Nasuno, S. aad Starr, M P. (1968) Extracellular and intracellular polygalacturonic acid trans-eliminase of *Erwinia carotovora* B37. Mol. Microbiol. **2,** 247-254.

Nasser, W., Reverchon, S., Condemine, G, and Robert-Baudouy, J. (1994) Specific interactions of the *Erwinia chrysanthemi* KdgR repressor with different operators of genes involved in pectinolysis. J. Mol.Biol. **236,** 427-440.

FLUORESCENT PARAMETERS OF TWO CUCUMBER CULTIVARS DIFFERING IN RESISTANCE TO MILDEW

V.A. Karavaev, M.K. Solntsev, T.P. Yurina*
Faculty of Physics, Faculty of Biology(*). Moscow State University. Vorobjovy, Moscow, 119899, Russia

The successful breeding of disease resistant plants requires knowledge concerning parameters characterising plant – pathogen interaction and correlating with non-specific resistance to the disease. The fluorescent method developed in this work for estimation of plant physiological state is based on the registration of so-called slow fluorescence induction (SFI) of plant leaves. Cvs. Izyashchnyi and Electron, which were used for comparative study, are both early-ripening, high-class cultivars but cv. Electron exhibits a combined resistance to a number of diseases, including mildew, gummosis, and bacteriosis, whereas cv. Izyashchnyi is resistant only to gummosis, and very susceptible to mildew.

Cucumber plants (*Cucumis sativus* L.) were grown in a greenhouse at 20-25'C under natural daylight supplemented with cool-white light. Light intensity was about $20 W/m^2$ at the plant level. Mildew infection was provoked by spraying with a spore suspension of the fungus *Sphaerotheca fuliginea* Poll. The cotyledonary or first true leaves of 20- or 30-day-old plants, both healthy and infected, were analysed. The photosynthetic activity and respiration rate were measured according to Warburg, and the chlorophyll content was measured with a colorimeter.

To measure SFI, the leaves were kept in darkness for 5 min and then exposed to blue light (15 W/m^2). Fluorescence was recorded at wave length 686 nm which corresponds to the maximum of chlorophyll fluorescence band. Switching on the actinic light induced the fast rise of fluorescence up to the maximal value F_M which was related to the photochemical reduction of primary electron acceptor Q. Then there was a slow decrease of fluorescence to the steady state level F_T. It is known that the transition F_M- F_T depends on the redistribution of excitation energy between photosystems II and I, alterations in trans-thylakoid proton gradient and carbon cycle activity. Earlier we have shown that the relative fluorescence quenching F_M / F_T positively correlates with the photosynthetic activity of the leaf (Karavaev et al., 1988). Table 1 shows that in healthy leaves this ratio was higher for cv. Izyashchnyi than for cv. Electron in accordance with the higher rate of photosynthesis on the chlorophyll basis in cv. Izyashchnyi than in cv. Electron (Table 2). In mildew-infected plants, the SFI curve retained the double-peak pattern and the same time courses for both maxima and steady-state fluorescence levels. But the F_M / F_T ratio was slightly lowered,

H.-W. Dehne et al. (eds.), Diagnosis and Identification of Plant Pathogens, 515–517.

especially in the leaves of susceptible cv. Izyashchnyi (Table 1). This indicates a more severe injury of chloroplast adaptability in a susceptible cv. A similar effect was described for the leaves of monogenic lines of wheat differing in their resistance to rust and mildew (Karavaev et al., 1996).

The mildew agent, a typical obligate parasite, provoked a decrease in the chlorophyll content and photosynthetic activity in susceptible cv. Izyashchnyi (Table 2). These changes were more evident with leaf senescence and infection development. However, in resistant cv. Electron, the early infection was accompanied by a slight chlorophyll accumulation and photosynthetic activation. Evidently, mildew metabolites, as an obligate parasite, can even activate some of the physiological processes in plants at the early stages of infectious development. This is more common for resistant plants.

TABLE 1. F_M to F_T ratio of slow fluorescence induction in the leaves of cucumber under mildew infection

Leaves	Cv.Izyashchnyi			Cv. Electron		
	Healthy	Infected	%	Healthy	Infected	%
			20-day -old plants			
Cotyledonary leaf	2.13	2.05	96	2.12	2.08	98
1st true leaf	2.29	2.18	95	2.10	2.03	97
2nd true leaf	2.15	1.80	84	1.87	1.78	95
			36-day -old plants			
4th true leaf	2.22	1.97	89	2.09	1.98	95
5th true leaf	2.55	1.83	72	2.16	2.05	95

TABLE 2. Effect of mildew infection on chlorophyll content and photosynthetic rate in the leaves of cucumber plants.

Cultivar	20-day -old plants			30-day -old plants		
	Healthy	Infected	%	Healthy	Infected	%
			Chlorophyll content [mg/g fr wt]			
Izyashchnyi	2.1 ± 0.07	2.0 ± 0.06	95	1.8 ± 0.05	1.3 ± 0.04	73
Electron	2.8 ± 0.10	3.1 ± 0.12	110	2.0 ± 0.08	1.9 ± 0.07	95
			Photosynthetic rate [μl O_2 / g fr wt]			
Izyashchnyi	617 ± 26	588 ± 25	96	560 ± 27	410 ± 17	74
Electron	703 ± 30	735 ± 32	104	600 ± 29	550 ± 25	91

All our data allow us to characterise the different types of plant response to pathogen in the leaves of cucumber cvs. The correlation between cucumber plant resistance to mildew and patterns of SFI changes under infection suggests a broad potential for the application of this rather simple spectroscopic method to elucidate the mechanisms of plant responses under infection and for the selection of resistant crop cvs..

References

Karavaev, V.A. and Shagurina, T.L. (1988). Slow fluorescence induction and CO2 exchange in bean leaves in the presence of various chemical agents, Soviet Plant Physiologv (Russian edition) 35, 962-972.

Karavaev, V.A., Solntsev, M.K., Kukushkina, M.A., Yurina, T.P, and Yurina, E.V.(1996) Estimation of plant resistance to powdery mildew and stem rust by the fluorcscent method, in H. Lyr, P.E. Russel and H.D. Sisler (eds.). Modern Fungicides and Antifungal Compounds, Intercept, Andover, pp. 251-255.

THE RATE CONSTANTS OF ELECTRON TRANSPORT IN RC PS2 DURING DISEASE DEVELOPMENT (STEM RUST, POWDERY MILDEW).

KUKUSHKIN ALEXANDER, EKOBENA F. HENRI P.
Faculty of Physics, Dept Biophysics, Moscow State University, Moscow, Russia

Introduction

The aim of this work was the study of the influence of plant diseases on the elementary constant of photosynthetic process at the level of PS2. We earlier used thermoluminescence (TL) method to study the development of plant infection [1,2,3] and have showed the possibility of TL to monitor the evolution of plant diseases. Previously [4] we developed the new theory of TL for application to the photosynthetic processes. The theory permitted us to evaluate elementary characteristics of states taking part in electron transport and the rate constants of these processes and, furthermore, to calculate theoretical TL curves for given rate constants of primary processes. Comparing them to experimental curves we can evaluate qualitatively the rate constants of primary processes in infected plants and monitor changes occurring while the evolution of plant diseases.

Materials and methods

The experiments were carried out on 7, 14, 21 and 28 days wheat leaves <u>*Triticum aestivum L.*</u> cv.Markiz resistant (with gene Sr-Tt) and non resistant (Sr-9B) to Pathogens *Puccinia graminis f. sp. tritici* and *Erysiphe graminis f. tritici* responsible to plant diseases powdery mildew and stem rust. TL curves of wheat leaves preliminary illuminated by far red light (707 nm) were registered after single flash illumination at 0 °C. We also analysed the experimental data obtained in works [1,2,3].

Theoretical calculations of TL were made according to the theory developed in the work [4]. Taking into account that the charges in the complex before the heating are supposed to be stable, we can reduce the number of reaction centre states to four described by the scheme:

$$D^+PA^- \underset{K_{-2}}{\overset{K_2}{\Leftrightarrow}} DP^+A^- \underset{K_{-1}}{\overset{K_1}{\Leftrightarrow}} DP^*A \overset{K_0}{\Rightarrow} DPA$$

TL caused by recombination of more complicated states corresponding with A, B, C maxima. The charge recombination, causing successively the peaks A (-40 ÷ 0 ^{0}C), B (0 ÷ 40 ^{0}C), C(40 ÷ 80 ^{0}C) occurs according to following scheme:

H.-W. Dehne et al. (eds.), Diagnosis and Identification of Plant Pathogens, 519–521.

$S_4^+PQ_A^- \Leftrightarrow S_3^+P^*Q_A$ lowest temperature component of peak A

$S_3^+PQ_A^- \Leftrightarrow S_2^+P^*Q_A$ component of peak A with maximum near -5 °C

$S_3^+PQ_B^{2-} \Leftrightarrow S_2^+P^*Q_B^-$ peak B_1 of B band

$S_2^+PQ_B^- \Leftrightarrow S_1^+P^*Q_B$ peak B_2 of B band

A more detailed scheme of recombination accounting for these peaks were given in work [4].

The corresponding TL intensity are given by the following formulae:

$$I_A = A_0 F_A(T)\exp(-\frac{1}{\beta}\int_{T_0}^{T} F_A(T)dT);$$

$$I_{B1} = F_{B1}(T)\exp(-\frac{1}{\beta}\int_{T_0}^{T} F_{B1}(T)dT)\left[B_{10} + K_1\int_{T_0}^{T} I_A(T)\exp(\frac{1}{\beta}\int_{T_0}^{T} F_{B1}(T)dT)dT\right];$$

$$I_{B2} = F_{B2}(T)\exp(-\frac{1}{\beta}\int_{T_0}^{T} F_{B2}(T)dT)\left[B_{20} + K_2\int_{T_0}^{T} I_{B1}(T)\exp(\frac{1}{\beta}\int_{T_0}^{T} F_{B2}(T)dT)dT\right];$$

Where

$$F_A(T) = \frac{\exp(-(E_{a1}+E_{d1}+E_{a2})/kT)}{\frac{1}{R_0}+\frac{1}{R_{a1}}+\frac{1}{R_{d1}}\exp(-\frac{E_{a1}}{kT})+\frac{1}{R_{a2}}\exp(-\frac{E_{a1}+E_{d1}}{kT})+\frac{1}{R_{d23}}\exp(-\frac{E_{a1}+E_{d1}+E_{a2}}{kT})};$$

$$F_{B1}(T) = \frac{\exp(-(E_{a1}+E_{d1}+E_{a2}+E_{d23}+E_{a32})/kT)}{\frac{1}{R_0}+\frac{1}{R_{a1}}+\frac{1}{R_{d1}}\exp(-\frac{E_{a1}}{kT})+\frac{1}{R_{a2}}\exp(-\frac{E_{a1}+E_{d1}}{kT})+\frac{1}{R_{d23}}\exp(-\frac{E_{a1}+E_{d1}+E_{a2}}{kT})+\frac{1}{R_{a32}}\exp(-\frac{E_{a1}+E_{d1}+E_{a2}+E_{d23}}{kT})};$$

similarly formula take place for $F_{B2}(T)$.

B_{10}, B_{20}- initial concentrations of states accounting for peaks B_1 and B_2, K_1, K_2- coefficients of connection between peaks A - B_1 and B_1 - B_2. $0 \le K_1 \le 1$, $0 \le K_2 \le 1$ depend on conditions. E_i - activation energy of i- transition, R_0 - fluorescence rate constant and R_i - rate constant of i- transition.

Results and discussions

The analyse of experimental data showed several typical TL curves characteristic for resistant and non resistant sort of wheat leaves during infection. We used the parameter $\beta = \left(\frac{S_B}{S_{A+B}}\right) / \left(\frac{S_B}{S_{A+B}}\right)_{control}$ to monitor changes occurring in the content of states $S_3^+PQ_B^{2-}$ and $S_2^+PQ_B^-$. Where S_i - light-sum of i- band of TL. The above parameter was decreasing from 115 down to 93 % for resistant and from 117 down to 73 % for non resistant. This certified that the amount of state $S_4^+PQ_A^-$ was increasing. However the same results could also be obtained theoretically if in the infected plants either the rate constant of electron transfer between primary and secondary electron acceptors of PS2 (Q_A and Q_B) shows 1.5-fold decrease or by decreasing the effectiveness of $S_3^+ \Rightarrow S_4^+$ transition of water splitting system. To clarify that point it is necessary to use additional data which can be got by slow and fast fluorescence induction method. In resistant plant leaves β was increasing with infection time as compare to the non resistant plant leaves. This may suggest the decrease of the PS2 activity in resistant plant leaves more than in non resistant (figure 1).

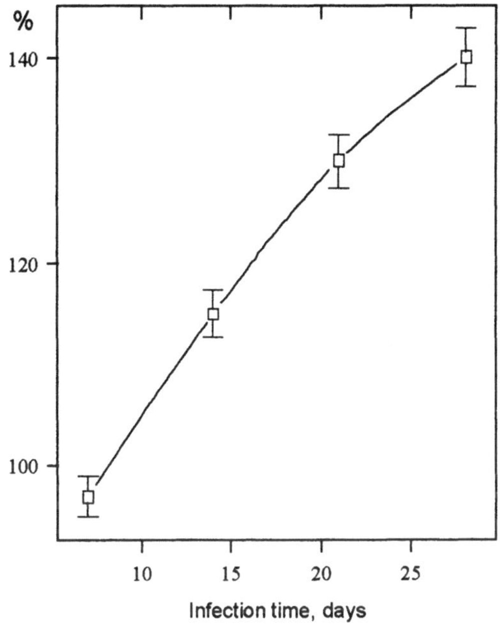

Figure 1: The dependance of $S_3{}^+PQ_B{}^{2-}$ and $S_2{}^+PQ_B{}^-$ concentration in resistant plant leaves, on infection time, as compared to the similary value in non resistant plants leaves

References.

1. Solntsev . M.K., Ekobena F. H.P., Tashish V., Yurina E.V., Yurina T.P.(1995) *Modern Fungicides and Antifungal Compounds,* Ed. by H. Lyz, P.E Russell and H.D. Sisler., Intercept Andover, pp. 257-262.
2. Yurina T.P, Karavaev V.A, Solntsev . M.K. (1993) Peculiarities of metabolism of two sort of cucumber with different resistance to powdery mildew, *Soviet Plant Physiology* **40,** 218-224.
3. Yurina T.P, Yurina E.V, Karavaev V.A.,. . Solntsev . M.K., Kukushkina M.A., Ekobena F. H.P. Physiological peculiarities of wheat leaves resistant and non resistant to powdery mildew, *Soviet Plant Physiology* **43,** 73-79.
4. Kukushkin A.K.,Novikova L.V. Theoretical investigation of thermoluminescence of higher plant, *Biofizika (USSR)* **36,** 640-647.

RAPID ASSESSMENT OF AGGRESSIVENESS OF ROOT ROT AGENTS OF ALFALFA (*MEDICAGO* L.)

R. MAČKINAITĖ
Institute of Botany
Žaliųjų ežerų 49, 2021 Vilnius, Lithuania

Root rot is the most widespread disease caused by phytopathogenic soilborne microorganisms. The disease has been detected in all regions of agriculture and it injures all cultivated plants. The spreading and severity of disease depends greatly on the aggressiveness of microorganisms which take part in the pathological process. In order to decrease the losses, caused by the disease, it is necessary not only to ascertain special composition of pathogens, but to evaluate their aggressiveness as well.

1. Materials and Methods

In order to evaluate the aggressiveness of 15 most widespread root rot agents of alfalfa (*Alternaria alternata* (Fr.:Fr.) Keissl., *Cylindrocarpon didymum* (Harting) Wollenw., *Cladosporium herbarum* (Pers.:Fr.) Link, *Fusarium avenaceum* (Fr.) Sacc., *F. culmorum* (Wm.G.Sm.) Sacc., *F. javanicum* Koord. var. *radicicola* Wollenw., *F. oxysporum* (Schltdl.) W.C.Snyder et H.N.Hansen var. *orthoceras* (Appel et Wollenw.) Bilai, *F. sambucinum* Fuckel var. *minus* Wollenw., *F. solani* (Mart.) Appel et Wollenw., *F. solani* (Mart.) Appel et Wollenw. var. *argillaceum* (Fr.) Bilai, *F. solani* (Mart.) Appel et Wollenw. var. *coeruleum* (Lib.) Bilai, *Phoma sclerotioides* G.Preuss ex Sacc., *Rhizoctonia crocorum* (Pers.:Fr.) DC., *Thielaviopsis basicola* (Berk. et Broome) Ferraris, *Verticillium albo-atrum* Reinke et Berthier), isolated from the naturally infected roots, the method for seedling inoculation in plates, described by Chi and Childers (1964) was applied. Two-days-old and the same (about 1 cm) length 15 seedlings of alfalfa (*Medicago varia* Martyn 'Augūnė II') were spread out around the 2-days-old colony of pathogen, which was grown on the medium of malt extract agar (MEA) in the centre of Petri plates. Seedlings were free of any infection and they were arranged so, that their root tips faced the margin of the developing colony. The aggressiveness of separate pathogens was ascertained according to the length of alfalfa seedlings and the intensity of injury after 4 days of their contact with the pathogen. It was developed five-graduate scale for evaluation of the intensity of injury of alfalfa seedlings: 0 - healthy seedlings, + - injured less than 1/3 of seedling, ++ - injured from 1/3 to 2/3 of seedling, +++ - injured more than 2/3 of seedling, ++++ - dead seedlings.

H.-W. Dehne et al. (eds.), Diagnosis and Identification of Plant Pathogens, 523–525.

In greenhouse tests alfalfa 'Augūnė II' was artificially infected by various strains of pathogens. The plants of alfalfa were cultivated in pots with sterile sand. The infection (15 days-old culture of pathogen, cultivated on MEA) was applied to the upper layer of sand before sowing, taking one Petri plate for one pot. 25 swollen seeds were sowed into every pot. In control variant plants were cultivated without any infection. The aggressiveness of tested pathogens was ascertained after two months, evaluating the percentage of injured plants and the average number of the intensity of their injury (Lubenec, 1964).

2. Results and Discussion

Our early investigations demonstrated, that root rot of alfalfa is caused by the whole complex of micromycetes, belonging to 40 genera (Mačkinaitė, Strukčinskas, 1992 a,b). Among them the most widespread are fungi of genus *Fusarium* Link:Fr. making over 50 % of the whole number of isolates.

This investigation indicated that fungi of the genus *Fusarium* were the most aggressive to the seedlings of alfalfa 'Augūnė II'. Their length has shortened twice or three times to compare with the control after the contact with these pathogens (Table 1). *F. avenaceum*, *F. oxysporum* var. *orthoceras* and *F. culmorum*

TABLE 1. Aggressiveness of the most widespread root rot agents to the seedlings of alfalfa 'Augūnė II'

The strain of fungus	Diameter of the colony of fungus	Injury of seedlings		Length of seedlings	
		amount	intensity	mm	% in comparison with control
Control (without fungus)	0	0	0	35.7	100.0
Alternaria alternata 6328	53.6	15.0	++	26.6	74.5
Cylindrocarpon didymum 6973	40.0	13.8	++	24.5	68.6
Cladosporium herbarum 3754	26.0	0	0	39.2	109.8
Fusarium avenaceum 5044	86.6	15.0	++++	10.6	29.7
F. culmorum 4743	110.0	15.0	++++	12.2	34.2
F. oxysporum var. *orthoceras* 6936	78.8	15.0	+++	15.3	42.9
Phoma sclerotioides 6955	38.4	15.0	+++	13.0	36.4
Rhizoctonia crocorum 5516	110.0	15.0	++	26.8	75.1
Thielaviopsis basicola 6349	27.8	12.8	+	33.3	93.3
Verticillium albo-atrum 4710	39.0	15.0	++	17.4	48.7

were the most aggressive among them. After the contact with these fusaria the length of alfalfa seedlings reached only 33.5, 43.3 and 44.5% to compare with the control (Table 2). *Phoma sclerotioides* and *Verticillium albo-atrum* appeared to be the most aggressive among the fungi of the other investigated genera. They have shortened the seedlings 'Augūnė II' by 22.7 and 18.3 mm respectively to compare with the control. The other fungi had much weaker influence on seedlings of alfalfa, while *Cladosporium herbarum* even slightly stimulated their growth (Table 1).

TABLE 2. Aggressiveness of fungi genus *Fusarium* to the seedlings of alfalfa 'Augūnė II'

The strain of Fusarium	Diameter of the colony of fungus	Injury of seedlings		Length of seedlings	
		amount	intensity	mm	% in comparison with control
Control (without fungus)	0	0	0	32.8	100.0
F. avenaceum 1316	85.6	15.0	++++	11.0	33.5
F. culmorum 1333	110.0	15.0	++++	14.6	44.5
F. javanicum var. *radicicola* 1145	54.2	14.4	+++	26.7	81.4
F. oxysporum var. *orthoceras* 1324	74.4	15.0	++++	14.2	43.3
F. sambucinum var. *minus* 1317	86.6	15.0	+++	26.9	82.0
F. solani var. *coeruleum* 1281	49.4	14.4	+++	25.5	77.4

The results of investigation on aggressiveness of fungi in laboratory experiments were confirmed by the results obtained by greenhouse tests. The artificial infection of alfalfa 'Augūnė II' seedlings with strains of 7 Fusarium species showed that the most aggressive to them were F. culmorum, F. oxysporum var. orthoceras and F. avenaceum. They injured 49.2, 40.8 and 38.6% of their roots, respectively (Fig. 1).

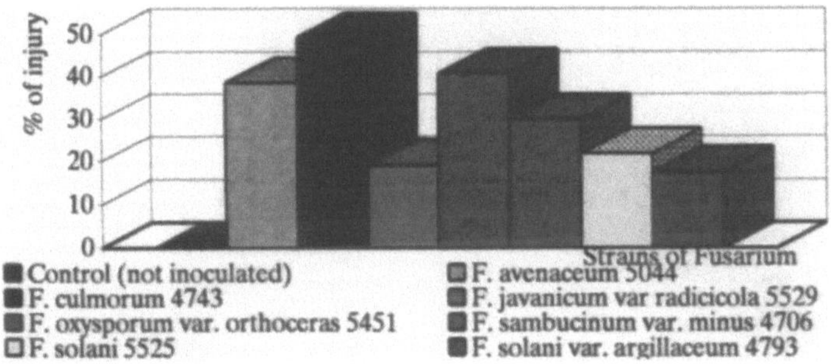

Fig. 1. Aggressiveness of fungi genus *Fusarium* to the alfalfa 'Augūnė II' after artificial inoculation under greenhouse conditions

For the rapid assessment of the aggressiveness of fungi - root rot agents, beside other methods, laboratory seedlings method, tested by us, should be used. This method speeds up the primary evaluation of the pathogen aggressiveness and does not require any special equipment.

3. References

Chi, C.C. and Childers, W.R. (1964) Penetration and infection in alfalfa and red clover by Pellicularia filamentosa, *Phytopathology* 54, 750-754.

Lubenec, P.A. (1964) Evaluation of collection of perennial grasses according to their resistance to root rots, *Selekcija i Semenovodstvo* 1, 60-64 (in russ.).

Mačkinaitė, R. and Strukčinskas, M. (1992 a) Micromycetes - Medicago L. root rot agents in Lithuania. 1. Root rot agents of seed alfalfa, *Ekologija*, 1, 98-106 (in russ.).

Mačkinaitė, R. and Strukčinskas, M. (1992 b) Micromycetes - Medicago L. root rot agents in Lithuania. 2. Root rot agents of wild alfalfa species and red clover growing beside, *Ekologija*, 4, 95-104 (in russ.).

THE ROLE OF PRUNUS SPINOSA L. IN EPIDEMIOLOGY OF PLUM POX VIRUS IN THE CZECH REPUBLIC

J. POLÁK
Research Institute of Crop Production
Division of Phytomedicine
161 06 Prague 6 - Ruzynì, Drnovská 507, Czech Republic

1. Abstract

Plum pox virus (PPV) was proved by ELISA in *Prunus spinosa* in the Czech Republic. The percentage of plants infected with PPV out of sixty eight tested was 17,6. Blackthorn plants infected with PPV showed distinct symptoms in leaves. Symptoms were very similar to those on plums. Blackthorn is suggested as secondary source of PPV infection. Its role in the epidemiology of PPV is probably very limited.

Prune dwarf virus (PDV) was proved in twenty five blackthorns out of sixty tested, i.e. 41,7% plants were infected. PDV was latent in blackthorns, occasionally were observed very mild symptoms. *Prunus spinosa* is an important source of PDV infection.

1. Introduction

Plants of blackthorn (*Prunus spinosa* L.) spontaneously infected with plum pox virus (PPV) were first detected by Šutic' and Pine (1968) and Jordovic' et al. (1969). The role of blackthorn as a natural host of PPV was later studied by Rankovic' and Dulic' Markovic' (1992) and found to be limited. Blackthorn is supposed to be an often symptomless host and important source of PPV infection in the Czech Republic (Ackermann, personal communication) being widely distributed. Polák (1988) proved the occurrence of natural infection of Prunus cerasifera ssp. myrobalana trees with PPV. PPV is also frequent in cultivated plums and wild plums in Central Bohemia.

The aim of our study is to prove the presence and distribution of PPV in blackthorn in the territory of the Czech Republic and to show its importance as a natural source of PPV infection.

H.-W. Dehne et al. (eds.), Diagnosis and Identification of Plant Pathogens, 527–530.

2. Material and Methods

2.1. SAMPLING OF PLANTS

The samples of *P. spinosa* flowers were collected mostly from different localities of Central Bohemia in May. Thirty flowers were usually taken from one plant. Leaves were collected from the same plants during the summer from June to August.

Plums in Central Bohemian areas used to be severely infected with PPV and the virus occurs also very often in wild Prunus cerasifera ssp. myrobalana trees.

2.2. SEROLOGICAL EVALUATION

PPV antiserum was prepared in our laboratory using a procedure published by Albrechtová et al. (1986). The DAS-ELISA method (Clark and Adams, 1977) was used. Samples for ELISA were prepared by grinding thirty flowers or 0,2 g of leaf blackthorn tissue in phosphate buffered saline pH 7,4 (ratio 1:20) with 2% polyvinylpyrrolidone and 0,2% of egg albumin. Samples were evaluated also for the presence of Prune dwarf virus (PDV). Commercial PDV antibodies (Bioreba) were used in DAS-ELISA. Plates were evaluated using MR 5000 reader (Dynatech) at 405 nm.

2.3. VISUAL EVALUATION

Presence of PPV and intensity of symptoms in leaves of *P. spinosa* were evaluated in June, July or August. Photo documentation of leaf symptoms was provided too.

3. Results

Results of serological detection of PPV and PDV in blackthorn trees from different localities of the Czech Republic are presented in Table 1. PPV was proved by ELISA in twelve blackthorns out of sixty eight tested while PDV in twenty five out of sixty tested. The percentage of plants infected with PPV and PDV in the total number of plants tested was 17,6% and 41,7% respectively.

Blackthorn plants infected with PPV showed distinct symptoms consisting of diffuse spots, rings and oak line patterns in leaves. Symptoms of necrotic rings and oak line patterns appeared in leaves of two PPV infected blackthorns.

Blackthorn plants infected with PDV were symptomless or showed individual mild pale green spots in leaves.

4. Discussion

Our results proved that only a part of blackthorns is infected with PPV in Central Bohemia. Brunt et al. (1966) reviewing references to PPV described *Prunus spinosa* as latent host of the virus. But we were able to prove distinct symptoms on leaves of PPV infected blackthorns. Fifteen to twenty percent of blackthorn shrubs were infected in

TABLE 1. Distribution of PPV and PDV by ELISA on *Prunus spinosa* L. in the Czech Republic.

No	Sampling site	ELISA results PPV	PDV
1	Praha - Troja	-	-
2	Praha - Bohnice	-	-
3	Praha - Dolní Chabry 1	+	-
4	Praha - Dolní Chabry 2	-	-
5	Kralupy 1	-	-
6	Kralupy 2	-	+
7	Zákolny	-	+
8	Dřešt'ovice	-	+
9	Hlubočepy 1	-	+
10	Hlubočepy 2	-	+
11	Slivenec 1	+	+
12	Slivenec 2	-	-
13	Lochkov	-	+
14	Rdaotín	-	+
15	Černošice	-	+
16	Černolice 1	-	nt
17	Černolice 2	-	nt
18	Třebenice 1	-	-
19	Třebenice 2	-	-
20	Košt'álov 1	+	-
21	Košt'álov 2	+	-
22	Kuzly	-	-
23	Velemyšleves	-	-
24	Kličín	+	-
25	Blšany	-	+
26	Bedlov	-	+
27	Šanov	-	-
28	Radovník 1	+	+
29	Radovník 2	-	+
30	Radovník 3	-	+
31	Vešín 1	+	nt
32	Vešín 2	+	nt
33	Pálava	-	nt
34	Nelahozeves	-	+
35	Praha - Smíchov	-	+
36	Mochov	-	-
37	Králíky	-	-
38	Holice	-	+
39	Hradec Králové	-	-
40	Hořice	-	-
41	Městec Králové	-	-
42	Poděbrady	-	-
43	Poběžovice	-	-
44	Myštěves	-	-
45	N. Bydžov	-	-
46	Kanice	-	-
47	Chvojenec	+	-
48	Dohalice 1	+	-
49	Dohalice 2	+	-
50	Miletín	-	-
51	Sulom	-	-
52	Žitenice	-	-
53	Podbořany 1	-	+
54	Podbořany 2	-	+
55	Drahonice	-	-
56	Lubenec	-	-
57	Žlutice	-	-
58	Bochov	-	-
59	Kněževes	-	+
60	Řeporyje	-	+
61	Mníšek	-	-
62	Zbelítov	-	+
63	Lety	-	+
64	Mikulov	-	+
65	Valtice	-	+
66	Strážnice	+	nt
67	Dubá	+	nt
68	Kokořín	-	nt

+: postive, -: negative, nt: not tested

the area with severe occurrence of PPV in plums. Our findings differ from Rankovic' and Dulic'-Markovic' (1992) who proved only 0,9% of blackthorns to be infected with PPV in Yugoslavia. Explanation for such a lower occurrence of PPV in blackthorns in Yugoslavia can be in higher temperatures during the vegetation period causing possible virus inactivation in aphid vectors. Nonpersistent PPV is probably transmitted by aphids to blackthorn more sporadically. Situation in Yugoslavia and also in the Czech Republic suggests blackthorn to be the only secondary source of PPV infection. Blackthorns surrounded by contaminated plums can be infected step by step. The role of *Prunus spinosa* in the epidemiology of PPV in the Czech Republic seems to be limited at present, but it can increase in future. We assume to verify our theory by next years. PPV occurs only occasionally in several marginal highland areas in the Czech Republic. Therefore our investigations will be concentrated to those marginal areas where we expect no incidence of PPV on blackthorns.

A different situation concerns the distribution of Prune dwarf virus. We proved more than forty percent blackthorns to be infected with PDV and confirmed results of Rankovic' and Dulic'-Markovic' (1992) who find 37,2% of PDV infected blackthorns. *Prunus spinosa* is an important source of PDV.

5. References

Albrechtová, L., Holubcová, L. and Jokeš, M. (1986) Purifikace viru šarky švestky a pøíprava antisera vhodného pro stanovení viru metodou ELISA (The purification of plum pox virus and the preparation of antiserum suitable for virus detection by ELISA). *Sb. ÚVTIZ-Ochr.rostl.* 22, 161-167.

Brunt, A.A., Crabtree, K., Dallwitz, M.J., Gibbs, A.J. and Watson M. (1996) Virus of Plants. *CAB International, Wallingford, United Kingdom*, 1484 pp.

Clark, M.F. and Adams A.N. (1977) Characteristics of the microplate method of enzyme-linked immunosorbent assay for the detection of plant viruses. *J. gen. Virol.* 34, 51-57.

Jordovic', M., Festic', H. and Rankovic', M. (1969) Ispitivanje *Prunus spinosa* L. kao domaèina virusa šarke (Preliminary studies of *Prunus spinosa* L. as a host of sharka virus). *Zaštita bilja* 105, 253-259.

Polák, J. (1989) Diagnosis of plum pox virus in infected symptomless trees of apricot peach and *Prunus cerasifera* ssp. myrobalana by ELISA and ISEM. *Acta Horticulturae* 235, 299-303.

Rankovic', M. and Dulic'-Markovic' I. (1992) Evaluation of *Prunus spinosa* L. as host of sharka and other viruses. *Acta Horticulturae* 309, 151-155.

Šutic', D. and Pine, S. (1968) Sharka (plum pox) disease. *Plant Disease Reporter* 52, 253-256.

Acknowledgements

The author thanks Mrs. J. Pívalová and Miss. M. Bartoòová for excellent technical assistance.

YIELD REDUCTION AND SEED TRANSMISSION RATE OF LITHUANIAN ISOLATE OF BROAD BEAN TRUE MOSAIC VIRUS

J. STANIULIS
Laboratory of Plant Viruses Institute of Botany
Žaliųjų ežerų 49, Vilnius, 2021, Lithuania

1. Introduction

Broad Bean True Mosaic Virus (Echtes Ackerbohnemosaik-Virus) (BBTMV) from Comovirus group, described for the first time in Germany by Quantz (1953). The virus was detected in many European countries and characterized in details by Gibbs and Paul (1970), Boswell and Gibbs (1983), Edwardson and Christie (1986). In Lithuania the virus was isolated in 1965 and preliminary identified on the basis of symptom expression, host range studies, morphology of particles, possibility of transmission by seeds of broad bean and not transmissibility by aphids (Staniulis, 1967; 1973). The incidence of BBTMV in fields of broad beans in Lithuania is not very widespread. It usually occurs within the range of 5 %, but it is found wherever broad bean is grown.

Recently more detailed characterization of local isolate of BBTMV from broad bean plants (Vicia faba L. cv. Aušra) expressing the symptoms of mosaic with slight distortion of leaves collected from Kėdainiai region (Dotnuva-Akademija) was undertaken (Staniulis, Elcševič, 1994). The host range of the isolate showed no deviation from typical isolate of the virus and was limited by species of Leguminosae family (Schmelzer, Wolf, 1971). Purified virus preparations after negative staining with 2 % of uranyl acetate consisted of isometric particles about 30 nm in diameter. Purified virus preparations as antigens and complete Freund's adjuvant were used for production of antiserum in rabbit. In gel double diffusion tests homologous titer was 1:1024. Identity or very close serological relatedness of our isolate of BBTMV with Dutch isolates of BBTMV (M96 and Rohloff) was confirmed. In gel double diffusion tests clear cut lines of precipitation between clarified sap of broad bean infected with Lithuanian isolate of BBTMV and antiserum to isolate BBTMV-M-96 were formed up till dilution of antiserum to 1:16 and up till 1:40 for isolate BBTMV-Rohloff. Both

H.-W. Dehne et al. (eds.), Diagnosis and Identification of Plant Pathogens, 531–533.

antisera were kindly provided by Dr. D. Maat from Institute for Plant Protection, Wageningen, The Netherlands.

2. Materials and methods

For investigation of the influence of BBTMV to yield of broad bean small trial in field conditions was undertaken. Small plots of broad bean cv. Aušra in three repeats were sown at appropriate time in spring of 1993. Emerging plants at 4 leaves stage and 8 leaves stage were dusted with carborundum and mechanically inoculated with purified virus preparations. Estimations were made taking into account plots of 100 plants. Number of pods, number of seeds per pod, number of seeds per plant and weight of seeds per plant were estimated.

For seed transmission experiment seeds of broad bean plants cv. Aušra, infected with BBTMV in field trial at four leaves stage in 1993 were sown in greenhouse conditions in 1994. Seed infection rate was evaluated according to appearance of typical symptoms for BBTMV infection, by serological reaction in gel, by electron microscopic observation of virus particles in negatively stained preparations and by back inoculation to broad bean plants.

3. Results

Broad bean plants inoculated with BBTMV at the four leaf of development stage were strongly reduced in growth and yielded considerable less seeds in comparison with noninoculated healthy plants (table 1).

TABLE 1. Influence of BBTMV to yield of broad bean plants inoculated at early stages of growth

Component of yield	Inoculation at 4 leaf stage / % from healthy plants	Inoculation at 8 leaf stage / % from healthy plants	Noninoculated plants
Number of pod per plant	4,55 ± 0,52 / 57,1	6,53 ± 0,96 / 82,1	7,95 ± 0,20
Number of seeds per pod	1,70 ± 0,0 / 85,0	1,87 ± 0,0 / 93,5	2,0 ± 0,0
Number of seeds per plant	7,80 ± 0,86 / 48,0	12,3 ± 1,5 / 76,7	16,0 ± 0,38
Weight of seeds per plant (g)	3,40 ± 0,45 / 51,9	5,14 ± 0,73 / 78,4	6,55 ± 0,23

As can be seen from table 1 main reason for the yield reduction in plants infected at early phases of development is very strong reduction in number of pods per plant (42,9% when infected at 4 leaf stage) and significant reduction of seeds per pod. Infection at 8 leaf stage still decreased the yield by nearly 22%. Infection just before flowering had very slight influence to yield of broad bean (data not presented in the table).

In greenhouse experiment for seed transmission rate of BBTMV seeds collected from plants infected at early stage of development in previous year were used. Germinated 3140 seedlings. At 8 leaves stage of development 133 plants showed characteristic symptoms of BBTMV infection. After additional examination of suspicious plants by electron microscopic and serological methods and by back inoculation additional 11 plants were attributed to plants with seed infection. Total number of plants with seed infection reached 144 or 4,59 %, that is within the limits indicated for this virus in other geographical locations.

4.References

Boswell K.F., Gibbs A.J. (1983) *Virus identification data exchange. Viruses of legumes.* The Australian University, Canberra.

Edwardson J.R., Christie R.G. (1986) *Viruses infecting forage legumes.* University of Florida, V. **1**, 89-142.

Gibbs A.J., Paul H.L. (1970) Echtes Ackerbohnemosaik-Virus. C.M.I./A.A.B. Descriptions of Plant Viruses, N 20, 1-5.

Quantz L. (1953) Untersuchungen über ein samensubertragbaren Mosaik Virus der Ackerbohne, *Phytopathologische Zeitschrift* **20**, 421-448.

Schmelzer K., Wolf P. (1971) Wirtspflanzen der Viren und Virusen Europas, *Nova Acta Leopoldina,* Suppl. **2**, N 36, 1-262.

Staniulis J. P. (1967) Some data for identification of virus diseases of leguminous crops in Lithuania, *Lietuvos TSR MA Darbai,* C serija **1(42)**, 3-18 (In Russian).

Staniulis J. (1973) Ankštinių augalų virusinės ligos, *Augalų virusinių ligų tyrimai Lietuvoje*, Vilnius, 49-72.

Staniulis J., Eleševič V. (1994) Broad bean true mosaic virus in Lithuania, *Annual report of the Institute of Botany*, Vilnius, 105-108.

CHARACTERIZATION OF *FUSARIUM OXYSPORUM* POPULATIONS BY GROWTH PARAMETER EVALUATION IN MICROTITER PLATES.

C. STEINBERG, V. EDEL, N. GAUTHERON, T. VALLEYS,
C. ALABOUVETTE.
INRA-CMSE , Laboratoire de Flore Pathogène du sol.
BV1540. F21034 DIJON.

1. Introduction

Trophic competition for the C source is the main mechanism involved in the interaction between pathogenic and non-pathogenic populations of *Fusarium oxysporum*. Therefore, a phenotypical characterization based on the growth characteristics of these populations will provide classical genotypical characterizations with additionnal information of ecological significance.

First, a reliable method, has been developed to estimate the growth characteristics of isolates of *F. oxysporum* on various C sources. This method is based on optical density (OD) changes of liquid culture in microtiter plates.

Second, this method was applied to a large collection of 60 isolates of *F. oxysporum* originating from a soil and from the rhizosphere of plants cultivated in this soil . This collection was previously characterized on a genotypical basis using several molecular methods (Edel *et al* 1995).

2. Experimental set up.

A *F.oxysporum* strain was previously grown in liquid mineral medium (LMM) supplied with a C source (glucose, galactose, sucrose, maltose) (1.6g C/l.) for 5 days on a rotary shaker at 25°C. Mycelium mat was discarded by filtration. The conidial suspension was adjusted to $4x10^3$ conidia/ml to inoculate 5 wells of a microtiter plate (50µl/well), each containing 150 µl of LMM supplied with the same C source that was used for the preculture. One well containing 200 µl of LMM + C was kept uninoculated for each set of 5 inoculated wells and used as the reference optical blank for this set. The plate was incubated at 25°C in the dark. OD measurements (λ=650nm) were performed at regular time intervals using a microtiter plate reader for around 120 hours.

Previously, the relevance of OD changes of a liquid culture to estimate the growth characteristics of a *F. oxysporum* strain was assessed: (i) because the variation of biomass within the well of a microtiter plate could not be directly measured, a relation between dry weight and protein content was first established using liquid culture

H.-W. Dehne et al. (eds.), Diagnosis and Identification of Plant Pathogens, 535–538.

(100ml) in 250ml conical flasks. Protein content was determined using the Biorad Kit (fig.1). (ii) A significant correlation (R=0.94) was then found between OD changes and protein content (fig. 2). These results together indicated that OD changes corresponded to biomass increases. (iii) The absence of a border effect was checked by comparing the *F. oxysporum* growth in inner wells and in border wells .
Measurements were highly reproducible within wells as shown in figure 3.

3. Applying to a collection.

The sigmoid shape of the growth curves could be described by a logistic equation:

$$dx/dt = K/1 + \exp((p-t)/R)$$

allowing for the estimate of 3 parameters for each replicate on each C source.
- growth rate 1/R,
- plateau value K,
- value of t for x= K/2, p. (p is related to the length of the lag phase).

A Principal Component Analysis (PCA) was performed on each set of growth parameter estimates within each C source. (60 strains x 5 replicates x 3 parameters = 900 values). The result of the PCA performed on growth parameter estimates on maltose is shown figure 4 as an example. Strains were represented on the figure by the genotypic group number to which they belonged to. K and 1/R were negatively correlated with the x axis, P was positively correlated with the y axis. Similar results were obtained with the 4 C sources but a better discrimination was obtained with the diholosides than with the monosaccharides. In none of the cases was P a good parameter to discriminate between the groups.

4. Conclusion

The 60 *F. oxysporum* strains have been tested for their growth characteristics on the 4 C sources. A high diversity has been found among these non-pathogenic strains. This diversity can be related to the genotypic diversity. Thus, genotypic group 4 included strains with poor growth characteristics, genotypic groups 1 and 3 were centered and included strains with various growth characteristics and genotypic groups 9 and 11 included strains with high growth characteristics.
This phenotypic characterization could then be used to discuss in terms of ecological significance changes in the structure of *F. oxysporum* populations.

5 References

Edel, V., Steinberg,C., Avelange, I. Laguerre, G., and Alabouvette, C. (1995) Comparison of three molecular methods for the characterization of *Fusarium oxysporum* strains, *Phytopathology* **85,** 579-585.

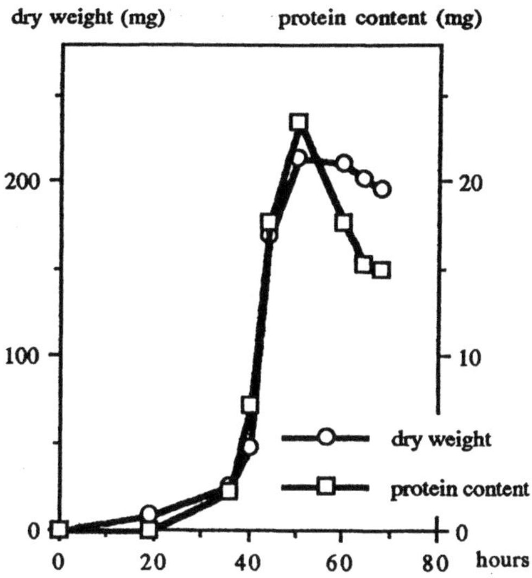

Figure 1. Relation between the dry weight and the protein content of a *F. oxysporum* culture in LMM + Glucose, in 250ml conical flasks.

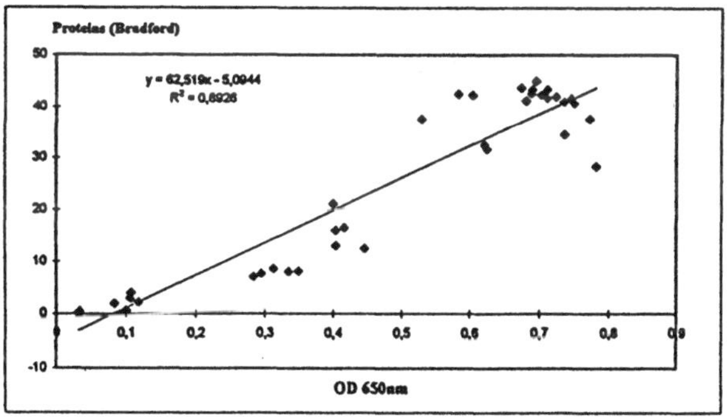

Figure 2. Relation between the OD and the rotein content of a *F.oxysporum* culture in LMM + Glucose in microtiter plates.

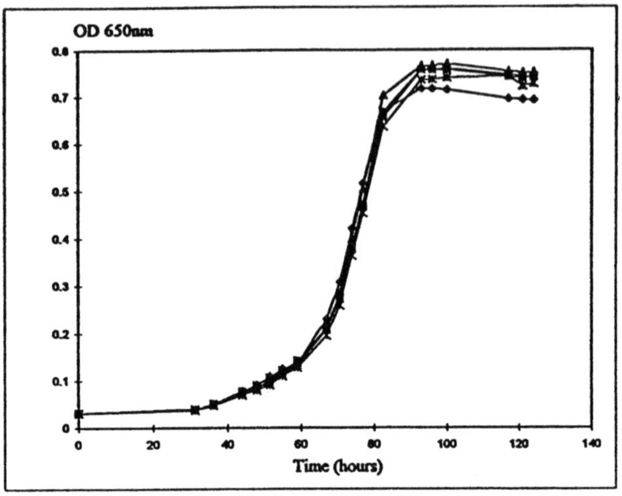

Figure 3. Growth curves of 5 replicates of a strain of F. oxysporum as assessed by OD changes in LMM + Glucose in microtiter plates. Measurements were highly reproducible within wells.

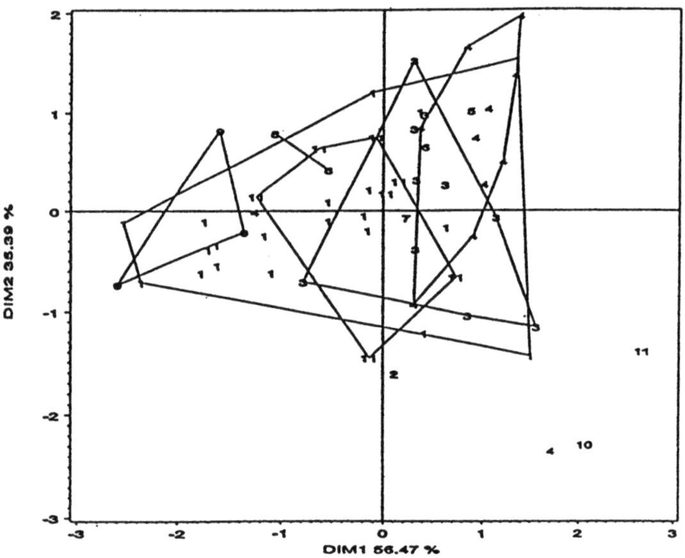

Figure 4. Result of the PCA performed on the growth parameter estimates of all the replicates of the 60 strains cultivated in LMM + maltose in microtiter plates. Strains with a high growth rate (1/R) and a high efficiency (K) are on the left part of the graph, strains with a high lag phase are on the upper part of the graph.

STUDY ON THE INFLUENCE OF *ASCOCHYTA FABAE* AND *BOTRYTIS FABAE* ON TWO GROWING TYPES OF FABA BEAN

ZAKRZEWSKA ELŻBIETA
Plant Breeding and Acclimatization Institute
Radzików, Poland

1. Introduction

Faba bean of the conventional indeterminate growth habit are the most common among existing cultivars so far. New cultivars of the determinate type were introduced during the last years in Poland. The information is needed on the relationship between growth habit and susceptibility to diseases.

Ascochyta blight and chocolate spot are severe diseases of faba bean (*Vicia faba* L.) in Poland. Variation in disease incidence following artificial inoculation with *Ascochyta fabae* Speg. has been observed among breeding lines of the conventional growth habit derived from English winter bean populations (Lockwood *et al.*, 1985). Variation in susceptibility of the spring cultivars to *A. fabae* was found in Canada (Kharbanda and Bernier, 1979). Marked variation in resistance to infection with *A. fabae* was observed among cultivars and strains of the conventional indeterminate and determinate growth habit in Poland (Zakrzewska, 1985, 1988).

The objective of this study was comparison of reaction of two morphological types of faba bean (*Vicia faba* L.) to inoculation with *A. fabae* or *Botrytis fabae* Sard. This research was made within the resistance breeding programme of faba bean in co-operation with Polish breeders.

2. Material and Method

The investigations have been conducted for seven years (1989 - 1995) at IHAR Radzików, Poland. The field experiments with faba bean cultivars and strains inoculated with *A. fabae* or *B. fabae* were carried out each year. The objective was to compare the reaction to infection with *A. fabae* of 210 entries of traditional growth habit and 164 entries of determinate growth habit and to compare the reaction to infection with *B. fabae* of 69 entries of traditional growth habit and 178 entries of determinate growth habit.

The plants were artificially inoculated by spraying with a suspension containing 10^6 *A. fabae* pycnospores per ml or 10^5 spores of *B. fabae*. Inoculum of *A. fabae* was prepared and inoculation was done as described by Zakrzewska (1985). The disease severity on each plant was estimated at maturity for experiments with *A. fabae* and approximately three weeks after inoculation for *B. fabae*. The incidence of disease on plants was

H.-W. Dehne et al. (eds.), Diagnosis and Identification of Plant Pathogens, 539–541.

assessed on a 1-9 scale, where 1 indicates no symptoms of infection and 9 indicates plant killed.

There were two standard cultivars. Cultivar Nadwiślański (Poland) over all years in the experiments with indeterminate type and cv. Tinova (Germany) in 1989-1990, cv. Tibo (Poland) in 1990-1993 and cv. Martin (Poland) in 1994-1995 for determinate type.

Results from 1990 were shown in figures as the most representative for the presented study.

3. Results and Conclusions

The stability of the reaction of cv. Nadwiślański to inoculation with *A. fabae* was observed over the years of study. Severity of ascochyta blight on this cultivar ranged on a moderate level of score and ranged from 4.52 in 1994 to 5.50 in 1995. Most cultivars and strains of the conventional growth habit were more susceptible to *A. fabae* in comparison to cv. Nadwiślański. Reaction of cv. Nadwiślański to *B. fabae* was more differentiated over the years and mean disease scores ranged from 3.87 in 1993 to 5.22 in 1992. Determinate type standard cv. Tibo was considerably more susceptible to both pathogens. The severity of infection of this cultivar to *A. fabae* ranged from 5.10 in 1993 to 6.33 in 1990. The reaction of cv. Tibo to infection to *B. fabae* was most various. Range from 4.20 in 1993 to 6.35 in 1990 was observed.

Significant differences in the resistance to infection with *A. fabae* between the traditional type of faba bean and determinate type were found over all years of the investigation. Most of the traditional type entries showed a higher level of resistance to *A. fabae* than the determinate forms (Fig. 1)

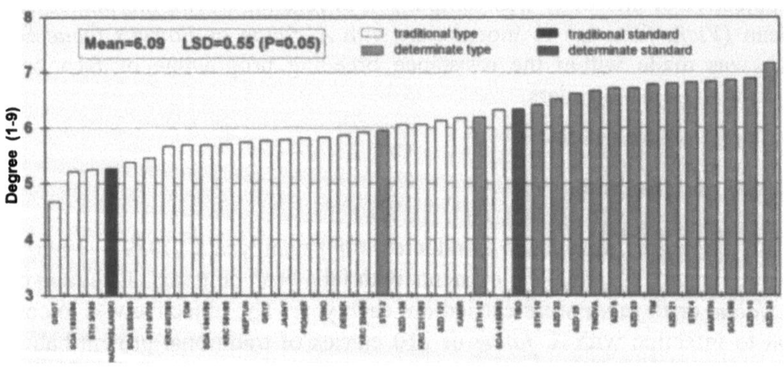

Figure 1. *Ascochyta fabae* infection of faba bean in 1990.

The *B. fabae* studies carried out in 1990 and 1994 demonstrated similar results as for *A. fabae* that most of the traditional type were higher resistant than the determinate entries. Significant differences in the resistance to *B. fabae* between entries evaluated in 1989 and 1991 were found, some determinate type entries of faba bean revealed a level of resistance similar to entries of the traditional growth habit (Fig. 2).

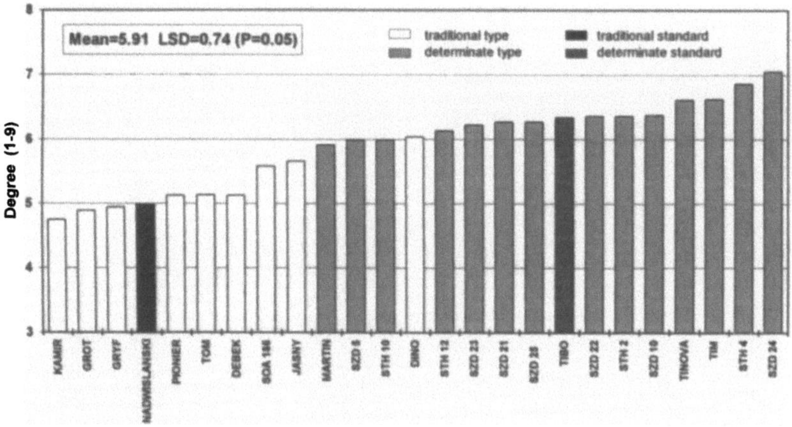

Figure 2. *Botrytis fabae* infection of faba bean in 1990.

Different range of variability of resistance to both diseases within entries in populations of faba bean of both conventional and determinate growth habit were observed over the years. But range of variability of resistance to *A. fabae* was higher than to *B. fabae* (Tab. 1).

TABLE 1. Reaction of faba bean entries to inoculations with *A. fabae* and *B. fabae* over seven years

Ascochyta fabae				Year	*Botrytis fabae*			
No. of entries	Range of reaction	Mean	$LSD_{0.05}$		No. of entries	Range of reaction	Mean	$LSD_{0.05}$
54	4.98 - 6.89	5.90	0.39	1989	34	4.64 - 5.84	5.31	0.73
38	4.67 - 7.16	6.09	0.55	1990	25	4.75 - 7.06	5.91	0.74
65	4.67 - 6.47	5.58	0.48	1991	40	4.80 - 5.86	5.34	0.48
52	4.58 - 6.57	5.83	0.55	1992	38	4.82 - 5.82	5.24	ns
50	4.40 - 6.41	5.28	0.66	1993	32	3.87 - 5.00	4.33	ns
62	4.24 - 5.86	5.05	0.69	1994	38	4.36 - 5.47	5.08	ns
53	4.74 - 6.83	5.92	0.61	1995	30	4.32 - 5.52	5.14	ns

ns - non significant.

The results obtained from this study demonstrated existence of thc association between growth habit and level of resistance of faba bean to *A. fabae*..

3. References

Kharbanda P. D. & Bernier C. C. (1979) Effectiveness of seed and foliar application of fungicides to control *Ascochyta fabae* of faba beans, Canadian Journal of Plant Science **59**, 661-666.

Lockwood, G., Jellis, J. G. and Aubury , R. G. (1985) Genotypic influences on the incidence of Infection by *Ascochyta fabae* in winter-hardy faba beans (Vicia faba), Plant Pathology **34**, 341-346.

Zakrzewska, E. (1985) Reakcja odpornościowa bobiku (*Vicia faba* L. var. minor Harz.) i bobu (*Vicia faba* L. var. *faba*) na porażenie grzybem *Ascochyta fabae* Speg. Materiały XXV Sesji Naukowej Instytutu Ochrony Roślin, Poznań, 295-318.

Zakrzewska, E. (1988) Variability in the resistance of *Vicia faba* L. to *Ascochyta fabae* Speg. Hodowla Roślin Aklimatyzacja i Nasiennictwo, **32**, 1/2, 311-317.

List of Contributors

Abad, P.
Depto. Producción Vegetal, Universidad Politécnica de Valencia, Spain

Adam, Guenther
Institute of Applied Botany, Dept. of Plant Protection, University of Hamburg, Germany

Ahmad, I.
Crop Disease Research Institute, PARC, Islamabad, Pakistan

Ahrens, Wilfried
Fachhochschule Weihenstephan, Abt. Triesdorf, 91746 Weidenbach, Germany

Alabouvette, C.
INRA-CMSE, Laboratoire de Recherches sur la Flore Pathogène dans le Sol, 17 rue Sully, 21034 Dijon Cedex, France

Albert, Guido
Cyanamid Forschung GmbH, P.O. Box 100, Zur Propstei, D-55270 Schwabenheim, Germany

Amelung, Dietrich
Bonhoeffer Straße 11, D-18051 Rostock, Germany

Andrina, M.
Di.Va.P.R.A. - Patologia vegetale, Università di Torino, Via P. Giuria 15, I-10126 Torino (Italy)

Arseniuk, E.
Plant Breeding and Acclimatization Institute, Radzików, 05-870 Błonie, Poland

Arts, M.S.J.
Plant Protection Service, P.O. Box 9102, 6700 HC Wageningen, The Netherlands

Baayen, Robert Paul
Research Institute for Plant Protection (IPO-DLO), P.O. Box 9060, 6700 GW Wageningen, The Netherlands

Bailey, John A.
IARC Long Ashton Research Station, Depart. of Agricultural Science, University of Bristol, Bristol, BS18 9AF, United Kingdom

Banks, J.N.
Central Science Laboratory, MAFF, Sand Hutton, York, YO4 1LZ, United Kingdom

Banniza, Sabine
International Mycological Institute, Bakeham Lane, Surrey TW20 9TY, Egham, United Kingdom

Barba, Marina
Istituto Sperimentale per la Patologia Vegetale - MiRAAF, ViaC.G. Bertero 22, 00156 Rome, Italy

Barbarossa, Loredana
Centro di Studio del C.N.R. sui Virus e le Virosi delle Colture Mediteranee, Via Amendola 165/A, 70126 Bari, Italy

Battilani, Paola
Istituto di Entomologia e Patologia vegetale, Università Cattolica S. Cuore, Via E. Parmense, 84 - 29100 Piacenza - Italy

Beckhove, Uli
Institut für Biologie I, RWTH Aachen, Worringer Weg 1, 52074 Aachen, Germany

Belli, G.
Centro CNR Miglioramento Sanitario Colture Agrarie, c/o Istituto di Patologia Vegetale, Università degli Studi di Milano, Via Celoria 2, 20133 Milan, Italy

Bianco, Piero A.
Centro CNR Miglioramento Sanitario Colture Agrarie, c/o Istituto di Patologia Vegetale, Università degli Studi di Milano, Via Celoria 2, 20133 Milan, Italy

Billings, C.
Central Science Laboratory, MAFF, Sand Hutton, York, YO4 1LZ, United Kingdom

Blakemore, Emily J.A.
National Institute of agricultural Botany, MBD Section, Huntingdon Road, Cambridge, CB3 0LE, United Kingdom

Bochow, Helmut
Humboldt University Berlin, Agric.-Hortic. Faculty, Div. Phytomedicine, Dorfstr. 9, D-13051 Berlin, Germany

Bonants, Peter J.M.
Research Institute for Plant Protection (IPO-DLO), Department of Detection, P.O. Box 9060, 6700 GW Wageningen, The Netherlands

Bosshard, Elisabeth
Swiss Federal Research Station, CH-8820 Wädenswil

Bottazzi, R.
Consorzio Provinciale Fitosanitario, Via Mazzini 40, 29100 Piacenza, Italy

Bouwen, Inge
DLO Research Institute for Plant Protection (IPO-DLO), Wageningen, The Netherlands

Briatore, E.
Di.Va.P.R.A. - Patologia vegetale, Università di Torino, Via P. Giuria 15, I-10126 Torino (Italy)

Bridge, P.D.
International Mycological Institute, Bakeham Lane, Surrey TW20 9TY, Egham, United Kingdom

Bruton, B.D.
Agricultural Research Service, U.S.D.A., Lane OK 74555, USA

Butgereitt, Anja
Institute for Biochemistry and Plant Virology, Federal Research Centre for Agriculture and Forestry, BBA, Messeweg 11/12, D-38104 Braunschweig, Germany

Büttner, Carmen
Institute of Applied Botany, Dept. of Plant Protection, University of Hamburg, Marseiller Strasse 7, D-22305 Hamburg, Germany

Cacciola, S.O.
Istituto di Patologia vegetale, University of Catania, Italy

Caffier, V.
INRA, Laboratoire de Pathologie Végétale, F-78850 Thiverval-Grignon, France

Camarasa, E.
Instituto Valenciano de Investigaciones Agrarias (IVIA), Depart. Virologia-Immunologia, Apartado Oficial, 46113 Moncada, Valencia, Spain

Cambra, Mariano
Instituto Valenciano de Investigaciones Agrarias (IVIA), Depart. Virologia-Immunologia, Apartado Oficial, 46113 Moncada, Valencia, Spain

Carling, D.E.
Agricultural and Forestry Experimental Station, University of Alaska Fairbank, 533 East Fireweed, Palmer, Alaska 99645

Casati, P.
Centro CNR Miglioramento Sanitario Colture Agrarie, c/o Istituto di Patologia Vegetale, Università degli Studi di Milano, Via Celoria 2, 20133 Milan, Italy

Černuško, Robert
Institute for Plant Pathology and Plant Protection, Georg-August-University Goettingen, Grisebachstr. 6, D-37077 Goettingen, Germany

Čeřovská, Noemi
Institute of Experimental Botany, Czech Academy of Sciences, Na Karlovce 1a, 160 00 Prague, Czech Republic

Cole, L.
School of Biological Sciences, Oxford Brookes University, Oxford, OX3 0BP, United Kingdom

Colinet, D.
Faculté Universitaire des Sciences Agronomiques, Passage des Deportes, 5030 Gembloux, Belgium

Contillo, R.
Istituto Sperimentale per il Tabaco, Via P. Vitiello 66, 84018 Scafati (SA), Italy

Cook, R.T.A.
Central Science Laboratory, MAFF, Sand Hutton, York, YO4 1LZ, United Kingdom

Cooke D.E.L.
Scottish Crop Research Institute (SCRI), Invergowrie, Dundee DD2 5DA, United Kingdom

Cox, S.
Scottish Crop Research Institute, Invergowrie, United Kingdom

Crepel, Caroline
Research Station for Plant Pathology, Centre for agricultural Research, B. van Gansberghelaan 96, 9820 Merelbeke, Belgium

Cubero, J.
Instituto Valenciano de Investigaciones Agrarias (IVIA), Depart. Virologia-Immunologia, Apartado Oficial, 46113 Moncada, Valencia, Spain

Czerwinska, A.
Plant Breeding Station Wiatrowo of the Poznań Plant Breeders, 62-100 Wagrowiec, Poznań

Dědič, Petr
Potato Research and Breeding Institute, Havličkův Brod, Czech Republic

Da Câmara Machado, A.
Institute of Applied Microbiology, University of Agriculture and Forestry, Nuszdorfer Lände 11, A-1190 Vienna, Austria

Dabkevicius, Zenonas
Lithuanian Institute of Agriculture, 5051 Dotnuva-Akademija, Kėdainiai, Lithuania

Dasi, M.A.
Instituto Valenciano de Investigaciones Agrarias (IVIA), Depart. Virologia-Immunologia, Apartado Oficial, 46113 Moncada, Valencia, Spain

Dehne, Heinz-Wilhelm
Institute for Plant Diseases, Rheinische Friedrich-Wilhelms-Universitaet Bonn, Nussallee 9, D-53115 Bonn, Germany

Dermastia, Marina
The National Institute of Biology, Vecna pot 111, 1000 Ljubljana, Slovenia

Dewey, Frances M.
Department of Plant Sciences, University of Oxford, South Parks Rd., Oxford, OX1 3RB, United Kingdom

Dijst, Gerda
DLO-Research Institute for Plant Protection (IPO-DLO),), P.O. Box 9060, 6700 GW Wageningen, The Netherlands

Dikshit, Anupam
Biological Product Laboratory, Department of Botany, Allahabad University, Allahabad - 211002, India

Dikshit, Shalini
Biological Product Laboratory, Department of Botany, Allahabad University, Allahabad - 211002, India

Dodd, J.C.
The International Institute of Biotechnology, P.O. Box 228, Canterbury, Kent CT2 7YW, United Kingdom

Dolenc, Jasna
Department of Biology, Biotechnical Faculty, University of Ljubljana, Slovenia

Duncan J.M.
Scottish Crop Research Institute (SCRI), Invergowrie, Dundee DD2 5DA, United Kingdom

Durán, N.
Instituto Valenciano de Investigaciones Agrarias (IVIA), Depart. Virologia-Immunologia, Apartado Oficial, 46113 Moncada, Valencia, Spain

Edel, Véronique
INRA-CMSE, Laboratoire de Recherches sur la Flore Pathogène dans le Sol, 17 rue Sully, 21034 Dijon Cedex, France

Ekobena, F. Henri P.
Faculty of Physics, Dept. Biophysics, Moscow Astate University, Moscow, Russia

Etienne, Laurent
Ciba-Geigy Ltd., Divison Crop Protection, CH-4002 Basel, Switzerland

Faggioli, Francesco
Istituto Sperimentale per la Patologia Vegetale - MiRAAF, ViaC.G. Bertero 22, 00156 Rome, Italy

Fessehaie, Anania
Institute for Plant Pathology and Plant Protection, Georg-August-University Goettingen, Grisebachstr. 6, D-37077 Goettingen, Germany

Figura, I.
Laboratory of Plant Pathology, Department of Biotechnology, University of Gdansk, Poland

Finetti Sialer, M.
Dipartimento di Protezione delle Piante dalle Malattie, Universitá degli Studi and Centro di Studio del C.N.R. sui Virus e le Virosi delle Colture Mediteranee, Via Amendola 165/A, 70126 Bari, Italy

Fokkema, Nyckle J.
DLO-Research Institute for Plant Protection (IPO-DLO),), P.O. Box 9060, 6700 GW Wageningen, The Netherlands

Frei, P.
Swiss Federal Research Station for Plant Production, CH-1260 Nyon, Switzerland

Frencel, Irina M.
Institute of Plant Genetics of the Polish Academy of Sciences, Strezyńska 34, 60-479 Poznań

Frey, Beatrice
Swiss Federal Research Station for Fruit-Growing, Viticulture and Horticulture, CH-8820 Wädenswil

Frey, Jürg E.
Swiss Federal Research Station for Fruit-Growing, Viticulture and Horticulture, CH-8820 Wädenswil

Gallitelli, D.
Dipartimento di Protezione delle Piante dalle Malattie, Universitá degli Studi and Centro di Studio del C.N.R. sui Virus e le Virosi delle Colture Mediteranee, Via Amendola 165/A, 70126 Bari, Italy

Gams, Walter
Centraalbureau voor Schimmelcultures, 3840 AG Baarn, The Netherlands

Garcia-Jimenez, J.
Depto. Producción Vegetal, Universidad Politécnica de Valencia, Spain

Garibaldi, A.
Di.Va.P.R.A. - Patologia vegetale, Università di Torino, Via P. Giuria 15, I-10126 Torino (Italy)

Gautheron, N.
INRA-CMSE, Laboratoire de Recherches sur la Flore Pathogène dans le Sol, 17 rue Sully, 21034 Dijon Cedex, France

Geistlinger, J.
Biocenter, University of Frankfurt, Marie-Curie-Str. 9/N200, D-60439 Frankfurt/Main, Germany

Giersiepen, Rainer
Institute for Plant Diseases, Rheinische Friedrich-Wilhelms-Universitaet Bonn, Nussallee 9, D-53115 Bonn, Germany

Gindrat, Daniel
Swiss Federal Research Station for Plant Production, CH-1260 Nyon, Switzerland

Giosue, S.
Istituto di Entomologia e Patologia vegetale, Università Cattolica S. Cuore, Via Parmense 84, 29100 Piacenza, Italy

Góral, Tomász
Plant Breeding and Acclimatization Institute, Radzików, 05-870 Błonie, Poland

Gorris, M.T.
Instituto Valenciano de Investigaciones Agrarias (IVIA), Depart. Virologia-Immunologia, Apartado Oficial, 46113 Moncada, Valencia, Spain

Grote, Dagmar
Institut für Gemüse- und Zierpflanzenbau Großbeeren / Erfurt e.V., Theodor Echtermeyer Weg 1, 14979 Großbeeren, Germany

Hack, Tim
Depto. Producción Vegetal (Patologia), Universidad Politécnica de Valencia, Camino de Vera s/n, 46022 Valencia, Spain

Hagenaar de Weerdt, M.
Research Institute for Plant Protection (IPO-DLO), Department of Detection, P.O. Box 9060, 6700 GW Wageningen, The Netherlands

Hahn, Renate
Institute for Plant Diseases and Plant Protection, University of Hannover, Herrenhaeuser Str. 2, D-30419 Hannover, Germany

Hakulinen, Johanna
Department of Biology, University of Joensuu, 80101 Joensuu, Finland

Hamacher, Joachim
Institute for Plant Diseases, Rheinische Friedrich-Wilhelms-Universitaet Bonn, Nussallee 9, D-53115 Bonn, Germany

Hanzer, V.
Institute of Applied Microbiology, University of Agriculture and Forestry, Nuszdorfer Lände 11, A-1190 Vienna, Austria

Harvey, P.
Cooperative Research Centre for Soil and Land Management and CSIRO, Division of Soils, Glen Osmond, PMB 2, South Australia 5064, Australia

Hau, Bernhard
Institut für Pflanzenkrankheiten und Pflanzenschutz, Universität Hannover, Herrenhaeuser Str. 2, D-30419 Hannover, Germany

Heinen, H.
Cyanamid Forschung GmbH, P.O. Box 100, Zur Propstei, D-55270 Schwabenheim, Germany

Heinze, Cornelia
Institute of Applied Botany, Dept. of Plant Protection, University of Hamburg, Germany

Heise, Silke
Institut für Biologie I, RWTH Aachen, Worringer Weg 1, 52074 Aachen, Germany

Heitefuss, Rudolf
Institute for Plant Pathology and Plant Protection, Georg-August-University Goettingen, Grisebachstr. 6, D-37077 Goettingen, Germany

Heppner, Claudia
Institute for Plant Pathology and Plant Protection, Georg-August-University Goettingen, Grisebachstr. 6, D-37077 Goettingen, Germany

Herdina
Cooperative Research Centre for Soil and Land Management and CSIRO, Division of Soils, Glen Osmond, PMB 2, South Australia 5064, Australia

Heupel, Monika
Landwirtschaftskammer Rheinland, Pflanzenschutzamt Bonn, Siebengebirgsstr. 200, D-53229 Bonn, Germany

Hilber, Urs W.
Swiss Federal Research Station, CH-8820 Wädenswil, Switzerland

Hilber-Bodmer, Maja
Swiss Federal Research Station, CH-8820 Wädenswil, Switzerland

Hindorf, Holger
Institute for Plant Diseases, Rheinische Friedrich-Wilhelms-Universitaet Bonn, Nussallee 9, D-53115 Bonn, Germany

Hoffstadt, Till
INRA, Laboratoire de Pathologie Végétale, F-78850 Thiverval-Grignon, France

Holderness, M.
International Mycological Institute, Bakeham Lane, Surrey TW20 9TY, Egham, United Kingdom

Honetšlegrová, J.
Institute of Plant Molecular Biology, Department of Plant Virology, Academy of Sciences, Branišovská 31, 370 05 České Budějovice, Czech Republic

Hooftman, R.
Inspection Service for Floriculture and Arboriculture (NAKB), Johan de Wittlaan 12, 2517 JR Den Haag, The Netherlands

Hughes, K.J.D.
Central Science Laboratory, MAFF, Sand Hutton, York, YO4 1LZ, United Kingdom

Hugouvieux-Cotte-Pattat, N.
Laboratory of Molecular Genetics of Microorganisms, URACNRS 1486, INSA Villeurbanne, France

Huttinga, Harm
DLO Research Institute for Plant Protection (IPO-DLO), P.O. Box 9060, 6700 GW Wageningen, The Netherlands

Iftikhar, S.
Crop Disease Research Institute, PARC, Islamabad, Pakistan

Ihrmark, K.
Department of Forest Mycology and Pathology, Swedish University of Agricultural Sciences, P.O. Box 7026, S-750 07 Uppsala, Sweden

Inman, Allan J.
Central Science Laboratory, MAFF, Sand Hutton, York, YO4 1LZ, United Kingdom

Jafra, S.
Laboratory of Plant Pathology, Department of Biotechnology, University of Gdansk, Poland

Janošová, Marta
Research Institute of Plant Production, Bratislavská cesta 122, 92168 Piešt'any, Slovakia

Janse, Jaap D.
Plant Protection Service,Depart. Bacteriology, P.O. Box 9102, 6700 HC Wageningen, The Netherlands

Jarosch, Birgit
Institut für Biologie I, RWTH Aachen, Worringer Weg 1, 52074 Aachen, Germany

Jayaratne, D.L.
Natural Resources Institute, Central Avenue, Chatham Maritime, Kent, ME4 4TB, United Kingdom

Jones T.
Scottish Crop Research Institute, Invergowrie, United Kingdom

Kahl, G.
Biocenter, University of Frankfurt, Marie-Curie-Str. 9/N200, D-60439 Frankfurt/Main, Germany

Kálmán, M.
Bay Zoltan Institute of Biotechnology, Derkovits fasor 2, H-6726 Szeged, Hungary

Kaminska, M.
Research Institute of Pomology and Floriculture, 96-100 Skierniewice, Poland

Kangasniemi, Aino
Department of Plant Biology, Plant Pathology Section, 00014 Helsinki, Finland

Karavaev, Vladimir A.
Faculty of Physics, Moscow State University, Vorobjovy, Moscow, 119899, Russia

Karjalainen, Reijo
Department of Plant Biology, Plant Pathology Section, 00014 Helsinki, Finland

Karlovsky, Petr
Institut für Pflanzenpathologie und Pflanzenschutz, Universität Göttingen, Grisebachstr. 6, 37077 Göttingen, Germany

Katinger, H.
Institute of Applied Microbiology, University of Agriculture and Forestry, Nuszdorfer Lände 11, A-1190 Vienna, Austria

Kessel, Geert J.T.
DLO Research Institute for Plant Protection (IPO-DLO), P.O. Box 9060, 6700 GW Wageningen, The Netherlands

Kiss-Tóth, E.
Bay Zoltan Institute of Biotechnology, Derkovits fasor 2, H-6726 Szeged, Hungary

Klusmann, J.
Technische Universität München, Lehrstuhl für Technische Mikrobiologie, 85350 Freising-Weihenstephan

Knapp, Elisabeth
Institute of Applied Microbiology, University of Agriculture and Forestry, Nuszdorfer Lände 11, A-1190 Vienna, Austria

Kogan, Nina
Department of Plant Pathology, ARO, The Volcani Center, Bet Dagan 50250, Israel

Kogel, Karl-Heinz
Institut für Biologie I, RWTH Aachen, Worringer Weg 1, 52074 Aachen, Germany

Köhl, Jürgen
DLO Research Institute for Plant Protection (IPO-DLO), P.O. Box 9060, 6700 GW Wageningen, The Netherlands

Kok, C.J.
DLO-Research Institute for Plant Protection (IPO-DLO),), P.O. Box 9060, 6700 GW Wageningen, The Netherlands

Kölber, Maria
Plant Health and Soil Conservation Station of Ministry of Agriculture, Budaörsi út 141-145, H-1118 Budapest, Hungary

Köllner, U.
AG Pflanzenviren, German Collection of Microorganisms and Cell Cultures, DSMZ, c/o BBA, Messeweg 11/12, D-38104 Braunschweig, Germany

Komorowska, B.
Research Institute of Pomology and Floriculture, 96-100 Skierniewice, Poland

Koopmann, Birger
Institut für Pflanzenpathologie und Pflanzenschutz, Universität Göttingen, Grisebachstr. 6, 37077 Göttingen, Germany

Körbler, M.
AG Pflanzenviren, German Collection of Microorganisms and Cell Cultures, DSMZ, c/o BBA, Messeweg 11/12, D-38104 Braunschweig, Germany

Kovács, N.Anna
Plant Health and Soil Conservation Station, Bacteriological Laboratory, Pécs, 7615 Pécs, P.O. Box 13, Hungary

Kraic, J.
Research Institute of Plant Production, Bratislavská cesta 122, 92168 Piešt'any, Slovakia

Krczal, Gabriele
Landesantsalt für Pflanzenbau und Pflanzenschutz, Essenheimerstr. 144, 55128 Mainz, Germany

Kreuzaler, Fritz
Institut für Biologie I, RWTH Aachen, Worringer Weg 1, 52074 Aachen, Germany

Krizbai, L.
Plant Health and Soil Conservation Station of Ministry of Agriculture, Budaörsi út 141-145, H-1118 Budapest, Hungary

Kukushkin, Alexander
Faculty of Physics, Dept. Biophysics, Moscow Astate University, Moscow, Russia

Lacourt I.
Scottish Crop Research Institute (SCRI), Invergowrie, Dundee DD2 5DA, United Kingdom

Lahoz, E.
Istituto Sperimentale per il Tabaco, Via P. Vitiello 66, 84018 Scafati (SA), Italy

Laimer da Câmara Machado, M.
Institute of Applied Microbiology, University of Agriculture and Forestry, Nuszdorfer Lände 11, A-1190 Vienna, Austria

Lane, Charles R.
Central Science Laboratory, MAFF, Sand Hutton, York, YO4 1LZ, United Kingdom

Latvala, Satu
Agricultural Research Centre, Institute of Plant Protection, 31600 Jokioinen, Finland

Lehto, Kirsi M.
Department of Biology, Laboratory of Plant Physiology and Molecular Biology, University of Turku, 20500 Turku, Finland

Lemmetty, Anne
Institute of Plant Protection, Agricultural Research Centre, 31600 Jokioinen, Finland

Leone, Gionata
DLO Research Institute for Plant Protection, Binnenhaven 5, 67oo GW Wageningen, The Netherlands

Lepoivre, Philippe
Faculté Universitaire des Sciences Agronomiques, Passage des Deportes, 5030 Gembloux, Belgium

Lesemann, Dietrich.-E.
Institut für Biochemie und Pflanzenvirologie, Biologische Bundesanstalt für Land- und Forstwirtschaft, Messeweg 11/12, D-38104 Braunschweig, Germany

Levi, Michal
Department of Plant Pathology, ARO, The Volcani Center, Bet Dagan 50250, Israel

Lewartowska, E.
Institute of Plant Genetics of the Polish Academy of Sciences, Strezyńska 34, 60-479 Poznań

Llop, P.
Instituto Valenciano de Investigaciones Agrarias (IVIA), Depart. Virologia-Immunologia, Apartado Oficial, 46113 Moncada, Valencia, Spain

Logrieco, Antonio
Istituto Tossine e micotossine da parassiti vegetali, C.N.R., Viale Einaudi 51, Bari, Italy

Lojkowska, Ewa
Laboratory of Plant Pathology, Department of Biotechnology, University of Gdansk, Poland

Lombaers-van der Plas, C.H.
DLO Research Institute for Plant Protection (IPO-DLO), P.O. Box 9060, 6700 GW Wageningen, The Netherlands

López, Maria M.
Instituto Valenciano de Investigaciones Agrarias (IVIA), Depart. Virologia-Immunologia, Apartado Oficial, 46113 Moncada, Valencia, Spain

Loreti, Stefania
Istituto Sperimentale per la Patologia Vegetale - MiRAAF, ViaC.G. Bertero 22, 00156 Rome, Italy

Mačkinaitė, Rimute
Institute of Botany, Žaliųjų ežerų 49, 2021 Vilnius, Lithuania

Maas, P.W.Th.
DLO-Research Institute for Plant Protection (IPO-DLO),), P.O. Box 9060, 6700 GW Wageningen, The Netherlands

Maes, Martine
Research Station for Plant Pathology, Centre for agricultural Research, B. van Gansberghelaan 96, 9820 Merelbeke, Belgium

Magan, N.
Cranfield Biotechnology Centre, Cranfield University, Cranfield, Bedfordshire, MK43 0AL, United Kingdom

Magnano di San Glio, Gaetano
Istituto di Deifesa delle piante, University of Reggio Calabria, Italy

Malinowski, T.
Research Institute of Pomology and Floriculture, 96-100 Skierniewice, Poland

Man in't Veld, W. and de Gruyter, H.:

Manulis, Shulamit
Department of Plant Pathology, ARO, The Volcani Center, Bet Dagan 50250, Israel

Masenga, Vera
Istituto di Fitovirologia Applicata, CNR, Delle Cacce 73, 10135 Torino, Italy

Mauler-Machnik, Astrid
Bayer AG, Agrochemicals Division, Research / Fungicides, D-51368 Leverkusen, Germany

McCartney, H.A.
IACR-Rothamsted, Department of Crop and Disease Management, Harpenden, Herts, AL5 2IQ, United Kingdom

Meijer, E.M.J.
DLO Research Institute for Plant Protection (IPO-DLO), P.O. Box 9060, 6700 GW Wageningen, The Netherlands

Merz, Ueli
Institute of Plant Sciences, Pathology Group, Federal Institute of technology, CH-8092 Zurich, Switzerland

Meyer, Linda
Department of Microbiology and Plant Pathology, University of Pretoria, Pretoria, 002 South Africa

Mierzwa, Z.
Potato Institute, Gdańsk, Poland

Migheli, Quirico
Di.Va.P.R.A. - Patologia vegetale, Università di Torino, Via P. Giuria 15, I-10126 Torino, Italy

Moravec, Tomáš
Institute of Experimental Botany, Czech Academy of Sciences, Na Karlovce 1a, 160 00 Prague, Czech Republic

Mörbel, J.
Landesantsalt für Pflanzenbau und Pflanzenschutz, Essenheimerstr. 144, 55128 Mainz, Germany

Mráz, Ivan
Institute of Plant Molecular Biology, Department of Plant Virology, Academy of Sciences, Branišovská 31, 370 05 České Budějovice, Czech Republic

Mühlbach, H.-P.
Institute of General Botany, Dept. of Genetics, University of Hamburg, Ohnhoststrasse 18, D-22609 Hamburg, Germany

Mulé, Giuseppina
Istituto Tossine e micotossine da parassiti vegetali, C.N.R., Viale Einaudi 51, Bari, Italy

Mumford, Richard A.
Natural Resources Institute, Central Avenue, Chatham Maritime, Kent, ME4 4TB, United Kingdom

Nabizadeh-Ardekani, Freidon
Institut für Pflanzenpathologie und Pflanzenschutz, Universität Göttingen, Grisebachstr. 6, 37077 Göttingen, Germany

Napierala, Anna
Polish Academy of Sciences, Institute of Dendrology, Parkowa 5, 62-035 Kórnik, Poland

Nel, L.H.
Department of Microbiology and Plant Pathology, University of Pretoria, Pretoria, 002 South Africa

Németh, J.
Plant Health and Soil Conservation Station, Bacteriological Laboratory, Pécs, 7615 Pécs, P.O. Box 13, Hungary

Nemeth, M.
Plant Health and Soil Conservation Station of Ministry of Agriculture, Budaörsi út 141-145, H-1118 Budapest, Hungary

Nicoletti, R.
Istituto Sperimentale per il Tabaco, Via P. Vitiello 66, 84018 Scafati (SA), Italy

Niessen, Ludwig
Technische Universität München, Lehrstuhl für Technische Mikrobiologie, 85350 Freising-Weihenstephan

Oerke, Erich-Christian
Institute for Plant Diseases, Rheinische Friedrich-Wilhelms-Universitaet Bonn, Nussallee 9, D-53115 Bonn, Germany

Oliva, A.
Istituto Sperimentale per il Tabaco, Via P. Vitiello 66, 84018 Scafati (SA), Italy

Olmos, A.
Instituto Valenciano de Investigaciones Agrarias (IVIA), Depart. Virologia-Immunologia, Apartado Oficial, 46113 Moncada, Valencia, Spain

Ophel-Keller, K.
Cooperative Research Centre for Soil and Land Management and CSIRO, Division of Soils, Glen Osmond, PMB 2, South Australia 5064, Australia

Ortega, Felipe
Institute for Plant Diseases, Rheinische Friedrich-Wilhelms-Universitaet Bonn, Nussallee 9, D-53115 Bonn, Germany

Oyarzun, P.J.
CIP, P.O. 17-2-1977 Quito, Ecuador

Pancaldi, D.
Dipartimento Protezione e Valorizzazione agroalimentare, Via Re 8, 40126 Bologna, Italy

Pane, A.
Istituto di Patologia vegetale, University of Catania, Italy

Pastrik, Karl-Heinz
Pflanzenschutzamt Hannover, Wunstorfer Landstr. 9, 30453 Hannover, Germany

Petrzik, K.
Institute of Plant Molecular Biology, Department of Plant Virology, Academy of Sciences, Branišovská 31, 370 05 České Budějovice, Czech Republic

Petzold, Ralf
Federal Ministry of Food, Agriculture and Forestry, Rochusstraße 1, 53123 Bonn, Germany

Polák, Jaroslav
Research Institute of Crop protection, Division of Phytomedicine, 161 06 Prague - Ruznì, Drnovská, Czech Republic

Prokinová, Evenie
Department of Plant Protection, Czech Agriculture University, Prague 6 - Suchdol, PC 165 21, Czech Republic

Pulawska, Joanna
Research Institute of Pomology and Floriculture, Pomolgiczna 18, 96-100 Skierniewice, Poland

Rabenstein, Frank
Federal Centre for Breeding Research on Cultivated Plants, Institute for Resistance Research and Pathogen Diagnostics, P.O. Box 1505, D-06435 Aschersleben, Germany

Racca, P.
Consorzio Emiliano Romagnolo Aziende Sperimentali (CERAS), Via Levante 18, 40026 Imola, Italy

Rainey, Fred
Deutsche Sammlung für Mikroorganismen DSM, Messeweg 11/12, D-38104 Braunschweig, Germany

Ravnikar, M.
The National Institute of Biology, Vecna pot 111, 1000 Ljubljana, Slovenia

Reeves, James C.
National Institute of agricultural Botany, MBD Section, Huntingdon Road, Cambridge, CB3 0LE, United Kingdom

Riedel, Dietmar
Institut für Biochemie und Pflanzenvirologie, Biologische Bundesanstalt für Land- und Forstwirtschaft, Messeweg 11/12, D-38104 Braunschweig, Germany

Rizvi, R.H.
Central Science Laboratory, MAFF, Sand Hutton, York, YO4 1LZ, United Kingdom

Roenhorst, J.W.
Plant Protection Service, P.O. Box 9102, 6700 HC Wageningen, The Netherlands

Roget, D.
Cooperative Research Centre for Soil and Land Management and CSIRO, Division of Soils, Glen Osmond, PMB 2, South Australia 5064, Australia

Roggero, Piero
Istituto di Fitovirologia Applicata, CNR, Delle Cacce 73, 10135 Torino, Italy

Román, M.P.
Instituto Valenciano de Investigaciones Agrarias (IVIA), Depart. Virologia-Immunologia, Apartado Oficial, 46113 Moncada, Valencia, Spain

Rossi, Vittorio
Istituto di Entomologia e Patologia vegetale, Università Cattolica S. Cuore, Via Parmense 84, 29100 Piacenza, Italy

Rudolph, Klaus
Institut für Pflanzenpathologie und Pflanzenschutz, Universität Göttingen, Grisebachstr. 6, 37077 Göttingen, Germany

Rutherford, M.A.
International Mycological Institute, Bakeham Lane, Surrey TW20 9TY, Egham, United Kingdom

Schiffer, Ruth
Institut für Biologie I, RWTH Aachen, Worringer Weg 1, 52074 Aachen, Germany

Schleier, S.
Friedrich-Schiller-Universität Jena, Institut für Mikrobiologie, Lehrstuhl für Allgemeine Mikrobiologie und Mikrobengenetik, Neugasse 24, D-07743 Jena, Germany

Schmechel, Detlef
IACR-Rothamsted, Department of Crop and Disease Management, Harpenden, Herts, AL5 2IQ, United Kingdom

Schoen, C.D.
DLO Research Institute for Plant Protection, Binnenhaven 5, 67oo GW Wageningen, The Netherlands

Schubert, J.
Federal Centre for Breeding Research on Cultivated Plants, Institute for Resistance Research and Pathogen Diagnostics, P.O. Box 1505, D-06435 Aschersleben, Germany

Schüepp, Hannes
Swiss Federal Research Station, CH-8820 Wädenswil, Switzerland

Schuld, P.
Institut für Pflanzenkrankheiten und Pflanzenschutz, Universität Hannover, Herrenhaeuser Str. 2, D-30419 Hannover, Germany

Schuster, Ralf-Peter
Institut für Pflanzenkrankheiten, Phytomedizin in Bodenökosystemen, Rheinische Friedrich-Wilhelms-Universität Bonn, Nussaallee 9, D-53115 Bonn, Germany

Schütze, Anja
Institute for Plant Diseases, Rheinische Friedrich-Wilhelms-Universitaet Bonn, Nussallee 9, D-53115 Bonn, Germany

Schwindt, Maya
Swiss Federal Research Station, CH-8820 Wädenswil

Seal, S.E.
Natural Resources Institute, Central Avenue, Chatham Maritime, Kent, ME4 4TB, United Kingdom

Shahi, Sushil
Biological Product Laboratory, Department of Botany, Allahabad University, Allahabad - 211002, India

Shukla, Amritesh Chandra
Biological Product Laboratory, Department of Botany, Allahabad University, Allahabad - 211002, India

Sikora, Richard A.
Institut für Pflanzenkrankheiten, Phytomedizin in Bodenökosystemen, Rheinische Friedrich-Wilhelms-Universität Bonn, Nussaallee 9, D-53115 Bonn, Germany

Simons, S.A.
International Mycological Institute, Bakeham Lane, Surrey TW20 9TY, Egham, United Kingdom

Šip, M.
Institute of Plant Molecular Biology, Department of Plant Virology, Academy of Sciences, Branišovská 31, 370 05 České Budějovice, Czech Republic

Sohn, M.
Institute of Applied Botany, Dept. of Plant Protection, University of Hamburg, Germany

Solntsev, Michail K.
Faculty of Physics, Moscow State University, Vorobjovy, Moscow, 119899, Russia

Staniulis, Juozas
Laboratory of Plant Viruses, Institute of Botany, Žaliųjų ežerų 49, 2021 Vilnius, Lithuania

Steden, C.
Ciba-Geigy Ltd., Divison Crop Protection, CH-4002 Basel, Switzerland

Steinberg, Christian
INRA-CMSE, Laboratoire de Recherches sur la Flore Pathogène dans le Sol, 17 rue Sully, 21034 Dijon Cedex, France

Steiner, Ulrike
Institute for Plant Diseases, Rheinische Friedrich-Wilhelms-Universitaet Bonn, Nussallee 9, D-53115 Bonn, Germany

Stenström, Elna
Department of Forest Mycology and Pathology, Swedish University of Agricultural Sciences, P.O. Box 7026, S-750 07 Uppsala, Sweden

Stevens, Elisabeth A.
National Institute of agricultural Botany, MBD Section, Huntingdon Road, Cambridge, CB3 0LE, United Kingdom

Stierl, Reinhard
Institute for Plant Diseases, Rheinische Friedrich-Wilhelms-Universitaet Bonn, Nussallee 9, D-53115 Bonn, Germany

Subr, Zdeno
Institute of Virology, Slovak Academy of Sciences, Dubravská cesta 9, 842 46 Bratislava, Slovak Republic

Susi, Petri E.
Department of Plant Production, University of Helsinki, P.O. Box 27 Viikki, 00014 Helsinki, Finland

Suter, J.
Ciba-Geigy Ltd., Divison Crop Protection, CH-4002 Basel, Switzerland

Suty, Anne
Bayer AG, Agrochemicals Division, Research / Fungicides, D-51368 Leverkusen, Germany

Tegel, Jukka
Plant Protection Inspection Centre, Plant Quarantine Laboratory, Box 42, 00501 Helsinki, Finland

Terhardt, Josef
Institut für Pflanzenkrankheiten, Phytomedizin in Bodenökosystemen, Rheinische Friedrich-Wilhelms-Universität Bonn, Nussaallee 9, D-53115 Bonn, Germany

Udupa, S.M.
ICARDA, P.O. Box 5466, Aleppo, Syria

Vaira, Anna Maria
Istituto di Fitovirologia Applicata, CNR, Delle Cacce 73, 10135 Torino, Italy

Valinsky, Lea
Department of Plant Pathology, ARO, The Volcani Center, Bet Dagan 50250, Israel

Valleys, T.
INRA-CMSE, Laboratoire de Recherches sur la Flore Pathogène dans le Sol, 17 rue Sully, 21034 Dijon Cedex, France

Van der Vlugt, René A.A.
DLO Research Institute for Plant Protection (IPO-DLO), Wageningen, The Netherlands

Van der Werf, Wobke
Department of Theoretical Production Ecology (TPE-WAU), Wageningen Agricultural University, Wageningen, the Netherlands

Van Gent-Pelzer, M.P.,
Research Institute for Plant Protection (IPO-DLO), Department of Detection, P.O. Box 9060, 6700 GW Wageningen, The Netherlands

Van Halteren, Paul
Plant Protection Service, P.O. Box 80048, 6700 HC Wageningen, The Netherlands

Van Zaayen, A.
Inspection Service for Floriculture and Arboriculture (NAKB), Johan de Wittlaan 12, 2517 JR Den Haag, The Netherlands

Vent-Schmidt, Katrein
Humboldt University Berlin, Agric.-Hortic. Faculty, Div. Phytomedicine, Dorfstr. 9, D-13051 Berlin, Germany

Vetten, H.J.
Institute for Biochemistry and Plant Virology, Federal Research Centre for Agriculture and Forestry, BBA, Messeweg 11/12, D-38104 Braunschweig, Germany

Vicedo, B.
Instituto Valenciano de Investigaciones Agrarias (IVIA), Depart. Virologia-Immunologia, Apartado Oficial, 46113 Moncada, Valencia, Spain

Vicente, M.J.
Depto. Producción Vegetal, Universidad Politécnica de Valencia, Spain

Vilhar, B.
The National Institute of Biology, Vecna pot 111, 1000 Ljubljana, Slovenia

Vilich, Vivian
Institut für Pflanzenkrankheiten, Phytomedizin in Bodenökosystemen, Rheinische Friedrich-Wilhelms-Universität Bonn, Nussaallee 9, D-53115 Bonn, Germany

Vlassova Tatjana
Moscow State University, Faculty of Biology, Department of Plant Physiology, Vorobjovy gory, 119899 Moscow, Russia

Vogel, R.F.
Technische Universität München, Lehrstuhl für Technische Mikrobiologie, 85350 Freising-Weihenstephan

Voigt, Kerstin
Friedrich-Schiller-Universität Jena, Institut für Mikrobiologie, Lehrstuhl für Allgemeine Mikrobiologie und Mikrobengenetik, Neugasse 24, D-07743 Jena, Germany

Waalwijk, C.
Research Institute for Plant Protection (IPO-DLO), P.O. Box 9060, 6700 GW Wageningen, The Netherlands

Walsh, John A.
Horticulture Research International, Wellesbourne, Warwick, CV35 9EF, United Kingdom

Weickel, Barbara
Institute for Plant Diseases, Rheinische Friedrich-Wilhelms-Universitaet Bonn, Nussallee 9, D-53115 Bonn, Germany

Weigand, Franz
ICARDA, P.O. Box 5466, Aleppo, Syria

Weiss, H.
Institute of Applied Microbiology, University of Agriculture and Forestry, Nuszdorfer Lände 11, A-1190 Vienna, Austria

Werner, Antoni
Polish Academy of Sciences, Institute of Dendrology, Parkowa 5, 62-035 Kórnik, Poland

Werner, Ralf
Institute of Applied Botany, Dept. of Plant Protection, University of Hamburg, Marseiller Strasse 7, D-22305 Hamburg, Germany

Werres, Sabine
Federal Biological Research Centre for Agriculture and Forestry, Institute for Plant Protection in Horticultural Crops, Messeweg 11/12, D-38104 Braunschweig, Germany

Wetzel, Thierry
Landesantsalt für Pflanzenbau und Pflanzenschutz, Essenheimerstr. 144, 55128 Mainz, Germany

Wieman, Liora
Department of Plant Pathology, ARO, The Volcani Center, Bet Dagan 50250, Israel

Winter, Stephan
AG Pflanzenviren, German Collection of Microorganisms and Cell Cultures, DSMZ, c/o BBA, Messeweg 11/12, D-38104 Braunschweig, Germany

Wiwart, Marian
Department of Plant Breeding and Seed Production, University of Agriculture and Technology, 10-724 Olsztyn, Poland

Wolf, Gerhard A.
Institut für Pflanzenpathologie und Pflanzenschutz, Universität Göttingen, Grisebachstr. 6, 37077 Göttingen, Germany

Wöstemeyer, Johannes
Friedrich-Schiller-Universität Jena, Institut für Mikrobiologie, Lehrstuhl für Allgemeine Mikrobiologie und Mikrobengenetik, Neugasse 24, D-07743 Jena, Germany

Wydra, K.
International Institute of tropical Agriculture, Cotonou, Benin

Yurina, Evdokija V.
Faculty of Biology, Moscow State University, Vorobjovy, Moscow, 119899, Russia

Yurina, Tatjana P.
Faculty of Biology, Moscow State University, Vorobjovy, Moscow, 119899, Russia

Zakrzewska, Elżbieta
Plant Breeding and Acclimatization Institute, Radzików, Poland

Zawadzka, B.
Research Institute of Pomology and Floriculture, 96-100 Skierniewice, Poland

Zoltanska, Ewa
University of Agriculture, Department of Phytopathology, ul. D'browskiego 159, 60-594 Poznañ, Poland

Zoon, F.C.
DLO-Research Institute for Plant Protection (IPO-DLO),), P.O. Box 9060, 6700 GW Wageningen, The Netherlands

Zu Xia, F.
UCRI, Guangshu, PR China

MIX
Papier aus verantwortungsvollen Quellen
Paper from responsible sources
FSC® C105338

If you have any concerns about our products,
you can contact us on
ProductSafety@springernature.com

In case Publisher is established outside the EU,
the EU authorized representative is:
Springer Nature Customer Service Center GmbH
Europaplatz 3, 69115 Heidelberg, Germany

Printed by Libri Plureos GmbH
in Hamburg, Germany